Introduction to
NEUROBEHAVIORAL TOXICOLOGY
Food and Environment

Pharmacology and Toxicology: Basic and Clinical Aspects

Mannfred A. Hollinger, Series Editor
University of California, Davis

Forthcoming Titles

CRC Handbook of Alcohol Addiction: Clinical and Theoretical Approaches, Gerald Zernig
Handbook of Mammalian Models in Biomedical Research, David B. Jack
Handbook of Theoretical Models in Biomedical Research, David B. Jack
Lead and Public Health: Integrated Risk Assessment, Paul Mushak
Molecular Bases of Anesthesia, Eric Moody and Phil Skolnick
Receptor Characterization and Regulation, Devendra K. Agrawal

Published Titles

Manual of Immunological Methods, 1999, Pauline Brousseau, Yves Payette, Helen Tryphonas,
    Barry Blakley, Herman Boermans, Denis Flipo, Michel Fournier
CNS Injuries: Cellular Responses and Pharmacological Strategies, 1999, Martin Berry
    and Ann Logan
Infectious Diseases in Immunocompromised Hosts, 1998, Vassil St. Georgiev
Pharmacology of Antimuscarinic Agents, 1998, Laszlo Gyermek
Handbook of Plant and Fungal Toxicants, 1997, Felix D'Mello
Basis of Toxicity Testing, Second Edition, 1997, Donald J. Ecobichon
Anabolic Treatments for Osteoporosis, 1997, James F. Whitfield and Paul Morley
Antibody Therapeutics, 1997, William J. Harris and John R. Adair
Muscarinic Receptor Subtypes in Smooth Muscle, 1997, Richard M. Eglen
Antisense Oligodeonucleotides as Novel Pharmacological Therapeutic Agents, 1997,
    Benjamin Weiss
Airway Wall Remodelling in Asthma, 1996, A.G. Stewart
Drug Delivery Systems, 1996, Vasant V. Ranade and Mannfred A. Hollinger
Brain Mechanisms and Psychotropic Drugs, 1996, Andrius Baskys and Gary Remington
Receptor Dynamics in Neural Development, 1996, Christopher A. Shaw
Ryanodine Receptors, 1996, Vincenzo Sorrentino
Therapeutic Modulation of Cytokines, 1996, M.W. Bodmer and Brian Henderson
Pharmacology in Exercise and Sport, 1996, Satu M. Somani
Placental Pharmacology, 1996, B. V. Rama Sastry
Pharmacological Effects of Ethanol on the Nervous System, 1996, Richard A. Deitrich
Immunopharmaceuticals, 1996, Edward S. Kimball
Chemoattractant Ligands and Their Receptors, 1996, Richard Horuk
Pharmacological Regulation of Gene Expression in the CNS, 1996, Kalpana Merchant
Experimental Models of Mucosal Inflammation, 1995, Timothy S. Gaginella
Handbook of Methods in Gastrointestinal Pharmacology, 1995, Timothy S. Gaginella
Handbook of Targeted Delivery of Imaging Agents, 1995, Vladimir P. Torchilin
Handbook of Pharmacokinetic Pharmacodynamic Correlations, 1995, Hartmut Derendorf
    and Guenther Hochhaus
Human Growth Hormone Pharmacology: Basic and Clinical Aspects, 1995,
    Kathleen T. Shiverick and Arlan Rosenbloom
Placental Toxicology, 1995, B. V. Rama Sastry
Stealth Liposomes, 1995, Danilo Lasic and Frank Martin
TAXOL®: Science and Applications, 1995, Matthew Suffness

Pharmacology and Toxicology: Basic and Clinical Aspects

Published Titles (*Continued*)

Introduction to
NEUROBEHAVIORAL TOXICOLOGY
Food and Environment

Edited by

R.J.M. Niesink, R.M.A. Jaspers, L.M.W. Kornet, J.M. van Ree, and H.A. Tilson

CRC Press
Boca Raton London New York Washington, D.C.

Library of Congress Cataloging-in-Publication Data

Introduction to neurobehavioral toxicology : food and environment /
    edited by Raymond J.M. Niesink ... [et al.]
        p.   cm. — (Pharmacology & toxicology)
    Includes bibliographical references and index.
    ISBN 0-8493-7802-8
    1. Behavioral toxicology.   2. Neurotoxic agents.   I. Niesink,
Raymundus Johannes Maria, 1953–   .   II. Series: Pharmacology &
toxicology (Boca Raton, Fla.)
    [DNLM: 1.   Nervous System—drug effects.   2.   Food—toxicity.
3. Environmental Pollutants—toxicity.   4.   Behavior—drug effects.
QV 76.5I615   1998]
RA1224.I55   1998
616.8′047—dc21
DNLM/DLC
for Library of Congress   98-39870
                                                                                                                                              CIP

© 1999 by CRC Press LLC and Open University of The Netherlands

No claim to original U.S. Government works
International Standard Book Number 0-8493-7802-8
Library of Congress Card Number 98-39870
Printed in the United States of America   1   2   3   4   5   6   7   8   9   0
Printed on acid-free paper

Preface

The brain has recently received more attention and consideration as evidenced by the 1990s being designated as "The Decade of the Brain" in the U.S., Europe, and many other parts of the world. Interest in the effects of exposure to drugs and chemicals on the development and functional integrity of the brain is on the rise. This book deals with the effects of chemicals on the central and peripheral nervous system and the subsequent changes in behavior. It provides an introduction to neurobehavioral toxicology.

The book contains essential information for students in neurotoxicology, psychiatry, and (neuro)psychology and for those professionally involved in the field of management of risk-assessment.

Neurobehavioral toxicology is the scientific study of unintended behavioral changes due to toxic agents. In this context, behavior is considered to form the ultimate integration of nervous functions at the level of the intact organism.

Initially, neurobehavioral toxicology was primarily practiced within occupational toxicology. Especially within this discipline, tests and methods have been developed to detect the undesired effects of toxicants on behavior. Meanwhile, it has also become clear that food components and environmental contaminants can influence our behavior. Neurobehavioral toxicology is important in studying the mechanisms underlying important Western diseases such as Alzheimer's disease, Parkinson's disease, and amyotrophic lateral sclerosis (ALS).

In this book, the neurobehavioral toxicity of food components and the behavioral effects of environmental toxicants will be discussed in particular. Emphasis will be on acute and chronic effects, reversible and irreversible consequences, functional disorders of the nervous system, neurobehavioral dysfunctions, and the underlying neurobiological mechanisms.

The book deals with the entire range of neurobehavioral toxicology in humans and animals, using a multidisciplinary approach to discuss all areas of the neuro- and behavioral sciences involved. Biological (physiological and biochemical), psychological, and social aspects of undesirable behavioral changes caused by toxicants will be reviewed.

This book is divided into three major sections. Part I provides the necessary background to understand the various topics. Relevant basic principles and processes stemming from different disciplines are described. These include: neurobehavioral toxicology and teratology, testing methods in humans and animals, basic neuroanatomical and neurophysiological functions, basic neuropathology, and genetic aspects of neurobehavioral toxicology and addiction. Parts II and III are devoted primarily to the neurobehavioral effects of natural food components, food contaminants, and food additives and to the effects of environmental chemicals such as metals, pesticides, and organic solvents on the nervous system. An extensive glossary that contains terms and concepts from the different disciplines, a list of commonly used abbreviations, and a list of tests used in neurobehavioral toxicology are provided at the end of the book.

Acknowledgments

The editors gratefully acknowledge the many people who helped organizing the original manuscript. A special thanks goes to Dr. Susan Boobis, for the editing, and Evelin Karsten, for typewriting and layout assistance for figures and tables. Jack Jamar and Vivian Rompelberg are greatly acknowledged for preparing the illustrations.

Edited by

Dr. R. J. M. Niesink, Ph.D., Open University of the Netherlands, Heerlen, and Rudolf Magnus Institute for Neurosciences, Utrecht University, The Netherlands

Dr. R. M. A. Jaspers, Ph.D., Ministry of Health, Welfare and Sport, Veterinary Public Health Inspectorate, Rijswijk, The Netherlands

Dr. L. M. W. Kornet, Ph.D., Waalre, The Netherlands

Prof. Dr. J.M. van Ree, Ph.D., Department of Medical Pharmacology, Rudolf Magnus Institute for Neurosciences, Faculty of Medicine, Utrecht University, The Netherlands

Dr. H.A. Tilson, Ph.D., U.S. Environmental Protection Agency, Health Effects Research Laboratory, Neurotoxicology Division, Research Triangle Park, North Carolina

Contributors

- Dr. K. Acuff-Smith, Ph.D., The Procter & Gamble Company, Ivorydale Technical Center, Professional and Regulatory Services Department, Laundry and Cleaning Products, Cincinnati, Ohio
- Dr. P. J. Bushnell, Ph.D., US Environmental Protection Agency, Neurotoxicology Division, Health Effects Research Laboratory, Research Triangle Park, North Carolina
- Dr. K. M. Crofton, Ph.D., Durham, North Carolina
- Dr. R. M. A. Jaspers, Ph.D., Ministry of Health, Welfare and Sport, Veterinary Public Health Inspectorate, Rijswijk, The Netherlands
- Dr. G. Jean Harry, Ph.D., National Institute of Environmental Health Sciences, National Institutes of Health, Research Triangle Park, North Carolina
- Dr. P. V. Kaplita, Ph.D., Boehringer Ingelheim Pharmaceuticals, Ridgefield, Connecticut
- Dr. B. M. Kulig, Ph.D., TNO Nutrition and Food Research Institute, Zeist, The Netherlands
- Dr. H. Lilienthal, Ph.D., Med. Inst. für Umwelthygiene, Abteilung Psychophysiologie, University of Düsseldorf, Düsseldorf, Germany
- Dr. R. C. MacPhail, Ph.D., U.S. Environmental Protection Agency, Neurotoxicology Division, Health Effects Research Laboratory, Research Triangle Park, North Carolina
- Prof. Dr. D. E. McMillan, Ph.D., Wilbur D. Mills Professor on Alcoholism and Drug Abuse Prevention, Director, Substance Abuse Treatment Clinic, Dept. of Pharmacology and Toxicology, University of Arkansas for Medical Sciences, Little Rock, Arkansas
- Dr. D. Mergler, Ph.D., CINBIOSE, Université du Québec à Montreal, Montreal, Quebec, Canada
- Prof. Dr. J. D. Mitchell, M.D., FRCP, Dept. of Neurology, Royal Preston Hospital, Fulwood, Preston,U.K.
- Dr. M. Panisset, M.D., The McGill Centre for Studies in Aging, St. Mary's Hospital Center, Montreal, Quebec, Canada
- Dr. D. C. Rice, Ph.D., Health Canada, Toxicology Research Division, Ottawa, Ontario, Canada
- Dr. S. L. Schantz, Ph.D., Institute for Environmental Studies, University of Illinois at Urbana-Champaign, Urbana, Illinois
- Dr. W. F. Sette, Ph.D., Office of Pesticide Programs, U.S. Environmental Protection Agency, Alexandria, Virginia
- Prof. Dr. I. C. Shaw, Ph.D., University of Central Lancashire, Preston, Lancashire, United Kingdom
- Dr. T. J. Sobotka, Ph.D., Neurobehavioral Toxicology Team, Division of Toxicological Research, Center for Food Safety and Applied Nutrition, Food and Drug Administration, Laurel, Maryland
- Dr. R. B. Stewart, Ph.D., Department of Psychiatry and Behavioral Sciences, U.T. Mental Science Institute, Substance Abuse Research Center, University of Texas, Houston
- Prof. Dr. C. V. Vorhees, Ph.D., Developmental Biology, Children's Hospital Research Foundation, Cincinnati, Ohio
- Prof. Dr. G. Winneke, Ph.D., Med. Inst. für Umwelthygiene, Abteilung Psychophysiologie, University of Düsseldorf, Düsseldorf, Germany

Table of Contents

Part I

Introduction

Contents Chapter 1

Neurobehavioral toxicology and addiction

Chapter 1

Neurobehavioral toxicology and addiction

D.E. McMillan

1   Basic principles of neurotoxicology

Neurotoxicology may be defined as the study of the adverse effects of chemicals on the structure and function of the nervous system. The subject matter includes chemical events at subcellular levels of the nervous system all the way to the study of chemical effects on the complex behavioral interactions of human organisms with their environment and with each other. Many chemicals have been shown to affect the nervous system. For example, psychotherapeutic drugs have been designed specifically for their effects on the brain. Although we do not usually think of such drugs as neurotoxicants, clearly these drugs can produce toxic "side effects" on the nervous system that can be defined as adverse effects.

Similarly, some chemicals have been developed to eliminate insects through their toxic effects on the nervous system. It should not be surprising that these insecticides can also produce toxicity to the nervous system of other organisms[9]. In some cases, chemicals that find their way into the environment have not been designed specifically for their effects on the nervous system, yet these chemicals can produce neurotoxicity. The range of chemicals that can produce neurotoxicity is enormous, and it is likely that we are aware of only a small number of the enormous number of unknown neurotoxicants[27].

1.1   NEUROTOXICOLOGY VERSUS NEUROBEHAVIORAL TOXICOLOGY

Some authors have used the term "neurotoxicology", or neurotoxicity, to mean much the same thing as our definition of behavioral toxicology[36]. Others make a distinction between neurotoxicology and behavioral toxicology. In this book, the term neurotoxicology will be used to refer to the study of those chemicals that produce structural damage to the nervous system, or that produce measurable biochemical changes in the cells of the nervous system. The term "neurobehavioral toxicology" will be applied to the study of the effects of chemicals on the behavioral interactions of the organism with its environment. Neurobehavioral toxicology developed as a hybrid of behavioral sciences and toxicology, because the field used behavioral methods to study the toxic effects of chemicals.

Probably all chemicals that produce behavioral effects do so by causing changes in cellular biochemistry; however, the distinction between neurotoxicity and neurobehavioral toxicity is made so that the direct effects of chemicals on the nervous system can be separated from those effects of chemicals that depend on the behavioral history of the organism and the interaction of the organism with its environment.

3

1.2 RELATIONSHIP BETWEEN NEUROTOXICITY AND SUBSTANCE ABUSE

Generally, when we think about exposure to neurotoxicants we think about the presence of these chemicals in our environment. Neurobehavioral toxicants may be in the air we breathe, the water we drink, or the food we eat. The exposure to these chemicals is "accidental" in that we do not volunteer to interact with environmental chemicals. In fact, government agencies in many countries are charged with the task of seeing that our exposure to such chemicals in air, water, and food is regulated. Parts II and III of the present volume will be devoted largely to neurobehavioral toxicants in food and in the environment.

In contrast, humans can choose to self-administer some drugs, because these drugs produce effects that the users find reinforcing. These self-administered chemicals produce effects that clearly meet the definition of neurobehavioral toxicity. When individuals "voluntarily" expose themselves to neurobehaviorally toxic drugs, such an action represents a particularly pernicious form of neurobehavioral toxicity, since the drug user actively seeks exposure to the chemical, sometimes in progressively higher doses with repeated exposure. Substance abuse can be a particularly difficult form of neurobehavioral toxicity to prevent, since the user frequently resists any attempts to terminate exposure to the chemical. Apart from the fact that these chemicals are addictive, some of them are also neurotoxic, especially when used chronically or in large doses. These chemicals may be particularly dangerous for women users because of the possibility that they might become pregnant while using, thereby also exposing the fetus to the drug.

In the case of the developing fetus, it is the mother who selects the level of exposure to an abused drug for both herself and for her fetus. In recent years, it has become apparent that many chemicals affecting the central nervous system (CNS) can affect the unborn fetus[40]. For example, fetal alcohol syndrome (FAS) has become widely recognized, and it appears that exposure to a number of psychoactive drugs *in utero* can also produce neurobehavioral toxicity in the offspring. The relationship of neurobehavioral teratology to substance abuse will be discussed in Chapter 2.

1.3 IMPORTANCE OF THE DOSE-RESPONSE RELATIONSHIP

To paraphrase Paracelsus, "The dose makes the poison." What Paracelsus was saying 500 years ago is a point that is sometimes neglected today. That point is that toxicity depends on the dose of the chemical to which individuals in the population are exposed. It is widely assumed that most chemicals have threshold levels below which no toxic effects can be observed. Conversely, all chemicals can probably produce toxicity of some type if the exposure level or dose is high enough. This is particularly true for neurobehavioral toxicology, where any chemical, if given at high enough doses, is likely to affect behavior. This fact has led to an area of contention between those promoting the use of behavioral data in making regulatory decisions about environmental exposure and those resisting the use of such data. For example, manufacturers of some chemicals have resisted legislation requiring that their chemicals be tested for possible effects using behavioral endpoints, arguing that "nonspecific" effects of certain chemicals on animal behavior might result in the misclassification of chemicals as neurotoxicants[33]. On the other hand, it can be argued that the behavior of an organism is a good reflection of the general well-being of the organism and that any change in behavior following exposure to a chemical is undesirable and reflects toxicity, although not necessarily neurotoxicity.

At least a partial solution to this problem becomes apparent when the relationship of dose to effect is considered. Those chemicals that produce neurobehavioral effects only at dose levels far above those to which any mammalian organism is likely to be exposed are not of primary concern to the neurobehavioral toxicologist. However, those chemicals that produce effects on behavior or other functions of the nervous system at doses at which environmental exposure is possible, or likely, are of great concern. This emphasizes the importance of studying a range of doses in any neurobehavioral toxicity study.

1.4   RELEVANCE OF CHRONIC EXPOSURE TO NEUROBEHAVIORAL TOXICOLOGY

Although occasionally neurobehavioral toxicity results from an acute poisoning episode, chronic exposure to lower levels of the offending chemical is more characteristic. With chronic exposure to a chemical, there are at least three possible outcomes, including no change in effect relative to acute exposure, increased effects (cumulative effects or increased sensitivity), or decreased effects (usually referred to as tolerance).

First, the chronic toxicity produced by repeated drug administration may mimic the acute toxicity, with few additional changes occurring with repeated exposure. This occurs when the chemical produces no lasting effects following each dosing, and the doses are spaced sufficiently far apart in time to prevent their accumulation in the body. If neither tolerance nor increased sensitivity to the drug occurs and the drug does not accumulate in the body under the schedule of repeated administration, there are few, if any, differences between acute and chronic exposure.

Second, it is possible that repeated exposure may result in progressively greater toxic effects. This is seen most clearly when the repeated exposure to the chemical occurs at short time intervals such that metabolic and excretory mechanisms cannot clear the chemical and the chemical accumulates in the body; however, some chemicals can produce increased effects on the CNS even when exposure occurs at intervals where the chemical does not accumulate. In some cases, the CNS effects may result from cumulative damage to the brain[25]; however, cumulative brain damage does not always explain increased sensitivity to drugs following chronic administration. For example, repeated administration of psychomotor stimulants such as cocaine can enhance many of their effects, although no specific changes in brain have been shown to be associated with this sensitization[21]. It has been suggested that sensitization to the stimulant effects of cocaine may be classically conditioned effects.

Accumulation of a drug may produce effects with chronic small doses that are similar to those seen with much higher acute doses. For example, the long-acting barbiturate, barbital, can accumulate with daily dosing to the point that it produces coma and finally death. This is because barbital is only slowly eliminated from the body by urinary excretion, and with successive administrations, blood and brain levels of the drug may continue to rise until toxic levels are reached, unless doses are widely spaced in time. This is illustrated in the figure below which shows plasma levels produced by five successive doses of a depressant such as barbital. Each dose of the drug is given at a single dose level that is not adequate to produce an effect initially, but at a frequency where drug accumulation occurs. The jagged curve represents the lowest and highest plasma concentration seen after each dose and the smooth curve represents the mean level obtained with each successive dose. The first dose does not produce the therapeutic effect, but the second one does. Following the administration of the third dose, a toxic level of drug has accumulated, and by the fourth or fifth dose a plasma level approaching that associated with a lethal dose has been achieved.

Third, it is possible that tolerance develops upon repeated exposure to the chemical. For example, exposure to cholinesterase inhibitors used as insecticides produces a neurobehavioral toxicity presumably due to the accumulation of acetycholine, but with repeated administration of some of these insecticides, the behavioral effects diminish despite continued inhibition of the enzyme[17]. With the repeated administration of abused drugs, the tolerance development can be quite large. For example, more than a 100-fold tolerance to morphine can develop, which means that to obtain the same effect (analgesia, euphoria) a dose 100 times higher will be needed. In the opiate addict, a dose lethal to the non-user produces only minimal pharmacological effects.

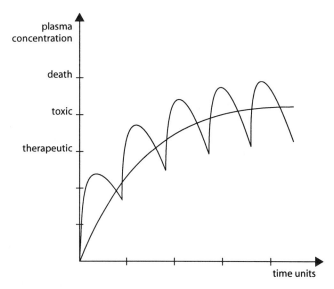

FIGURE 1.1

Relationship of plasma drug concentration to pharmacological effects with repeated drug administration at time intervals that result in the accumulation of the drug. The smooth curve shows the increase in mean drug concentration, and the wavy curve shows the minimum and maximum plasma concentrations with repeated drug administration. The labels death, toxic, and, therapeutic show the approximate plasma levels at which these pharmacologic effects are expected.

In Figure 1.1 above, which compares the effects of acute dosing with those of chronic dosing, it was illustrated that small cumulative doses of a drug can produce plasma blood levels associated with toxicity if the doses are given frequently; however, if the repeated administration of the drug also results in the production of tolerance, these high blood levels may not result in toxicity.

This discussion of the consequences of repeated exposure to toxic chemicals has assumed that the effects of the chemicals in question are reversible. After the administration of the chemical, the body has metabolized and/or excreted the chemical, and it has produced no lasting effect on the tissue. This is not always the case with neurotoxicants. A chemical that destroys neural tissue is likely to produce lasting effects. Chronic administration of such chemicals can produce cumulative damage even when the drug itself does not accumulate with repeated administration. It is also clear that there is a degree of plasticity in the nervous system and that compensation can occur following exposure to some neurotoxicants, although in many cases only partial recovery is observed.

1.5 EFFECTS OF CHEMICALS IN VULNERABLE POPULATIONS

Populations exposed to neurotoxic chemicals are not homogeneous. People exposed to environmental chemicals differ in genetics, age, gender, health, nutrition, and many other ways. Although the study of individual differences in response to neurobehavioral toxicants is only beginning, there are a few areas where enough work has been done that some generalities can be stated. For example, it has long been accepted that exposure of the immature organism is more likely to produce toxic insult than is a similar exposure in the adult organism. Presumably, the greater sensitivity of immature organisms to toxic insult results from less well-developed metabolic systems, a lowered excretory capacity, an immature blood-brain barrier, different binding capacities of tissue and serum protein, and other chemical distributional differences[29]. Perhaps the extreme example of the special vulnerability of immature organisms to neurobehavioral toxicants is best illustrated by fetal exposure to chemicals. For example, in fetal alcohol syndrome pregnant women consuming modest amounts of alcohol, perhaps less than 3 oz of ethanol (85 g of absolute ethanol) daily, produce offspring with a variety of somatic and developmental problems. These children show moderate to severe dysfunction of the CNS with mental retardation apparent in most cases[13]. Other drugs have also been suggested to produce neurobehavioral toxicity in children exposed to them *in utero*, including cocaine and the widely used anticonvulsant drug, phenytoin.

There is considerable evidence that environmental chemicals can also produce neurobehavioral toxicity to the developing organism. For example, lead is thought to produce greater effects in children than in adults, both because lead exposure is often greater in children, and because of the special sensitivity of the developing brain to lead. Generally these effects occur at exposure levels that produce minimal effects in the adult[8]. Other environmental chemicals that produce developmental neurobehavioral toxicity include the polychlorinated biphenyls (effects on motor and cognitive function) and methylmercury (sensory function deficits, motor disturbances and adverse effects on cognitive function). Interest in this area has given rise to a new field of study, that of behavioral teratology.

Just as young organisms can be especially vulnerable to some chemical insults, so are aging organisms. As in the child, excretory and metabolic capacity is reduced in the aged. In addition, as the brain ages the number of neurons in the brain decreases, memory function declines, learning capacity is reduced and reaction time slows[14]. Against this declining level of functioning, the impact of neurobehavioral toxicants could be increased.

In addition to the natural decline in function with aging, the elderly are more likely to suffer from disease. The development of diseases such as Alzheimer's disease, or Parkinsonism in the elderly further complicates the picture for older individuals who are exposed to neurobehavioral toxicants. Even very subtle chemical effects can be significant to an individual whose nervous system function is already compromised. Similarly, liver and kidney disease may also limit the ability of the elderly patient to metabolize and excrete neurotoxic chemicals from the body, thereby increasing the potential for toxic effects even further. Table 1.1 compares the half-life of antipyrine in children, geriatric subjects, and adult controls as estimated from figures reviewed by Alveres and Pratt[1]. Although there are some differences in the estimates of the half-life of antipyrine in control adults, the table illustrates the general point that the metabolism of chemicals is decreased in both the very young and the old.

It has become increasingly apparent that there are genetic determinants of our response to chemicals. At present, most of the genetic differences that have

TABLE 1.1

Age-dependent metabolism of antipyrine in different groups of subjects

Group	Estimated half life (Hrs)	Ratio to control group
Adult controls*	6.3	
Children	13.8	2.2
Adult controls*	12.1	
Geriatric adults	16.3	1.3

* Data in the table are from two different studies, each using its own
control group. Ages of subjects in these control groups varied which
may account for the differences in values in the two control groups

been demonstrated in response to environmental chemicals have involved genetic differences in metabolism[31]. Examples of these, and other genetic determinants of differences in neurobehavioral responses to toxicants will be given in Chapter 6. The area of genetic influences on the neuro-behavioral toxicity of environmental toxicants remains in its infancy.

2 Issues in neurobehavioral toxicology

2.1 EXTRAPOLATION FROM ANIMALS TO HUMANS

Almost from its inception, the field of neurobehavioral toxicology has been involved in, perhaps even driven by, the needs of regulatory agencies. Early concern about protecting the public from exposure to potential neurobehavioral toxicants was stimulated by reports from the former Soviet Union indicating that some chemicals produced effects on conditioned behavior at very low doses[28]. These exposure levels were frequently much lower than those permitted in the United States. This provided a strong stimulus for increased research in neurobehavioral toxicology related to hazard identification and subsequent assessment of human risk.

In the risk assessment process, determining risk frequently depends on epidemiological and other data from human populations and on data derived from animal exposure, since humans cannot be exposed to potential neurobehavioral toxicants deliberately in controlled laboratory settings. Although epidemiological and other studies with humans provide data on the most relevant species, human beings, there are frequently difficulties with such studies. When an acute poisoning occurs there is usually a lag varying from days to months before scientists can begin to study the exposed population. By this time it is almost impossible to determine the level and duration of exposure to the chemical, and the pattern of onset of any toxicity. Furthermore, when the public becomes informed about the exposure, placebo responses and imagined toxicity can occur. For some potential neurotoxicants the extent of human exposure is unknown. Therefore, data derived from animal models have become increasingly important for estimating human risk, since it is often the only data available.

2.1.1 Species and strain differences

The use of animals presupposes that it is possible to extrapolate the results across species. Although such extrapolation has been generally accepted for other forms of toxicity, there has been some reluctance to accept extrapolation across species for nervous system function, especially behavior. Critics of neurobehavioral toxicology point to the complexity of human behaviors, citing reading, mathematics, music, and other higher functions that apparently occur only in human beings, and suggest that because they cannot be modeled successfully in animals, widespread requirements for neurobehavioral toxicity testing of chemicals are not appropriate. Although even the most clever neurobehavioral toxicologist is unlikely to be able to teach calculus to rats, it seems likely that there are analogous models of most human behaviors in lower organisms. For example, neurotoxic chemicals can produce alterations in sensory, motor and cognitive functions such as learning and memory. All of these functions can be measured in both human and animal species. Under this assumption neurobehavioral toxicology is developing along the same lines as other areas in toxicology have done. As with the study of toxic effects in other organ systems, the determination of effective doses of toxicants in animal models can be used to develop a data base to which safety factors can be applied to protect for sensitive individuals within a population and to provide for differences across species[27].

It might be expected that higher organisms would be more sensitive to neurobehavioral toxicants than would lower organisms, both because of slower rates of drug metabolism in higher organisms and higher brain to body weight ratios. There is a linear relationship between body weight and the half-life of at least some drugs, with the plasma half-life of the drug increasing with the body weight[34]. This means that these drugs reach higher plasma concentrations and take a longer period to leave the body with increasing body weight. Similarly, animals high on the phylogenetic scale have larger brains and a higher ratio of brain weight to body weight. Although there are differences among species in response to chemicals that affect the CNS, the correlation between position on the phylogenetic scale and sensitivity to chemical effects is not perfect[24]. For example, pigeons appear to be more sensitive to the effects of tetrahydro-cannabinol on behavior than squirrel monkeys, despite the lower placement of the pigeon on the phylogenetic scale.

Even within species, large individual differences are sometimes observed. One approach to minimizing individual animal differences in response to chemicals is to use inbred animal strains for neurobehavioral toxicity testing. A possible advantage of comparing inbred strains for testing is that they are likely to show a lower within-group variability than outbred strains[12], although considerably more work will be required to document this contention convincingly for neurobehavioral toxicology. Inbred strains could provide an opportunity to study the genetic mechanisms underlying differences in neurobehavioral responses to chemical insult.

2.1.2 *Dosing differences*

Ideally, studies that use animal models for toxicity testing should employ the same doses, the same routes of administration, and the same duration of chemical exposure to which human beings are exposed. In practice this is rarely feasible. As was discussed in the previous section, different species may differ in their sensitivity to chemicals. A patient suffering from postoperative pain may obtain significant relief from a 10 mg dose of morphine (about 0.14 mg/kg). The tail-flick test in rats, which is often used to screen for narcotic analgesics, may require 3-10 mg/kg to show an analgesic effect of morphine. With this difference in sensitivity in mind, should a chronic study with morphine in rats be done at the human analgesic dose, or at the dose shown to be effective in rats when the investigator is interested in extrapolating safe dose levels from the rat to humans? In toxicity testing most investigators probably would choose to compare doses with similar effects in animals and humans, and then apply a safety factor to the animal data to attempt to predict "safe levels" for humans. It is obvious that the basis for determining this safety factor is crucial in such studies.

Another problem is posed by differences in the life span of animals. Humans can be exposed to environmental chemicals for a lifetime, which averages about 70 years in the western world. How can chronic toxicity in humans be predicted from lifetime exposure data from rats, which have a life span of only 2 years? The solution often taken to this problem is to expose the animals to relatively high dose levels under the assumption that an extrapolation can be made from data gathered from more intense short-term exposure to the situation where longer but less intense exposure occurs.

Humans are exposed to environmental chemicals by several routes, often over different time periods. For example, humans are exposed to lead in the drinking water, air, and food. In addition, children may ingest lead paint fragments. Some of the lead exposure can be as inorganic salts and some as organic lead compounds. These complex differences in dosing patterns are nearly impossible to replicate in the laboratory and are rarely attempted. This constitutes another barrier to the extrapolation of animal data to humans.

Despite all of these differences and problems in attempting to extrapolate toxicity data from animals to humans, reviews of almost any segment of the literature would show that successful extrapolation is the rule and not the exception. Chemicals that produce neurotoxicity in animal models almost always do so in humans. Drugs that act as reinforcers in animals almost always do so in humans. The list of examples of successful extrapolations would be very lengthy. While there are also some exceptions to these successes where attempts to extrapolate from animals to humans have been misleading, the list of exceptions would be much shorter.

2.1.3   *Validity of animal models*

Ideally, animal models would be most useful when they have construct validity, that is, they produce identical effects in the animal model and in the human being, and the mechanism underlying these effects is the same. To establish construct validity for an animal model of neurobehavioral toxicity, not only would it be necessary to show that the chemical produced very similar effects on the behavior in animals and humans, but also that the detailed mechanisms at the cellular and subcellular level that underlie these behavioral changes are also identical. True construct validity is difficult to establish, primarily due to a lack of knowledge about many of the physiological and biochemical mechanisms underlying complex human and animal behaviors. When construct validity has been established for an animal model of neurobehavioral toxicity, it means that the model produces the same behavioral effects in animals and man and for the same reasons. When this occurs, great confidence can be placed on any extrapolation from that model to the human situation.

More frequently, we must be satisfied with other forms of validity, such as face validity and predictive validity. An example of face validity can be found in screening for antianxiety drugs using punishment procedures. In punishment procedures the responses of an animal are suppressed by scheduling these responses to produce punishing stimuli. It has been known for some years that a variety of drugs that have anti-anxiety effects clinically are effective in restoring responding suppressed by punishment[26]. Because the activities of the animal during the punishment procedure are consistent with the interpretation that the animal is experiencing some level of fear and anxiety which is reduced by the anti-anxiety drug, the procedure has face validity. Too heavy a reliance on face validity can lead to dangerous generalizations. If a monkey appears to follow with his eyes moving objects that are really not there (or at least not observable to the investigator) following administration of a chemical, then this may appear to the investigator to be "hallucinatory behavior", but it certainly does not establish that the animal is really experiencing a hallucination.

Because punished responding has been so successful as a procedure for predicting clinical anti-anxiety activity, the procedure can also be said to have predictive validity. That is, the animal model predicts the effects of a chemical on human behavior. For an animal model to have predictive validity, it need not have construct validity, nor face validity, although it may have both. Predictive validity only means that the test predicts the response that it has been designed to predict. For example, Seiden and his colleagues have studied the effects of drugs on animals trained to space their responses apart in time by some minimum duration[23,32]. Many antidepressant drugs tested under this procedure decrease the proportion of unreinforced responses under this schedule. Most investigators have considered this schedule to be a measure of an animal's ability to differentiate time durations. It is not apparent why a drug that affects an animal's ability to differentiate time would have antidepressant activity, since depressed people are not particularly characterized by problems in their time estimations. Nevertheless, the animal model is useful since it accurately predicts the activity of at least some kinds of antidepressant drugs, even if the basis of the prediction is unclear.

How do these concepts of validity relate to neurobehavioral toxicity testing? Obviously, it would be best if all of our animal tests had construct validity, since this would assure us that we were measuring the exact same effects across species. As discussed previously, this is rarely the case. When attempting to establish an animal test for regulatory puposes it makes sense to begin with a behavioral test that at least appears to be similar to some aspect of human behavior, or in other words, has face validity. If the test is predictive of the human response it can be said to have predictive validity and as such it becomes very useful; however, considerable study of the mechanisms of action with the chemical of concern at both the animal and the human level will be required to establish construct validity.

In addition to being valid, a neurobehavioral toxicity test should also be reliable. Although there are mathematical formulae for measuring various types of reliability, what we usually mean by reliability is whether or not our test results are reproducible. A behavioral test may be an excellent predictor of neurobehavioral toxicity in humans in one test, but subsequently the results may not be reproducible. A low reliability of a test is usually due to the presence of uncontrolled variables that add to the true variance of a predictive test in a random manner. A useful test for predicting neurobehavioral toxicity will be characterized by both high reliability and high validity.

2.1.4 *Differences imposed by measuring systems*

In the previous section a comparison was made between the doses of morphine that produced "analgesia" in rat and human. The point was made that much higher doses were required in the rat. Why? One possibility is that there are pharmacokinetic differences among species in the way that morphine is handled. Another possibility is that there may be differences in the anatomical substrates upon which morphine produces its characteristic effects (e.g., differences in morphine receptors across species). However, it is also likely that we are measuring something very different in a rat and a human in the response to morphine, despite giving both phenomena the verbal label of analgesia. In such cases we may impose species differences by measuring different things while assuming that they are the same thing. It is very tempting to assume two behaviors are similar or identical because they appear to be so, but like most temptations, it should be avoided.

2.2 DEFINING "ADVERSE" EFFECTS OF NEUROBEHAVIORAL TOXICANTS

Toxicant equals poison. When we use the word toxicant we are implying that the chemical has adverse consequences; however, defining the term adverse in neurobehavioral toxicology is not always easy. This is less of a problem for some other measures of neurotoxicity, as almost everyone would agree that a chemical that produces a lesion in the brain has produced an adverse effect, even when the functional consequences of the lesion are not readily apparent. With behavioral measurements disagreements can arise quickly. For example, as part of the registration and reregistration process, a number of chemicals are now subject to neurobehavioral toxicity testing in the United States under the Federal Insecticide, Fungicide and Rodenticide Act (FIFRA). The general requirements under the Act require that studies on these chemicals be conducted to determine their effects in a functional observational test battery and on spontaneous motor activity in rats. The question arises as to what constitutes an "adverse" effect on a behavior such as spontaneous motor activity. If exposure to a chemical slightly increases motor activity should this be considered to be an adverse effect? Are large changes required? Should increases and decreases in motor activity be given the same degree of consideration in determining adverse effects? Several years ago some unpublished experiments in our laboratory suggested that trimethyltin at a low dose exposure slightly reduced the number of errors made by the test animal in a response-acquisition procedure. Should this behavioral change be considered to be adverse, even though it represents an "improved" performance? These considerations have led to disagreements between those who would define any behavioral change as an adverse effect and those who would attempt to define adversity.

2.2.1 *Direct versus indirect neurobehavioral toxicants*

A criticism sometimes leveled at neurobehavioral toxicity studies is that behavioral changes produced by a chemical do not always reflect effects on the CNS. Anyone ever stung by a bee knows that the incident can produce immediate behavioral changes, yet few would argue that these changes were mediated by transport of the bee venom into the brain. Obviously, the sting has produced a marked peripheral effect which has indirectly produced effects in the CNS and subsequent changes in behavior. Should bee stings be labeled as producing behaviorally toxic effects? Certainly a bee sting fits the definition of an event that results in a profound change in behavior.

Similar problems arise when considering toxic chemicals to which we are environmentally exposed[33]. A chemical ingested by the animal may produce a transient nausea, and subsequently a change in behavior. Is it appropriate to label this chemical as a neurobehavioral toxicant? Certainly such chemicals do not qualify as neurotoxicants, but they may be behavioral toxicants. The behavior of an organism can be considered to be a measure of the well being of that organism. Behavioral changes suggest that in some way the well-being of the organism has changed and all such data need to be considered seriously.

2.2.2   *Are all chemicals behavioral toxicants at high doses?*

The importance of studying the relationship between dose and response has already been discussed. It is probably true that the administration of enough of any chemical will produce a behavioral change, but it is only through a description of the dose-effect curve that we are able to solve the dilemma that this problem produces. If a chemical does not produce effects on behavior until very large doses are given, perhaps doses well in excess of those likely to be encountered environmentally, then that chemical may not need to be considered to be a neurobehavioral toxicant.

2.2.3   *Quality of life issues*

Some chemicals do not produce strong effects on the brain, nor any obvious adverse effects on peripheral receptors, yet we find exposure to them to be unpleasant. For example, many chemicals smell bad, but otherwise produce few pharmacological or toxicological effects. People frequently react to unpleasant chemical odors by attempting to avoid further contact with them. The finest houses are not located immediately downwind from the paper mill, because the output of the mill smells bad. Chemical odor has influenced behavior and it lowers the quality of life for those forced to live near the mill. As such it might be said that the chemical with an unpleasant odor meets the criteria of a neurobehavioral toxicant.

3   Basic principles of addiction

3.1   BEHAVIORAL DEPENDENCE, PHYSICAL DEPENDENCE, AND TOLERANCE

The word addiction is difficult to define precisely. The term could refer to an individual whose entire life revolves around drug seeking and drug taking; however, it may be a term that is better to avoid than to define. The words drug dependence and drug tolerance have more precise meanings. Tolerance occurs when a pharmacological effect produced by a given dose of a drug decreases with repeated administration of the drug. Usually, the pharmacological effect can be reinstated by giving a higher dose of the drug. An extremely powerful demonstration of tolerance occurs when a dose-effect curve is established for a drug and then increasing doses of the drug are repeatedly administered and the dose-effect curve is shifted. If a marked tolerance has occurred, the whole dose-effect curve shifts to the right on the abscissa of a graph plotting dose (on the abscissa) against response (on the ordinate). The degree to which tolerance has occurred can be inferred by determining the degree to which the dose-effect curve has shifted during chronic drug administration. For example, repeated dosing with morphine causes a large shift to the right in the dose-effect curve for the analgesic effects of morphine, perhaps more than 100-fold (the original curve can be reproduced at doses more than 100 times those that originally established the dose-effect curves). Most other drugs, however, do not produce shifts in the dose effect curve of this magnitude. For example, the repeated administration of phencyclidine can only shift the dose-response curve about 2-fold[4].

The hypothetical data in Figure 1.2 compare two drugs in tolerance development. The drug in the frame on the left is phencyclidine-like, in that its dose-effect curve is shifted only slightly to the right after chronic administration, while the drug in the frame on the right is morphine-like because its dose-effect curve is shifted far to the right following repeated administration.

13

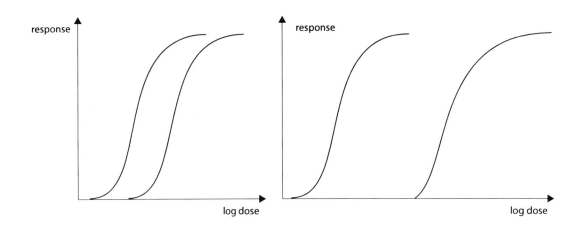

FIGURE 1.2

Effects of repeated administration of two different drugs showing differences in tolerance development

Abscissa: dose, log scale. Ordinate: percentage of maximum effect. The curve on the left in each frame shows the original dose-effect curve and the curve on the right shows the dose-effect curve after repeated administration of the drug. The hypothetical drug in the left frame resembles phencyclidine and the hypothetical drug in the right frame resembles morphine.

Several mechanisms have been implicated as a basis for tolerance development. Some drugs produce tolerance by increasing the rate of their own metabolism. It has been known for many years that barbiturates stimulate the liver microsomal enzyme system to increase liver size and the amount of enzyme[1]. This decreases the duration of action of barbiturates metabolized by this system. However, enzyme induction is not a major factor in the development of tolerance to drugs such as morphine. At the biochemical level, many mechanisms have been proposed to explain tolerance development, including changes in neurotransmitters, receptors, and second messengers, but at this time the exact mechanisms underlying some very large tolerances to abused drugs are not well understood[6]. It also appears that animals can learn tolerance, that is, they learn to counteract the effects of a drug when it is administered repeatedly. This has been shown in experiments in which the degree of tolerance development is greater when animals have the opportunity to practice the task while under the influence of the drug. This has led to an interesting theory which suggests that in instances when the drug effect is costly to the animal in its interaction with the environment, tolerance is most likely to develop. When the drug effect is less "costly" to the animal, tolerance is less likely to occur[38].

Drug dependence is an inferred state which is based on what happens when drug use is discontinued. There are two types of drug dependence, physical dependence and behavioral dependence. Physical dependence is inferred when physical signs appear after a drug is withdrawn. For example, during withdrawal from chronic morphine administration a variety of physical signs appear including vomiting, diarrhea, piloerection, cardiovascular changes, and many others. Often the signs and symptoms of withdrawal are opposite in direction from the acute pharmacological effects that the drug produces.

Behavioral dependence is also an inferred state based on what happens when a drug is withdrawn. The major symptom of behavioral dependence is drug-seeking behavior. Such behavior often leads to the exclusion of other behaviors, such as eating or working at a job. When the addict does not have the drug he is attempting to obtain it, often to the exclusion of other behaviors. Behavioral dependence occurs because abused drugs are powerful reinforcers. They produce effects that users find to be very pleasant. The repeated delivery of a powerfully reinforcing drug following successful drug-seeking behavior strengthens this behavior until it becomes exceedingly difficult to disrupt it or eliminate it. Furthermore, drugs differ in their reinforcing properties. Hallucinogens such as LSD are only weak reinforcers compared to drugs like cocaine, therefore, cocaine maintains drug-seeking behavior to a much greater extent than LSD does.

As mentioned above, some drugs produce physical dependence so that when the drug is not available the chronic user shows physical withdrawal signs and symptoms. Thus one of the motivators of continued drug use is the avoidance of this syndrome (termed a withdrawal syndrome) when the drug is not available. Although it is clear that the avoidance of withdrawal illness is an important factor in maintaining drug use on a regular schedule, it is also clear that some drugs that are widely abused produce only a mild physical withdrawal syndrome. In some cases, the positive reinforcing effects of drug use can be more important in maintaining drug-seeking behavior than is the avoidance of withdrawal. In fact, it is not unusual for users of abused drugs that produce a marked physical dependence to choose to undergo voluntary withdrawal from their preferred drug, only to relapse to drug use at a later time, well after physical dependence signs and symptoms have disappeared.

3.2   THE SELF-ADMINISTRATION MODEL

The patterns of drug self-administration can be studied in human addicts; however, there are many variables that cannot be manipulated easily in such studies. Furthermore, it is obviously unethical to study the acquisition of human drug abuse patterns in a laboratory setting. Therefore, we have come to depend on animal models to answer a number of questions about addiction. Such models have proven useful in screening for the abuse liability of new drugs, developing new drugs for the treatment of drug abuse, and for the study of the stimulus properties of abused drugs.

One of the most fruitful models for the study of addiction has been the drug self-administration model developed by Weeks[43] and subsequently used by a host of investigators. In this model a cannula is placed in a major blood vessel of the experimental animal (usually a monkey, although rats, dogs, and other animals have been used), and the other end of the cannula exits the body at a convenient site where it is then attached to a motor-driven syringe. The animal is then given control of the operation of the motor-driven syringe, such that an operant response on a lever activates the motor resulting in the delivery of a measured amount of the fluid from the syringe into the animal's bloodstream. If a drug solution such as cocaine hydrochloride is placed in the syringe, the animal quickly learns to press the lever to deliver the drug. When this occurs, the drug is said to be a positive reinforcer, since it establishes and maintains the operant responding that produces the drug.

This procedure has been very useful in studying drug self-administration. It is a valid procedure for predicting the abuse liability of drugs in humans since most drugs that humans readily abuse are self-administered by animals, including cocaine, amphetamines, barbiturates, benzodiazepines, opiates, ethanol, phencyclidine, and many others[22]. An exception may be the class of hallucinogenic drugs. Drugs such as LSD, mescaline and other hallucinogenic drugs are not readily self-administered by animals. In fact, there are some suggestions that these drugs may be aversive in animals[18]. The psychoactive ingredients (tetrahydrocannabinols) of marijuana, one of the most widely abused drug preparations in the world, are usually not self-administered by animals. Nevertheless, the self-administration procedure has successfully predicted the abuse liability of many drugs.

The technology of the self-administration model has also been extended to other routes of administration. Falk and colleagues[10] have shown convincingly that animals can be trained to self-administer a drug by the oral route, and other investigators have established the self-administration of volatile chemicals by inhalation. These technical advances have made it possible to study drug self-administration in animal models by the same routes of administration used by human drug users.

3.3 DRUG DISCRIMINATION

Drug self-administration procedures can be used to determine whether or not a drug functions as a reinforcer. The procedure does not indicate which chemical class of reinforcers the drug belongs to. Drug discrimination techniques can provide this classification. Drug discrimination techniques can tell us if a drug is of the amphetamine type, opiate type, barbiturate type, and so on.

The drug discrimination procedure is usually conducted by training an animal to respond on two different manipulanda (e.g. response levers). On some days the animal is injected with the training drug before the training session, while on other days the animal receives a placebo or no drug. Responses on only one of the two manipulanda are reinforced, usually with food, during each training session. The only cue that the animal has as to which response lever will produce the reinforcer during that session is the drug state. For example, the animal may undergo discrimination training with responses on the left lever reinforced on days after pentobarbital administration and responses on the right lever reinforced on days after saline administration. After a number of training sessions, the animal acquires the discrimination such that responses are almost always made on the left lever after pentobarbital and on the right lever when pentobarbital has not been given. Once the animal has learned to discriminate the presence and absence of the training drug, a host of experiments become possible, including determination if other doses of the training drug will generalize to the training dose, and if other drugs from the chemical class of which the training drug is a member will generalize to the training drug. It has been found that there is a high degree of specificity across drug classes[37]. For example, in animals trained to discriminate morphine from saline, the administration of other opiate agonists that act primarily at the mu receptor produce responding on the drug key, but the administration of non-opiate depressants do not. Data from drug discrimination experiments in animals have been found to correlate highly with data from subjective reports in human beings. Humans find the subjective effects of barbiturates, benzodiazepines, and ethanol to be very similar. Animals trained to discriminate any one of these drugs from saline show cross generalization to the others[2].

3.3.1 Dose-effect curves in drug discrimination

As is always the case in pharmacology and toxicology it is important to determine full dose-effect curves in drug discrimination studies. At low doses of the training drug, subjects typically respond on the key associated with the non-drug state. As the dose increases the subject will switch to the drug-associated key. Under some conditions the shift is quite abrupt, while under other conditions the subject may distribute responses on both keys at intermediate dose levels[41]. The conditions under which graded verses abrupt shifts occur are beyond the scope of this chapter. The use of drug discrimination procedures to study behavioral toxicity has not been employed widely in the study of environmental toxicants, although Cory-Slechta and colleagues[5] have begun to apply the procedure in behavioral toxicology. Her experiments have been designed to study the role of neurotransmitter mechanisms in the behavioral toxicity of lead. Drug discrimination is established for drugs considered to exert their effects through neurotransmitter mechanisms and then dose-response curves are established for these drugs before and after exposure to lead. The data suggest that chronic postnatal exposure to lead produces a supersensitivity to dopamine agonists[5].

Not only has the drug discrimination procedure been very useful in identifying drugs with similar discriminative stimulus effects, but it also has been helpful in determining the mechanisms underlying the stimulus properties of drugs. For example, many drugs are thought to exert their stimulus effects by binding to specific receptors in brain. Much of the evidence to support the idea that the stimulus properties of drugs arise from the ability of drugs to bind to specific receptors in the brain comes from the high correlation between the affinity of a series of drugs for a particular receptor and the relative potency of these drugs as discriminative stimuli in animals trained to discriminate one of them. The table below shows the affinity of a series of drugs for the phencyclidine (PCP) receptor based on the data of Quirion *et al.*[35] and the potency of these same drugs in animals trained to discriminate PCP from saline[39]. The correlation coefficient for receptor binding affinity and discriminative stimulus potency is 0.99. Such data strongly suggest that the stimulus effects of PCP-like drugs are mediated through actions at the PCP receptor.

TABLE 1.2

Relative potency of drugs as PCP-like stimuli and as PCP-receptor ligands

Drug	Relative potency	Relative affinity
Phencyclidine	1.0	1.0
Ethyl derivative	4.84	6.0
Thienyl derivative	1.46	1.7
Morpholine derivative	0.91	1.4
Thienyl/morpholine derivative	0.07	0.05
Dimethyl derivative	0.5	0.3
Ketamine	0.1	0.1

3.3.2 *Chronic exposure and drug discrimination*

As has been mentioned previously, exposure to most drugs by humans is chronic. Very little work has been done on the effects of chronic exposure to drugs on drug discrimination. For drugs to which marked tolerance has long been recognized, there is some evidence that with repeated exposure to the training drug, tolerance also develops to the discriminative stimulus properties of the drug, although not necessarily to the same degree as tolerance to other effects of the drug.

The repeated administration of some drugs results in an increased sensitivity of the organism to the effects of these drugs (sensitization), as has been discussed previously. Questions regarding sensitization have not received much attention in drug discrimination research.

4   Issues in addiction research

In the previous section on neurotoxicology it was pointed out that animal testing and data from human studies have formed the basis for the risk assessment process. In this process, animals typically have been exposed to potential neurobehavioral toxicants and dose-responsive effects have been determined. From these data a lowest observed effect level or a no-observed effect level has been calculated and then safety factors have been applied to determine an acceptable daily intake (ADI) or a "reference dose" for humans. Animal tests are playing a similar role in the substance abuse field, although under less formal rules than those employed for regulation of environmental chemicals. For example, animal tests have played an important role in the search for psychotherapeutic agents of many types[15].

4.1   EXTRAPOLATION FROM ANIMALS TO MAN

In drug abuse research the basic question about extrapolation of data from animals to humans must be answered, as was the case with neurobehavioral toxicology. The question is whether or not animal tests can be used as predictors of the development of tolerance, physical dependence, behavioral dependence and the subjective effects of drugs in humans. The answer to this question seems to be a qualified yes.

Using behavioral baselines, many investigators have studied the development of tolerance to abused drugs. Generally, when tolerance occurs to a drug in an animal model, a similar tolerance can be observed in humans. There are a few exceptions to this generalization, especially when the drugs produce somewhat different effects across species when given acutely. For example, morphine produces predominantly stimulant effects on motor activity in species such as mice and horses, while morphine produces depressant effects on motor activity in rats, dogs, monkeys, and humans. Tolerance seems to develop to a far greater degree to the depressant effects of morphine on motor activity than it does to the stimulant effects. Thus the rat would be a better predictor of the human response to repeated morphine administration than would the mouse. Similarly, some species receiving drugs chronically may reflect the human development of physical dependence more than others. As an example, the repeated administration of morphine in high doses to a pigeon does not seem to produce many overt signs of physical dependence when drug administration is discontinued. In contrast, rats, monkeys, and dogs show marked signs of withdrawal from this drug after it has been administered chronically. Thus the extrapolation can be species dependent.

Drug self-administration by animals has become an exceedingly popular technique for determining the abuse liability of drugs. Animals generally self-administer opioids, sedative-hypnotic drugs, stimulants, volatile solvents, and ethanol. So do humans, however they also self-administer hallucinogenic drugs such as LSD, mescaline, and marijuana, none of which has been well established as good reinforcers that will maintain self-administration in animal models. Thus the self-administration model is excellent for predicting the self-administration of drugs by humans for most drug classes, but there are exceptions.

Similarly, the drug discrimination procedures have been used to model the subjective effects of drugs in humans. Although animals cannot provide verbal descriptions of their subjective response to a drug, drug discrimination procedures do permit us to classify drugs as being either similar to or different from a drug that has been established as a discriminative stimulus. Animals trained to discriminate amphetamine from saline will respond on the drug key when given cocaine, but not when given a variety of depressant drugs. Humans will report that cocaine and amphetamine have similar subjective effects that are quite different from those of depressant drugs.

4.2 VULNERABILITY TO DRUGS OF ABUSE

Today, many young people experiment with illegal drugs. For many of them the experience terminates after some experimentation, but for others the frequency of use, the doses, and the number of kinds of drugs used increases to the point where some form of addiction is evident. Why some individuals use drugs to the point that it becomes the most important aspect of their lives, while others can walk away from drug use is one of the most important questions in the field.

Several explanations have been made to explain such phenomena. Historically it has been argued that the drug user suffers from some form of psychopathology, or is attempting to escape from a life filled with tension and personal difficulties, or has been seduced into drug usage by a friend, or by a peer group. Although these events probably influence the initial phase of drug use, most research has focused on the ancient nature/nurture question.

It has been clear for some time that hereditary factors play an important role in drug abuse. The sons of alcoholics have a high probability of becoming alcoholics[15]. In identical twins reared apart, there is evidence that hereditary factors are very important. Various rat and mouse strains have been inbred to accept or not accept alcohol. There are well recognized differences among ethnic groups in alcohol metabolism and the genetic subtypes for these differences have been well described. Although there is general agreement that genetics plays a very important role in the development of alcoholism, the gene, or genes, involved continue to elude us despite some interesting possibilities[3]. The evidence for a hereditary component for the abuse of drugs other than alcohol is limited, but evidence is beginning to accumulate.

The degree to which drug abuse vulnerability cannot be explained by genetic variables must be a function of the behavioral history of the animal, especially the attempt of the animal to adjust to the contingencies of reinforcement presented by the environment. A very obvious aspect of such environmental reinforcement contingencies is the lifetime availability of the drug. A genetic vulnerability to a drug, unlike an inherited metabolic disease, cannot be expressed unless the drug is available. Some cultures have prevented access to certain drugs so effectively that few members of the culture are ever exposed to the drug. A better understanding of how historical events and environmental contingencies lead to drug use remains an important research priority today.

Although hereditary factors clearly play a role in at least some forms of substance abuse, it is clear that environmental contingencies are also important determinants of drug use. All animals, including humans, will develop physical dependence on opiates if administered adequate doses for a sufficient time period. Similarly, all monkeys successfully implanted with an intravenous cannula can be trained to self-administer cocaine. Obviously, the access to abused drugs is a crucial environmental factor that leads to drug usage.

Much literature has discussed the role of other environmental factors in the etiology of drug abuse. Stress, peer pressure, and other environmental and historical factors have frequently been cited as risk factors for the development of drug abuse problems; however, it should be emphasized that the prediction of drug abuse on the basis of environmental factors is far from an exact science[19]. Most of the studies are retrospective and we are not able to explain why two family members with very similar behavioral backgrounds can differ in their vulnerability to substance abuse.

Personality theories have traditionally looked for individual differences in needs, drives, and motives that might be related to substance abuse in some, but not others. For instance, an array of personality evaluation tests has been given to alcoholic and prealcoholic individuals in an attempt to identify factors that may contribute to alcohol abuse. In general, factors of impulsiveness, unconventionality, and social adroitness are currently thought to contribute to the development of alcoholism[7]. Similarly there has been much recent interest in "dual diagnosis" which refers to the coexistance of drug abuse and a psychiatric disorder. There have been suggestions that personality disorders are more closely associated with drug abuse problems than are other types of psychiatric disorders[7]. Again, most studies on putative relationships between personality types, or psychiatric disorders and drug abuse, are based on retrospective analyses and we are as yet unable to predict with any confidence whether or not an individual with a certain personality type, or a particular psychopathology will become a drug abuser.

4.3 PREVENTION OF DRUG ABUSE

If drug abuse is determined by genetic, environmental, and historical factors then it follows that these are the areas upon which prevention efforts should focus. Although the techniques of modern biology do not yet allow altering gene expression in vulnerable subjects, identification of those with a high risk for the disease might provide possible benefits (and perhaps dangers) for prevention. Similarly, as we begin better to understand the environmental determinants of drug abuse we are in a better position to change the environment to prevent drug abuse patterns from developing. The science of drug abuse prevention is in its infancy, but many investigators are devoting their research efforts to this area. With many illnesses, prevention of the development of the disease is far more cost effective than the treatment of those who have contracted the disease. The same is likely to be true for substance abuse problems as well.

4.4 TREATMENT OF DRUG ABUSE

4.4.1 *Twelve-step programs*

Twelve-step programs, as exemplified by Alcoholics Anonymous (AA), have been very successful programs for the treatment of alcoholics. In recent years a number of related programs have spun off from AA, including Narcotics Anonymous and Cocaine Anonymous. In a 12-step program, group processes are used to help the user "work the steps", beginning with a recognition of loss of control over drug use and proceeding sequentially to helping other users to abstain from drug use, then practicing the 12-step principles in all of their daily activities. The 12-step process includes elements of group therapy, behavioral modification, and modification of the environment, although these are not terms that AA members use in describing the program. Although controlled outcome studies for the 12-step approach are fairly limited, it is clear that 12-step programs are effective for many alcoholics attempting recovery and these programs have expanded rapidly. For example, in 1989 there were almost 50,000 AA groups in the United States and Canada[30]. The approach has several limitations including a sometimes fanatical rigidity to the steps and traditions of the organization by some members that some drug users find offensive. Furthermore, the user must have reached a stage of drug or alcohol use in which there is a desire to discontinue use before treatment can be effective. Nevertheless, the programs have reached tremendous numbers of drug and alcohol users with little drain on the health care dollar[42].

Structured group therapy has also proven to be effective in the treatment of drug and alcohol users. Different types of groups employ different treatment approaches, but all of them have in common that they provide a sense of belonging, social reinforcement and feedback, behavioral limit setting, role modeling, confrontation, and so on[42]. They are also very time efficient and cost effective. As with most drug-abuse treatments outcome data is very limited, especially in terms of long term follow-up of patients undergoing treatment.

Therapeutic communities and halfway houses perhaps represent the ultimate in control of the drug users environment. They provide the recovering addict with strong social reinforcement and usually employ a variety of treatment approaches. The administrators and therapists of such programs frequently consist of former drug users who have shared many experiences that the recovering addict has and will encounter.

Behavioral modification has been found to be increasingly useful in treating addicts. Behavioral modification is based on the behavioral principles of reward and punishment developed by Skinner[16] and other behaviorists. Recent studies have shown that abused drugs function as reinforcers to maintain the behaviors leading to their self-administration. Treatment research has shown that the reinforcement of drug-free behaviors can play an important role in the treatment of drug abuse problems[16].

Psychotherapy has traditionally been a medical approach to the treatment of the drug abuser. Although it is undeniably effective in the treatment of some drug abusers under some conditions, it is not very cost effective given the large numbers of drug abusers and the intensive labor required for this treatment. Some 12-step programs have been extremely hostile toward psychotherapeutic approaches, suggesting that psychotherapy has done little to help the alcoholic and addict. The role of psychotherapy in the treatment of drug abusers is not well defined at this time.

Recently, there has been considerable interest in pharmacotherapy for the drug abuser. Some types of pharmacotherapy have been used for many years. For example, disulfiram has been used to prevent relapse in the recovering alcoholic. If the alcoholic uses alcohol while taking disulfiram, the user becomes sick, thereby receiving punishment for drug-taking behavior. Also methadone has been used to treat opiate abusers. Because it has long lasting opioid activity, it can substitute for the use of other opioid drugs to block craving for these drugs and to prevent the withdrawal discomfort from them. Furthermore, if the addict uses these drugs, due to the cross tolerance between methadone and other opioids, the customary reinforcing effects of the drug are less likely to be experienced. Unfortunately, there are drawbacks to these approaches and appropriate pharmacotherapy is not available for all classes of abused drugs.

5 Some final comments

Neurobehavioral toxicity and drug abuse have many aspects in common. Whether we are exposed to dangerous chemicals because they are present in the environment and in our food, or we actively seek exposure to recreational drugs, the behavioral consequences may be similar. Chronic exposure to such chemicals decreases the ability of the organism to function in its environment, which can decrease the quality of life.

In the next five chapters the reader will learn about principles and techniques for measuring changes in the structure and function of the nervous system produced by chemical exposure. Chapters 7 through 10 will focus on the neurobehavioral toxicity of chemicals in our food, while Chapters 11 through 14 will discuss some of the major classes of environmental toxicants. Impaired function due to chemical exposure, be it voluntary exposure through drug abuse or involuntary exposure to environmental chemicals, represents one of the world's most important public health problems. The present volume should begin to shed light on this subject.

6   Summary

This chapter provides an introduction to neurobehavioral toxicology and to substance abuse. The overview of some of the basic principles in these fields indicates that they have much in common. The major site of action of both neurobehavioral toxicants and drugs of abuse is the brain and exposure to either of these groups of chemicals produces marked effects on behavior, frequently with serious consequences to the life of the person exposed. Both neurobehavioral toxicology and drug abuse stress the importance of the route of exposure, the frequency and pattern of contact with the chemical, and exploration of dose-response relationships. Both fields struggle with similar problems, such as the extrapolation of animal data to human populations, the degree to which chemical exposure should be regulated by the law, and the rehabilitation of those whose lives have been harmed by exposure. Perhaps the major difference between environmental neurobehavioral toxicants and abused drugs is that exposure to abused drugs is usually voluntary, while exposure to neurobehavioral toxicants is not, but even this distinction sometimes breaks down. Hopefully, this brief overview will set the stage for a more detailed discussion of some of these issues in the chapters that follow.

References

1.   Alvares, A.P. and Pratt, W. (1990) Pathways of drug metabolism. In: Pratt, W.B. and Taylor, P. (Eds.) *Principles of Drug Action*. Churchill Livingstone, New York, pp 365–422.

2.   Ator, N.A. (1990) Drug discrimination and drug stimulus generalization with anxiolytics. *Drug Dev. Res.* **20**: pp 189–204.

3.   Blum, K., Noble, E.P., Sheridan, P.J., Montgomery, A., Richie, T., Jagadeeswaran, P., Nogami H., Briggs, A.H., Cohn, J.B. (1990) Allelic association of human dopamine D2 receptor gene in alcoholism. *JAMA* **263**(15): pp 2055–2060.

4.   Brocco, M.J., Rastogi, S.K., McMillan, D.E. (1983) Effects of chronic phencyclidine administration on the schedule-controlled behavior of rats. *J. Pharmacol. Exp. Ther.* **226**: pp 449–454.

5.   Cory-Slechta, D.A. and Widzowski, D.V. (1991) Low level lead exposure increases sensitivity to the stimulus properties of dopamine D1 and D2 agonists. *Brain Res.* **553**(1): pp 65–74.

6.   Cox, B.M. (1990) Drug tolerance and physical dependence. In: Pratt, W.B. and Taylor, P. (Eds.) *Principles of Drug Action*. Churchill and Livingstone, New York, pp 639–690.

7.   Cox, W.M. (1988) Personality theory. In: Chaudron, C.D. and Wilkinson, D.A. (Eds.) *Theories on Alcoholism*. Addiction Research Foundation, Toronto, p 143.

8.   Davis, J.M. (1990), Risk assessment of the developmental neurotoxicity of lead. *Neurotoxicology* 11: pp 285–292.

9.   Ecobichon, D.J., Davies, J.E., Doull, J., Ehrich, M., Joy, R., McMillan, D., MacPhail, D., Reiter, L.W., Slikker, W., Jr., Tilson, H. (1990) Neurotoxic effects of pesticides. In: Baker, S.R. and Wilkinson, C.F. (Eds.) *The Effect of Pesticides on Human Health*. Princeton Scientific Publishing Co., Springfield, VA., pp 131–199.

10. Falk, J.L., Samson, H.H., Winger, G. (1972) Behavioral maintenance of high concentrations of blood ethanol and physical dependence in the rat. *Science* **177**: pp 811–813.

11. Ferster, C.B., Skinner, B.F. (1957) *Schedules of Reinforcement*. Appleton-Century-Crofts, New York.

12. Festing, M.F.W. (1990) Use of gentically heterogenous rats and mice in toxicological research. A personal perspective. *Toxicol. Appl. Pharmacol.* **102**(2): pp 197–204.

13. Finnegan, L.P. and Kandall, S.R. (1992) Maternal and neonatal effects of alcohol and drugs. In: Lowinson, J.H., Ruiz, P., Millman, R.B., Langrod, J.G. (Eds.) *Substance Abuse: A Comprehensive Textbook*. Williams & Wilkins, Baltimore, pp 628–656.

14. Gambert, S.R. (1992) Substance abuse in the elderly. In: Lowinson, J.H., Ruiz, P., Millman, R.B., Langrod, J.G. (Eds.) *Substance Abuse: A Comprehensive Textbook*. Williams & Wilkins, Baltimore, pp 843–851.

15. Goodwin, D.W., Schulsinger, F., Hermansen, L., Guze, S.B., Winokur, G. (1973) Alcohol problems in adoptees raised apart from alcoholic biological parents. *Arch. Gen. Psychiatry* **28**: pp 238–242.

16. Grabowski, J., Stitzer, M.L., Henningfield, J.E. (Eds.) (1984) *Behavioral Intervention Techniques in Drug Abuse Treatment*. NIDA Research Monograph 46, U.S. Government Printing Office, Washington, D.C.

17. Ho, I.K. and Hoskins, B. (1987) Biochemical and pharmacological aspects of neurotoxicity from and tolerance to organophosphorus cholinesterase inhibitors. In: Haley, T.J., Berndt, W.O. (Eds.) *Handbook of Toxicology*. Hemisphere, Washington, D.C., pp 44–73.

18. Hoffmeister, F., Wuttke, W. (1976) Psychotropic drugs as negative reinforcers. *Pharmacol. Rev.* **27**: pp 419–428.

19. Holden, C. (1994), A cautionary genetic tale: the sobering story of D_2. *Science* **264**: pp 1696–1697.

20. Iversen, S.D. and Iversen, L.L.,(1981) *Behavioral Pharmacology*. Oxford University Press, New York.

21. Johansson, C.E., Fischman, M.W. (1989) The pharmacology of cocaine related to its abuse. *Pharmacol. Rev.* **41**: pp 3–52.

22. Johansson, C.E. and Schuster, C.R. (1981) Animal models of drug self administration. In: Mello, N.K. (Ed.) *Advances in Substance Abuse: Behavioral and Biological Research*. JAI Press, Greenwich, Conn., pp 219–297.

23. McGuire, P.S. and Seiden, L.S. (1980) The effects of tricyclic anti-depressants on performance under differential reinforcement of low rates schedule in rats. *J. Pharmacol. Exp. Ther.* pp 635–641.

24. McMillan, D.E. and Owens, S.M. (1995) Extrapolating scientific data from animals to humans in behavioral toxicology and behavioral pharmacology. In: Chang, L.W. and Slikker, Jr., W. (Eds.) *Neurotoxicology Approaches and Methods*. Academic Press, New York, pp 329–332.

25. McMillan, D.E. and Wenger, G.R. (1986), Neurobehavioral toxicology of trialkyltins. *Pharmacol. Rev.* **37**: pp 365–379.

26. McMillan, D.E. (1975) Determinants of drug effects on punished responding. *Fed. Proc.* **34**: pp 1870–1879.

27. McMillan, D.E. (1987) Risk assessment for neurobehavioral toxicity. *Environ. Health Perspect.* **76**: pp 155–161.

28. Medved, L.I., Spynu, E.L., Kagan, E.I. (1963) Effects of chemicals on classically conditioned responses. *Res. Rev.* **6**: pp 42–74.

29. Mirkin, B.L. (1978) Pharmacodynamics and drug dispositon in pregnant women, in neonates, and in children. In: Melmon, M.L., Morrelli, H.F. (Eds.) *Clinical Pharmacology: Basic Principles in Therapeutics*. MacMillan Publishing, New York.

30. Nace, E.P. (1992) Alcoholics anonymous. In: Lowinson, J.H., Ruiz, P., Millman, R.B., Langrod, J.G. (Eds.) *Substance Abuse*. Williams & Wilkins, Baltimore, pp 486–495.

31. Nebert, D.W., Weber, W.W. (1990) Pharmacogenetics. In: Pratt, W.B. and Taylor, P. (Eds.) *Principles of Drug Action*. Churchill Livingston, New York, pp 469–531.

32. O'Donnell, J.M., Seiden, L.S. (1983) Differential reinforcement of low rate 72-sec schedule selective effects of antidepressant drugs. *J. Pharmacol. Exp. Ther.* **224**: pp 80–88.

33. O'Donoghue, J.L. (1994) Defining what is neurotoxic. In: Weiss, B., O'Donoghue, J.L.,(Eds.) *Neurobehavioral Toxicity: Analysis And Interpretation*. Raven Press, New York, pp 19-33.

34. Owens, S.M., Hardwick, W.C., Blackall, D. (1987) Phencyclidine pharmacokinetic scaling among species. *J. Pharmacol. Exp. Ther.* **242**: pp 96–100.

35. Quirion, R., Hammer, Jr., R.P., Herkenham, M., Pert, C.B. (1981) Phencyclidine (angel dust)/ sigma "opiate" receptor: visualization by tritium sensitive film. *Proc. Natl. Acad. Sci.* **9**: pp 5881–5885.

36. Rees, D.C., Francis, E.Z., Kimmel, C.A. (1990) Qualitative and quantitative comparability of human and animal developmental neurotoxicants: a workshop summary. *Neurotoxicology* **11**: pp 257–270.

37. Schuster, C.R. and Balster, R.L. (1977) The discriminative stimulus properities of drugs. In: Thompson, T. and Dews, P.B. (Eds.) *Advances in Behavioral Pharmacology* (vol. 1). Academic Press, New York.

38. Schuster, C.R., Dockens, W.S., Woods, J.H. (1966) Behavioral variables affecting the development of amphetamine tolerance. *Psychopharmacologia* **9**: pp 170–182.

39. Shannon, H.E. (1981) Evaluation of phencyclidine analogs on the basis of their discriminative stimuls properties in the rat. *J. Pharmacol. Exp. Ther.* **216**: pp 543–551.

40. Slikker, Jr., W., (1994) Placental transfer and pharmacokinetics of developmental neuro-toxicants. In: Chang, L.W. (Ed.) *Principles of Neurotoxicology*. Marcel Dekker, New York, New York, pp 659–680.

41. Snodgrass, S.H. and McMillan, D.E. (1991) Effects of schedule of reinforcement on a pentobarbital discrimination in rats. *J. Exp. Anal. Behav.* **56**: pp 313–329.

42. Washton, A.M. (1992) Structured outpatient group therapy with alcohol and substance abusers. In: Lowinson, J.H., Ruiz, P., Millman, R.B., Langrod, J.G. (Eds.) *Substance Abuse*. Williams &Wilkins, Baltimore, pp 508–519.

43. Weeks, J.R. (1962) Experimental morphine addiction: method for automatic intravenous injection injections in unrestrained rats. *Science* **138**: pp 143–144.

Contents Chapter 2

Neurobehavioral teratology

0-8493-7802-8/99/$0.00+$.50
© 1999 by CRC Press LLC

Chapter 2

Neurobehavioral teratology

Karen D. Acuff-Smith and Charles V. Vorhees

1   Neurobehavioral teratology: definition and relationship to neurobehavioral toxicology

The development of an organism from the time of conception through death is a complex and intriguing process. There are numerous times during development, especially prenatally, in which the organism is vulnerable to insult. Neurobehavioral teratology is the study of the effects of prenatal exposure to agents, e.g., environmental, physical, and chemical, on the development of the central nervous system (CNS) and includes the evaluation of postnatal behavior and functioning. There are four characteristics of the developing CNS that contribute to its susceptibility to injury including: (1) its structural complexity, (2) the extended time over which development occurs, (3) the slow maturation of the blood-brain barrier, and 4) delayed development of metabolic mechanisms to detoxify various agents[103]. There are several other closely related and slightly overlapping terms which also describe this type of research and have similar meaning including developmental neurotoxicology, behavioral teratology, and neuroteratology. These terms may be used interchangeably throughout this chapter.

The related field of neurobehavioral toxicology also studies the adverse effects of agents on behavior and functioning as has been described in detail in Chapter 1. The two fields, neurobehavioral teratology and neurobehavioral toxicology, share many common interests and technologies, yet differ in one major concept. Neurobehavioral toxicology has come to refer to the adverse effects of an agent on behavior and functioning in adult animals and neurobehavioral teratology is the study of postnatal behavior, that is the consequence of prenatal or early postnatal exposure to an agent during development. This distinction exists because the developmental basis of the two fields, embryology and maturity, are unique disciplines.

2   Origins of neurobehavioral teratology

2.1   BIRTH OF NEUROBEHAVIORAL TERATOLOGY

In discussing the beginnings of neurobehavioral teratology one must consider the importance of the parent field, teratology, and the critical role it played in the birth (and rebirth) of neurobehavioral teratology. As pointed out in a recent review[50], teratology is formally defined as "that division of embryology and pathology which deals with abnormal development and congenital malformation"[26]. In 1973, James Wilson[106] defined teratology as "the study of the adverse effects of environment on developing systems, that is, on germ cells, embryos, fetuses, and immature postnatal individuals". Wilson also suggested four types of outcomes that might result from prenatal exposure to various agents: death, malformation, growth alterations, and functional deficits. The origins of and influences on neurobehavioral teratology have been reviewed in greater detail[50,96] and will be briefly presented herein. The birth of neurobehavioral teratology can be traced back to 1963. The term behavioral teratology was coined by Werboff and Gottlieb to refer to the postnatal effects on behavior of prenatal exposure to drugs.

2.2   INFLUENCE OF EARLY RESEARCHERS

In the early research on effects of prenatal drugs many essential elements were not considered. An important principle in teratology that deals with the relationship between the stage of development at the time of exposure to an agent and the type of effects produced in the offspring was not investigated for behavioral effects observed in the offspring. Another pitfall of the early research is that the agents studied were not established teratogens in the sense that they did not produce malformations in humans or experimental animals, and biochemical/CNS end points were not examined. Many teratologists were skeptical of the validity and importance of behavior as a subspeciality of teratology.

The work of some early researchers, including Hamilton and Heeb, influenced the birth of neurobehavioral teratology. Heeb (1947)[41] studied early postnatal effects of differing environmental conditions on later maze behavior. Hamilton and Harned (1944)[39] investigated the effects of prenatal sodium bromide treatment on activity, problem solving, and seizure susceptibility. Hamilton emphasized three findings that have become well known throughout the field of neurobehavioral teratology. These observations include: (1) the irreversible nature of prenatal drug-induced damage, (2) that drug exposed offspring show differences in the mean level of performance as well as increased variability, and (3) dose-response effects appeared on some tests and not on others; this may be related to differences in task difficulty and the level and type of impairment (lack of a direct relationship).

During the later 1950s and early 1960s many researchers studied the biological effects of X-irradiation and attempted to relate behavioral changes to alterations in brain structure (see ref 42 for a review of this work). The concept of recovery of function or functional plasticity was consistently noted in this research. That is, animals could perform normally in many tasks despite large amounts of CNS destruction. This also suggests that considerable compensation occurs after early CNS insult and emphasizes the resilience of the developing brain instead of its vulnerability, as later researchers were to stress[96].

In "Prenatal Determinants of Behavior" (1969) Justin Joffe[53] reviewed undernutrition, hormone, X-ray, and drug research, and he critiqued the shortcomings. Joffe made many recommendations on experimental design and control procedures. This work had an important influence on subsequent work. There was a gap of about 10 years between the initial work of Werboff (1963)[96] and the field of neurobehavioral teratology as we know it today. Factors contributing to this gap include the fact that Werboff changed the focus of his research, partial lack of funding for this new area, and some resistance and skepticism from some investigators in the field of teratology.

The re-emergence of neurobehavioral teratology began between 1969 and 1973 and included a change of perspective. Several investigators were influential in the rebirth of the field and are reviewed in greater detail[96], highlights of which are summarized below. Among the first to examine behavioral teratogenesis from a teratological or toxicological perspective were Haddad and co-workers[38] who studied prenatal methylazoxymethanol. Other important reports followed on the prenatal effects of hyperphenylalaninemia, hypervitaminosis A, salicylate, methylmercury, and 5-azacytidine (see reference 12, 13, and 96 for details and additional references).

The change of perspective which accompanied the newer research differed from the early reports by focusing on principles from teratology for the design of the experiments. The changes included: (1) addressing the relationship between manifestations of postnatal CNS dysfunction, impaired growth and malformations, (2) investigating specificity of effects as a function of stage of embryofetal development, and (3) establishing effects with agents that are documented CNS teratogens at higher doses.

The work of several prominent morphologists from the field of teratology (James Wilson, Joseph Warkany, and Harold Kalter) sparked the interest of several neurobehavioral teratologists. The one key person who played a role in the development of neurobehavioral teratology was James Wilson, who was an early supporter of this subspecialty field of teratology. From Wilson's work one can see that he recognized early that teratology involved the study of more than just abnormal structure. A growing realization was that embryotoxicity embraced a range of effects far broader than that represented by death or malformation alone. Many of these manifestations involve effects on function that cannot be detected as malformations, but that might be detectable during postnatal development as dysfunctions of growth, behavior, or other physiological systems. In 1974, Wilson helped to form the Committee for the Investigation of the Effects of Prenatal Insults on Postnatal Development, which brought together most of those interested in behavioral teratogenesis, leading morphologists to discuss how the entire area of behavioral effects might be addressed.

2.3   REBIRTH OF NEUROBEHAVIORAL TERATOLOGY

Finally, probably the single most important event in behavioral teratology was the discovery, or rediscovery, of the prenatal effect of alcohol on development. In 1973, Jones and Smith coined the term fetal alcohol syndrome (FAS)[54], which immediately gained widespread attention in the scientific, political and public arenas. The characteristic features of FAS are organized into four clusters. These involve effects on psychological functioning, growth (pre- and postnatal), facial morphology and structural development of some major organs. The psychological effects involve mild to moderate retardation, abnormal reflexes, irritability, and state lability in infancy, hyperactivity and learning disabilities in childhood and adulthood[87]. The growth effects involve pre- and postnatal growth retardation and microcephaly. The facial effects involve a series of distinct dysmorphologic features. The structural effects are major malformations; FAS children have an increased incidence of birth defects, most commonly reported to involve the heart, joints, and the palate. By 1976, a rat model was produced that demonstrated the functional effects of prenatal exposure to alcohol. FAS emerged as the largest and most pervasive line of research in neurobehavioral teratology and remains a dominant area of study in the field. FAS represents the largest environmental cause of neurobehavioral teratogenesis to date, with an estimated incidence rate in humans of 1/700 to 1/1,000 births in North America and Europe. FAS has ignited the growth of research in neurobehavioral teratology just as the demonstration of thalidomide embryopathy did in structural teratology.

Another event that stimulated growth of research in this relatively new field of neurobehavioral teratology stemmed from the heroin epidemic during the 1960s and early 1970s in the United States. (For a review see ref. 50.) In this case, neither heroin nor methadone produces structural malformations. However, prenatal exposure is associated with at least transitory neurobehavioral and physiological alterations in the offspring and in neurobehavioral effects that might persist in postnatal development and childhood.

3   Principles, suppositions and central issues of neurobehavioral teratology

The basic principles of teratology proposed by James Wilson[106,107] were generalized sufficiently so they may include behavioral teratogenic responses as well. These principles have previously been extended specifically to apply to the developing CNS[92,96] and will be reviewed below. The first six principles of neurobehavioral teratology closely parallel the six principles of teratology proposed by Wilson. Some of the remaining points are suppositions to encourage research that will address central issues and questions in neurobehavioral teratology and will provide direction for future research.

The *first principle* of behavioral teratogenic response is directly analogous to Wilson's first principle of teratogenic response of genetic determination.

1   The type and magnitude of behavioral teratogenic effects depend on the genetic composition of the organism.

Species differences have been demonstrated for structural teratogens (e.g., thalidomide and acetazolamide) and there are examples that suggest different species show differing responses to various behavioral teratogens (e.g., alcohol and phenytoin). Therefore, it appears that this principle applies just as much to behavioral responses as to other teratogenic responses. Different strains or stocks of animals of the same species may also respond in different degrees to the same agent (trypan blue and hypervitaminosis A,[107]).

The *second principle* of behavioral response is the principle of critical periods of vulnerability and parallels Wilson's second principle:

2   The type and magnitude of behavioral teratogenic effects depend on the stage of development when the organism was affected.

The period of susceptibility to these effects is isomorphic with CNS development, namely, the period of neurogenesis.

The four major periods of general development include preimplantation, organogenesis, histogenesis, and functional organization. These periods overlap in time and each one has its own vulnerability to insult. Since the brain and CNS are the target organ (system) in behavioral teratogenic insults, stages of CNS development are mapped on the developmental timeline in relation to the events of general development and embryonic, fetal, and postnatal periods (Figs. 2.1 and 2.2).

It has been repeatedly demonstrated that organogenesis is the most vulnerable period for behavioral as well as for structural defects. It is important to note that nothing in this principle states that after neurogenesis, during later phases of CNS development, damage cannot be produced. In fact, the principle implies that the CNS is vulnerable at all stages of its development.

Only a few behavioral alterations have been reported from exposure to an agent exclusively from conception to implantation (for review see ref. 44). Furthermore, the few effects reported have not been replicated or confirmed. Exposures during the stage of organogenesis are well known to cause malformations; however, exposures affecting behavioral development are most important when they are at or below minimal malforming doses. Some agents produce behavioral effects as a result of exposure during histogenesis and/or functional organization. Examples of such agents include vitamin A,

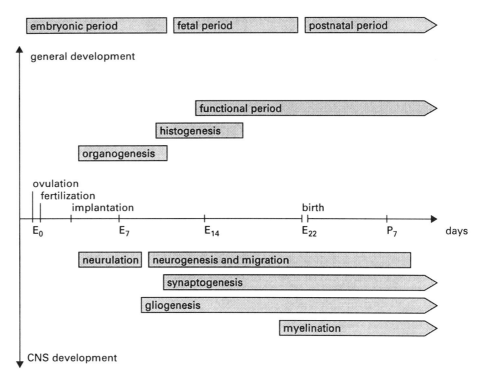

FIGURE 2.1

Major features of rodent development

Depicted are the approximate starting and stopping of various developmental events. The broken line around the time of birth for the events of neurogenesis and migration reflects the slowing down of these processes during the perinatal period. The neurogenesis of most major brain structures is completed prenatally, but postnatal neurogenesis occurs in some structures (cerebellum, olfactory bulbs, and dendate gyrus). The events of general development were adapted from Wilson (1973). The events of CNS development were adapted from Rodier (1980). In the rat, implanation occurs on about Days 6, organogenesis ends on E 15-17, and birth occurs on E 21-22 (the day on which sperm are detected is counted as E 0). Gliogenesis generally succeeds neurogenesis region by region, except for radial glia, which appear earlier and guide neuronal migration. Synaptogenesis includes dendritic and axonal development (arborization), synaptic thickening, and the processes of cell death and, later, synapse reduction. E = embryonic day; P = postnatal day. (Adapted from Vorhees, C.V. (1986) In: Riley, E.P., Vorhees, C.V. (Eds.), *Handbook of Behavioral Teratology*. Plenum Press, New York, Chapter 2, p 25.)

5-azacytidine and methylazoxymethanol (see ref. 96 for a more detailed discussion of this topic).

The *third principle* of behavioral teratogenic response is the principle of specific mechanisms.

> 3   *Agents that are behaviorally teratogenic act on the developing nervous system by specific mechanisms to alter development.*

Regardless of the specific mechanism involved in any type of developmental injury, there is a finite set of final common pathways that mediate all defects. Examples of common pathways include increased or decreased cell death, impaired morphogenetic movements, impaired cell-cell interactions, reduced production of essential cell products, or mechanical injury to cells.

The *fourth principle* of neurobehavioral teratology is the principle of behavioral teratogenic response.

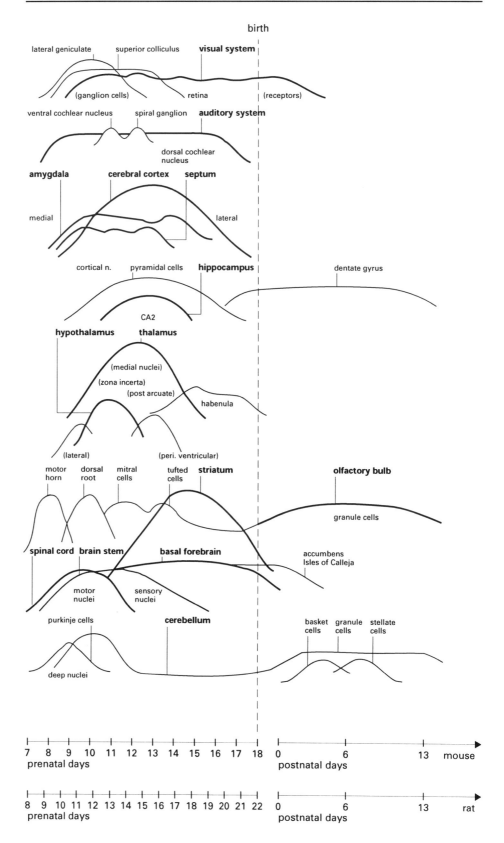

FIGURE 2.2

Stages of neurogenesis in different regions of rodent brain by ³H-thymidine incorporation

(Adapted from Rodier ;[80] reprinted here to include modifications from Vorhees.[103])

4 *Behavioral teratogenesis is expressed as impaired cognitive, affective, social, arousal, reproductive, and sensorimotor behavior, delayed maturation of these capacities, or related indices of compromised behavioral competence.*

The principle of target access is the *fifth principle* of neurobehavioral teratology.

5 *The type and magnitude of behavioral teratogenic effects depend on the type of agent and its access to the developing nervous system.*

Most prenatal drug and chemical exposures are routed through maternal circulation and must either alter maternal function or cross the placenta. Agents which cross the placenta may do so unchanged as the parent compound or be biotransformed as a metabolite. Placental transfer is a unique feature of pregnancy that depends primarily on the size, charge and lipophilic character of the agent and if active or passive transport mechanisms occur at the placental-fetal interface. The same pharmacological principles of absorption, biotransformation, distribution, binding, and elimination apply to the embryo or fetus that apply to the mother, except that the pharmacodynamics of the conceptus may be markedly different because of the immaturity of functions such as those of liver microsomal enzyme systems and of the blood-brain barrier. Infectious agents must also pass through the maternal organism and cross the placenta. When they reach the CNS, brain damage is the most common adverse outcome.

The *sixth principle* of dose-response relationships parallels Wilson's sixth principle of teratogenic response.

6 *The type and magnitude of behavioral teratogenic effects depend on the dose (and actual exposure) of the agent reaching the developing nervous system.*

There are four major endpoints of a teratogenic response, including embryolethality (death), malformations, growth alterations, and functional teratogenesis. The relationship among these endpoints are graphically depicted in Figure 2.3.

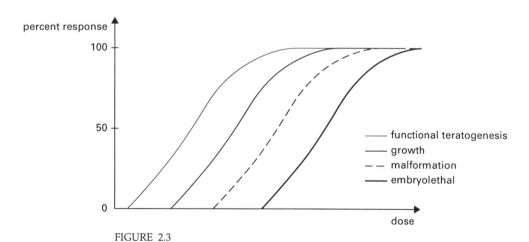

FIGURE 2.3

Idealized dose-response curves for the major manifestations of teratogenesis
The slope and spacing of the curves are dependent on the agent under investigation. The position of the curves is valid only in their ordinal relationship to one another, and if, and only if, the agent in question is capable of producing all four types of embryotoxicity shown. (Adapted from Vorhees, C.V. (1986) In: Riley, E.P. and Vorhees, C.V. (Eds.), *Handbook of Behavioral Teratology*. Plenum Press, New York, Chapter 2, p 29.)

This conceptualization originally included a no effect level, lethality, and malformation, and was formulated by Wilson. Later it was extended by Vorhees and Butcher[92] and Vorhees[96] to include growth and behavior. The theoretical malformation dose-response curve sits to the left of the curve for embryolethality, and the curve for functional (behavioral) effects sits to the left of malformations. The growth curve is often between the curves for malformation and for dysfunction. This suggests that behavioral or functional indices of abnormality are more sensitive measures of teratogenic response than malformation or lethality. This is based on a critically important factor, that the agent is capable of producing behavioral effects. There are drugs that can produce malformations but will produce no behavioral dysfunction. Obviously, in these cases behavior would not be the most sensitive index of embryo toxicity.

There has been some debate about whether or not dose-response relationships clearly exist in neurobehavioral teratology[70]. A number of studies report a dose-response relationship for alcohol, methylazoxymethanol, hydroxyurea, and phenytoin to name a few[93]. When it is not found the situation seems to be one in which the dose-response curve for behavioral effects is very steep, allowing only a very narrow window for the induction of dysfunction.

In designing a study to evaluate dose-response relationships, the two most commonly used approaches to choosing test doses are to approach the issue from either a toxicological or pharmacological perspective. From a toxicological perspective one begins by establishing first a minimal teratogenic dose and then a sub-teratogenic dose, the latter being used as the highest dose for the behavioral analyses. An alternative approach has been to approach dose level from a pharmacological perspective, using doses matched to human therapeutic values or some other pharmacological end point. Toxicological doses are often justified by demonstrating clear dose-response effects in moderately sized laboratory groups as the basis for extrapolations to lower doses and larger groups (population estimates).

The *seventh principle* seems particularly important for neurobehavioral teratology and may be called the concept of environmental determination:

7 *The type and magnitude of behavioral teratogenic effects depend on the environmental influences on the organism, including both prenatal and postnatal environmental factors.*

There are many environmental variables that might be controlled in developmental research. One variable that has received much attention is maternal rearing influence. The fundamental purpose of fostering offspring to a parent other than the biological parent is to sort out the direct effect of the agent on the offspring from the indirect effects. These indirect effects could occur if the agent affects the dam and alters her rearing behavior substantially enough so that it alters the offspring behavior. Researchers are in agreement that this distinction is important; however, many are divided about the priority of determining its role in early investigations. A noteworthy fact is that for all of the drugs reported in the literature to be behavioral teratogens that were given only prenatally and did include fostering/crossfostering procedures, there is not a single one that has been shown to be mediated entirely through postnatal maternal influences. Hence, for some researchers serious questions are raised about the real gain to be made by using fostering/crossfostering procedures indiscriminately. Vorhees (1986)[96] argues against using fostering techniques in initial studies of behavioral teratogenicity. If one decides to study a drug that has not been examined for behavioral teratogenesis before, should the design include a fostering/crossfostering procedure? Starting out with a complex and resource intensive design may not be efficient because it uses many animals, is expensive, and it expends technical effort against the proposition that the drug is a behavioral teratogen even before this is proved. A more reasonable design would include administering the drug prenatally,testing the offspring when reared by their biological dam, and determine if the drug is a behavioral teratogen. The resources required to do a fostering/crossfostering study from the onset could be used to examine several dose levels or replications of the study or even to include more behavioral tests, thus producing more informative data. Fostering/crossfostering experiments have been done with some of the major known behavioral teratogens (e.g., vitamin A, aspirin, phenylalanine, and alcohol) and in all cases, the influence of postnatal maternal variables on the offspring's behavior were either not significant or only accounted for a small minority of the treatment related variance; the majority being attributable to prenatal causation. It seems reasonable that if an agent appears to be a behavioral teratogen based on initial studies, fostering needs to be addressed only during later phases of research[96]. The role of maternal rearing when more subtle agents are administered also needs to be addressed.

The *eighth principle* is strictly an empirical one and is not theoretically based, the principle of types of behavioral teratogens.

> 8 *Only those agents that are CNS teratogens (producing structural malformations such as neural tube defects) or, if not CNS teratogens, those that are psychoactive, are capable of producing behavioral teratogenic effects.*

Vorhees[96] based this conclusion on the observation that thus far only two general groups of agents have been found to produce behavioral effects after prenatal exposure. No evidence has yet appeared in the literature showing any CNS teratogen not also to be a behavioral teratogen (affecting behavior or functioning of the CNS) at lower doses.

The *ninth principle* stems from the principle of dose-response relationships and states that the curve for behavioral dysfunction lies reliably to the left of that for malformations even though the curves may and usually do overlap. To our knowledge there are no exceptions.

> 9 *Behavioral teratogenic effects are demonstrable at doses below those causing overt structural malformations, if the agent is capable of producing adverse behavioral effects.*

Such effects are presumed to be based on lesions observable at a histological or ultrastructural level but are usually not apparent by standard pathological examination.

The *tenth principle* has been indirectly discussed before but bears remembering for the sake of completeness.

> 10 *Not all agents that are capable of producing malformations or other types of teratogenic effects are necessarily behavioral teratogens.*

The agent that best represents this principle is acetazolamide (a potent teratogen for producing limb deformities), but there are undoubtedly others. It should also be noted that none of the principles described imply that behavioral teratogenesis is defined as behavioral effects in the absence of other abnormalities, although some believe these might represent the most important instances of behavioral injury. These have been termed "pure behavioral teratogens". This designation was intended to convey the large region of non-overlap between the doses having functional effects and those having major structural effects. The search for pure behavioral teratogens is important but so is the search for behavioral teratogens with overlapping defects. Fetal ethanol, methylmercury (Minamata disease), retinoids, and fetal anticonvulsants, the most significant behavioral teratogenic agents yet identified in human beings, all have closely associated growth and dysmorphic effects. The fetal alcohol and anticonvulsant syndromes also reflect another important feature of behavioral abnormalities, that the behavioral effects of some teratogens, even when present with physical defects, may be the most significant, devastating and uncorrectable of all the effects observed with the syndrome.

The *eleventh principle*, the principle of preconceptional and transgenerational effects, includes the concept of male-mediated behavioral teratogenesis. This effect is seen when male animals are exposed to a toxin, usually for a discrete interval, and are withdrawn from the exposure and various intervals are allowed to transpire before they are bred with untreated females. The offspring of these matings are then tested during postnatal development and as adults for behavioral competence.

11 *Some agents can induce transmissible behavioral damage to their offspring by exposures that occur before conception, that is, by the exposure of germ cells to the toxin either as developing spermatozoa or as unfertilized ova.*

The *twelfth principle* of behavioral teratology is the principle of pattern of exposure.

12 *Some agents produce a behavioral teratogenic response as a result of total exposure (area under the dose-time curve), while others produce their effects based on peak concentration irrespective of cumulative exposure.*

4 Methods in neurobehavioral teratology and developmental neurotoxicology

4.1 ADVANTAGE OF BEHAVIORAL ASSESSMENT

There are numerous tests that can be used to assess behavior and functioning of the CNS. There are several inherent advantages of using behavioral assessments to evaluate the functional integrity of the CNS[103].

First, behavioral assessment reflects the integrated output of the CNS, therefore, if behavior is abnormal it is reasonable to infer that some underlying brain process is disrupted. Second, behavior can be used to measure the magnitude of a dysfunction regardless of its location within the brain and provides quantitative information concerning the severity of the effect and suggests qualitative information on the type of dysfunction that is present. Third, since functional assessment is noninvasive, it can be measured more than once. To obtain a complete understanding of neurotoxic effects, other techniques are used to provide additional data as suggested by Hartman (1988)[40] and Vorhees (1992)[100].

On the other hand, despite the advantages of behavioral analyses for identifying and characterizing neurotoxic agents, there are also limitations to this approach. One drawback with behavioral testing is that of variability, which presents some intrinsic problems for reliably detecting changes in function. Another problem is that alterations in behavior do not usually provide information as to the specific location or site of injury. Finally, the interpretation of functional changes is sometimes difficult to translate across species.

Assessment of the functional integrity of the CNS is important for determining the potential for neurotoxic effects of drugs, chemicals, and toxins in both developing and adult organisms. As discussed earlier, neurobehavioral teratology evaluates many of the same categories of behavior and often employs similar test procedures as neurobehavioral toxicology. The major differences between these fields of study are the stage of maturity and the time of exposure and testing. A few points are noteworthy when studying developmental effects following exposure during gestation or early in the postnatal period. In evaluating behavioral screening batteries the following points should be considered when studying early brain damage[79].

The CNS of developing animals possesses a greater plasticity which suggests that a tremendous amount of recovery may be possible following a toxic insult during gestation. Another point to be considered is whether or not the severity of effect may be influenced by the amount of time between exposure (insult) and testing. The earlier testing is performed the more likely it is that damage will be detected, before recovery is complete. This point exemplifies the importance of testing neonates and young organisms. However, it should also be emphasized that neonatal testing alone is not sufficient. Some forms of behavioral dysfunction may not be detectable until the animal reaches a critical age, or until more complex behaviors have developed. Even if a recovery process does occur it does not indicate that agent is not dangerous. After the recovery process the effects may be revealed in more subtle behavioral changes or under conditions in which the organism is challenged[79].

4.2 SENSITIVITY, RELIABILITY AND VALIDITY

Three important features of scientific testing in all fields, including neurobehavioral teratology, are adequate sensitivity, reliability, and validity. For a test to be reliable it should produce both a consistent pattern of results within the same laboratory concurrently, over time (intralaboratory reliability), and across different laboratories (interlaboratory reliability)[99].

Within the context of developmental toxicity, the sensitivity of behavioral test methods may relate to: (1) CNS functional integrity in relation to other methods of detecting developmental toxicity, (2) the ability of the test to respond to particular types of dysfunctions, or (3) the detection sensitivity of measurements based on error variance (i.e., how small a change the test can detect as significant)[99].

Anastasi (1982)[8] and Cronbach (1970)[20] describe three basic types of validity which are common to test development in all fields. These include construct validity, criterion validity, and predictive validity. The term validity refers to two aspects of a test, including the parameter that the test measures and how well the test measures the parameter. Construct validity is a term that refers to what biological functions can be measured by a given test method. Test development in neurobehavioral toxicology and teratology has proven more efficient when there is a concept for what should be measured and when there is some understanding about which test methods will provide accurate measurement of the observed trait. The ability of a test to measure a characteristic that can be independently defined is referred to as criterion validity. The criterion selected in neurobehavioral toxicology and teratology has often been based on the ability of tests to detect behavioral dysfunction after exposure to known neurotoxic agents. This method takes advantage of the use of positive controls. Predictive validity refers to the ability of a test to predict effects from an incomplete or partial data set. It is equally important that well chosen tests correctly label those agents that do not produce CNS damage as negative[98,99].

4.3 CATEGORIES OF BEHAVIOR

In developing screening procedures for behavioral teratogenesis a battery approach is advantageous. The battery used must be designed to systematically evaluate a variety of CNS functions. Based on current knowledge of the numerous CNS functional domains that require monitoring and the test methods available for measuring these functions, there are at least five categories of behavior which should be included in a reasonable test battery. These categories include: (1) reflex (neuromotor) development, (2) locomotor activity, (3) reactivity, (4) learning and memory, and (5) sensorimotor functions[5,22,98].

Table 2.1 contains examples of behavioral tests which are used to monitor the above categories of behavior in laboratory animals. Some of the tests which are commonly used for assessing behavioral capacities/deficits in human teratology research are listed in Tables 2.2 (from refs 11, 33 and 52) and 2.3 (from ref. 76).

TABLE 2.1

Measures of various categories of behavior

I. Reflex (Neuromotor)	Surface righting
	Negative geotaxis
	Swimming ontogeny
	Pivoting
	Mid-air righting
	Rotarod
	Forelimb grip strength
	Jumping down to home cage
II. Locomotor Activity	Open field
	Figure 8
	Hole board
	Video-tracking systems
III. Reactivity	Startle
IV. Learning & Memory	Biel or Cincinnati maze
	M-maze
	Active or passive avoidance
	Auditory startle habituation
	Morris maze (spatial navigation)
	Olton maze (radial arm maze)
	Lashley III maze
	Hebb-Williams maze
	Spontaneous alternation
V. Sensorimotor Function	Olfactory orientation
	Acoustic startle
	Tactile startle
	Brightness discrimination M or Y-maze

The lists are not complete, but rather provide a sample. A detailed discussion of these tests will not be provided due to the similarity of these testing procedures with those used to detect neurotoxic effects in adult animals (behavioral toxicology). Many of these procedures will be described in the next chapter (see Chapter 3) which discusses methods used in behavioral toxicology.

TABLE 2.2

Assessing behavioral capacities/deficits in human teratology research

Infant Assessment	Childhood Assessment

– Brazelton's Neonatal Behavioral Assessment Scale (measures habituation, orientation, motor activity, state regulation)	– McCarthy Scales of Children's Abilities
	  Verbal
	  Perceptual-performance
	  Quantitative
	  Memory
	  Motor

– Bayley's Scales of Infant Development
  Mental Development Index
  Psychomotor Development Index
  Infant Behavior Record
	– Stanford Binet IQ Test
	  Verbal
	  Abstract/Visual
	  Quantitative
	  Short term memory

– Fagan Visual Recognition Memory Test (assesses infant information processing)	– Reaction Time Paradigms (assesses cognitive processing efficiency within specific domains) – Measurements of Sustained Attention – Measurements of Activity Levels

TABLE 2.3

An example of a neuropsychological testing battery appropriate for use in young children

Domain Measured	Tests*	Age Range of Norms
Language-Based	Vocabulary (WISC-III) Comprehension (WISC-III) Similarities (WISC-III) Word Attack (Woodcock-Johnson)	6-16 6-16 6-16 6-18
Executive Control	Digit Span Symbol Search (WISC-III) Wisconsin Card Sorting Test Continuous Performance Tasks	6-16 6-16 6-12 5-adult
Spatial Cognition	Block Design (WISC-III) Object Assembly (WISC-III) Beery Visuomotor Integration	6-16 6-16 4-15
General Mental Ability	Based on above WISC-III subtests plus Picture Completion, Picture Arrangement, Coding, and Arithmetic	6-16
'Academic Achievement'	Reading Recognition (PIAT) Reading Comprehension (PIAT) Spelling (PIAT and WRAT) Math (PIAT) Spelling (WRAT)	6-18 6-18 6-18 6-18 6-18

* WISC-III, Wechsler Intelligence Scale for Children; PIAT, Peabody Individual Achievement Test; Woodcock-Johnson Psycho-educational Battery-Revised.

4.4   OTHER APPROACHES

Other approaches can also be used when designing assessments to detect neurotoxic effects in young or developing animals following early insult. One approach is to study the appearance and disappearance of age-related behaviors. Another approach is to evaluate the emergence of adult behaviors which occur in a specific ontological pattern. These two approaches along with examples of each will be briefly discussed.

4.4.1 Age-related behaviors

There are many behaviors which appear transiently during development, the appearance of the response being dependent on the developmental state of the subject. These transiently expressed behaviors exist in humans (sucking reflex and crawling) and rodents. A well-documented example in rats is ambulatory development.

There are consistent stages which include head elevation, shoulder elevation, pelvic elevation, pivoting, crawling, and finally full quadrupedal forward ambulation[7] (see Figure 2.4). Spontaneous forelimb movements become frequent by postnatal (P) days 4 and 5. Pivoting results when the hind limbs do not support the movement of the forelimbs in a coordinated fashion and the pelvis remains on the ground. Pivoting is frequent by the end of the first week (see Figure 2.5). The next stage of development is crawling, which results when the paws make a paddling motion along with some propulsion by the hind limbs. Individuated control over the hind limb digits is evident by P10 or 11. The quadrupedal posture becomes the main style by P12 or 13 with a sudden increase in proficiency around P15.

Olfactory-guided homing (orientation) is another example of an age specific behavior. Researchers have shown that infant rats respond to olfactory cues found in their home environment. Gregory and Pfaff (1971)[37] have described the time course of development of the behavioral response by preweanling rats to orient and move toward the scent of bedding from their home cage. The authors report that by P8, 50% of rats choose the home bedding side over the clean bedding side. The next day (P9) offspring begin to show a preference for the home side and by P12 over 90% of subjects choose the home bedding side. A similar preference is shown for the scent of littermates[6].

Rats also exhibit a variety of nursing behaviors including establishing maternal contact, nipple locating and attachment, sucking, milk withdrawal,

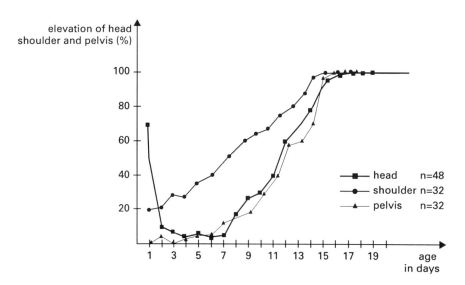

FIGURE 2.4

Mean percentages of the time when head, shoulder and pelvis were raised off the ground during daily 3-min observation periods in the open field

Raising of the shoulder matured earlier than raising of the head and pelvis. By day 15 the entire body was lifted off the ground during the entire period in virtually all animals. *n* in this and the following figures indicates the number of pups tested. (Redrawn from Altman, J. and Sudarshan, K., (1975) *Anim. Behav.* **23**: pp 896–920.)

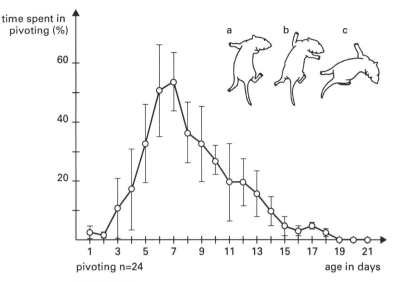

FIGURE 2.5

Mean percentage of time spent in pivoting (either punting or treading)
The decline in pivoting is associated with the emergence of co-ordinated quadruped progression with the entire trunk raised off the ground. Inset: tracings from motion-picture of the three essential steps in "punting". (A) the head is turned to the right; (B) the right forelimb is pulled out from under the head and placed to the right; (C) the shoulder region is pushed to the right by the left forelimb. Vertical lines, 1 SD. (Redrawn from Altman, J. and Sudarshan, K., (1975) *Anim. Behav.* **23**: pp 896–920.)

and cessation of sucking[10]. Several aspects of these behaviors are developmentally specific, so they increase during early developmental stages and then decrease with later development.

Components of swimming behavior serve as another example of a transiently expressed behavior. The emergence of the various stages of swimming behavior have been well-documented. The components have been defined and later refined and expanded by several researchers. The components include head position (body angle), usage of limbs (paddling), and swimming direction[82,91,96]. In the present context, limb usage is particularly interesting because it is a behavior which exhibits patterns only seen during early development. On day 6, rats placed in water often exhibit asynchronous movement of the limbs that may result in small turning movements, but seldom produce forward motion. Within a few days, however, they begin to paddle with all four limbs in an alternating coordinated pattern that propels them forward. Paddling with all limbs begins to disappear around day 16 and animals switch to a behavior termed forelimb inhibition. This marks the emergence of more adult-style swimming in which only the hind limbs are used for forward propulsion and the forelimbs are held stationary tucked under the jaw except during turning, at which time lateral stroking motions of the forelimbs are seen.

4.4.2 *Emergence of adult behaviors*

Another approach used in the assessment of animals for developmental neurotoxicity is the study of the emergence of adult behaviors. Because the unfolding of most adult forms of behavior occurs in stages, the rates of attainment of the adult skills can be used as markers of developmental disruption[103].

A classic example of the emergence of an adult behavior is the developmental pattern associated with the maturation of ambulation. Quadrupedal locomotion emerges in rats on about day 10–12 of life, then quantitatively increases until it peaks between 16–20 days . After the peak there is a rapid decline in the amount of ambulation with advancing age. This phase is followed by a gradual decline that plateaus to nearly asymptotic levels during adulthood. Furthermore, other aspects of motor activity, including nose-poke frequency and rearing, show progressive increases with advancing age, from day 16 until adult levels are reached. Thus, the progressive increase in various components of motor activity represents a series of stages which can be profiled after treatment and examined for shifts in the slope or position of the plotted normal developmental curve over time.

Another example of an adult behavior that emerges in a documented pattern is the mammalian startle reflex. Evaluation of the startle reflex has come to play an important role in neurotoxicity testing. The startle response measures an animal's reactivity and provides a rapid, objective and quantitative assessment of sensory function. The startle response emerges around day 12, increasing in amplitude with age. Small changes in the sensory environment, known as prepulses which can be auditory, visual or tactile and precede the startle stimulus produce highly reproducible changes in the response (inhibits the magnitude of the startle response). Prepulse inhibition develops from postnatal day 12–30 and can be used with young animals to detect defects in sensory abilities. In a single simple paradigm startle amplitude, habituation, ontogeny of startle, and sensory thresholds can be determined.

The dynamic or air righting reflex follows a similar developmental time course to that seen with startle. It has been widely used as a measure of developmental impairment following prenatal and/or neonatal exposure to chemical or physical agents. Basically, animals are held with all four plantar surfaces of their feet pressed gently against the underside of a clear acrylic plate mounted 30 cm above a padded landing surface. The animals are released and evaluated for their ability to right themselves, landing on their feet. The beginnings of the reflex can be detected as early as postnatal day 9 in rats and

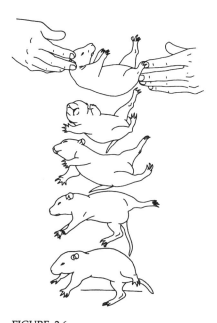

FIGURE 2.6

Tracings of righting in mid-air

Speed 64 frames per s. Rat aged 17 days. (Redrawn from Altman, J. and Sudarshan, K., (1975) *Anim. Behav.* **23**: pp 896–920.)

the response is fully present by day 18. During development not only does the latency to begin the response dramatically shorten with advancing age, but also the speed of rotation accelerates. The most rapid phase of development of the reflex occurs during a window between days 10–14. The animal's ability to perform this response can be compared to the normal developmental curve over time (Figure 2.6).

An ontological pattern has been established for spontaneous alternation behavior[28]. Rats have an innate tendency to alternate direction when placed in the same choice-making environment repeatedly. Rats alternate less than 50% of the time early in life, during the second week of postnatal development (i.e., they make the same direction choice repeatedly). Fifty percent alternation frequency emerges around the time of weaning, and alternation in the adult range (67–75%) is evident by day 30. Thereafter only small additional increases are seen up to 100 days and then no further changes occur.

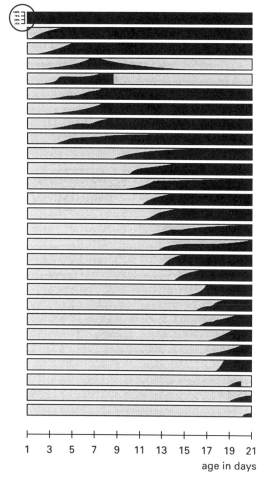

FIGURE 2.7

Summary diagram of the emergence of different postural, locomotor, and related skills

In the majority of instances performance level (0, 25, 50, 75 and 100 per cent) refers to the percentage of animals successful in the full display of the response. In a few instances the reference is to level of performance with respect to asymptotic response frequency. (Redrawn from Altman, J. and Sudarshan, K., (1975) *Anim. Behav.* **23**: pp 896–920.)

TABLE 2.4

Summary of chemical agents for which positive neurobehavioral teratogenicity have been found

Drugs	Pesticides	Solvents	Heavy metals
Drugs of abuse	– Aldrin[60]	– Benzene[71]	– Cadmium[71]
– Alcohol	– Baygon[71]	– t-Butanol[71]	– Lead[71]
– Cocaine	– Carbaryl[71]	– Benzene[71]	– Manganese[71]
– Heroin	– Carbofuran[71]	– Carbon disulfide[72]	– Mercury (organic)[71]
– Marijuana[2]	– Chlordane[71]	– Chloroform[71]	– Tin (organic)[71]
– Methadone	– Chlordecone[60]	– 2-Ethoxyethanol (2-)[71]	
– Methamphetamine[48]	– DDE[71]	– formaldehyde[71]	*Others*
– Morphine	– DDT[60]	– 2-Methoxyethanol[71]	– Halogenated hydro-
– Nicotine	– Diazinon[71]	– Methyl chloroform[71]	carbons
	– Dieldrin[60]	– Methyl-n-butyl ketone[71]	• PBBs[71]
Anti-anxiety drugs	– Endrin[60]	– Methylene chloride[71]	• PCBs[71]
– Chlordiazepoxide[97]	– Kepone[71]	– Paint thinner[71]	– Halothane[71]
– Diazepam[97]	– Maneb[71]	– Perchloroethylene[71]	
	– Mirex[71]		
Anticonvulsants[94]	– Nitrofen[71]		
– Barbital[93, 97]	– 2,4,5-T[71]		
– Carbamazepine[97]	– Toxaphene[60]		
– Phenytoin[93, 95, 97]			
– Primidone[97]			
– Trimethadione[93, 97]			
– Valproate[97]			

[2, 48, 60, 71, 72, 93, 94, 95, 97] see Reference List.

Almost every adult behavior emerges from a normal ontological pattern of progressive stages. The behaviors discussed above are not intended to be comprehensive but rather are meant to serve as examples of how behaviors can be developmentally mapped and used as an index of developmental neurotoxicity by comparing normal to deviant patterns of maturation.

There is one last point with regard to methods used in neurobehavioral teratology which is a limitation of current behavioral screens. This shortcoming is the lack of assessment of complex behaviors which are unique to the neonate[79]. A number of behaviors unique to neonates have been discussed earlier, however, few of these involve learning. There are several ways to test young animals for learning that are under-utilized in current behavioral screens including classical and instrumental conditioned (taste, olfactory, or odor aversion) learning. Furthermore, the reader is referred to two excellent reviews on age-specific behaviors as markers of neurotoxicity which will provide additional examples not included in this discussion[84,85].

5   Examples of neurobehavioral teratogens

Chemicals from different classes of substances can act as neurobehavioral teratogens or developmental neurobehavioral toxicants. In Table 2.4 classes of these substances have been summarized and the major representatives of the different classes are mentioned. The reader is referred to Chapters 7–10 for food contaminants and Chapters 11–14 for environmental contaminants.

In the following sections, the neurobehavioral teratology of some popular drugs of abuse will be dealt with. For the neurobehavioral teratology and developmental toxicity of representatives of the other chemical classes, the reader is referred to Section II and III of this volume in which these aspects are covered in the chapters on the neurobehavioral toxicology of specific chemicals.

5.1   BEHAVIORAL TERATOLOGY OF
        SOME POPULAR DRUGS OF ABUSE

The abuse of drugs is not a new problem in our society as history is replete with diverse examples of drug epidemics which tend to come and go in a cyclic pattern. The issue of women abusing illicit drugs received little attention until the realization that women taking drugs during pregnancy may place the developing fetus at risk for congenital malformations and abnormal behavioral development. It is a basic concept of biology and teratology that immature or developing organisms are more susceptible to change than are mature ones. This implies that heightened susceptibility extends throughout the span of development from fertilization to postnatal maturation, although not necessarily at a constant rate.

The types of drugs that are abused are about as diverse as the persons using them and often vary with demographic and cultural characteristics. This section will summarize the current literature of several popular drugs of abuse: (1) methamphetamine (MA), (2) cocaine, (3) opioids, (4) nicotine, and (5) alcohol.

5.1.1   *Methamphetamine*

5.1.1.1   Behavioral teratological studies
         on methamphetamine in humans

MA is the N-methyl analog of amphetamine (AMPH). The amphetamines have been abused in many cultures over a long period of time due to their CNS stimulant properties. The manifestations of MA include psychomotor stimulation, euphoria, increased alertness and self-esteem, decreased fatigue, and loss of appetite. The CNS stimulatory effects can occur with a dose as low as 1 mg/kg. As with some drugs of abuse, tachyphylaxis or acute tolerance develops with repeated MA use. Long-term users tolerate higher doses with few symptoms. Chronic abusers sometimes increase their dose to as much as 5,000–15,000 mg/day.

In the past several years there has been a dramatic increase in the illicit production and use of MA. The renewed popularity is appearing in a new dosage form that is a pure crystalline product of the hydrochloride salt, known on the street as "ice". This form of the drug is smokable and sufficiently volatile to vaporize in a pipe and be inhaled. This results in rapid absorption which produces a rapid onset of action; the duration of effects lasts several times longer than cocaine due to the longer half-life of MA (8–24 hours compared to 30 minutes for cocaine). The advantage to the user of smoking MA is that it avoids the problems associated with i.v. administration, including the risk of HIV or hepatitis B infection, yet achieves the same rapid onset of effect. The amphetamines are known to affect the presynaptic terminals of catecholamine (CA) neurons. They are indirect agonists, exerting their effect primarily through the release of CAs rather than acting directly on the receptors. MA causes the release of neurotransmitters, inhibits the breakdown, and prevents the re-uptake resulting in an increased amount of neurotransmitter being active in the synaptic cleft. Despite the rising popularity of MA use and the estimated number of women using the drug who are of childbearing age, the clinical literature on prenatal MA exposure is limited.

Little *et al.* (1988)[56] evaluated a group of 52 neonates born at Parkland Memorial Hospital in Dallas that were exposed to MA *in utero*. The drug-using mothers used other substances along with MA including tobacco, marijuana, cocaine, and alcohol. No differences in pregnancy or neonatal complications were found. Maternal MA abuse was associated with significantly decreased birth weight, birth length, and head circumference. There was no increase in major anomalies.

Another group of neonates was retrospectively studied in 1987 by Oro and Dixon[75]. This study included infants born in San Diego county with urine positive for MA (n = 28) or cocaine (n = 13) in addition to a control group of 45 infants. They also found growth retardation and decreased head circumference. In addition, they note an increase in preterm deliveries with fetal distress. Using the Finnegan scoring system[30] MA exposed infants showed abnormal sleep patterns, tremors, hypertonia, poor feeding, vomiting, high pitched crying, frantic sucking, and tachypnea. Three MA-exposed infants required gavage feeding due to lethargy and poor feeding. Statistical analyses were used that evaluated the role of demographic and economic factors, such as poverty, poor nutrition, and lack of prenatal care. The results indicate that only drug exposure contributed to the adverse outcome.

A later study by Dixon[24] evaluated 103 infants exposed to cocaine (41%), MA (30%), opiates (18%), and combined cocaine and opiates (22%). The cocaine and MA users were distinct groups. The cocaine users were mostly inner-city minorities and poor, while 77% were black. On the other hand, MA users were predominantly white and came from all areas. Little *et al.* (1988) also found that MA abuse occurred predominantly among whites (90%) with a mean age of 24 years. Dixon performed a series of Brazelton Neonatal Assessment Scale (BNAS) examinations on the infants and found alterations in visual processing, poor quality of alertness, tremulousness, and altered startle reflex. The visual-motor system in both stimulant groups was particularly impaired.[24] Interestingly, during the first year of life, the MA-exposed infants appeared less impaired than the cocaine group. However, other symptoms including oculomotor apraxia, a parkinsonian dystonia, severe tactile-elicited dystonia, intention tremor, severe active hypotonia, and a hemiparesis developed later in the MA-exposed children.

A prospective study was reported by Dixon and Bejar in 1989[25] in which echoencephalography (ECHO) was performed on or before day 3 on a total of 74 infants prenatally exposed to cocaine (n = 32), MA (n = 24), or stimulant and narcotic drugs (n = 18). These three drug groups formed Group 1. Group 2 represented a group of ill neonates with hypoxic-ischemic encephalopathy. Group 3 was comprised of normal neonates. Of the neonates in Group 1, 35% had abnormal ECHO results. Cocaine and MA combined contained 39% with abnormal ECHO patterns suggestive of hemorrhagic or ischemic injury with subsequent cavitation in the white matter in the drug-exposed infants but not the other two groups. Lesions were concentrated in the frontal lobes and basal ganglia. Intraventricular hemorrhage occurred in 12.1% of Group 1 and 5.7% of Group 2 and was not present in Group 3. Subependymal and subarachnoid hemorrhages were also noted in 10.8% and 13.5% of the Group 1 infants respectively, but only 3.4% and 10.3% in the hypoxic group, and 5.3% and 0% in the control group. Ventricular dilation was present in 9.5% of Group 1 neonates, but was present in only 6.9% of Group 2, and was not present in any of the control infants[25].

The children with evidence of cerebral injury did not have any specific neurobehavioral marker in the newborn period that would differentiate them clinically from other drug-exposed infants. Damage in the frontal lobes and basal ganglia may become evident after the first year, when more complex visual-motor and social cognition tasks are required of preschool and school-age children[25]. The results from the above discussed studies are summarized in Table 2.5

TABLE 2.5

Studies on neurobehavioral teratologic effects in children prenatally exposed to methamphetamine

Study	Ref.	Subjects	(n)	Effect
Little et al., 1988	56	MA *in utero*; polydrug exposure	52	↓ birth weight/length ↓ head circumference no major anomalies
Oro & Dixon, 1987	75	MA *in utero*	28	↓ birth weight ↓ head circumference increase in preterm deliveries ↑ withdrawal symptoms
Dixon, 1989	24	MA *in utero* (i.v.)	52	BNAS: altered visual processing ↓ alertness tremulousness *frontal lobe dysfunction?* altered startle reflex ↓ visual motor system
Dixon & Bejar, 1989	25	MA (prospective)	24	Abnormal ECHO results. Brain hemorrhages, ventricular dilation, atrophy

BNAS = Brazelton Neonatal Assessment Scale; ↓ decrease in/or impairment of; ↑ significantly more than in controls

5.1.1.2 Behavioral teratological studies on methamphetamine in rodents

In a study by Sato and Fujiwara (1986)[81] MA was administered to pregnant rats (2.0 mg/kg) s.c. for 21 days during gestation. The authors reported a decrease in overall activity with an increase in vertical activity following prenatal MA exposure. The offspring also showed an altered reactivity to sound as evidenced by reduced suppression of activity to the auditory signal diurnally and increased vertical activity to the signal nocturnally. There was no change in catecholamine concentrations in the striatum, a motor control region of the brain. However, there was a significant decrease in dopamine concentration in the frontal cortex. A serotonin receptor assay showed a decrease binding of a serotonin agonist in the frontal cortex. In contrast, prenatal exposure to MA did not alter striatal binding of other drugs active at specific receptor sites. The decrease in serotonin receptor binding may have resulted from a down regulation of serotonin receptors.

Cho *et al.* (1991)[16] administered MA by s.c. injection at doses of 1.0–4.5 mg/kg on gestation days 7–20 to rats. Doses of MA (2.0 mg/kg and greater) caused a reduction in maternal weight gain. The highest dose of MA (4.5 mg/kg) proved to be embryotoxic; only 1 out of 13 pregnant females delivered live pups. Developmental delays were noted for lower incisor eruption, eye opening and testes descent. Delays were found in negative geotaxis and air righting development. A significant decrease in spontaneous motor activity was reported for males. In later testing, MA subjects showed higher avoidance response rates and decreased latency times on a conditioned avoidance response.

Martin and coworkers have done several studies investigating the effects of methamphetamine. Martin (1975)[63] administered 1, 3, or 5 mg/kg of methamphetamine s.c. twice per day to pregnant rats throughout gestation. The 5.0 mg/kg dose of MA resulted in four out of seven dams failing to deliver, and one litter had only one offspring. MA reduced maternal weight gain, shortened gestation length, reduced litter size, and delayed eye opening. One dam in the 3 mg/kg group cannibalized her litter after birth. Few effects were found in the lower dose groups. In another study, Martin *et al.* (1976)[64] extended the dosing regimen and administered 5.0 mg/kg MA s.c. x 2 per day throughout pregnancy and the suckling period. Confirming earlier findings, MA dams gained less weight during gestation and had a shortened gestation length. Birth and neonatal weights of offspring were also decreased compared to controls. In contrast to the previous study where 5.0 mg/kg MA was maternally toxic, this study administered the same dose of stimulant to 25 dams and it resulted in 23 delivering litters with live pups. Offspring were tested for activity in running wheels once a month until 39 months of age. MA offspring were significantly lighter and more active in the running wheel throughout the experiment. This finding of hyperactivity persisted even when body weight was used as a covariate in the analyses[66]. These authors also report that exposure to MA changed the pattern of saccharin preference but not the overall level consumed.[67] The MA animals were significantly lighter in body weight from birth to 5 months.

Weissman and Caldecott-Hazard (1993)[105] administered 2 or 10 mg/kg MA to female rats before and throughout pregnancy. They report a decrease in open-field locomotor activity in the high dose groups. The MA-treated offspring did not differ from controls in the ability to find the goal on a spatial navigation learning and memory task.

In another developmental study[14], *d,l*-MA 5.0 mg/kg s.c. was administered twice per day to rats on days 12–20 of gestation. All litters were cross-fostered to non-treated females. The authors report that male offspring had significantly shorter crown rump length and lower body weights (24% decrease). MA offspring showed deficits in the acquisition of a food foraging task. The results of an open field locomotor activity test demonstrated there was no difference in baseline nose-poke performance between the MA-exposed and the controls. The MA-exposed rats showed an increase in interleukin-2 stimulated natural killer cell cytotoxicity; however, no differences were found in T- or B-cell mitogenesis. Finally, a greater stress-induced increase in medial frontal cortex serotonin turnover was noted in the MA-treated rats compared to the vehicle controls.

Acuff-Smith *et al.* (1992)[3] administered 50 mg/kg *d,l*-MA or distilled water twice daily by subcutaneous (s.c.) injection on days 7–12 of gestation. The offspring exposed prenatally to MA showed developmental delays such as impaired performance on an olfactory orientation task and were slightly less active than controls as preweanlings. The drug-exposed progeny exhibited hyperreactivity and shortened response time on a test of the acoustic startle reflex. No differences were revealed in response to a pharmacological challenge or on performance in a complex water maze. Furthermore, prenatal MA induced eye defects (small or missing eyes and folded retinas) in more than 16% of the offspring.

Vorhees *et al.* (1993)[101,102] administered 30 mg/kg d-MA s.c. twice daily to neonatal rats on days 1–10 or 11–20 and evaluated the effects of MA on postnatal functioning. Postnatal exposure to this dosing regimen resulted in increased preweaning mortality in the early exposed pups, altered startle reactivity, impaired learning in a complex water maze and in a spatial navigation task.

In another developmental study[4] *d*-MA (5–20 mg/kg) was given to pregnant rats during early (days 7–12) or late pregnancy (days 13–18). The results indicate that prenatal MA is developmentally neurotoxic at higher doses and the effects were stage dependent, organogenesis being more susceptible than fetogenesis. Exposure to MA during early pregnancy induced more behavioral alterations (deficits on development of early locomotion and memory impairments). MA administered late in gestation produced more deleterious effects on survival, body weight and brain neurochemistry (reduction of serotonin in the nuclues accumbens). Ocular defects described previously were also present.

A summary of the behavioral effects of MA in the existing literature is provided in Table 2.6. Overall, it appears that MA has effects on the offspring following prenatal exposure. It is probable that the issues raised earlier, including dose of drug, route of administration, exposure period, and behavioral testing measures employed all interact to produce a complex outcome that warrants further attention.

TABLE 2.6

Summary of behavioral teratology studies on MA

Authors	Dose (mg/kg)	Exposure Period (Gestation days)	Effect
Martin et al. 1975 (63)	1, 3, 5 (x 2/day	E1-21	↑ maternal toxicity ↑ 2 way avoidance ↓ body weights
1976 (64) 1981 (66) 1979 (65)	5 (x 2/day)	E1-21 + P2-21	Trends for delayed devt. ↓ body weights ↑ activity in running wheel No gross pathology
Sato & Fujiwara 1986 (81)	2.0	E1-21	↓ total motor activity Altered reactivity to stimuli ↓ in DA level (cortex) ↓ in binding
Cho et al. 1991 (16)	1.0-4.5	E7-20	Delays in physical devt. ↓ incisor eruption, eye opening and testes decent Altered functional reflexes ↓ neg. geotaxis and air righting ↓ motor activity ↓ latency on C.A.R.
Acuff-Smith et al. 1992 (3)	50	E7-12	↓ olfactory orientation ↑ reactivity in acoustic startle ↑ eye defects
Acuff-Smith et al. 1995 (4)	5, 10, 15, 20	E7-12 or E13-18	↓ maternal and pup weight ↓ devt. of early locomotion Memory impairments ↓ 5-HT nucleus accumbens ↑ eye defects
Cabrera et al. 1993 (14)	5.0	E12-20	↓ body weights ↓ open field activity ↑ 5-HT turnover with stress
Weissman & Caldecott-Hazard 1993 (105)	2 or 10	Before conception +E0-22	↓ open field activity No differences on spatial nav.
Vorhees et al. 1993 (101, 102)	30	P1-10 or 11-20	↑ mortality (P1-10) Altered startle reactivity Impaired learning (complex water maze & spatial nav.)

E = embryonic day, P = postnatal day, C.A.R. = conditioned avoidance responses

5.1.2 Cocaine

Another popular drug of abuse is cocaine, a powerful, short-acting cerebrocortical stimulant. It is available in two forms: cocaine hydrochloride and purified cocaine alkaloid (free base). Cocaine holds a unique place in history and was at one time an ingredient in Coca-Cola in the 1890s. (For a review of the history of cocaine over the past century see ref. 51). Cocaine has been the leading drug abused for the past decade in the United States. Prior to the latest epidemic, cocaine was used mostly by the rich, famous, and educated upper middle class. During the mid-1980s the pattern shifted and the drug was abused by a diverse group, including the poor; this marked the beginning of the recent cocaine epidemic. Epidemiological data reveal that 30 million Americans have tried cocaine and up to 8 million have become regular users, with 5–15% being women of childbearing age. The stimulatory and pharmacological effects of cocaine are similar to those described for MA; the major difference is the duration of the effects. With cocaine, the desired effects last approximately 30 minutes. Cocaine can be administered in a variety of ways, including orally, intravenously, intranasally, and inhalation (smoking) with the latter two being the most popular routes. The route of administration will determine the amount of drug entering the body and the onset of these effects.

The upsurge in cocaine popularity was accompanied by increased media coverage of all sorts, including newspaper, television news and special reports, and popular weekly magazines covering the dangerous side effects. The degree of concern escalated even more in 1985 when the first reports surfaced on the effects of maternal use of cocaine during pregnancy and the resulting effects on the developing fetus and newborn.

Studies from the mid-1980s reported that cocaine was teratogenic, producing genitourinary tract defects, congenital heart abnormalities, gastrointestinal anomalies, and skeletal defects in man and rodent. These reports also described spontaneous abortion, abruptio placenta, premature labor, intrauterine growth retardation, reduced head circumference, SIDS (sudden infant death syndrome), craniofacial defects, and missing digits following cocaine exposure. Possibly even more devastating were the neurobehavioral effects noted to occur with prenatal cocaine exposure. These findings included increased irritability, tremulousness, altered state regulation, abnormal reflexes, impairment of orientation and motor function, poor interactive behaviors, abnormal sleep patterns, and poor feeding. It was speculated by many professionals that these babies would suffer from permanent and severe neurological damage.

These findings were generally accepted without much reservation by the public and many scientific groups, leaving these babies and their mothers negatively labeled as severely impaired and hopeless. The enormous amount of media coverage no doubt played an important role in this rush to judgment. However, some scientists were not confirming these reports in their own research. One researcher summed it up quite well, "The crack baby has become a media star, with multiple misconceptions arising from journalistic sensationalism"[74].

Over the past couple of years there has been a movement toward more critical analysis of these studies and of the possible deleterious effects which raised such great concern throughout society. Several factors are surfacing which reinforce the need for more careful research with appropriate design features and controls for confounding variable (e.g., polydrug abuse, undernutrition). Several investigators have reviewed the literature on the effects of prenatal cocaine exposure and are questioning the results and the rush to judgment that occurred[51,73,74,90] in an attempt to provide a balanced and objective view of this complex picture.

The fate of abstracts submitted to the Society for Pediatric Research related to prenatal cocaine was examined[55]. They report the presence of a selection bias against studies in which no adverse effects were found with cocaine exposure; only 11% of the negative submitted abstracts were accepted compared to 58% acceptance for abstracts reporting serious effects.

Another attempt at critical analysis of the current reproductive/teratological literature was made by Lutiger and co-workers (1991)[59]. They performed a meta-analysis which is a statistical technique used to determine an overall evaluation of the effects of prenatal exposure across a number of previously published smaller studies. This method allows one to pool data from many studies which share similar design features. This process does not, however, overcome the limitations of each study. Nevertheless, random errors that are present in any particular study tend to cancel out. The results of this meta-analysis (20 studies were pooled) indicate that only a significant increase in genitourinary tract malformations was found to be associated with cocaine use. The results did not confirm an increase in other types of defects (e.g., cardiovascular, limb reduction defects, craniofacial defects) suggested by earlier studies.

In a review by Hutchings (1993)[51], four well-controlled studies were summarized which provide important additional information on the developmentally toxic potential of prenatal cocaine exposure. These studies will be briefly reviewed herein. First, Richardson and Day (1991)[78] studied infants who had been exposed to cocaine (n = 34) and evaluated them for growth, morphology, and behavioral function using the BNAS (Brazelton Neonatal Assessment Scale) on day 2 after birth. The BNAS evaluates habituation, orientation, motor activity, and state regulation. No differences were noted between the cocaine-exposed and the control infants.

Second, Neuspiel et al. (1991)[73] studied 51 infants exposed to cocaine (72% of mothers smoked crack). They measured growth, behavior (BNAS at two time points) and mother-infant interactions during feeding. Their findings suggest intrauterine growth retardation (IUGR), evidenced by lower birth weight, length, and head circumference. No behavioral deficits were revealed.

Third, Coles et al. (1992)[18] reported similar effects on growth and behavior as above. The neonates exposed to cocaine prenatally had decreased growth and showed abnormal reflexes at 28 days and altered state regulation at 2 and 4 weeks compared to control infants, but were still within the normal range.

The fourth study by Chasnoff and associates (1992)[15] reports a decrease in birth weight, length, and head circumference. The first two growth parameters were within the normal range by 1 year, but the microcephaly remained through 2 years. Intense follow-up is imperative with these cocaine exposed babies because some drug effects may not appear until after infancy when the child is more challenged cognitively.

These four studies summarized in Table 2.7 provide some optimism for the babies that are exposed to cocaine during development. It may turn out that initial reports overestimated the severity and duration of the effects present at birth. Two important issues dealing with the neurotoxic effects of cocaine raised by Hutchings (1993)[51] are the dose-response effects of cocaine and the possible additive/synergistic effects stemming from polydrug abuse.

The message from these newer studies should not be taken to mean that cocaine is an innocent and safe drug; quite the contrary. Current investigations suggest an association of prenatal cocaine and decreased fetal growth; behavioral effects are certainly possible by direct or indirect mechanisms. The available data, especially animal data, suggest an association between maternal cocaine use and neurobehavioral deficits (see Table 2.8, for a summary of the positive effects of cocaine on postnatal functioning), but the type of effects, duration and long-term implications of these findings remain to be determined[74,83].

TABLE 2.7

Studies on neurobehavioral teratologic effects in children prenatally exposed to cocaine

Study	Ref.	Subjects*	(n)	Effect
Richardson & Day, 1991	78	infants prenatally exposed to cocaine	34	– no effect on behavior (BNAS)
Neuspiel et al., 1991	73	infants prenatally exposed to cocaine (incl. crack; 72%)	51	– no effect on behavior (BNAS) – lower birth weight; smaller head circumference
Coles et al., 1992	18	infants prenatally exposed to cocaine	106	– no effect on behavior (BNAS) – lower birth weight; smaller head circumference – abnormal reflexes (P28) – altered state regulation (P14 and P28)
Chasnoff et al., 1992	15	infants prenatally exposed to cocaine and usually marijuana and/or alcohol	107	– lower birth weight and length; smaller head circumference

BNAS = Brazelton Neonatal Assessment Score; P = postnatal day

Currently, the available data on the developmental neurotoxic potential of cocaine is not conclusive. In closing, an important question to answer is whether or not conventional tests of neurologic and cognitive development are sufficient to detect potential impairments in infants exposed prenatally to drugs. Neurologic and behavioral functions include attention, arousal, motivation, cognition, and social interactions, yet some of these are not easily quantifiable; research may require new test methods and years of follow-up to detect functional impairment. Without this scrutiny there also exists the possibility that we will seriously mis-estimate the adverse effects of cocaine on development[90].

5.1.3   Opiates

Another class of compounds which have a high abuse potential are the opiates, which readily cross the placenta. The opiates include exogenous agents (codeine, morphine, heroin and methadone) while the opioids are endogenous in nature (enkephalins and endorphins). The focus of this section will be on the opiates. Heroin (diacetylmorphine) is the major drug of choice among opiate addicts, likely due to its greater solubility and because it enters the CNS more rapidly than morphine. Heroin was first manufactured in 1874 as an analgesic and was originally thought to be a non-addictive alternative to morphine. In the late 1960s heroin became a popular drug of abuse in the US and was administered by a number of different routes, with intravenous injection being the most common. Methadone is a synthetic opiate and was introduced in 1945. The major use nowadays for methadone (oral administration) has been for the treatment of heroin addiction. Methadone is by far the most extensively studied opiate in pregnancy in both humans and animals.

TABLE 2.8

Summary of positive behavioral teratologic experiments on prenatal cocaine 1982-1993

Article**	Route	Dose	Exposure	Effect
Smith et al. 1989	sc	10	3–17*	↓ Spont. alt. freq. (males only) ↓ Open-field act. (males only) ↑ DRL-20 resp. rate ↓ Tail-flick lat. ↑ T-water mz. lat. (M, early trls. only) ↓ Shock sensitivity
Spear et al. 1989	sc	40	7–19*	↓ Odor conditioning ↓ Shock-induced wall climbing ↑ Act. during shock-ind. wall climb.
Church/Overbeck (II) 1990	sc	20, 30, 40, 50	7–20	↑ Left bias in spont. alt. (non-D-dep.) ↓ Pass. av. ret. at 50 mg/kg
Henderson/McMillen 1990	sc	15	1–birth	↓ Surface righting dev. ↑ Loco. act. at P30; – at P60
Heyser et al. 1990	sc	40	7–19*	↓ Sens. precond., P8, P12; not at P21 ↓ 1st order cond., P8; not at P12, 21
Raum et al. 1990	sc	10, 30	15–20	↓ Scent marking in males ↓ Lat. to intromission
Sobrien et al. 1990	sc	20	14–20*	↑ Surface righting and cliff avoid. dev. ↑ Startle dev. ↓ Res. on loco. act. to d-A and coc.
Church et al. 1991	sc	30	7–20	↓ Loco. act.
Heyser et al. 1992	sc	40	7–19*	↓ Cond. place preference ↑ No. chamber entries
Johns et al. (I) 1992	sc	15	1–20 2–3, 8–9 14–15, 19–20	↑ Loco. act. P30 1st 15' (Coc-D grp.) ↓ Loco. act. P30. dark cycle (Coc-I grp.)
Johns et al. (II) 1992	sc	15	Ibid	↑ Open-field non-entries (Coc-D grp.) ↑ Open-field act. (Coc-I grp.)
Bilitzke & Church 1992	sc	40	7–20	↓ Immobil. time on Porsolt swim test
Goodwin et al. (II) 1992	sc	40	7–19*	↓ Odor cond. P7; coc-coc, 2-4 train. trls. fos/coc 2 and 3 train. trials only – aud. cond. P17; – Odor cond. P17 ↓ Lat. to 1st attack shock-induc. aggr. in coc-coc and fos-coc grps.
Meyer et al. 1992	sc	20	10–19*	Effects tested using coc. challenge: ↑ wall climb. all ch. doses in AL and PF ↑ wall climb. high dose only in coc. gp.
Hutchings et al. 1989	po	30, 60	7–21*	↑ Loco. act. P20 and 23 only at 60 mg/kg

* Adjusted for evidence of conception as embryonic (F.) day 0.
** See Reference 104 for original references.

Opiate metabolism and disposition is highly species specific with the half-life of methadone being 90 min, 3–5 h, and 24 h for non-gravid rats, gravid, rats, and humans, respectively. The continued use of opiates results in the development of tolerance (e.g., it takes an increasingly higher dosage of drug to produce the same desired effect). Furthermore, a physical dependence also occurs and abstinence from narcotic use will lead to withdrawal symptoms. In a pregnant woman this withdrawal syndrome will occur in both the mother and the neonate after birth.

The opiates do not produce an increase in overt structural malformations in animals or humans[43,57,62]. Exposure to opiates during pregnancy does result in adverse effects on development and reproduction. For example, Hutchings and associates (1976)[43] report a dose related increase in maternal mortality, resorptions, and stillbirths after oral administration of methadone (5-20 mg/kg)

TABLE 2.9

Neonatal withdrawal syndrome

(From Hutchings, D.E. and Dow, E.D. (1991) In: Chasnof, I.J. (Ed.), *Chemical Dependency and Pregnancy: Clinics in Perinatology*. W.B. Saunders, Philadelphia, pp 1–22.)

Drug exposure ⟶ Tolerance and physical dependence
  Compound withdrawn ⟶ Rebound hyperexcitability
  Neonatal Symptoms
    CNS Irritability
      Tremors
      Poor state regulation
      Incessant shrill crying
      Sleeplessness
      Hypertonicity
      Hyperreflexia
      Seizures
    Gastrointestinal
      Hyperphagia
      Vomiting
      Diarrhea
    Other
      Lacrimation
      Rhinorrhea
      Yawning
      Perspiration

to rats on days 8–22 of gestation. Opiates are associated with fetal wastage and growth retardation. Clinical studies also report small for gestational age (SGA) babies; infants show decreased birth weight, length, and head circumference. However, in both animals and humans the effects on growth appear to be transient and reversible, since later the narcotic-exposed subjects do not appear different from controls. Methadone substitution for heroin use in pregnancy leads to improved outcome.

Of the neurobehavioral effects that can be produced by the opiates, the most serious effects identified are a set of complex symptoms referred to as neonatal withdrawal (or abstinence) syndrome[49]. This cluster of symptoms occurs in approximately 60–80% of prenatally exposed neonates. The symptoms appear at birth or soon after, usually peaking at about 3–6 weeks of life and then slowly subside, usually by 4–6 months.

Table 2.9 (taken from ref. 49) summarizes the neonatal symptoms; most of these represent a disturbance of physiologic control mechanisms. Typically the infants' alertness, visual responsiveness, and sleep regulation are affected. Some infants show less severe symptoms of subacute withdrawal (restlessness, agitation, tremors, and sleep disturbances) which usually resolve in several months.

To evaluate the long-term behavioral effects induced by prenatal opiate exposure, there are two major sources of data: 1) clinical studies, and 2) animal research. Each will be briefly summarized.

There have been a number of clinical studies, a few longitudinal, which evaluate the neurobehavioral status of infants exposed *in utero* to heroin and/or methadone. The studies typically report no significant deficit in cognitive development (IQ) but there appear to be impairments in organizational and perceptual processing along with low persistence and goal directedness. The narcotic-exposed children are hyperactive, have attention deficits (especially in a structured environment), poor fine motor coordination, and temper (emotional) rages. Another consistent finding is disturbances in sleep patterns or state regulation. Dinges (1980)[23] reported a dose dependent effect of heroin/methadone on sleep patterns. Both low and high dose heroin groups and a polydrug narcotic group exhibited less quiet sleep time, and the latter group also displayed more indeterminate sleep (disorganized pattern of sleep that does not fit into any standard sleep pattern).

The animal data corroborate many of the behavioral findings described in infants exposed *in utero* to opiates. Hutchings and associates (1979)[45,46,47,48] have done a lot of research on the effects of methadone on behavioral outcome. They evaluated rats exposed prenatally to methadone in several operant learning tasks. These measures failed to reveal any learning or inhibitory deficits. The only behavioral effect noted was elevated response rates. This find of increased rates of responding on learning tasks was confirmed by Middaugh and Simpson (1980)[68]. Hutchings *et al.* (1979)[46] developed a test to measure changes in a young rats' rest-activity cycle. The methadone-exposed offspring were more active than the control offspring. In addition, the progeny showed poor state regulation; their normal sleep pattern was severely disrupted and they displayed frequent shifts from low to high activity[45]. These reported findings closely parallel observations noted in human neonates.

In summary, while opiates do not appear to be structurally teratogenic, prenatal exposure is associated with neonatal abstinence syndrome, developmental delays, and persistent problems with fine motor coordination. Furthermore, a potent effect on the newborn is the disruption of normal sleep (rest-activity) patterns and a risk for developing an attention deficit disorder, especially in a structured setting.

5.1.4   *Maternal cigarette smoking*

The use of the previous drugs (methamphetamine, cocaine and opioids) which have been discussed is illegal. Use of the two drugs, cigaretts and alcohol, is not illegal and they are considered non-medicinal "social drugs". Maternal use of nicotine or alcohol during pregnancy can result in adverse and detrimental outcomes in the developing conceptus. The neurobehavioral teratogenicity of cigarettes and alcohol will be briefly reviewed.

There are over 3,800 products formed in the smoke of a single cigarette. There are 10 biologically active components including (in descending concentration): (1) total particulate matter, (2) carbon monoxide, (3) nicotine, (4) acetaldehyde, (5) acetic acid, (6) nitrogen oxides, (7) formic acid, (8) hydrogen cyanide, (9) acetone, and (10) methanol[61]. The majority of studies investigating the developmental effects of cigarette smoke employ prenatal nicotine or carbon monoxide exposure, since they are present in the highest concentrations.

Cigarette use by pregnant women in non-ghetto urban areas has been found to range from 22–28%[36]. Smoking during pregnancy has been associated with a variety of adverse pregnancy outcomes in a large number of well-controlled studies. There is no significant increase in the risk for major malformations following exposure to cigarettes (maternal or paternal) *in utero*, after controlling for confounding factors. However, many studies report maternal smoking to increase significantly the rate of spontaneous abortion with the relative risks ranging from 1.2 to 1.8. In addition, there is a strong association between maternal smoking during pregnancy and fetal growth retardation, manifested as a lowered birth weight. Most of the effects of smoking on fetal growth take place after 30 weeks of gestation, suggesting the later stages of development are more sensitive to the effects of cigarette smoking on growth. The negative effect that smoking has on fetal growth occurs in a dose-dependent fashion (e.g., the greater the number of cigarettes smoked during pregnancy the greater the growth retardation). It has been estimated that the risk for a low birth weight baby increases by a factor of 1.5 for every 10 cigarettes smoked. The lowered birth weight is not the result of a shortened gestation period. Furthermore, the growth retardation is not a long term effect, and it does not correlate with the neurobehavioral deficits detected.

There have been several studies which evaluated infants that were exposed to cigarette smoke *in utero*. A prospective study from Ottawa[32,34,35] is well-designed and it collected subjects over a number of years. A unique feature of this study is the length of time of follow-up, more than 4 years. Data from this study will be summarized briefly. In the first month after birth, those infants exposed to cigarette smoke prenatally exhibited symptoms suggestive of nervous system excitation. These symptoms included increased motor reflexes, muscle tone, and fine tremors. Additionally, the neonates showed poor auditory habituation. Hyperactivity, poor orientation, and impaired attention were also noted in the babies. The babies showed a dose dependent decrease on a mental development index (Bayley Scale) and also a negative association between auditory related items on an infant behavior record.

Evaluation of these subjects at 24 months of life also showed impairments in auditory related behaviors[32]. Maternal smoking did not contribute to the variance of the cognitive deficits after considering the confounding factors.

However, the children of heavy smokers scored more poorly on all measurements compared to children of lighter smokers. The home environment was a very important contribution to the deficits in cognition at this age. The relationship between lowered cognitive skills in childhood and maternal smoking is relatively consistent; a point to consider is that the tests used to evaluate cognitive functioning at this age were not sensitive enough to detect more subtle effects.

The same sample of children from the Ottawa study was evaluated at 36 months[34]. The results showed a significant negative dose-response relationship between maternal smoking and many subscales on the McCarthy Scales of Children's Abilities (verbal, perceptual, quantitative, cognitive index, and memory), with the verbal component being the most severely affected. The children were again evaluated at 48 months for behavioral and cognitive competence[34]. The performance of the smoke-exposed group was negatively associated with cognitive measures in a dose-dependent manner. Three different language associated tests differentiated the heavy, light, and non-smokers to the greatest degree, with the heavy smokers showing the largest impairment. The findings reported for these younger children persisted in 6 year olds, with deficits in auditory and language performance as well as cognitive functioning[35].

The approach taken when investigating the effect of cigarette smoke in a rodent model is based on a primary component exposure. Animals are usually exposed to one of the major components found in cigarette smoke, mainly nicotine or carbon monoxide. Nicotine has been administered by several routes including subcutaneous delivery, addition to the drinking water, or infusion via an osmotic pump. Carbon monoxide is typically administered through inhalation methods, which are more difficult. The reader is referred to a review[61] of the literature on both nicotine and carbon monoxide in animals for details of the reproductive and behavioral effects. Basically, the animal literature supports the clinical data for the effects of maternal cigarette smoking during pregnancy. Prenatal exposure to nicotine or carbon monoxide results in increases in maternal and perinatal mortality and altered body weights, corroborating the growth effects reported in the clinical data. Furthermore, there are reports of delayed development and impairments on learning tasks for animals of mothers treated with nicotine or carbon monoxide during gestation. Many studies also report alterations in the concentration and/or uptake of some of the neurotransmitters found in various regions of the brain (see ref. 61 for details).

The data from both clinical and animals studies suggest that there are adverse reproductive and developmental effects that result from exposure to cigarette smoke during development. There are consistent findings of fetal growth retardation, increased spontaneous abortions and consistent long-term behavioral deficits involving cognitive functioning, auditory performance, and verbal/language deficits[69]. These findings warrant concern for the developing conceptus when maternal smoking occurs during pregnancy.

5.1.5 *Fetal Alcohol Syndrome*

Another popular legal "social" drug is alcohol. The adverse effects associated with alcohol exposure during pregnancy were already noted in the Bible. In early Carthage there was a prohibition against a bride and groom drinking on their wedding night for fear of producing a defective child (for a historical perspective, see reference 87).

In 1968, Lemoine and associates[58] described 127 offspring of women who drank alcohol during pregnancy; they noted the similarity of distinct facial features, growth retardation, and psychomotor disturbances[87]. In 1973, Jones and Smith[54] and co-workers independently described eight children born to alcoholic women. The children displayed the distinct facial characteristics, growth deficiency, and mental retardation; the term fetal alcohol syndrome (FAS) was coined by Jones and Smith to describe this set of symptoms. The incidence rate for FAS is approximately 1/1,000 to 1.9/1,000 live births. If the criteria for a FAS diagnosis are not met but some features are present it is often classified as fetal alcohol effects (FAE), which is milder and not as encompassing as the full syndrome. The incidence rate for FAE is even higher than for FAS (1/300 to 1/400 live births)[86]. The actual rates for both FAS and FAE are probably higher due to cases which are undiagnosed. For pregnant alcoholic women, the risk of having a baby with FAS is approximately 6%, but for the second offspring if the previous child has FAS the risk increases to 70%[1]. To date there have been over 2,000 articles reporting the adverse effects of alcohol on development.

In order for the diagnosis of FAS to be made three teratogenic characteristics of FAS must be present including: (1) characteristic cluster of facial features, (2) growth deficiency, and (3) central nervous system damage[19,21,87]. Alcohol is a known teratogen; intrauterine exposure to alcohol increases the risk for congenital anomalies, including birth defects and dysmorphia, particularly of the face. The facial features of children born to alcoholic women are distinctive and recognizable.

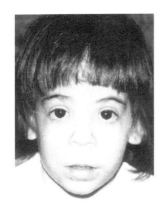

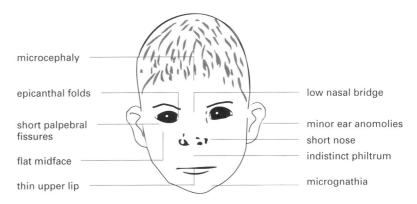

FIGURE 2.8

This figure describes the typical facial features in fetal alcohol syndrome

The features on the *left* are those that are the most characteristic and delineating; those on the *right* are less frequently observed and are not as differentiating. (Adapted from Sterling (1981) and Streissguth et al.[88])

A schematic representation of an FAS face is shown in Figure 2.8[19,21,87,88]. The facial characteristics in FAS include short palpebral fissures (eye slits), low nasal bridge, epicanthic folds, short nose, indistinct philtrum (the ridges running between the nose and mouth), narrow upper lip, small chin (micrognathia), flat midface, and minor ear anomalies. Minor joint anomalies, altered palmar crease patterns and minor genital anomalies have also been reported. In addition to the unique facial characteristics, other physical teratogenic problems include cardiac and urogenital systems, immunologic and skeletal abnormalities, visual and hearing impairment, and dental abnormalities[19,87]. A maternal drinking history together with malformations in the absence of the full FAS are known as alcohol related birth defects (ARBDs).

A second teratogenic characteristic of FAS is growth retardation. The growth deficits begin *in utero* and are typically symmetric, including head circumference, birth length, and birth weight. The extent of growth retardation is related to the extent (dose and duration) of maternal alcohol use. In children who have FAS, the prominent growth deficiencies are persistent, particularly short stature and microcephaly (small head), despite adequate nutrition[19,87,89].

The third teratogenic characteristic of FAS is evidence of damage to the central nervous system. As with other teratogens that are capable of producing behavioral abnormalities, the neurobehavioral effects of prenatal alcohol are produced at lower doses (exposures) than the morphologic defects or growth deficits. In most cases, the behavioral impairments are the more devastating and long-term effects to the offspring[19,21,87,88,89].

In most studies which determine the effects of prenatal exposure to alcohol, neonates are evaluated. There are a much smaller number of studies that follow the children in a longitudinal study. At birth, neonates exposed to alcohol *in utero* show increased irritability and tremulousness, altered state control, poorer habituation, hyperactivity, and eating and sleeping disturbances. Later in life, there are reports of abnormalities in gait, persistence of primitive reflexes, and delays in motor and speech abilities, plus attentional deficits.

As the children who were prenatally exposed to alcohol develop, more complex assessments can be administered for behavioral and cognitive functioning. Many studies have reported persistent cognitive deficits in subjects exposed prenatally to alcohol, with most showing performance in the range of mild to moderate mental retardation. The intelligence quotient (IQ) scores span a wide range varying from 20–108, with the average being about 68. Several research groups note significant decreases in mental and motor scores, as well as long-term deficits in academic achievement. Even at older ages, adverse psychomotor effects are persistent, including speech deficits, hyperactivity, and behavioral disorders. Specific deficits were noted in academic and adaptive functioning with problems involving attention, judgment, and abstraction. Researchers have reported impairments in short-term memory processing, which is mediated by subcortical structures in the brain including the hippocampus (an area susceptible to the effects of alcohol). This impairment leads to deficits in ability to encode visual or auditory information. Deficits were also present in math skills and prereading identification of words and letters[17,79,88,89].

In addition to the long-term effects on cognition and academic performance, 58% of adolescents and adults with FAS or FAE were classified as having significant levels of maladaptive behavior including poor concentration and attention, social withdrawal, impulsivity, dependency, bullying behavior, and elevated anxiety.

Researchers have also investigated the effects of prenatal alcohol exposure in pregnant laboratory animals. The data from these studies corroborate the clinical findings reported in humans. Alcohol has an adverse effect on reproduction and development, evidenced by increased resorptions and reduced litter size. The progeny also show characteristic growth retardation and increased incidences of congenital defects including eye defects, neural abnormalities, digit anomalies, malformations of the cardiovascular and urogenital systems, head malformations, and delayed skeletal ossification. Delays in development as well as neurobehavioral deficits (i.e., learning and cognition) are also documented in animal models of FAS. In addition, there are adverse neurochemical effects that result from transplacental exposure to alcohol. These effects include decreased protein synthesis, delayed myelination, and decreased serotonin in the brain[87].

Heavy maternal drinking of alcohol during pregnancy clearly places the fetus at risk for increased neonatal anomalies, including the three classic teratogenic effects of dysmorphia (especially unique facial characteristics), growth retardation, and long-term persistent effects on cognition. The adverse effects are later manifested as intellectual impairments, problems in learning, attention and memory deficits, fine and gross motor abnormalities, and difficulty with organization and problem solving. An important fact to remember is that the effects that result from prenatal exposure to alcohol follow a continuum of effects. Each adverse outcome is directly related to the dose and duration of maternal drinking during pregnancy.

As long as women continue to drink alcohol during pregnancy FAS, FAE and related developmental problems will continue to be a major public health concern. The impact of genetic susceptibility and postnatal home environment on adverse alcohol-induced effects remains to be clearly elucidated.

6 Regulations for neurobehavioral teratology in government agencies

Exposure of the developing nervous system to various drugs and chemicals may produce effects at lower doses than in the adult, and may result in long-term or permanent effects. Guidelines for developmental neurotoxicity safety assessments of pharmaceuticals were introduced in 1974 by Japan and in 1975 by Great Britain; this was accomplished by amending existing regulations to include behavioral effects on the offspring following prenatal and prenatal/neonatal exposure to test substances. The United States made no effort at that time to add testing for behavioral teratogenesis. This is in sharp contrast to structural teratology where the United States was clearly a leader and set the standard. It has been 20 years since the initiation of behavioral testing in some countries. Until recently, evidence that developmental exposure to chemicals can cause neurotoxicity has had little impact on actions taken by some US regulatory agencies.

In general, reproductive and developmental toxicity assessment has traditionally been divided into three segments. The segmental approach divides ontogeny into the following phases: segment I includes reproduction through term segment II includes organogenesis and segment III includes fetogenesis and neonatal development. Behavioral assessment was added to existing segments but not uniformly. The status of regulations for developmental neurotoxicity safety assessments in Japan, Great Britain, and US agencies, and the proposed international harmonization guidelines will be briefly reviewed.

6.1 REGULATIONS IN JAPAN

There can be little doubt that from a regulatory standpoint, the Japanese have been leading the way in neurobehavioral teratology. The Japanese government requires behavioral assessments during segment II and III testing. In segment II testing they require two-thirds of the litters for standard morphological examination on a day late in pregnancy and the remaining one-third to be left alone and allowed to deliver. The latter offspring are examined postnatally for growth and development (including behavior), abnormal symptoms, and reproductive function. The requirement for segment III testing is almost identical. The method of observation in their regulations states, "behavioral observation can be made by a series of specific methods concerning motor, learning sensibility, or emotion". In 1986, this language was revised to state simply that tests of behavior were required. More specific methods of observation have not been officially established in segment II and III studies, indicating that Japanese authorities are not prepared to provide detailed guidance.

While Japan has been progressive in the implementation of developmental neurotoxicity safety assessments for pharmaceuticals, no comparable regulations exist for behavioral assessments of new food additives or environmental chemicals.

6.2   REGULATIONS IN GREAT BRITAIN AND OTHER COUNTRIES OF THE EUROPEAN UNION (EU)

On January 1, 1975, the British government amended its reproductive guidelines for new drugs to include a requirement that animals be tested for "auditory, visual and behavioral impairment". In Great Britain behavioral testing is required in the fertility and perinatal studies, segments I and III. The British have kept the language for behavioral assessments general. The British regulatons state that vision and audition must be evaluated; however, it is not clear what is intended by the phrase "and behavioral impairment".[9]

The behavioral guidelines of Great Britain like those of Japan apply only to new drugs. No comparable guidelines exist to regulate neurotoxicity testing for food additives or industrial, or for environmental chemicals.

In 1983 the EU proposed a set of guidelines for all of its member nations. The guidelines included evaluations for reproductive and developmental toxicity testing, and encompassed behavioral teratogenesis testing. The EU guidelines followed the design of Britain, requiring behavioral assessment in the fertility (segment I) and perinatal studies (segment III). The guidelines specify "late effects of the drug on the progeny in terms of auditory, visual, and behavioral impairment should be assessed". In the summer of 1985, the guidelines were officially adopted by the EU. The requirement of behavioral teratogenesis evaluations for new drugs now includes essentially all of Europe.

6.3   INTERNATIONAL COMMITTEE ON HARMONIZATION (ICH) HARMONIZED TRIPARTITE GUIDELINE

National regulatory guidelines vary somewhat from country to country. These variations make it increasingly difficult to strictly follow every country's requirements using a single study design. It has been recognized that many of these variations do not represent substantial differences in the information obtained on a substance's potential for adverse developmental effects. A large amount of time and effort have been put forth by a committee initiated by the European Union to develop a uniform set of international reproductive and developmental guidelines. The document, entitled "Guideline on Detection of Toxicity to Reproduction for Medicinal Products" [29], has been supported by the International Federation of Teratology Societies (IFTS), the World Health Organization (WHO), and related groups[100]. The objective of this guideline is to eliminate some of the redundancy currently present in the safety assessment arena among the countries. The goal is that the international guidelines will be accepted by regulatory authorities from Japan, the EU, and the US. This would help to eliminate repetitive studies, reduce costs, and reduce the number of animals used in research without compromising the safety assessment of new pharmaceuticals.

The guideline states that the studies should allow exposure of mature adults and all stages of development from conception to sexual maturity. The typical three study design is recommended. The combination of studies should include evaluation of effects on:

1. Fertility and early embryonic development to implantation
2. Embryo-fetal development (organogenesis)
3. Pre- and postnatal development (from implantation to weaning)

The guideline has many features in common with guidelines of other agencies. Some of the design details include the recommendation of using rats as the primary species, the use of three dose groups and controls, and an exposure route that mimics the expected human route. The guideline has a simplified narrative in which the main themes and aims of each segment are outlined. Many of the specific details and alternatives have been placed in an extensive set of notes which accompany the guideline.

Assessment of neurobehavioral function was included in both segment II and III of an earlier version of the guidelines, however, now such assessment only occurs in the segment III study. This change occurred because the embryo-fetal study has been reduced to a strict structural malformation study and the original postnatal cohort has been deleted. In the segment that includes assessment of neurotoxicity, much less information is provided on which functions should be assessed. In a previous version, the guideline specified assessments of "sensory function, reflexes, motor activity, and learning/ memory". The final guideline states that physical development (note 21), sensory functions and reflexes (note 21), and behavior (note 21) should be assessed. Note 21 includes comments on functional tests but does not suggest which behavioral tests should be used; it only encourages investigators to find methods that will assess sensory function, motor activity, learning, and memory. The optimal situation would be for enough guidance to be provided to ensure appropriate evaluation and still include some flexibility in the final choices of tests.

6.4 REGULATORY AGENCIES OF THE UNITED STATES AND CANADA

Despite the problems mentioned in the previous section, one major advantage of the ICH-Harmonized Tripartite Guideline is that it would be applied to all new pharmaceuticals throughout the industrialized world. The results of this effort would be that even drugs submitted to the US Food and Drug Administration (FDA) and the Canadian Health and Welfare Agency would be required to include nervous system functional assessment, which is a major improvement over the present requirements in the United States and Canada. Pragmatically, most new drugs registered in the US do have behavioral testing because it is required for registration elsewhere in the world and companies do not want to have to repeat testing programs.

6.4.1 *The Food and Drug Administration (FDA)*

The Food and Drug Administration (FDA) is the American governmental authority responsible for the regulation of safety assessment for food/color additives as well as new drugs. From 1974 to 1986, different entities within the FDA took separate positions on the potential importance of developmental neurotoxicity assessments. Since 1986, the need for neurotoxicity testing has been under internal review at the FDA.

The Center for Food Safety and Applied Nutrition (CFSAN) of the FDA has recently proposed new guidelines on reproduction and developmental toxicity studies which included adult and developmental neurotoxicity assessments for direct food and color additives. The CFSAN proposes a structured process of tiered testing in evaluating neurotoxicity. Each tier would focus on a different objective. The CFSAN suggests the tiered testing include screening (initial identification of chemicals that exhibit any potential for adversely affecting the nervous system), characterization of effects (determining the scope of nervous system involvement), and dose-response (determining dose-response kinetics, including finding no-observed-effect-level, NOEL).

As described in the guidelines, "An effective and comprehensive basic neurotoxicity screen would include both (1) a specific histopathological examination of tissue samples representative of all major areas and cellular elements of the brain, spinal cord, and peripheral nervous system, in conjunction with (2) a systematic examination of experimental animals inside and outside of their cages using a clearly defined functional evaluation battery of clinical tests and observations to provide a general assessment of the primary neurological, behavioral and physiological functions of the nervous system."

As indicated in the proposal, the "characterization of neurotoxicity should include information about the severity of effect, the temporal pattern of onset of effects, and duration of effects. The neurofunctional assessment at this level should include a core battery of behavioral and physiological tests designed to detect adverse changes to the primary subfunctions (e.g., cognitive, sensory, motor, and autonomic) of the nervous system in mature and developing nervous systems".

The NOEL is a critical element in determining a chemical's neurotoxic hazard. Accurate and reliable determination of dose-response and dose-time relationships are imperative in defining the NOEL.

The FDA is receiving praise for the inclusion of neurotoxicity testing within the reproduction and developmental toxicity studies for food and color additives. It seems appropriate to extend the application of these guidelines to the safety assessment of pharmaceuticals in FDA guidelines directly, in addition to the indirect incorporation through the ICH guidelines.

6.4.2   *The Environmental Protection Agency (EPA)*

In 1986, the US Environmental Protection Agency (EPA) took the lead in the US regulatory arena and issued a protocol for developmental neurotoxicity testing of glycol ethers which was later expanded and revised (1989, 1991) to form a guideline for developmental neurotoxicity testing of other chemicals and pesticides. The guideline provides a basic protocol for assessing the potential effects of prenatal exposure to a variety of chemicals or environmental toxins. The EPA has the authority to require testing of agents under two acts including the Federal Insecticide, Fungicide and Rodenticide Acts (FIFRA) and the Toxic Substances Control Act (TSCA). Under TSCA, two sections (4 and 5) give the EPA the authority to require adult and developmental neurotoxicity testing for existing and new chemicals, respectively. The EPA's Office of Toxic Substances (OTS) carries out the mandates of TSCA and developed the first systematic approach to developmental neurotoxicity testing by a regulatory office[77].

The functioning of several integrated systems are assessed in this approach including sensory systems, neuromotor development, reactivity and/or habituation, and learning and memory. The progeny are evaluated for the presence of physical developmental landmarks (e.g., preputial separation and vaginal patency). Neuropathological assessments are included in the guideline at two different ages.

The protocol promulgated by the EPA to be used in screening for developmental neurotoxicity is summarized in Table 2.10[27]. There are several unique features of the guideline which make it different from any of the regulatory guidelines of other agencies and countries. One of the unique features is that the EPA guideline is truly developmental in nature. The ontological framework allows the development and functioning of the CNS to be followed over time to determine the persistence of effects. The longitudinal design requires that each male/female pair be tested at different ages on the same test. This approach allows several windows of development to be assessed which would be helpful in identifying transient and/or delayed neurotoxic effects. Another attractive feature of the EPA guideline is the specification of procedural requirements. Clear categories of CNS functional assessment are provided along with key criteria for each test, for example demonstration of an acquisition curve in learning tests and of habituation in automated activity monitoring. This guideline is superior to others with regard to the amount of detail included for the assessment.

TABLE 2.10

EPA guidelines for developmental neurotoxicity screening (March 1991)
(Adapted from Driscoll, C.D. (1991) *J. Am. Coll. Toxicol.* **10**: pp 697–703.)

Number of litters	≥ 20/Dose Group
Dose levels	≥ 3 + Vehicle Control
Dosing period	GD6 – PND10
Route	Oral (gavage)
Gestation day 0 (GD0)	Sperm/Plug Positive
Postnatal day 0 (PND0)	Parturition
Body weights, clinical signs	
Dams	GD0, GD6-PND10, PND21
Pups	PND0, 4, 11, 17, 21 and biweekly thereafter
Development landmarks	
Males	Preputial Separation
Females	Vaginal Opening
Motor activity	PND13, 17, 21, 60 ± 2
Auditory startle	PND22, 60 ± 2
Learning & memory	PND21-24, 60 ± 2
Neuropathology & brain weights	PND11, 60+

There are also some areas of concern with regard to the guideline. The guideline allows for the assessment of clinical signs in a categorical manner; however, such observations may be too blunt to be of use in evaluating signs of neurotoxicity beyond detecting gross effects. The EPA guideline for assessment of developmental neurotoxicity is viewed as "all purpose" but in reality it is not. The guideline is adequate for the detection of developmental neurotoxicity following exposure to some chemicals but may not be capable of detecting neurotoxicity for others. A concern of the EPA guideline is that it is a separate entity from the main guideline for detecting developmental toxicity. The current approach assumes that if an agent is neurotoxic in adults, it should be tested for developmental neurotoxicity. If a chemical produces structural effects or death at higher dosages, then this might trigger further testing which would be adequate for safety evaluation. If, on the other hand, developmental neurotoxicity testing is only required when prior information shows CNS effects, then one runs the danger of missing potential behavioral teratogens. One possibility is for the EPA to reorganize the developmental studies, merging the developmental toxicity and developmental neurotoxicity guideline into one unified protocol.

In the past few years, there has been an increasing effort among US regulatory agencies to include assessment for potential developmental neurotoxicity with food/color additives and some chemicals (i.e., pesticides). While there is still work to be done in this area, progress has been made. The nervous system is a critically important area of development and functioning and should not be overlooked by risk assessors when determining the potential neurotoxicity of drugs and chemicals.

7 Conclusions

It has been nearly 25 years since neurobehavioral teratology emerged from its parent discipline, teratology. The brain has recently received more attention and consideration as evidenced by the 1990s being designated as "The Decade of the Brain" by the US Congress and subsequently in many other countries, e.g., in Europe. Overall interest in the effects of *in utero* exposure to drugs, chemicals, and environmental and physical agents on the development and functional integrity of the brain and CNS is on the rise. This increase is also reflected by the actions which are being initiated by government regulatory agencies requiring safety testing on some chemicals.

The history and fundamental principles of neurobehavioral teratology have been presented as well as approaches and methods used to detect neurotoxicity following prenatal exposure to a variety of agents. The literature was reviewed for the developmental effects of several popular drugs of abuse (methamphetamine, cocaine, opiates, cigarette smoking, and alcohol). Recent actions of regulatory agencies and the new guidelines specifying more safety testing in an attempt to detect and prevent alterations in brain development and function are finally moving forward to ensure protection of this long neglected, but vital, organ system.

References

1. Abel, E. and Sokol, R. (1987) Incidence of fetal alcohol syndrome and economic impact of FAS-related anomalies. *Drug Alcohol Depend.* **19**: pp 51–70.

2. Abel, E.L., Rockwood, G.A,, Riley, E.P. (1986) The effects of early marijuana exposure. In: Riley, E.P. and Vorhees, C.V. (Eds.), *Handbook of Behavioral Teratology.* Plenum Press, New York, Chapter 12.

3. Acuff-Smith, K.D., George, M., Lorens, S.A., Vorhees, C.V. (1992) Preliminary evidence for methamphetamine-induced behavioral and ocular effects in rat offspring following exposure during early organogenesis. *Psychopharmacology* **109**: pp 255–263.

4. Acuff-Smith, K.D., Schilling, M.A., Fisher, J.E., Vorhees, C.V. (1995) Stage specific effects of prental *d*-methamphetamine exposure on eye and behavioral development in rats. *Neurotoxicol. Teratol.* (in press).

5. Adams, J. (1986) Methods in behavioral teratology. In: Riley, E.P. and Vorhees, C.V. (Eds.) *Handbook of Behavioral Teratology.* Plenum Press, New York, pp 67–97.

6. Altman, J., Brunner, R.L., Bulert, F.G., Sudarshan, K. (1974) The development of behavior in normal and brain damaged infant rats, studied with homing as motivation. In: Vernadakis, A. and Weiner, N. (Eds.), *Drugs and the Developing Brain.* Plenum Press, New York, pp 321–348.

7. Altman, J. and Sudarshan, K., (1975) Postnatal development of locomotion in the laboratory rat. *Anim. Behav.* **23**: pp 896–920.

8. Anastasi, A. (1982) *Psychological Testing*, 5th edition. Macmillan, London.

9. Barlow, S.M. (1985) United Kingdom: regulatory attitudes toward behavioral teratology testing. *Neurobehav. Toxicol. Teratol.*

10. Blass, E.M. and Teicher M.H. (1980) Suckling. *Science* **210**: pp 15–22.

11. Brazelton, T.B., Nugent, J.K., Lester, B.M. (1987) Neonatal behavioral assessment scale. In: Osofsky, J.D. (Ed.) *Handbook of Infant Development*, 2nd edition. Wiley-Interscience, New York, NY, pp 780–817.

12. Butcher, R.E., Vorhees, C.V., Berry, H.K. (1970) A learning impairment associated with induced phenylketonuria. *Life Sci.* **9**: pp 1261–1268.

13. Butcher, R.E. (1970) Learning impairment associated with maternal phenylketonuria in rats. *Nature* **226**: pp 555–556.

14. Cabrera, T.M., Levy, A.D., Li, Q., Van De Kar, L.D., Battaglia, G. (1993) Prenatal methamphetamine attenuates serotonin mediated renin secretion in male and female rat progeny: evidence for selective long-term dysfunction of serotonin pathways in brain. *Synapse* **15**: pp 198–208.

15. Chasnoff, I.J., Griffith, D.R., Freier, C., Murray, J. (1992) Cocaine/polydrug use in pregnancy: two year follow up. *Pediatrics* **89**: pp 284–289.

16. Cho, D., Lyu, H., Lee, H., Kim, P., Chin, K. (1991) Behavioral teratogenicity of methamphetamine. *J. Toxicol. Sci.* **16** (Suppl.) pp 37–49.

17. Coles, C.D., Brown, R.T., Smith, I.E., Platzman, K.A., Erikson, S., Falek, A. (1991) Effects of prenatal alcohol exposure at school age: I. Physical and cognitive development. *Neurotoxicol. Teratol.* **13**: pp 357–366.

18. Coles, C.D., Platzman, K.A., Smith, I., James, M.E., Falek, A. (1992) Effects of cocaine and alcohol use in pregnancy on neonatal growth and neurobehavioral status. *Neurotoxicol. Teratol.* **14**: pp 23–34.

19. Coles, C.D. (1993) Impact of prenatal alcohol exposure on the newborn and the child. *Clin. Obstet. Gynecol.* **36**: pp 255–265.

20. Cronbach, L.J. (1970) *Essentials of Psychological Testing*, 3rd edition. Harper & Row, New York.

21. Day, N.L. and Richardson, G.A. (1991) Prenatal alcohol exposure: a continuum of effects. *Semin. Perinatol.* **15**: pp 271–279.

22. Dietrich, K.N. and Bellinger, D. (1994) The assessment of neurobehavioral development studies of the effects of prenatal exposure to toxicants. In: Needleman, H.L. and Bellinger, D. (Eds.) *Prenatal Exposure To Toxicants: Developmental Consequences*. John Hopkins University Press, Baltimore, MD, pp 57–88.

23. Dinges, D.F., Davis, M.M., Glass, P. (1980) Fetal expsoure to narcotics: neonatal sleep as ameasure of nervous system disturbance. *Science* **209**: pp 619–621.

24. Dixon, S.D. (1989) Effects of transplacental exposure to cocaine and methamphetamine on the neonate. *West J. Med.* **150**: pp 436–442.

25. Dixon, S.D. and Bejar, R. (1989) Echoencephalographic findings in neonates associated with maternal cocaine and methamphetamine use: incidence and clinical correlates. *J. Pediatr.* **115**(5): pp 770–778.

26. *Dorland's Medical Dictionary* (1957) 23rd edition. Saunders, Philadelphia.

27. Driscoll, C.D. (1991) Screening for developmental neurotoxicity: approaches and controversy. *J. Am. Coll. Toxicol.* **10**: pp 697–703.

28. Egger, C.J., Livesey, P.J., Dawson, R.G. (1973) Ontogenetic aspects of central cholinergic involvement in spontaneous alternation behavior. *Dev. Psychobiol.* **6**: pp 289–299.

29. European Drafting Group (1992) *Guideline on Detection of Toxicity to Reproduction for Medicinal Products*. Draft No. 13, July 1992. International Federation Teratology Societies, Berlin.

30. Finnegan, L.P., Connaughton, J.F., Kron, R.E., Emich, J.P. (1975), Neonatal abstinence syndrome: assessment and management. *Addict. Dis.* **2**: pp 141–158.

31. Francis, P.L., Self, P.A., Horowitz. F.D. (1987) The behavioral assessment of the neonate: an overview. In: Osofsky, O.J. (Ed.) *Handbook of Infant Development*, Second edition. Wiley-Interscience, New York, NY, pp 723–779.

32. Fried, P.A. and Waktinson, B. (1988) 12 and 24 month neurobehavioral follow-up of children prenatally exposed to marijuana, cigarettes and alcohol. *Neurotoxicol. Teratol.* **10**: 305-313.

33. Fried, P.A. (1989) Cigarettes and marijuana: are there measurable long term neurobehavioral teratogenic effects. *Neurotoxicology* **10**: pp 577–584.

34. Fried, P.A. and Watkinson B. (1990), 36 and 48 month neurobehavioral follow-up of children prenatally exposed to marijuana, cigarettes and alcohol. *Dev. Behav. Pediatr.* **11**: pp 49–58.

35. Fried, P.A., O'Connell, C.M., Watkinson, B. (1992) 60 and 72 month follow-up of children prenatally exposed to marijuana, cigaretttes, and alcohol: Cognitive and language assessment. *Dev. Behav. Pediatr.* **13**: pp 383–391.

36. Fried, P.A. (1993) Prenatal exposure to tobacco and marijuana: effects during pregnancy, infancy, and early childhood. *Clin. Obstet. Gynecol.* **36**(2): pp 319–337.

37. Gregory, E.H. and Pfaff, D.W. (1971) Development of olfactory guided behavior in infant rats. *Physiol. Behav.* **6**: pp 573–576.

38. Haddad, R.K., Rable, A., Laqueur, G.L., Spatz, M., Valsamis, M.P. (1969), Intellectual deficit associated with transplacentally induced microcephaly in the rat. *Science* **163**: pp 88–90.

39. Hamilton, H.C. and Harned, B.K. (1944) The effect of the administration of sodium bromide to pregnant rats on the learning ability of the offspring: III. Three table test. *J. Psychol.* **18**: pp 183–195.

40. Hartman, D.E. (1988) *Neuropsychological Toxicology: Identification and Assessment of Human Neurotoxic Syndromes.* Pergamon Press, New York.

41. Heeb, D.O. (1947) The effects of early experience on problem solving at maturity. *Am. Psychol.* **2**: pp 306–307.

42. Hicks, S.P. and D'Amato, C.J. (1978) Effects of ionizing radiation on developing brain and behavior. In: Gottlieb, G. (Ed.) *Studies on the Development of Behavior and the Nervous System: Early Influences,* Vol. 4, Academic Press, New York.

43. Hutchings, D.E., Hunt, H.F., Towey, J.P., Rosen, T.S., Gorinson, H.S. (1976) Methadone during pregnancy in the rat: dose level effects on maternal and perinatal mortality and growth in the offspring. *J. Pharmacol. Exp. Ther.* **197**: pp 171–179.

44. Hutchings, D.E. (1978) Behavioral teratology: embryopathic and behavioral effect of drugs during pregnancy. In: Gottlieb, G. (Ed.) *Studies on the Development of Behavior and the Nervous System: Early Influences,* vol. 4. Academic Press, New York.

45. Hutchings, D.E., Feraru, E., Gorinson, H.S. (1979) Effects of prenatal methadone on the rest-activity cycle of the pre-weanling rat. *Neurobehav. Toxicol.* **1**: pp 33–40.

46. Hutchings, D.E., Towey, J.P., Gorinson, H.S., Hunt, H.F.. (1979) Methadone during pregnancy: assessment of behavioral effects in rat offspring. *J. Pharmacol. Exp. Ther.* **208**: pp 106–112.

47. Hutchings, D.E. (1982) Methadone and heroin during pregnancy: a review of behavioral effects in humans and animal offspring. *Neurobehav. Toxicol. Teratol.* **4**: pp 429–434.

48. Hutchings, D.E. and Fifer, W.P. (1986) Neurobehavioral effects in humans and animal offspring following prenatal exposure to methadone. In: Riley, E.P. and Vorhees, C.V. (Eds.) *Handbook of Behavioral Teratology.* Plenum Press, New York, Chapter 6.

49. Hutchings, D.E. and Dow, E.D. (1991) Animal models of opiate, cocaine and cannabis use. In: Chasnof, I.J. (Ed.), *Chemical Dependency and Pregnancy: Clinics in Perinatology.* W.B. Saunders, Philadelphia, pp 1–22.

50. Hutchings, D.E. (1993) A contemporary overview of behavioral teratology. In: Kalter, H. (Ed.) *Issues and Reviews in Teratology* **6**: 12-167, Plenum Press, New York.

51. Hutchings, D.E. (1993) The puzzle of cocaine's effects following maternal use during pregnancy: are there reconcilable differences? *Neurotoxicol. Teratol.* **15**(5): pp 281–286.

52. Jacobson, J.L. and Jacobson, S.W. (1991) Assessment of teratogenic effects on cognitive and behavioral development in infancy and childhood. *NIDA Res. Monogr.* **114**: pp 248–261.

53. Joffe, J.M. (1969) Prenatal determinants of behavior. Pergamon Press, Oxford.

54. Jones, K.L. and Smith, D.W. (1973), Recognition of hte fetal alcohol syndrome in early infancy. *Lancet* **2**: 999-1001.

55. Koren, G., Shear, H., Graham, K., Einarson, T. (1989) Bias against the null hypothesis: the reproductive hazards of cocaine. *Lancet* **2**: pp 1440–1442.

56. Little, B.B., Snell, L.M., Gilstrap, L.C. (1988) Methamphetamine abuse during pregnancy outcome and fetal effects. *Obstet. Gynecol.* **72**: pp 541–544.

57. Little, B.B., Snell, L.M., Klein, V.R., Gilstrap, L.C., Knoll, K.A., Breckenridge, J.D. (1990) Maternal and fetal effects of heroin addiction during pregnancy. *J. Reprod. Med.* **35**: pp 159–162.

58. Lemoine, P., Harrousseau, H., Borteyra, J.P. (1968) Les enfants de parents alcoliques. Anomalies observees a propos de 127 cas. *Ouest Med.* **21**: pp 476–482.

59. Lutiger, B., Graham, K., Einarson, T.R., Koren, G. (1991) Relationshiop between gestational cocaine use and pregnancy outcome: a meta-analysis. *Teratology* **44**: pp 405–414.

60. Mactutus, C.F. and Tilson, H.A. (1986) Psychogenic and neurogenic abnormalities after perinatal insecticde exposure. In: Riley, E.P. and Vorhees, C.V. (Eds.) *Handbook of Behavioral Teratology.* Plenum Press, New York, chapter 15.

61. Mactutus, C.F. (1989) Developmental neurotoxicity of nicotine, carbon monoxide, and other tobacco smoke constituents. *Ann. N.Y. Acad. Sci.* **562**: pp 105–122.

62. Markham, J.K. (1971) Teratogenicity studies of methadone HCl in rats and rabbits. *Nature* **233**: pp 342–343.

63. Martin, J.C. (1975) Effects on offspring of chronic maternal MA exposure. *Dev. Psychobiol.* **8**: pp 397–404.

64. Martin, J.C., Martin, D.C., Radow, B., Sigman, G. (1976) Growth, development and activity in rat offspring following maternal drug exposure. *Exp. Aging Res.* **2**: pp 235–251.

65. Martin, J.C., Martin, D.C., Radow, B., Day, H.E. (1979) Life span and pathology in offspring following nicotine and methamphetamine exposure. *Exp. Aging Res.* **5**: pp 509–522.

66. Martin, J.C. and Martin, D.C. (1981) Voluntary activity in the aging rat as a function of maternal drug exposure. *Neurobehav. Toxicol. Teratol.* **3**: pp 261–264.

67. Martin, J.C., Martin, D.C., Sigman, G., Day-Pfeiffer, H. (1983) Saccharin preferences in food deprived aging rats are altered as a function of perinatal drug exposure. *Physiol. Behav.* **30**: pp 853–858.

68. Middaugh, L.D. and Simpson, L.W. (1980) Prenatal maternal methadone effects on pregnant C57BL6 mice and their offspring. *Neurobehav. Toxicol.* **2**: pp 307–313.

69. Naeye, R.L. (1992) Cognitive and behavioral abnormalities in children whose mothers smoked cigarettes during pregnancy. *Dev. Behav. Pediatr.* **13**(6): pp 425–428.

70. Nelson, B.K. (1981) Dose/effect relationships in developmental neurotoxicology. *Neurobehav. Toxicol. Teratol.* **3**(3): pp 255.

71. Nelson, B.K. (1985) Developmental neurotoxicology of environmental and industrial chemicals (1985). In: Blum, K. and Manzo, L. (Eds.) *Neurotoxicology.* Marcel Dekker, New York, Chapter 8.

72. Nelson, B.K. (1986) Behavioral teratology of industrial solvents. In: Riley, E.P. and Vorhees, C.V. (Eds.) *Handbook of Behavioral Teratology.* Plenum Press, New York, Chapter 16.

73. Neuspiel, D.R., Hamel, S.C., Hochberg, E., Greene, J., Campbell, J. (1991) Maternal cocaine use and infant behavior. *Neurotoxicol. Teratol.* **13**: pp 229–233.

74. Neuspiel, D.R. and Hamel, S.C. (1991) Cocaine and infant behavior. *Dev. Behav. Pediatr.* **12**(1): pp 55–64.

75. Oro, A.S. and Dixon, S.D. (1987) Perinatal cocaine and methamphetamine exposure: maternal and neonatal correlates. *J. Pediatr.* **111**(4) pp 571–578.

76. Paule, M. and Adams, J. (1994) Interspecies comparison of the evaluation of cognitive developmental effects of neurotoxicants in primates. In: Chang, L. (Ed.) *Handbook of Neurotoxicology. 1. Basic Principles and Current Concepts.* Marcel Dekker, New York.

77. Rees, D.C., Francis, E.Z., Kimmel, C.A. (1990) Scientific and regulatory issues relevant to assessing risk for developmental neurotoxicity: an overview. *Neurotoxicol. Teratol.* **12**: pp 175–182.

78. Richardon, G.A. and Day, N.L. (1991) Maternal and neonatal effects of moderate cocaine use during pregnancy. *Neurotoxicol. Teratol.* **13**: pp 455–460.

79. Riley, E.P., Hannigan, J.H., Balaz-Hannigan, M.A. (1985) Behavioral teratology as the study of early breain damage: considerations for the assessment of neonates. *Neurobehav. Toxicol. Teratol.* **7**: pp 635–638.

80. Rodier, P.M. (1980) Chronology of neuron development: animal studies and their clinical implications. *Dev. Med. Child. Neurol.* **22**: pp 525–545.

81. Sato, M. and Fujiwara, Y. (1986) Behavioral and neurochemical changes in pups prenatally exposed to methamphetamine. *Brain Dev.* **8**: pp 390–396.

82. Schapiro, S., Salas, M., Vukovich, K. (1970) Hormonal effects on the ontogeny of swimming ability in the rat: assessment of central nervous system development. *Science* **193**: pp 146–151.

83. Slutsker, L. (1992) Risks associated with cocaine use during pregnancy. *Obstet. Gynecol.* **79**(5): pp 778–789.

84. Spear, L.P., Enters, E.K., Linville, D.G. (1985) Age specific behaviors as tools for examining teratogen induced neural alterations. *Neurotoxicol. Teratol.* **7**: pp 691–695.

85. Spear, L.P. (1984) Age at the time of testing reconsidered in neurobehavioral teratology research. In: Yanai, J. (Ed.) *Neurobehavioral Teratology.* Elsevier, Amsterdam, pp 315–328.

86. Streissguth, A.P. (1978) Fetal alcohol syndrome: an epidemiologic perspective *Am. J. Epidemiol.* **107**(6): pp 467–478.

87. Streissguth, A.P., Landesman-Dwyer, S., Martin, J.C., Smith, D.W. (1980) Teratogenic effects of alcohol in humans and laboratory animals. *Science* **209**: pp 353–361.

88. Streissguth, A.P., Sampson, P.D., Barr, H.M. (1989) Neurobehavioral dose-response effects of prenatal alcohol exposure in humans from infancy to adulthood. In: Hutchings, D.E. (Ed.) *Prenatal Abuse of Licit and Illicit Drugs.* New York Academy of Sciences, New York, 562: pp 145–158.

89. Streissguth, A.P., Aase, J.M,, Clarren, S.K., Randels, S.P., LaDue, R.A., Smith, D.F. (1991) Fetal alcohol syndrome in adolescents and adults. *JAMA* **265**(15): pp 1961–1967.

90. Volpe, J.J. (1992) Effect of cocain on the fetus. *N. Engl. J. Med.* **327**(6) pp 399–407.

91. Vorhees, C.V., Brunner, R.L., Butcher, R.E. (1979), Psychotropic drugs as behavioral teratogens. *Science* **205**: pp 1220–1225.

92. Vorhees, C.V. and Butcher, R.E. (1982) Behavioral teratogenicity. In: Snell, K. (Ed.) *Developmental Toxicology.* Praeger, New York.

93. Vorhees, C.V. (1983) Fetal anticonvulsant syndrome in rats: dose and period response relationships of prenatal diphenylhydantoin, trimethadione, and phenobarbital exposure on the structural and functional development of the offspring. *J. Pharmacol. Exp. Ther.* **227**: pp 274–287.

94. Vorhees, C.V. (1986) Developmental effects of anticonvulsants. *Neurotoxicology* **7**: pp 235–244.

95. Vorhees, C.V. (1986) Fetal anticonvulsant exposure: effects on behavioral and physical development. *Ann. N.Y. Acad. Sci.* **477**: pp 49–62.

96. Vorhees, C.V. (1986) Origins of behavioral teratology (Chapter 1); Principles of behavioral teratology (Chapter 2). In: Riley, E.P. and Vorhees, C.V. (Eds.) *Handbook of Behavioral Teratology.* Plenum Press, New York.

97. Vorhees, C.V. (1986) Behavioral teratology of anticonvulsant and antianxiety medications. In: Riley, E.P. and Vorhees, C.V. (Eds.) *Handbook of Behavioral Teratology.* Plenum Press, New York, Chapter 10.

98. Vorhees, C.V. (1987) Methods in behavioral teratology screening: current status and new developments. *Congr. Anom.* **27**: pp 111–124.

99. Vorhees, C.V. (1987) Reliability, sensitivity and validity of behavioral indices of neurotoxicity. *Neurotoxicol. Teratol.* **9**: pp 445–464.

100. Vorhees, C.V. (1992) Developmental neurotoxicology. In: Tilson, H.A. and Mitchell, C.L. (Eds.) *Neurotoxicology: Target Organ Toxicology Series.* Raven Press, New York, pp 295–329.

101. Vorhees, C.V., Ahrens, G.A., Acuff-Smith, K.D., Schilling, M.A., Fisher, J.E. (1993) Methamphetamine exposure during early postnatal development in rats: I. Acoustic startle augmentation and spatial learning deficits. *Psychopharmacology* **114**: pp 392–401.

102. Vorhees, C.V., Ahrens, G.A., Acuff-Smith, K.D., Schilling, M.A., Fisher, J.E. (1993) Methamphetamine exposure during early postnatal development in rats: II. Hypoactivity and altered responses to pharmacological challenge. *Psychopharmacology* **114**: pp 402–409.

103. Vorhees C.V. (1994) Behavioral and functional ontogeny: biomarkers of neurotoxicity. In: Chang, L.W. (Ed.) *Principles of Neurotoxicology.* Marcel Dekker, New York, pp 733–763.

104. Vorhees, C.V. (1994) Long-term effects of developmental exposure to cocaine on learned and unlearned behaviors. In: Finnagan, L. (Ed.) NIDA Res. Monogr. Series: *Behaviors of Drug-Exposed Offspring: Research Update.* Dept. of Health and Human Services, National Institutes of Health, Washington D.C. (in press).

105. Weissman, A.D., Caldecott-Hazard, S. (1993) In utero methamphetamine effects: I. Behavior and monoamine uptake sites in adult offspring. *Synapse* **13**: pp 241–250.

106. Wilson, J.G. (1973) Principles of teratology. In: *Environment and Birth Defects.* Academic Press, New York, pp 11–34.

107. Wilson, J.G. (1977) Current status of teratology: general principles and mechanisms derived from animal studies. In: Wilson, J.G. and Fraser, F.C. (Eds.), *Handbook Of Teratology, General Principles And Etiology*, Vol. 1. Plenum Press, New York.

Photograph
The photograph in illustration 2.8 is reproduced by permission from an article by Sterling K. Clarren (1991) *JAMA*, **245:** pp 2436–2439.

Contents Chapter 3

Assessment techniques for detecting neurobehavioral toxicity

0-8493-7802-8/99/$0.00+$.50
© 1999 by CRC Press LLC

Chapter 3

Assessment techniques for detecting neurobehavioral toxicity

Beverly M. Kulig and Rob M.A. Jaspers

1 Introduction

1.1 THE SCOPE OF THE PROBLEM

Neurotoxic effects of chemicals as reflected by behavioral changes have been known through overexposure of humans and animals since antiquity. The neurotoxic properties of lead, for example, were identified as early as 200 BC and the first recorded regulatory measures in industrial hygiene ever taken was aimed at protecting workers from the neurological effects of mercury exposure in the Idrian mines in 1665[14].

Industrial activities over the last 100 years have significantly added both to the number of neurotoxic chemicals in existence and to the number of persons potentially exposed. Occupational exposure limits for approximately 25% of all chemicals governmentally regulated in the United States, for example, have been based at least in part on nervous system effects, and in the US alone, an estimated 14 million persons are occupationally exposed to these chemicals[2].

Much of our knowledge regarding the potential neurotoxicity of specific chemicals has not originated in the laboratory, but rather has come from outbreaks of neurotoxic disease due to environmental and industrial overexposures as a consequence of accident or ignorance. In the last century, for example, exposure to mercury in the manufacture of felt hats produced a neurotoxic syndrome characterized by tremors and other CNS disturbances. One well-known case is shown in Figure 3.1. Further, the ingestion of food inadvertently contaminated with substances containing neurotoxic compounds has been the cause of major outbreaks of neurologic disease involving thousands of persons. In addition, over-the-counter medications and prescription drugs with unrecognized neurotoxic effects as well as drugs of abuse have also produced severe, sometimes irreversible, neurological and behavioral impairment[29].

Episodes of neurological injury have often occurred in the occupational setting where exposure levels tend to be high. Recently, for example, the solvent 2-*tert*-butylazo-2-hydroxy-5-methylhexane (BHMH) was introduced as a foaming catalyst for the manufacture of plastic bathtubs. Within 2 weeks of its introduction, a microepidemic began when workers directly exposed to BHMH developed a syndrome consisting of weight loss, sensorimotor neuropathy, loss of vision, and impaired memory. Follow-up studies of these patients two years later indicated that although affected workers showed some improvement, residual weakness, peripheral nerve abnormalities, and sensory dysfunction were still evident[13].

71

The Mad Hatter

In the 1800s, hatters were exposed to large amounts of mercury in the production of felt hats. Poor industrial hygiene led to permanent nervous system injury in hundreds of workers. Lewis Caroll's *The Mad Hatter* is one of the better known clinical cases of occupational mercury poisoning.

FIGURE 3.1

A well-known example of occupational neurotoxic exposure
(From Dodgson, O.L. (1976) *The Complete Works of Lewis Carroll.* Vintage Books, Random House, New York.)

Following this outbreak, animal studies were undertaken which showed that exposure to BHMH in rats for as short a period as 3 weeks also produced weight loss, corneal opacity, and hindlimb weakness progressing to severe signs of peripheral neuropathy. These clinical signs were accompanied by axonal degeneration of the optic tracts, ascending and descending spinal tracts, and peripheral nerves as well as degeneration of nerve fibers in the brain and cerebellum[30]. Thus, in animals a syndrome remarkably analogous to the human clinical picture in terms of rapidity of onset, severity of clinical signs and nature of clinical neurotoxic effects could easily be reproduced.

It is perhaps surprising that animal testing examining the neurotoxic effects of chronic exposure to BHMH was not conducted prior to its introduction for use in industry. However, this state of affairs is presently the rule, rather than the exception. For most of the 65,000 chemicals currently in commerce as well as the 2,000 new chemicals which are introduced on to the market each year, relatively few have been tested for neurotoxicity. Table 3.1 summarizes the results of a study estimating the percentage of chemicals by chemical category which have been tested for effects on the nervous system. As the table shows, even for compounds such as pesticides, which in many cases are designed to act through toxic effects on the nervous system of lower animals, less than 10% have been sufficiently examined to determine their possible consequences to the mammalian nervous system. Surprisingly, even drugs and food additives have not been studied very well. As a result, we cannot estimate with any degree of accuracy the number of neurotoxic chemicals that exist nor can we always predict which substances may produce neurotoxic disease.

TABLE 3.1

The number of chemicals for which there is sufficient information to judge neurotoxic potential
(Adapted from McMillan, D.E. (1987) *Environ. Health Perspect.* **76**: p 155.)

Chemical category	Chemicals with minimal neuro-toxicity data
Pesticides and pesticide ingredients	< 10%
Drugs	25-29%
Food additives	< 20%
Chemicals in commerce (production: < 1 million lbs/yr)	< 1%
Chemicals in commerce (production: > 1 million lbs/yr)	< 1%

In response to this lack of information, there have been increasing regulatory activities aimed at developing testing guidelines for neurotoxicity assessment at the animal level. The United States Environmental Protection Agency (USEPA) has recently published generic neurotoxicity testing guidelines for evaluating commercial chemicals and pesticides and similar guidelines have been proposed by the Organization of Economic Cooperation and Development (OECD). In addition, the extension of neurotoxicity testing to drugs and food additives has also been proposed by the US Food and Drug Administration[25].

1.2   THE RATIONALE FOR THE USE OF BEHAVIORAL END-POINTS

Historically, morphological and classical toxicologic methods have been used to provide evidence of neurotoxicity. However, there has been increasing emphasis on the use of behavioral methods for assessing neurotoxic effects[36]. The rationale for the use of behavioral end-points is based on the fact that behavior represents the net sensory, motor, and integrative outputs of the central, peripheral, and autonomic nervous systems. As such, behavioral changes can be used to provide a sensitive index of chemically induced changes in nervous system function. In addition, behavioral methods are noninvasive and can be used in chronic studies to track the progressive development of adverse effects on the nervous system during long-term exposure.

The experimental study of behavior has been the subject matter of psychologists for over a hundred years. Many of the methods used by behavioral toxicologists have been adopted from experimental psychology, psychopharmacology, and behavioral neuroscience. However, it was not until the 1980s that behavioral methods came into increasingly widespread use in toxicology. Thus, behavioral toxicology, as a scientific discipline, is relatively young and its methodology is still evolving.

The purpose of the present chapter is to acquaint the reader with some of the basic principles involved in evaluating the effects of chemical exposures on behavior, and to survey some of the methods currently being used in behavioral toxicology. Section 2 of the chapter deals with a discussion of animal behavioral methods for evaluating neurotoxicity, including current methods for neurotoxicity screening in adult animals and more sophisticated techniques for evaluating subtle changes in motor, sensory and cognitive changes. Section 3 provides a short survey of testing techniques which are being used at the adult human level for evaluating neurobehavioral changes in persons exposed to neurotoxic compounds. Finally, the application of behavioral methods to evaluate a special type of behavioral toxicity, namely the potential of drugs and chemicals to produce addiction and psychological dependence, is also discussed.

1.3   CLASSES OF BEHAVIOR

Many different paradigms have been used to study the effects of drugs and chemicals on behavior. Some paradigms require the animal to learn to do some type of task, while others require no learning on the part of the animal. Before beginning a discussion of the methods used for studying chemically induced behavioral effects, it is important to understand what behavioral scientists mean when they use the term behavior and to very briefly review the different general classes of behavior, which can be distinguished.

Broadly speaking, behavior refers to the movement or response of an animal or its parts which can be operationally defined, observed, and measured. In the experimental analysis of behavior, the primary focus is on the functional relationship between an organism's behavior and those aspects of the environment which affect it.

The basic units of behavior are termed responses. Events occurring in both the internal and external environment are termed stimuli. Responses fall into two general categories of behavior, *respondent* and *operant*, based on the functional relationships that control their occurrence. Respondent behaviors depend on the occurrence of a prior stimulus event which serves to elicit a specific response. In contrast, operant behaviors are controlled by their consequences and these responses are emitted by the animal. Further, both respondent and operant behaviors can be modified by conditioning (i.e., learning). These different classes of behavior are summarized in Table 3.2.

Respondent behaviors typically involve smooth muscles, glandular secretion, autonomic responses, or environmentally elicited effector responses. Examples of unconditioned respondent behaviors include simple and complex reflexes such as the startle reflex in response to a sudden loud stimulus, or salivation in response to the introduction of a morsel of food into the mouth. When these reflexes occur naturally in response to the stimuli that normally elicit them, they are unlearned and are referred to as unconditioned respondent behaviors.

Respondent behavior, however, is also subject to conditioning. That is, they can be elicited by a stimulus which does not normally elicit them by means of learning. Pavlov, for example, demonstrated that by repeatedly pairing the presentation of food (the unconditioned stimulus) into the mouth of a hungry dog with the ringing of a bell (conditioned stimulus), the ringing of the bell itself will eventually elicit salivation. In respondent conditioning paradigms, the pairing of two stimuli, the unconditioned stimulus and the conditioned stimulus, eventually leads to the learning of a conditioned response.

In contrast, operant behaviors are those which are emitted by an organism, rather than being reliably elicited by a specific stimulus. Although unlearned operants are to some degree species-specific, common examples include loco-motor activity, exploratory behavior, and the production of different vocalizations. Animals and humans, for example, will emit locomotor

TABLE 3.2

Classes of behavior

(Adapted from Tilson, H.A. and Harry, G.J. (1982) In: Mitchell, C.L. (Ed.), *Nervous System Toxicology*. Raven Press, New York, Chapter 1.)

	Respondent	Operant
Unconditioned	1. Response is elicited by an identifiable stimulus	1. Response is emitted with no known eliciting stimulus
	2. Responses typically include those of smooth muscle, glandular secretions, autonomic responses, and environmentally elicited effector responses	2. Responses typically are mediated by CNS and involve skeletal muscle movements
	3. Data are measures of response magnitude, probability, latency, or are related to intensity of eliciting stimulus	3. Measures are typically response probability, frequency or speed
	Classical conditioning	Instrumental conditioning
Conditioned	Conditioned response (CR) is elicited by a particular stimulus as the result of close temporal pairing of that stimulus (CS) with a stimulus (US) which naturally elicits the response (UR)	Operant response (R) changes in frequency of occurrence as a function of the consequences of the response

movements and movements of different groups of skeletal muscles in the absence of any identifiable stimulus or learning.

The rate at which operant behaviors are emitted is dependent upon the consequences which they have on the organism. In operant conditioning, the likelihood that a given behavioral response will be emitted by an organism depends on the consequences of that behavior. That is, learning involves the pairing of a particular behavioral response with the consequences of that response, rather than the pairing of two stimuli as in respondent conditioning. Some well-known examples are the hungry rat which learns to depress a lever (emitted response) in order to obtain a food reward, or the dog that comes to the call of its owner for a pat on the head.

When the consequences which occur as the result of a given behavior serve to maintain or increase a given response, they are considered to be reinforcing stimuli. In operant conditioning paradigms, reinforcing stimuli occur contingent upon a specific response, and they increase the probability that the particular response will be emitted. The two examples above are illustrations of rewarding stimuli which serve to reinforce a particular response and increase the probability of that particular response being emitted.

Rather than providing the occasion to obtain reward, however, some stimuli can also be aversive. When an aversive stimulus is contingent upon a response, it can serve as a punishment. In that case, the aversive stimulus will serve to decrease the probability that a response will be emitted. If the rat in the above example was given a painful electric shock instead of a food reward each time it depressed the lever, the rat's bar pressing would decrease, rather than increase.

75

Aversive stimuli, however, do not always produce decreases in responding. If a lever press, for example, has the result of terminating or postponing the shock stimulus, the probability of eliciting the lever press response will increase. In this case, the shock stimulus is still aversive, but the postponement of shock, which is contingent on the response, acts as a reinforcer.

Many different behavioral paradigms employing both conditioned and unconditioned respondent and operant techniques have been developed for studying the effects of drugs and chemicals on behavior. Unconditioned respondent behaviors such as the pupil reflex in response to light, or the startle reflex in response to a sudden noise and the unconditioned operant, spontaneous motor activity, for example, have been included in screening batteries designed to identify chemicals with possible neurological and behavioral effects. For more in-depth studies that are designed to characterize the effects of a compound or to establish no-effect levels, other paradigms have been developed which employ operant methodologies for examining a wide range of sensory, motor and cognitive effects.

2   Behavioral methods for assessing neurotoxicity in animals

The nervous system is an extremely complex organ so different chemicals have different targets within the nervous system. Methanol, for example, affects the eye and produces visual disturbances and blindness; aminoglycosides affect the hair cells of the cochlear and produce permanent deafness; acrylamide attacks sensory peripheral nerves and leads to changes in the sense of touch as well as the control of movement; manganese destroys neurons in the basal ganglia, producing Parkinson-like symptoms; the organometal trimethyl tin destroys cells in the hippocampus, a brain structure involved in learning and memory. As a result of the multitude of different sites in the brain sensitive to different compounds, neurotoxic effects can be expressed in a myriad of ways (Table 3.3). For this reason, neurotoxicologists generally advocate the use of batteries of behavioral tests designed to measure different aspects of nervous system functioning for evaluating potential neurotoxicity of compounds. Because of the large number of chemicals already in existence, it is unlikely that each one of them can be thoroughly evaluated for possible neurotoxic effects. For this reason, a number of expert scientific groups have advocated a "tiered" approach to neurotoxicity assessment consisting of a stepwise progression from general screening using simple observational methods and neuropathology to more comprehensive assessment using more specific and sophisticated behavioral techniques.

2.1   NEUROTOXICITY SCREENING METHODS

2.1.1   *Observational methods*

The approach which has been most frequently advocated for screening compounds that have little or no neurotoxicity data available is the use of a functional observational battery (FOB). An FOB consists of standardized observational ratings designed to assess different aspects of nervous system function as well as several noninvasive tests to evaluate sensory, motor, and autonomic dysfunction (Table 3.4).

Usually observations of the animal, which in most studies is the rat, are carried out both while the animal is in its home cage and while it is moving freely about in a test arena for a specified period of time. Animals are observed for changes in arousal, reactivity to handling, presence and type of convulsions, alterations of gait and mobility, and autonomic signs. In addition, simple tests

TABLE 3.3

Types of behavioral and neurological effects reported in humans

Functional domain	Effects
Affective/personality	Depression, euphoria, irritability, sleep disturbances, restlessness, delirium, hallucinations
Motor effects	weakness, incoordination, ataxia, reflex abnormalities, paralysis, tremor, convulsions
Sensory effects	blindness, color vision disturbances, hearing loss, abnormalities of taste and smell, parathesias, loss of the sense of touch, changes in pain perception
Cognitive changes	attention deficits, memory loss, psychomotor slowing, visuomotor disturbances, reduced intelligence (IQ), deficits in problem solving, dementia
Autonomic effects	sweating abnormalities, changes in appetite and body weight, abnormalities in control of cardiovascular and gastrointestinal function, difficulties in urination, sexual dysfunction

TABLE 3.4

Examples of types of observations and simple tests included in a functional observational battery
(Adapted from Moser, V.C. (1989) *J. Am. Coll. Toxicol.* **8**: p 885.)

Neuromuscular	Observation and ranking of gait abnormalities
	Landing foot splay measurement
	Ranking of righting reflex
	Forelimb grip strength measurement
	Hindlimb grip strength measurement
Arousal	Observation of posture in the home cage
	Ranking of ease of removal from the home cage behaviors
	Ranking of ease of handling
	Observation of vocalizations
	Observation of piloerection
CNS hyperexcitability	Observation and ranking of clonic movements
	Observation and ranking of tonic movements
Physiological	Measurement of body weight
	Measurement of body temperature
Sensorimotor reactivity	Ranking of rat's reaction to:
	– approach of a blunt object
	– being touched with an object
	– a sound stimulus
	– a tail pinch
Bizarre behavior	Observations of bizarre behaviors such as walking backwards, self-mutilation, stereotyped motor behavior
Autonomic	Ranking of salivation
	Ranking of lacrimation
	Observation of response of pupil to light stimulus
	Observation and ranking of palprebral closure
	Observation and ranking of degree of urination
	Observation and ranking of degree of defecation

of sensory function such as the pupil reflex to light, reaction to a sudden noise stimulus (see Figure 3.2), tactile stimulation and tail pinch, are also conducted. Furthermore, simple tests of motor function such as the righting reflex and landing foot splay are also usually included. To measure landing foot splay, the hindpaws of the rat are inked and the rat is dropped from a horizontal position on to a blank sheet of paper covering the cushion. The distance between the two resulting ink spots on the paper provide a measure of "landing foot splay" which tends to be increased in rats with motor dysfunction.

FIGURE 3.2

One measure of sensory reactivity in a functional observational battery (FOB) evaluates a rat's response to an auditory stimulus.
(Photo credit: John O'Donaghue, Eastman Kodak Co.)

One of the advantages of using a standardized battery of observations with measures that are operationally defined is that semi-quantitative estimates of the severity of impairment on a particular measure can be made. As Figure 3.3 demonstrates, quite orderly dose-response relations can be obtained with standardized observational techniques to assess the effects of a particular exposure. In the example presented in the figure, acute exposure to the pesticide cypermethrin produced clear-cut dose-related effects on a number of different functions including autonomic function, landing foot splay, convulsive movements, and decreased reactivity to tail pinch when tested shortly following exposure which are still observable 24 hours later. Further, some residual effects on reactivity to tail pinch, and to a lesser degree on landing foot splay, were still measurable one week later, suggesting that some of the effects of a single high exposure may be irreversible.

The human eye is a very powerful means of detecting neurotoxicity and observational assessments are essential in documenting the clinical effects of neurotoxic agents. The most serious drawback of observational methods, however, is that they are subjective in nature and depend on the experience and acumen of the observer. For this reason, a number of investigators have proposed that screening for neurotoxicity should also include objective, automated measures of motor activity as part of a screening battery.

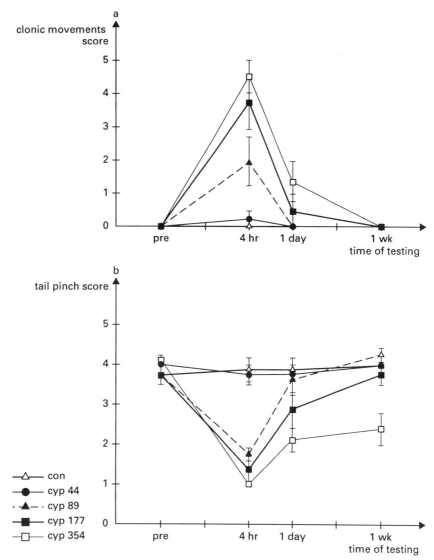

FIGURE 3.3

The pesticide cypermethrin administered in doses ranging from 0 (con) to 354 mg/kg (cyp 354) produces many different effects on neurobehavioral functioning

Some effects, e.g., clonic movements (upper panel), do not seem to persist after 24 hours. Other effects such as decreased sensitivity to tail pinch (lower panel) can be more permanent. (From Kulig, unpublished data.)

2.1.2   *Motor activity assessment*

Similar to observational methods, motor activity assessment requires no prior learning on the part of the animal, thus it is quite suitable for neurotoxicity screening. Further, because the technique has been used extensively in behavioral pharmacology for many years, quite a bit is known about the sensitivity of activity measurements to the effects of different drugs and brain damage as well as the advantages and limitations of different measurement devices.

Different approaches to detecting the movement of animals include field detectors, activity wheels, photocell-based systems, and video-based devices.

Although all of these systems detect movement, there are very large differences in the type of "motor activity" that they measure. One example of a field detection device, for example, is based on the generation of a capacitive field around the test chamber. In this system, an adjustable oscillator supplies high-frequency current to an input coil creating a field around the test chamber. Movement within the test chamber produces momentary changes in the voltage in the output coil which are digitized and reported as "activity counts". One of the disadvantages of field systems is that any movement large enough to activate the coil may be detected. This means that movements other than spontaneous locomotion, such as body-part movements, grooming and even tremors and convulsive movements will be included in the measurement, making the interpretation of effects on locomotor activity *per se* difficult.

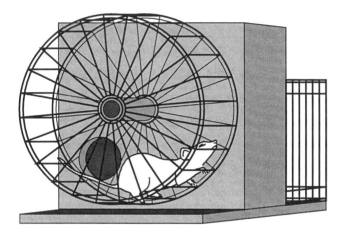

FIGURE 3.4

Activity wheel shown with access from the home cage
(From Silverman, P. (1978) *Animal Behavior in the Laboratory*. Chapman & Hall. With permission.)

In contrast to field detection devices, activity wheels tend to measure a very specific type of ambulation under very specific circumstances. Activity wheels are electromechanical-based devices which consist of an enclosed wheel attached to the animal's home cage (see Figure. 3.4). When the animal enters the wheel, they tend to "run" so the measure of activity is the number of revolutions which the animal makes in the wheel. All other activities including eating, drinking, grooming, exploring, etc. occur in the home cage portion of the apparatus. The major drawback of activity wheels is that the movement itself provides feedback to the rat which can modify running rate in a rather unspecified manner. This is perhaps one of the reasons underlying the large differences among rats in this type of activity measurement.

In order to obtain a more refined measure of locomotor activity in a stationary environment, detection methods employing either photocells or computerized video imaging techniques are typically employed. In photocell-based systems, the photocells are positioned in such a way that they are primarily sensitive to horizontally directed movement which is measured in either simple test environments, such as a rectangular box or circular enclosure, or in complex maze-type environments, such as the figure-8 maze or the residential maze.

In video-based systems, a camera positioned above the test apparatus receives a video image of the white rat on a black background which is digitized by computer into a series of X-Y coordinates, providing information on the amount of ambulatory movement in terms of "total meters run", the

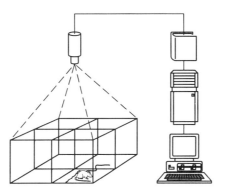

FIGURE 3.5
Motor activity of more than one animal can be measured simultaneously by
using computer-based videoimaging devices

distribution of movement at different speeds, and the amount of locomotor
activity in different locations within the test chamber (see Figure 3.5).

Although one might expect that motor activity would decrease as the result
of drug or chemical exposure, this is not the case for all compounds.
Barbiturate-like drugs and organic solvents, for example, tend to have
stimulating properties at low doses which produce an increase in locomotor
activity, whereas at high doses sedation occurs, leading to a decrease in activity.

2.2   QUANTIFYING CLINICAL SIGNS OF MOTOR IMPAIRMENT

Exposure to many different classes of substances including metals, solvents,
pesticides, gases, and drugs has been associated with toxicant-induced motor
disturbances. In some cases, functional effects are the result of damage to
peripheral nerve, while in others, structures within the central nervous system
are involved. Not surprisingly, the types of motor effects which have been
reported in the literature for different compounds are equally diverse ranging
from specific deficits such as ataxia, paresis, tremor, and other types of
dyskinesias to more generalized effects such as activity changes and changes in
psychomotor performance.

One technique in widespread use to examine the effects of chemical
exposures on neuromuscular function is the measurement of grip strength. The
general principle of the technique is depicted in Figure 3.6. The apparatus
consists of a push-pull strain gauge attached to a T-bar positioned at the end of
a specially built platform. To measure forelimb grip strength, the rat is placed
with its forepaws on the T-bar and gently pulled backwards by the tail until it
engages the bar. It is then pulled back further until its grip is broken. To
measure hindlimb grip strength, the rat is placed on the platform facing away
from the T-bar and a similar procedure is employed.

Figure 3.7 demonstrates the effects of inhalatory exposure of rats to the solvent n-hexane. As the figure demonstrates, exposure for 3 months produces
progressive decreases in grip strength in both the upper and lower limbs, with
more marked effects seen in hindlimb grip strength. The gradual and insidious
loss of neuromotor function with greater involvement of the lower limbs has
also been described in human cases of chronic n-hexane poisoning and is due to
the selective vulnerability of long, large diameter axons in the spinal cord and
periphery to the n-hexane metabolite, 2,5-hexanedione. Thus, there is good
agreement between the nature of effects seen in human cases and those
measured in rats with this technique.

a b

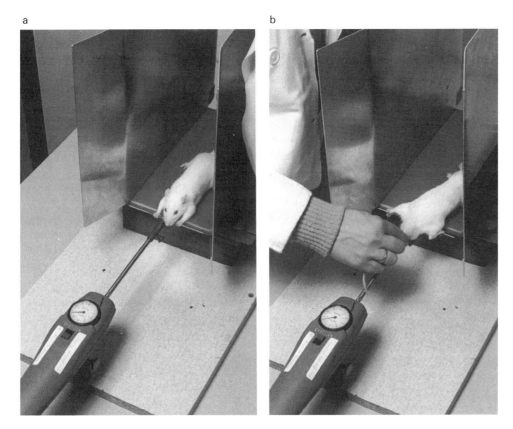

FIGURE 3.6

A simple technique for assessing changes in neuromuscular strength uses strain gauges to measure fore- and hindlimb grip strength

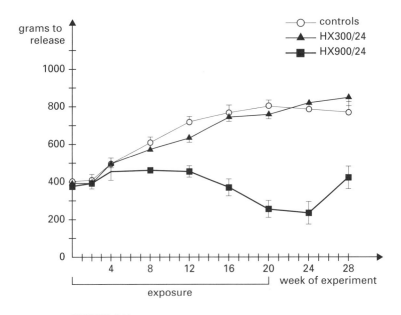

FIGURE 3.7

Chronic exposure to *n*-hexane at 900 ppm for 24 hours per day (HX900/24) produces a gradual loss of neuromuscular strength particularly in the hindlimbs which can be measured using grip strength measurement techniques

Exposure at 300 ppm (HX300/24) was ineffective in this experiment

In addition to neuromuscular weakness, exposure to many chemicals and drugs also produces disturbances in coordinated movement. In pharmacology, one of the most common techniques used to evaluate the effects of drugs on coordination is the rotorod (see Figure 3.8). With this technique, a quantitative assessment of coordinated movement is obtained by placing the rat or mouse on a motor-driven rotating rod and measuring the time it takes the animal to lose its footing and fall off. Although this simple technique is quite good for measuring uncoordination following acute drug administration, its application is limited in studies in which testing is repeatedly conducted over weeks or months. With repeated testing, control rats have a tendency to start jumping off the rotating rod, making the test unsuitable for long-term studies.

FIGURE 3.8

The rotorod has been extensively used to assess the effects of acute exposures on motor coordination however, responding by control rats is often unreliable with this technique.

To counteract this problem, several automated techniques using video imaging have recently been developed (see Figure 3.9). In one such test, the apparatus is comprised of a motor-driven wheel with cross-pieces attached to the floor of the wheel which rotates at a constant speed, a color video camera, and a computer for the automatic analysis of different aspects of hindlimb movement (e.g., stride length, time that a hind paw rests on the rungs, mis-stepping, etc). When the wheel is in motion, normal rats tend to walk on the tops of the rungs. When rats are motorically impaired, however, they either fail to stay in motion resulting in walking failure, or they make missteps in which their hindpaw drops between the rungs. In a test as short as 90 seconds, a quantitative measure of coordinated movement can be obtained for evaluating the dose-response effects of acute exposures as well as the development of motor deficits in subchronic studies.

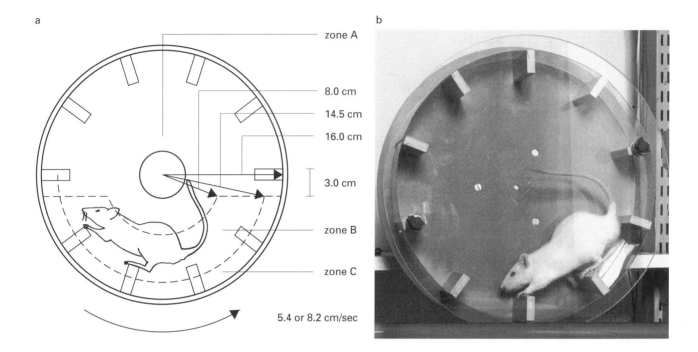

FIGURE 3.9

Automated video-based analysis of hindlimb placement and movement can be used to measure quantitatively coordination deficits

(A) Movements of the dyed hindpaw are recorded and digitized by computer. (B) Normal rats walking on tops of rungs.

In addition to methods designed to quantify loss of neuromuscular strength and both ataxic and paretic movement dysfunction, specialized tests have also been developed for studying the effects of chemicals and drugs which produce tremor. Tremor refers to rhythmic, involuntary oscillations of the whole body, or of parts of the body, and is seen as a clinical sign in several neurological diseases or as a consequence of exposure to particular therapeutic drugs (e.g., drugs used in the treatment of schizophrenia), chemicals, and pesticides. Some of the most dramatic instances of human neurotoxic disease — from mercury-induced tremor in the hatting industry (see Figure 3.1) to a more recent outbreak in chlordecone-exposed workers in a pesticide manufacturing plant — have involved occupational exposure to tremorogens.

The need to characterize chemically induced tremor in laboratory animals has led to the use of a number of different types of transducers, (e.g., force-displacement transducers, load-cell transducers, strain gauges, etc.) capable of converting the mechanical energy of tremor to electrical energy for the purpose of quantification. Most paradigms involve the measurement of whole-body tremor in which the rat is placed in a cage mounted on an appropriate transducer. Although care must be taken to avoid confounding the measurement of the animal's tremor with resonating oscillations from the cage itself, measurement of whole-body tremor can provide extremely sensitive and specific measures of tremorogenic effects.

2.3   EVALUATING SENSORY FUNCTIONS IN ANIMALS

Although a considerable amount of importance has been placed on the motor effects of toxic exposures, possible effects on sensory function are also of growing concern. It has been recently reported, for example, that approximately 44% of chemicals which possess neurotoxicant effects have effects on sensory function.

Basically, two different types of behavioral paradigms have been described for evaluating the effects of chemicals, those based on operant conditioning and those using reflex modification techniques. Instrumental or operant conditioning techniques have included the use of both active avoidance paradigms as well as psychophysical operant discrimination methodologies in both rodents and primates.

One behavioral technique using avoidance learning to evaluate sensory function is the multisensory conditioned avoidance paradigm (see Figure 3.10). In this test, animals are first trained on an avoidance conditioning task with different sensory cues, for example, light, tones of different frequencies, or mild shock as the conditioned stimuli. Sensory impairments, fsuch as auditory threshold changes are apparent when animals fail to make avoidance responses to tones while continuing to make responses to stimuli in other modalities. Using a multisensory conditioned avoidance paradigm such as this, investigators have recently uncovered a neurotoxic effect of organic solvents not previously measured in laboratory animals with other methods, namely the ability of some organic solvents to produce irreversible hearing loss.

Psychophysical operant discrimination techniques provide a very elegant approach to the evaluation of neurotoxicant-induced specific sensory deficits. In this technique, animals are first trained to emit a specified response in the presence or absence of a stimulus of a particular modality. That is, the animal is rewarded for "reporting" whether or not it can detect a stimulus by making a particular response. To study the effects of acrylamide on vibratory sensation, for example, Maurissen and his co-workers trained monkeys to hold down a response lever with one hand and to let go of the lever if a vibratory stimulus was detected to the fingertip of the other hand. Each trial was signalled by a tone and the animal was rewarded if it released the lever when the vibratory stimulus was delivered. To control for "guessing", trials on which no vibratory stimulus was delivered were also presented during the test session and the monkey was also rewarded if it did not release the lever during the trials where no stimulus was presented[21]. Similar methodologies have also been adapted to study the effects of acrylamide on the visual system as well as the sensory deficits produced by developmental methylmercury exposure[23,27].

One of the principal drawbacks usually cited in discussions of operant sensory testing paradigms is the relatively long periods of time required to achieve stable baseline levels of responding. Usually months of training are required before animals respond in a reliable fashion. Because of this limitation, several authors have investigated the use of reflex modification of the startle response as a possible tool both in terms of screening and in characterization of sensory deficits, particularly with respect to the auditory system.

Reflex modification refers to the change in the startle response caused by a perceptible change in the sensory environment occurring shortly before the stimulus which produces the startle response. The basic paradigm of the reflex

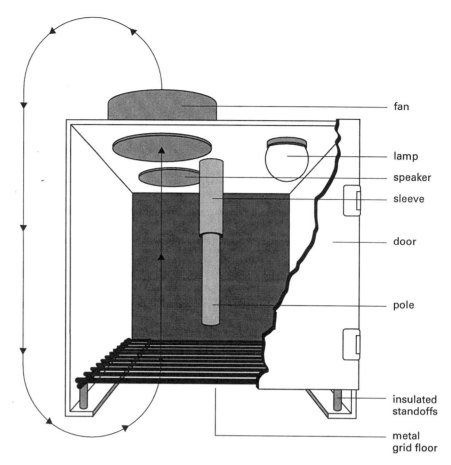

fan

lamp

speaker

sleeve

door

pole

insulated
standoffs

metal
grid floor

FIGURE 3.10

Apparatus used for automated multisensory conditioning testing used to assess losses in hearing, vision, and somatosensory function

The technique utilizes a pole-climbing avoidance task with light. tone, and mild foot-shock as discriminative stimuli. (From Pryor, G.T. et al. (1983) *Neurotox. Teratol.* **5**:1991. Copyright 1983 by Elsevier Science. With permission.)

modification of the startle response is illustrated in Figure 3.11. The left side of the figure shows the response to a loud sudden auditory stimulus (S2) which elicits the startle response reflexively. The strength of this response can be measured quantitatively using a force transducer or similar device. The right hand side of the figure shows the effects of a reflex-modifying prepulse stimulus (S1) presented shortly before the startle stimulus. If the animal is capable of perceiving S1, the effect of S1 will be to inhibit the startle response and reduce the amplitude of the startle reflex elicited by S2.

One principle of this phenomenon is the linear relationship between the intensity of S1 and the decrease in the amplitude of the startle response. This relationship makes it possible to obtain empirically derived sensory thresholds.

Figure 3.12 illustrates the auditory threshold for tones of 4 Khz measured in a single rat. In this type of study, the rat received approximately 300 trials to obtain a hearing threshold. For some of the trials, no prepulse was presented and on others the S2 was preceded by a 4 Khz prepulse presented at 13 different intensities. With very low intensities, the amplitude of the startle response was unaffected. However, as the prepulse increased in intensity, the effects on startle amplitude became increasingly pronounced. By applying a mathematical

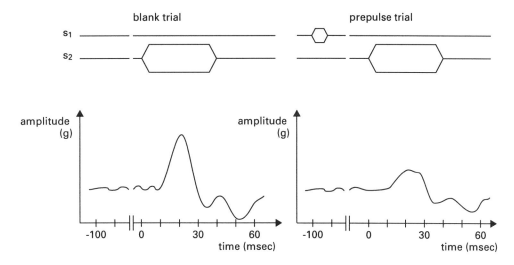

FIGURE 3.11

Reflex modification of the startle response

Left: Response amplitude on a trial in which only the reflex-eliciting (S2) is presented. Right: Reduction in response amplitude on a trial in which S2 is preceded by a prepulse (S1). (Adapted from Crofton, K.M. (1990) *Neurotoxicol. Teratol.* **12**:461–168. With permission.)

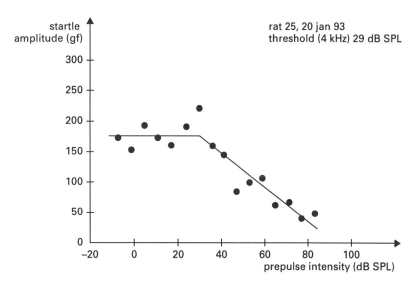

FIGURE 3.12

Hearing threshold for 4 Khz tones in a single rat using reflex modification procedures

(From Muijser and Kulig, unpublished data.)

model to these data, the hearing threshold at specific frequencies, for example 29 dB in this particular rat, can be determined both before and following a particular exposure.

In Figure 3.13, for example, hearing thresholds for 5 and 20 kHz tones were determined in rats prior to exposure to trichloroethylene and for 6 weeks following the end of exposure. As the figure illustrates, auditory thresholds for the lower frequency were unaffected by the solvent. However, at the higher frequency a significant threshold shift of approximately 20 Db could be detected in the first week following the end of exposure which appeared to be permanent. The permanent loss of hearing just for higher frequencies is probably due to the selective destruction of sensory hair cells in the cochlear.

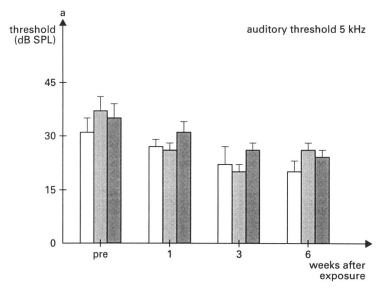

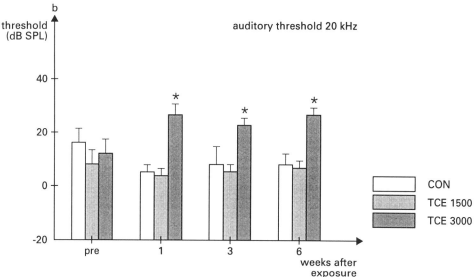

FIGURE 3.13

Frequency-specific effects of short-term exposure to trichloroethylene at 1500 ppm (TCE 1500) or 3000 ppm (TCE 3000) on auditory thresholds
(a) No effects seen for 5 kHz hearing. (b) Persistent effects at 20 kHz. (From Jaspers, R.M.A. et al. (1993) *Neurotox. Teratol.* **15**:407–412. Copyright 1993 by Elsevier Science. With permission.)

Although most studies with reflex modification have been directed to the study of audition, the phenomenon of reflex inhibition is not limited to auditory perception. Studies have demonstrated that the amplitude of the startle response can be modified, not only by tones, but by light stimuli, shock, noise, and gaps in a continuous background of noise. To what extent reflex modification procedures can be applied to the study of chemical effects on other modalities, however, awaits further investigation.

2.4   EVALUATING CHANGES IN COGNITIVE BEHAVIORS

Behavioral impairments indicative of cognitive changes have been associated with exposure to a number of chemicals. Developmental exposure to lead, methylmercury, and polychlorinated biphenyls (PCBs), for example, have been causally related to delayed development and intellectual impairments in children. Further, in adults, chronic exposure to organic solvents has been associated with the development of toxic encephalopathy characterized by memory loss and cognitive impairments. Given the fact that intellectual abilities related to memory and learning capacity are of such importance in successfully adapting to changes in the environment, it is not surprising that concern has been raised regarding the need to include measures of learning and cognition in evaluating the health effects of drugs and chemicals.

Learning is not a unitary phenomenon, so a result many models have evolved to evaluate different aspects of learning and other higher order functions in animals. Some studies have concentrated on studying the effects of chemical exposures on the performance of learned behaviors, using, for example, free operant or discrete trial techniques, while others have attempted to develop models to study acquisition and memory. Furthermore, a variety of different types of testing environments such as two-compartment shuttle-boxes, mazes, and operant chambers have been used with different behavioral paradigms including avoidance learning, reversal learning, repeated acquisition, and delay tasks. Although the list is far from exhaustive, Table 3.5 summarizes some of the more frequently used techniques to study the effects of chemicals on different aspects of cognitive behavior.

TABLE 3.5

Several examples of different conditioning paradigms used to study the effects of neurotoxicants on cognitive behaviors

Testing environment	Behavioral paradigm	Cognitive function
Two-compartment test chamber	– passive avoidance learning	– short-term memory
	– shuttlebox avoidance learning	– learning acquisition, memory, performance
Mazes		
T-maze	– learning	– acquisition
	– delayed alternation	– short-term memory
	– discrimination reversal	– learning
8-arm radial maze	– spatial memory testing	– short-term memory
Morris water maze	– spatial learning	– short-term memory, learning
Operant chamber	– intermittent schedules of reinforcement	– learned performance
	– discrete-trial operant discrimination	– learned performance
	– delayed alternation	– short-term memory
	– discrimination reversal	– learning
	– repeated acquisition	– learning

2.4.1 *Performance testing*

The potential to affect the performance of learned behavior is an important consideration in evaluating the behavioral toxicity of drugs and chemicals. Two different approaches which have been used to assess changes in learned performance are free-operant tasks using intermittent schedules of reinforcement and discrete-trial operant discrimination tasks. Both approaches utilize operant chambers as the test environment. A typical operant chamber, such as the one depicted in Figure 3.14, consists of a test box equipped with one or two levers and a device to deliver water or food reinforcement. Animals are first trained to depress the lever in order to obtain the reinforcement. Once learning of the bar press response has been accomplished, the rat is switched over to the task chosen by the experimenter and given additional training until a high level of stable performance has been reached. Once a stable baseline performance has been obtained, the animal is then exposed to the compound in order to evaluate its effects on performance.

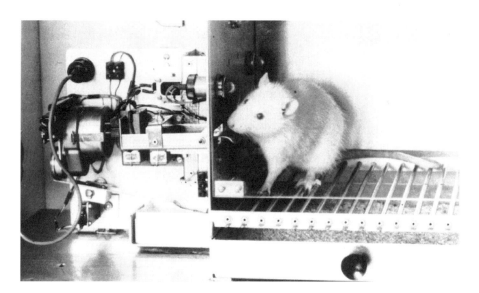

FIGURE 3.14

Operant chamber apparatus

In this example, a thirsty rat has been trained to depress the lever in order to obtain a drop of water from a dipper which can be automatically introduced into the chamber.

2.4.1.1 Intermittent schedules of reinforcement

Although reinforcement following a response leads to an increase in the probability that the response will be subsequently emitted, it is not necessary that every response an animal (or human) makes be reinforced in order to maintain or increase responding of that response. In fact, higher rates of responding are typically achieved when responses are intermittently reinforced than when they are reinforced continuously each time they are made. In the operant chamber situation, for example, the rate at which a rat will depress a lever to obtain a single reinforcement is much higher when the rat is required to make 20 lever presses for each reinforcement compared to the situation where each lever press is followed by reinforcement.

Reinforcement can be scheduled to occur on the basis of either time or on the number of responses required. Those that are based on the number of responses required for the delivery of a single reinforcement are referred to as ratio schedules and in the above example, the schedule of reinforcement being employed was a fixed-ratio 20 (or FR20), signifying that 20 responses are required for each reinforcement. Reinforcement can also be scheduled to occur after a certain amount of time has passed since the last reinforcement. Such schedules are referred to as interval schedules. For example, in a fixed-interval 1-min (or FI 1-min) schedule, the first lever press occurring after at least one minute has passed since the last reinforcement is reinforced. Although these examples utilize a fixed number of responses or a fixed interval of time, schedules can also be variable. In a variable-ratio 20 (VR20) schedule, for example, the rat receives reinforcement on average once every 20 lever presses. However, the ratio requirement between reinforcements might vary from 10 to 30. Likewise, in a variable-interval 1 min schedule (or VI 1-min), the intervals separating reinforcements would vary from interval to interval with an average length of one minute[5].

As Figure 3.15 illustrates, FI and FR schedules produce different patterns of responding. In FI schedules, delivery of reinforcement produces a pause in responding, known as the post-reinforcement pause, immediately after the

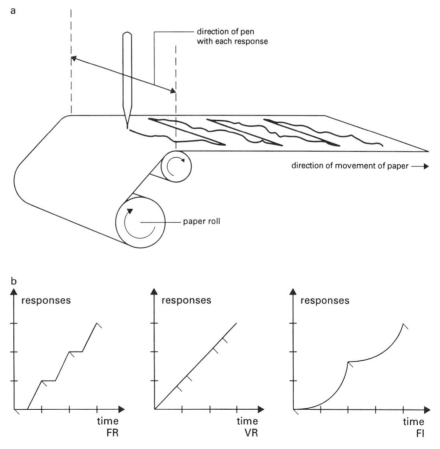

FIGURE 3.15

Upper diagram: A drawing of a cumulative recorder

Every response moves the pen a small notch up the paper in the direction of the arrow. At the top of the paper it resets to the bottom in a single motion. Reinforcements are usually indicated by diagonal slash marks. Commercially available software programs are available with present-day programming equipment for generating cumulative records off-line from computer-stored data files. Lower diagram: Typical cumulative records for an animal (or human) responding on FR, VR, and FI reinforcement schedules. The diagonal slash indicates the delivery of reinforcement. (Redrawn from McKim, W.A. (1986) *Drugs and Behavior*. Prentice Hall, New York.)

delivery of reinforcement. This post-reinforcement pause is then followed by low rates of responding during the early part of each interval which gradually increase as the availability of reinforcement approaches. This pattern of responding produces a scalloped-shape curve which is characteristic of FI behavior. In FR schedules, delivery of reinforcement also produces a post-reinforcement pause, but in this case, the animal resumes responding at a high rate immediately thereafter, producing a step-like pattern of responding. Ratio and interval schedules can be used alone or in combination to evaluate treatment effects on both ratio and interval schedules in the same animal.

Dependent measures in fixed ratio schedules include response rate, measures of post-reinforcement time, running rate (i.e., mean time from the end of the post-reinforcement pause to the delivery of the next reinforcement), and inter-response times (i.e., the distribution of times between responses calculated across a test session). For FI schedules, mathematical models for quantifying changes in the shape of the scallop-like patterning of responding during an interval have also been developed. In test sessions of sufficient duration, e.g., 1 hour, the rat can produce many thousands of responses which can be used to provide very reliable estimates of responding in individual animals across time. This characteristic of intermittent schedules of reinforcement translates into the need for fewer animals to evaluate compound effects, which is obviously an important consideration in the choice of a particular technique.

Studies using different schedules of reinforcement indicate that chemical effects on schedule-controlled operant behavior are dependent not only on the particular substance and dose, but also on the type of schedule controlling the behavior. Lead, for example, typically produces a dose-effect curve on FI responding which has an inverted U-shape. That is, at lower levels of lead exposure, FI response rates are increased whereas higher levels decrease responding. This appears to be the case regardless of the developmental stage at which exposure occurs. That is, different studies using prenatal, postnatal, postweaning or adult exposure to inorganic lead all show a similar inverted-U dose response effect on FI responding. In contrast to effects of FI responding, FR responding is relatively insensitive to lead exposure, and when effects do occur, responding is typically decreased[5].

In psychopharmacology, schedule-controlled operant behavior has been used extensively for evaluating the psychoactive properties of therapeutic drugs for the last thirty years, and this technology is being increasingly advocated in the study of chemical effects as well. In the US, regulatory guidelines for the inclusion of chemical testing using intermittent schedules of reinforcement have been published and are currently being mandated for the study of a number of organic solvents.

2.4.1.2 Discrete-trial operant discrimination

In contrast to the intermittent schedules of reinforcement described above, discrete-trial discrimination tasks using the same type of operant chamber have also been used to examine the effects of chemicals, particularly organic solvents. In some ways, such an approach is analogous to the two-choice reaction time tasks used to study drug and chemical effects on human subjects. In human studies, increases in response latency in reaction time tasks is a common finding as a consequence of human solvent exposure.

In the discrete-trial paradigm, a rat is trained in an operant chamber equipped with two levers, two stimulus lights located above the levers, and a water (or food) reinforcement device located equidistant between the two levers. On each trial, one of the lights above the levers is illuminated in a semi-random fashion and the rat's task is to depress the lever under the illuminated light in order to receive a drop of water. Trials are separated by an inter-trial interval (ITI) of, for example, 10 s. To ensure that responding occurs primarily during the light-cued trial, lever responses made during the inter-trial interval postpones the start of the next trial.

Such a two-choice discrete-trial discrimination task is easily acquired and in daily sessions of, for example, 100 trials, very reliable baselines of responding can be achieved within 10–15 sessions. Responding is not only stable with respect to the accuracy of the discrimination, but also in the speed with which the rat makes the two-choice discrimination response[17]. In addition, this measure also appears to be quite sensitive to the effects of inhaled organic solvents and studies using this approach have demonstrated the response-slowing effects of a number of inhaled organic solvents, including styrene, trichloroethylene, white spirits, and perchloroethylene in rats at relatively low concentration levels.

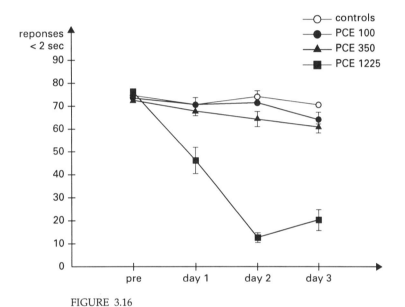

FIGURE 3.16

Effects of exposure to perchloroethylene at 0 (Controls), 100 (PCE 100), 350 (PCE 300) or 1225 (PCE 1225) ppm on the mean number of short-latency correct responses made by rats working on a discrete-trail operant visual discrimination task
(From Kulig, B.M. unpublished data.)

Figure 3.16, for example, shows the acute effects of inhalatory exposure to perchloroethylene measured shortly after each daily exposure for three days. As the figure demonstrates, the effects of perchloroethylene were cumulative over the three-day period. The number of short-latency trial responses occurring within 2 s of light onset progressively decreased over the three-day exposure period while long-latency pauses increased. This cumulative effect is probably due to the slow metabolism of this solvent and the gradual build-up of concentrations of the solvent in the brain.

2.4.2   *Techniques to evaluate memory and learning*

In addition to evaluating the effects of drugs and chemicals on the performance of learned behavior, different paradigms have also been used to evaluate the effects of a particular treatment on the acquisition and retention of information, i.e., learning and memory. Below is a short survey of several methods which have been used to study neurotoxicant effects on learning and memory processes.

2.4.2.1 Passive and active avoidance learning

Two very different types of avoidance learning have been used to study learning and memory: passive avoidance learning and active avoidance learning. In passive avoidance learning paradigms, the animal is trained to *withhold* a response in order to avoid punishment, while in active avoidance learning the animal is required to *make* a response in order to avoid a shock stimulus.

One standard procedure for studying passive avoidance learning utilizes a two-compartment test chamber comprised of a brightly illuminated compartment and a darkened one. The darkened compartment is also equipped with a grid floor which can be activated to produce an electric shock to the paws of the rat. The technique utilizes the fact that normal rats typically prefer an environment that is dimly illuminated. In the one-trial learning procedure, the animal is first given a training trial in which it is placed in the lighted compartment and allowed to enter the darkened compartment. Upon entering the darkened compartment, the rat immediately receives a painful electric shock. The test trial consists of subsequently reintroducing the rat into the lighted compartment and measuring the latency to enter the darkened compartment.

Although passive avoidance learning paradigms have been extensively used in the past to study the effects of drugs on short-term memory processes, they have come under heavy criticism in recent years[8]. Despite the advantages of being technically easy and rapid, passive avoidance procedures also tend to produce highly variable results and are sensitive to a number of variables which are not related to learning *per se* such as activity level and sensitivity to footshock. Effects on motivational variables, arousal level, and sensory changes can obviously influence the acquisition, retention and/or performance of any learned task. However, many other types of paradigms provide measures to check for these types of effects, a feature which is lacking in passive avoidance procedures. Furthermore, passive avoidance learning cannot be carried out repeatedly in the same animal to evaluate the cumulative effects of repeated chemical exposures on memory and learning processes which is an important consideration in assessing the health effects of prolonged low-level chemical exposures.

In active avoidance paradigms, a two compartment test chamber is also typically used, but in this situation an active response is required from the animal. In one standard procedure, the rat is placed in one of the two compartments. Each trial is signalled by the presentation of a tone and the rat is required to move to the other compartment to avoid being shocked. If the response is not made within a specified period of time, the grid floor is electrified and the rat is allowed to escape the shock by moving to the other compartment. In the one-way avoidance paradigm, the rat is returned to the original starting compartment and the procedure is repeated. Since this usually involves handling the animal after each trial, a two-way active avoidance paradigm is often employed. In this paradigm, the animal remains undisturbed in the apparatus between trials. Upon presentation of the tone on each successive trial, the rat must learn to return to the compartment which had been shocked on the preceding trial.

Dependent measures of active avoidance learning include the number and latency of avoidance responses, the number and latency of escape responses, the number of trials to a predetermined criterion, and the number of inter-trial crossings from one compartment to another. High levels of correct avoidance responding can be achieved quite rapidly within several days and is quite stable with repeated testing. Furthermore, in contrast to passive avoidance techniques, the inclusion of measures of escape latencies and inter-trial interval responding can help in interpreting if deficits in avoidance learning in treated animals are truly effects on learning processes, or if changes in sensory-motor effects or activity levels may also play a role.

Active avoidance procedures, however, also have a number of disadvantages. First, by their nature avoidance paradigms must employ aversive stimuli such as electric shock to maintain behavior. Secondly, they cannot be used to track the time course of effects on learning and memory over time. When they are used repeatedly during and following exposure, they become measures of learned performance rather than of acquisition and memory.

2.4.2.2   Maze and operant techniques to study working memory

Mazes are one of the oldest types of apparatus which have been used by psychologists to study the rate of acquisition and a number of techniques involving discrimination of spatial or sensory cues (visual, auditory, or olfactory) have been developed. Studies of spatial or visual discrimination learning often use simple T-mazes or more complex variants, such as the Hebb-Williams maze, which is a series of joined T-mazes.

When mazes are used to evaluate changes in learning, i.e., the rate of acquisition of a particular response, they, too, cannot be used to track changes in memory or learning capacity over the course of time in the same animal. There are, however, several other maze paradigms which might prove more useful in this regard, for example, delayed alternation techniques using the simple T-maze or the use of the more complex eight-arm radial maze.

In the T-maze situation, the food-deprived rat is first trained to turn into the right or left arm of a maze shaped in the form of a "T" in order to obtain a food reward. Only one of the arms is baited with food on a given trial and the location of the reinforcement changes from left to right to left, etc. on each successive trial. There are no stimulus cues in this task to indicate which arm contains reinforcement on a given trial. If the animal makes a correct response, e.g., entering the right arm, it is allowed to consume the reinforcement and then removed from the apparatus. Following a delay, the rat is re-introduced into the start box, and this time the rat is required to choose the left arm of the maze in order to obtain reinforcement. Thus, in order to respond correctly, the rat must remember which arm contained the reward on the previous trial and choose the opposite arm on the next trial.

Delayed alternation in the maze situation requires only a minimal investment in equipment and has been shown to be sensitive to a number of different drugs and different types of brain lesions. However, delayed procedures of this type are extremely time-consuming and are not easily automated. Furthermore, they involve a considerable amount of handling of the animals during the test session which may be difficult to carry out in a standardized manner for each rat. For those reasons, techniques employing other maze configurations or delayed procedures carried out in operant chambers have also come in to greater use.

The eight-arm radial maze, depicted in Figure 3.17, consists of eight alleys or arms connected to a center platform. At the end of each arm is located a reinforcement delivery device and, in some paradigms, doors located at the end of arm close to the center platform are also included. The technique makes use of the fact that, under normal circumstances, the rat will forage for food, consuming the available food in one location and, if still hungry, going to other locations in search of more food. In one eight-arm radial maze paradigm, for example, a hungry rat is placed on the center platform with free access to all eight arms. The end of each arm is baited with a single food pellet which is not visible to the rat until it reaches the end of the arm. Once the food is consumed in that arm, the rat must travel to another arm which it has not yet visited in order to obtain food reward.

FIGURE 3.17

An eight-arm radial maze apparatus used to measure working memory in the rat

(Photo credit: Gerhard Winneke, Univ. of Düsseldorf.)

Untreated rats are remarkably efficient in this task, typically showing no or only one or two arm re-entries when obtaining the eight reinforcements. Further analysis of arm entries from session to session indicates that the rat does not employ a fixed or stereotyped strategy in obtaining the food pellets, such as always going to the arm to the right of the arm last visited. By using doors which automatically confine the rat to the center platform between choices, delays can also be introduced to increase the difficulty of the task and to evaluate changes in memory gradients.

In addition to maze environments, operant chambers are also suitable environments for evaluating working memory. Similar to the right-left delayed alternation technique employed in simple mazes for studying working memory, for example, an analogous approach can also be utilized in the operant chamber situation with the rat being required to alternate between the right and left lever on each trial in order to obtain reinforcement. Unlike the maze situation, no handling of the rat is necessary between trials so the technique can be fully automated.

Furthermore, interesting variations in delay-type tasks for non-maze test environments, i.e., delayed-matching-to-sample and non-matching-to sample tasks have also been developed for testing both rodents and primates. In one type of delayed matching task, the animal receives a number of paired trials.

On the first trial of each pair, only one response is possible and provides information as to which response is correct for that trial-pair. Then, after a delay, the animal is presented with a two-choice trial in which the correct choice "matches" that of the first trial in the pair. One example using rodents is the delayed-matching-to-position task. In this technique, the operant chamber is equipped with retractable levers which can be automatically extended and retracted. On the first trial of either pair, only one of the levers is extended into the chamber, e.g., the right lever, and a lever press results in reinforcement and retraction of the right lever. After a delay, both the right and left levers are extended. Correct responding depends on the rat's ability to choose the same lever as that presented on the first trial of the pair. In addition to spatial cues such as position, visual discrimination cues can also be used.

As one might expect, delayed-matching tasks are quite difficult for rats to master and they usually require extensive training. Thus, although they are quite interesting from a theoretical point of view, they may prove somewhat impractical for testing chemicals on a wide scale.

2.4.2.3 Approaches to study learning ability

As mentioned above, it is quite difficult to evaluate changes in learning capacity in the same animal over time using behavioral paradigms which continually present the same problem to the animal since, once the problem is mastered, what is being assessed is performance rather than learning ability.

To address this problem, several general approaches have been developed using both maze and non-maze environments. These approaches include procedures for evaluating discrimination reversal learning and repeated acquisition of response chains.

In discrimination reversal learning, animals are first trained to respond to one of two stimuli on a two-choice discrimination task. First, animals are trained to a high level of correct responding on the original problem. Once criterion performance has been reached, the discrimination is reversed. If the original problem, for example, is to choose the right arm of a T-maze, choosing the left arm would be the correct response on the reversal problem. Following the first reversal, the problem can be again reversed each time the animal reaches some pre-set criterion of performance.

Studies across a wide variety of species indicate that reversal learning shows a typical pattern which indicates that reversal learning requires processes other than those needed to learn the original discrimination. First, acquisition of the first reversal requires many more trials to reach criterion than those needed for the original discrimination. Secondly, various CNS lesions, particularly those of the limbic system, frequently disrupt reversal learning while having no effect on the learning of the original discrimination. Finally, repeated serial reversal testing leads to a type of steady-state of acquisition in which successive reversals are learned more quickly than the original problem. Once a steady-state of reversal learning has been achieved its baseline can be used to obtain a learning curve for each test session.

A further approach to the study of learning capacity is the use of repeated acquisition of response chains. In such a paradigm, three levers may be present in the operant chamber and the sequence in which they must be depressed in order to obtain reinforcement is signalled by the illumination of a particular cue light. In each test session, different cues are associated with different to-be-learned sequences and the animals are required to learn a new specific response sequence with each problem. Although repeated acquisition paradigms have successfully been used in rodents to evaluate the effects of compounds such as trimethyl tin on learning ability, these techniques, similar to delayed-matching-to-sample tasks, require extensive pretraining to achieve stable levels of performance and may be better suited for use in primates than in rodents.

In summary, many different types of techniques have been used to study the effects of drugs, brain lesions, and chemical exposures on learning and memory. However, no consensus has yet emerged as to which approach is the most appropriate. All of the techniques described above have both advantages and disadvantages, and there is as yet no sufficient database of comparable studies using these different models to determine which approach is best suited for evaluating chemical effects on processes related to learning and memory.

3   Behavioral methods for assessing neurotoxicity in humans

Different approaches derived from both clinical and experimental psychology have been used to evaluate the neurobehavioral effects of toxic exposures in humans. These include questionnaires designed to document subjective symptoms associated with neurotoxic exposures, clinical neuropsychological test methods, automated psychological test batteries, and specialized tests of sensory and motor function. These different approaches are not mutually exclusive and are often used in combination with each other for assessing human neurotoxic effects as well as with neurological and electrophysiological methods. The choice of assessment methods depends to some extent on what is known regarding the effects of the compound, the purpose of the evaluation, and the feasibility of different methods in different types of studies (Table 3.6).

TABLE 3.6

Some examples of different types of behavioral assessment studies

Study type	Purpose	Approach used
Patient evaluation	establish differential diagnosis, evaluate degree of neurological and psychological impairment	neurological examination, structured interview, work and exposure questionnaire, clinical neuro-psychological testing
Epidemiological studies A. Case control studies	examine neurobehavioral functioning in a group of exposed persons compared to that of non-exposed matched controls	clinical neuro-psychological tests or automated test battery, symptom questionnaire, and/or specialized sensory-motor test
B. Field studies	examine relationship between level of neurobehavioral effects and level of exposure on the same day in field situations	automated neuro-psychological test battery and measurement of levels of chemicals or metabilites in blood or urine
Experimental exposure studies	evaluate dose-response effects of experimental exposure to neurotoxicants (usually organic solvents)	automated neuro-psychological tests or questionnaires, usually performed in conjunction with metabolic studies

3.1   EVALUATION OF CLINICAL CASES

The purpose of clinical evaluations of individual patients is to establish a differential diagnosis of neurotoxic disease and to rule out other possible etiologies. Such evaluations typically involve both a neurological examination and neuropsychological evaluation.

The purpose of the neurological examination is to assess peripheral and central nervous system function by evaluating a number of specific and global measures designed to examine the functional integrity of particular neuroanatomical and neurophysiological systems. As Table 3.7 demonstrates, such an examination typically includes examination of cranial nerve function, motor changes, gait, reflexes, and somatosensory changes.

TABLE 3.7
Items in a clinical neurological examination

Adapted from Johnson, B.L. (1987) *Prevention of Neurotoxic Illness in Working Populations*. John Wiley & Sons, New York.

General appearance	tremor – arms outstretched and at rest ability to sit still speech – clear, slurred gums, nails, color of skin
Vital signs	
Cranial nerves	I sense of smell tested II fundoscopic examination, disk margins, visual acuity, visual fields III pupil size and reactivity to light; extraocular movements – nystagmus IV extraocular movements V pin and touch over face, corneal reflex VI extraocular movements VII symmetry of facial movement VIII hearing acuity, vestibular function IX-X presence of gag reflex, ability to swallow XI symmetry of shoulder bulk and movement XII Tongue: midline presence of atrophy or abnormal movements
Motor examination	presence of atrophy, fasciculations tone: resistance to passive movement ability to hold arms outstretched with eyes closed grip strength: scored on scale of 0 to 5 (normal strength) deep knee bend hopping on each foot, walking on heels and toes finger-tapping extensor plantar response
Tendon reflexes, scored as o (absent), 1 (decreased), 2 (normal), 3 (increased), 4 (grossly exaggerated)	biceps triceps knee ankle jerk
Coordination	tandem gait finger-to-nose pointing rapid alternating movement foot-tapping
Gait	stance, presence of arm swing, rapid turning ability to walk on toes and heels walk quickly run
Sensory	vibration and pin-testing in arms and legs position sensation

The neurological examination may be extended to include evaluation of other functions or the use of more quantitative methods. For example, some compounds (e.g., methanol, quinine, nitrobenzene, and methyl mercury) produce visual loss in humans. Thus, more extensive assessment of visual function including evaluation of the extent of visual field defects would be called for in these cases. Furthermore, the quantification of a particular sign may be particularly helpful in tracking a patient's recovery of function following the termination of exposure. Figure 3.18, for example, demonstrates the use of behavioral methods in quantifying the degree of tremor in a female patient occupationally exposed to elemental mercury in the course of calibrating pipettes. To obtain a quantitative estimate of the degree of tremor, the patient was asked to exert a prescribed force, designated by lights corresponding to the upper and lower limits, on a strain gauge which was fed to a digital computer. As a comparison of the upper and lower tracing in Figure 3.18 shows, the degree of tremor upon admission to hospital for treatment was quite marked compared to that measured nine months later, with no intervening exposure. The application of spectral analysis techniques to these data allows for a quantitative assessment of the degree of power at different tremor frequencies in a manner similar to that used in the quantification of tremor in animals discussed above.

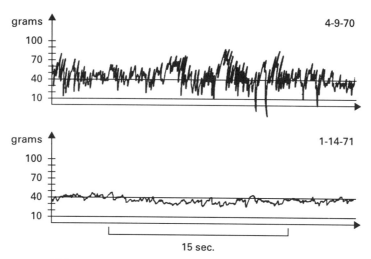

FIGURE 3.18

Tremor tracing from a worker exposed to mercury vapor during pipette calibration measured when the worker entered the hospital for treatment (upper tracing) and 9 months later (lower tracing)

(From Weiss, B. (1990) *Appl. Occup. Environm. Hyg.* **5**: 587. With permission.)

In addition to the neurological examination, extensive neuropsychological testing is also often carried out, especially in those cases where there is an indication of cognitive or affective changes. Similar to the neurological examination, neuropsychological testing also helps in ruling out other etiologies as well as establishing the extent of psychological impairment. Although different clinical neuropsychologists use different tests for evaluating neurotoxic effects, in general, the choice of tests is aimed at assessing different aspects of verbal function, memory function, attention, cognitive tracking and flexibility, and psychomotor abilities. Since one of the purposes of the clinical evaluation is the differential diagnosis of the patient's condition but the patient may possibly have a neurological condition unrelated to toxic exposure, clinical testing usually involves the use of tests which can be expected to be affected by neurotoxic exposures as well as those that are less likely to be affected. Such testing obviously is quite extensive and may require several hours to complete. An example of a clinical neuropsychological test battery is presented in Table 3.8.

TABLE 3.8

Some tests which may be included in a clinical neuropsychological test battery to examine possible neurotoxic cases

(Adapted from White, R.F. et al. (1992) In: White, R.F. (Ed.) *Clinical Syndromes in Adult Neuropsychology.* Elsevier Science Publishers, Chapter 1.)

Alberts Famous Faces Memory Test[1]
Benton Visual Recognition TEst[1]
Boston Naming Test[2]
Continuous Performance Test[1]
Controlled Word Association Test[1]
Finger Tapping Test[1]
Milner Facial Recognition Test[1]
Minnesota Multiphasic Personality Inventory[1]
Profile of Mood States[1]
Santa Ana Formboard Test[1]
Trail Making Test[1]
Selected subtests from the Wechsler Adult Intelligence Scale Revised (WAIS-R)[1]
Wechsler Memory Scale[1]
Wide Range Achievement Test-Revised[2]
Wisconsin Card Sorting Test[1]
Writing Sample[2]

1) Test may be sensitive to central nervous system effects of toxic exposure
2) Test usually qualifies as a 'hold' test and is relatively insensitive to chemical effects.

* Appendix 2 gives a short discription of the tests mentioned in this table and other tests mentioned in this text book.

Not all patients presenting with CNS toxic disorders demonstrate the same types or degree of neuropsychological changes. For this reason, a general diagnostic scheme has been devised for classifying neuropsychological changes in CNS function on the basis of both the duration of exposure and the persistence of effects. As Table 3.9 shows, acute exposure to some compounds may produce adverse pharmacological effects which are readily reversible and produce no residual effects. Examples of such agents include short-term exposure to organic solvents, anaesthetics, and psychopharmacological agents. If very high exposure levels occur, effects may be more severe.

With respect to organic mental disorders related to chronic exposure, three syndromes have been described (see Table 3.9). Organic affective syndrome may develop over days or weeks in susceptible individuals exposed to neurotoxicants, and involves primarily mood disturbances which are not usually accompanied· by cognitive changes as measured by formal neuropsychological testing. In mild chronic encephalopathy, subjective symptoms of cognitive and affective changes are accompanied by cognitive deficits on formal testing. Abnormal test performance is usually seen on tests of attention and short-term memory, psychomotor function and perceptual speed, and mood. Visuoconstructive abilities may also be affected. In contrast, verbal and nonverbal reasoning rarely are affected. Finally, in cases of severe chronic toxic encephalopathy, a similar pattern of cognitive deficits are also seen, but to a much greater degree, with changes sufficiently severe to interfere with daily living. For example, both verbal and non-verbal short-term memory may be so affected that the patient is no longer able to remember appointments or organize daily activities.

TABLE 3.9

Diagnostic system for classifying CNS disorders related to toxic exposures
(Adapted from White, R.F. et al. (1992) In: White, R.F. (Ed.) *Clinical Syndromes in Adult Neuro-psychology*. Elsevier Science Publishers, Chapter 1.)

I. Organic mental disorders related to acute exposures	**A. Acute intoxication** 1. Symptoms: CNS depression or attentional deficits 2. Duration of effects: minutes to hours 3. Residual effects: none **B. Acute toxic encehalopathy** 1. Symptoms: confusion, coma, seizures 2. Pathophysiology: cerebral edema, CNS capillary damage, hypoxia 3. Residual effects: permanent cognitive deficit may occur
II. Organic mental disorders related to chronic exposures	**A. Organic affective syndrome** 1. Symptoms: mood disturbances (depression, irritability, fatigue, anxiety) 2. Duration of effects: days to weeks 3. Residual effects: none **B. Mild chronic toxic encephalopathy** 1. Symptoms: fatigue, mood disturbances, cognitive complaints 2. Cognitive deficits seen on testing: may include attentional impairment, motor slowing or incoordination, visuospatial deficits, short-term memory loss 3. Course: insidious onset 4. Duration: weeks to months 5. Residual effects: improvement may occur in absence of exposure, but mild cognitive deficits may persist **C. Severe chronic toxic encephalopathy** 1. Symptoms: cognitive and affective changes sufficient to interfere with daily living 2. Cognitive deficits seen on testing: same as mild toxic encephalopathy, but more severe 3. Neurological changes: abnormalities seen on some neurophysiologic or neuro-radiologic measures 4. Course: insidious onset 5. Duration: irreversible 6. Residual effects: permanent cognitive dysfunction

Such severe cognitive deficits can occur as the result of long-term, high level exposure to some organic solvents in which the person voluntarily inhales the solvent vapor, often from a plastic bag, for its euphoric effects. Under such conditions, exposure levels of up to 10–20,000 ppm can be reached. Figure 3.19 presents an example of a such a patient's performance on the visual reproduction subtest of the Weschsler Memory Scale. In this case, reproductive memory was clearly disturbed and was most likely accompanied by evidence of attentional deficits and disturbances in verbal memory as well.

Although neurological and neuropsychological examinations together with information regarding the types and levels of exposure can be effectively used to diagnose neurotoxic disorders in clinical patients showing signs and symptoms of neurotoxic disease, an equally important concern is the possibility that neurobehavioral changes may be occurring in exposed populations at levels which are more typically encountered in, for example, the occupational setting. Limits of workplace exposures are often based on the results of animal studies using safety factors to adjust for extrapolation from high to low dose and from one species to another. To determine whether such levels are without effect, epidemiological studies can be used to evaluate the effects of chronic exposure over many years. In addition, field studies and experimental exposure studies can help determine if the exposure limits deemed to be safe are, in fact, low enough to avoid acute neurobehavioral effects. Experimental and epidemiological studies, however, require the use of tests which can be administered in a much shorter period of time than that needed for a full clinical evaluation, but are equally or more sensitive in detecting neurotoxic effects. Several approaches to the use of neurobehavioral methods in these types of studies are presented below.

3.2 BEHAVIORAL METHODS IN EPIDEMIOLOGICAL AND EXPERIMENTAL STUDIES

One approach to evaluating changes in neurobehavioral functioning in epidemiological studies of exposed workers involves the use of a shortened battery of clinical neuropsychological tests focusing on those effects most commonly seen in CNS toxic disorders. Such an approach was taken in one of the first studies reporting neurobehavioral changes in carbon disulfide-exposed workers conducted at the Finnish Institute of Occupational Health by Helena Hänninen[9]. Subsequently, this approach has been adopted by a number of investigators and a core battery of tests has been published by the World Health Organization for use in epidemiological studies. Tests included in the WHO core test battery are presented in Table 3.10.

TABLE 3.10

WHO Core Test Battery

(Adapted from Anger, W.K. (1990) *Neurotoxicology* **11**: p 629.)

Perceptual coding	– Digit symbol (WAIS-R)
Attention and short-term memory	– Simple reaction time – Digit span (WAIS-R) – Benton visual recognition
Psychomotor performance	– Aiming motor – Santa Ana test
Mood and affect	– Profile of mood states questionnaire

One of the advantages of such a battery is that it does not require complicated equipment and can be administered virtually anywhere that it might be needed. One disadvantage of these tests, however, is that they require administration by an experienced examiner and a considerable amount of interaction between the examiner with the person being tested.

With the advent of low-cost computer technology, however, it is possible to use automated methods for administering tests and recording the person's response accurately and objectively. At present, one of the most widely used automated test batteries is the Neurobehavioral Evaluation System (NES). In the NES battery, test administration and the recording of responses is accomplished using a personal computer. Before the beginning of each test, the subject receives instructions via the computer monitor explaining each test. Then after several practice trials, the test is administered. Because such an automated approach allows for the exact timing of stimulus presentation and speed of response, quite subtle effects on the accuracy, speed, and patterning of responding can be measured.

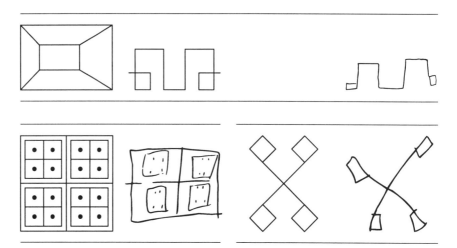

FIGURE 3.19

Performance on the memory for designs subtest of the Wechsler Memory Scale by a patient with a history of solvent abuse

For the information of the reader, the correct designs are provided on the left side of each diagram. (From Hartman, D.E. (Ed.) (1988) *Neuropsychological Toxicology: Identification and Assessment of Human Neurotoxic Syndromes.* Pergamon Press, New York. With permission.)

Figure 3.20 demonstrates a subject working on the pattern recognition subtest of the NES which is a test designed to measure visuospatial memory. Other NES tests include an automated version of the digit span test for evaluating short-term memory and the symbol-digit test for examining perceptual coding, as well as other tests of reaction time, attention, and verbal memory. Performance tests such as these are not only sensitive to the chronic effects of neurotoxic exposure, but acute effects can also be detected using these techniques, making them applicable in field studies and experimental exposure studies as well.

Furthermore, because one of the earliest indications of neurotoxic exposures is often an increase in subjective symptoms, questionnaires are also typically included in epidemiological studies. A number of questionnaires designed to obtain information in a standardized manner regarding subjective complaints have been described which typically include items designed to evaluate the prevalence of complaints related to sleep disturbances, sensory and motor changes, depression and other affective changes, and concentration and memory disturbances[12]. Examples of some typical types of questions included in neurotoxic symptom questionnaires are presented in Table 3.11.

Finally, in addition to measures of cognitive and affective functioning, specialized behavioral tests are also being used in epidemiological studies to monitor sensory changes in workers exposed to particular substances known to affect certain sensory functions. For example, there is increasing evidence for changes in color vision in workers exposed to organic solvents so a number of investigators have used methods for measuring color vision in these workers. Also, several neurotoxicants are known to produce a central-peripheral axonopathy producing damage to primarily long, large diameter sensory fibers.

TABLE 3.11

Some examples of the types of questions included in neurotoxicity questionnaires
(Adapted from Hogstedt, C. et al. (1984) In: Aito, R., Riihimaki, K., Vaino, H. (eds.) *Biological Monitoring and Surveillance of Workers Exposed to Chemicals*. Hemisphere Publishing, Washington.)

1. Do you have a short memory?
2. Have you ever been told that you have a short memory?
3. Do you often have to make notes about what you must remember?
4. Do you often have to go back and check things that you have done, such as turned off the stove, locked the door, etc?
5. Do you generally find it hard to get the meaning from reading newspapers and books?
6 Do you often have problems with concentrating?
7. Do you often feel irritated without any particular reason?
8. Do you often feel depressed for any particular reason?
9. Are you abnormally tired?
10. Are you less interested in sex than what you think is normal?
11. Do you have palpitations of the heart even when you don't exert yourself?
12. Do you sometimes feel a pressure in your chest?
13. Do you perspire without any particular reason?
14. Do you have a headache at least once a week?
15. Do you often have painful tingling in some parts of your body?

For this reason, some investigators have advocated the use of vibration sensitivity testing as a means of monitoring workers exposed to chemicals such as acrylamide. Figure 3.21, for example, illustrates the use of psychophysical methods for examining changes in vibration thresholds in human volunteers. In this technique, the subject is asked to report the presence or absence of a sensation of vibration to the left forefinger which is resting on a metal plate. Holes drilled in the plate allow for the protrusion of small pins which vibrate at specified frequencies.

FIGURE 3.20

Computerized methods are being increasingly used in epidemiological studies
In this example, the subject is working on a visual memory task from the Neurobehavioral Evaluation System test battery.

FIGURE 3.21

The application of psychophysical techniques for evaluating somatosensory changes in vibration thresholds in humans

The subject is wearing sound-insulating headphones to prevent any possible sound cues during the presentation of the vibratory stimulus. (From Marissen, J.P.J. et al. (1979) *Neurobehav. Toxicol.* **1**(Suppl. 1): pp 23–31. With permission.)

In general, the purpose of epidemiological studies, both those designed to assess sensory changes as well as those aimed at identifying cognitive and/or affective changes, is to detect individuals at an early (and reversible) stage who are beginning to show the first signs of overexposure. Such evaluations can be used as a basis for the decision to remove the person from the exposure, thus avoiding more serious effects, or to conduct testing in workers in the same location who may also be experiencing mild neurobehavioral changes.

In conclusion, similar to animal methods there is a wide range of techniques available for assessing the effects of neurotoxicants on the human nervous system. Such tests are currently being used by a number of laboratories throughout the world to examine and monitor exposed populations. There is, however, one effect of chemical exposures which was alluded to in our discussion of chronic toxic encephalopathy which deserves special consideration and that is the potential of certain drugs and chemicals to be abused by humans. In the next section, an overview of several methods used to study this uniquely behavioral neurotoxic phenomenon is presented.

4   The application of behavioral methods to the assessment of the abuse liability of drugs and chemicals

Although issues related to addiction are rarely discussed in textbooks of toxicology, the potential to produce addiction or patterns of abuse is a serious health consequence of exposure to a number of different compounds. Analgesic opiates, alcohol, cocaine, and nicotine are well-known examples of drugs which are abused by humans. In addition, some industrial chemicals are also potential "drugs" of abuse as well. For example, acute high-level exposure to inhaled solvents can produce feelings of euphoria, so intentional inhalational solvent abuse has become a serious health problem, particularly among children and adolescents. The most commonly abused products include glues, paints and paint thinners, gasoline, lighter refills, and cleaning fluids. Although toluene-containing products appear to be particularly popular, the chemical diversity of the different compounds which are abused suggests that the potential for abuse is not limited to a single compound, but may be a general property of different types of solvents.

There is an operational distinction which is usually drawn between physical dependence liability and abuse liability. Physical dependence refers to a drug's ability to produce signs of withdrawal following a period of exposure, whereas abuse liability is related to a drug's ability to maintain self-administration regardless of whether or not this is associated with evidence of physical dependence. In the section below, several examples of the types of behavioral methods being used to evaluate the dependence and abuse liability of drugs and chemicals at the animal and human level are presented.

4.1   TECHNIQUES FOR ASSESSING PHYSICAL DEPENDENCE

Removal of a drug following its chronic use may trigger a constellation of physiological and behavioral manifestations referred to as a withdrawal syndrome which can be evaluated using standardized observational techniques similar to those discussed in the beginning of this chapter. Most drugs of abuse from the opiate and depressant classes have the capacity to produce withdrawal signs upon the termination of drug treatment (Table 3.12). Three different paradigms are typically used to assess physical dependence: a) direct dependence studies, b) substitution studies, and c) precipitation withdrawal studies[4].

TABLE 3.12

Withdrawal abstinence signs associated with morphine-like dependence in the rat

(Adapted from Aceto, M.D. (1990) In: Adler, M.W. and Cowan, A. (Eds.) *Testing and Evaluation of Drugs of Abuse: Modern Methods in Pharmacology*, vol. 6. Wiley-Liss, New York, p 67.)

Autonomic	Psychomotor	Behavioral
Hypothermia	Rearing	Squealing and/or hostility when handled
Chromodacryorrhea	Jumping	Exaggerated startle reaction
Ptosis	Writhing	
Lacrimation	Rubbing (face)	Decreased eating and drinking
Rhinorrhea	Teeth-chattering	
Urination	Wet-dog-shakes	Loss of body weight
Diarrhea	Decreased seizure threshold	

In direct dependence studies, the compound is administered for a period of several weeks to several months. Then, after a suitable period of exposure, the test substance is abruptly stopped. Drug-treated animals are then evaluated at preselected time points over several days following drug termination and compared to vehicle-treated controls. Obviously, the choice of time points is critical and usually requires some knowledge regarding the rate of elimination of the substance or its active metabolites from the body. Withdrawal signs can be evaluated over a much shorter period of time by administering an antagonist which serves to precipitate the withdrawal syndrome in dependent animals beginning several minutes after administration. For example, the opiate antagonist naloxone will produce clear-cut withdrawal signs such as wet-dog shakes, escape jumping, writhing, teeth chattering, hypothermia, and diarrhea in morphine-dependent rats which can be assessed within a 30-minute observation period following naloxone treatment.

In substitution studies, animals are typically fitted with an indwelling catheter and chronically infused with a model compound such as morphine. Following a period of time long enough to produce dependence to the model compound, the animals are switched over to the infusion of the test compound and observed for signs of morphine withdrawal. If animals show no signs of withdrawal from the model compound when the test compound is infused, then the test compound is judged to substitute for, e.g., morphine. In that case, it is likely that the compound under study is as equally capable of producing physical dependence as the model compound.

A further approach to evaluating the physical dependence potential of a test substance involves studies designed to determine whether the test substance can itself precipitate withdrawal signs in animals made dependent on a model compound. For example, in the evaluation of opiates, animals are first made dependent on morphine and then administered the test compound several hours after the last morphine administration. If the test substance precipitates a withdrawal syndrome, the compound would be judged to possess antagonistic properties. Evidence for antagonistic effects of opiates would suggest a lower potential for abuse than compounds which possess full agonist effects, since they are less likely to support dependence in drug abusers.

4.2 EVALUATING THE REINFORCING PROPERTIES OF DRUGS AND CHEMICALS

The most important property of drugs leading to their abuse by humans is related to their ability to function as reinforcers of drug-taking behavior. To study this phenomenon, animal models have been extensively developed for evaluating the potential of drugs to act as reinforcers. That is, just as food and water can serve as positive reinforcers for maintaining high rates of operant behavior in deprived animals, drugs, too, can function as positive reinforcers to maintain high rates of drug self-administration. Although techniques have been developed whereby animals consume the drug solution orally, the most successful techniques involve the surgical placement of intravenous catheters through which animals are able to self-administer the drug under study by making some arbitrary response, such as pressing a lever.

The techniques for studying intravenous self-administration of drugs in animals have been used to evaluate a wide range of different psychoactive compounds and there appears to be good agreement between a drug's ability to maintain self-administration and its potential for abuse in humans (see Table 3.13). Two basic paradigms have typically been employed in self-administration studies: unlimited access and drug substitution. In unlimited access studies, animals with no history of drug exposure are placed in an operant chamber equipped with a lever. Depression of the lever results in the automatic delivery of the drug solution via the indwelling catheter. Typically, a fixed ratio-1 (FR1) schedule in which each lever press results in delivery of the drug solution is employed to establish self-administration behavior. Once responding has been established, animals are typically switched to schedules with other response requirements. e.g., FR20 or interval schedules. In unlimited access studies, care must be taken to insure that the total dose which the animal is able to administer within a single test session does not exceed tolerable levels. That is, similar to (drug-dependent) humans, animals are also capable of overdosing. With some compounds such as cocaine, for example, monkeys will self-administer lethal doses of the drug if they are allowed to do so.

TABLE 3.13

Intravenous drug self-administration in animals

(Adapted from Sanger, D.J. (1991) In: Willner, P. (Ed.) *Behavioural Models in Psychopharmacology.* Cambridge University Press, Cambridge, Chapter 18.)

(1) Drugs which unequivocally support self-administration	Opioids: morphine, heroin, codeine Stimulants: cocaine, amphetamine Barbiturates: pentobarbital, secobarbital Dissociative anaesthetics: phencyclidine, ketamine Ethanol
(2) Drugs which support self-administration under certain limited conditions	Nicotine Caffeine Benzodiazepines: diazepam, triazolam
(3) Drugs which do not support self-administration	Antidepressants Neuroleptics Cannabis Hallucinogens: LSD, mescaline Some anorectics: fenfluramine, phenyl- propanolamine Some anxiolytics: buspirone

For volatile compounds such as inhaled anaesthetic gases and organic solvents, analogous methods using the inhalatory route have also been devised[35]. In these studies, reinforcement involves the delivery of the inhalant into the test chamber for several seconds following a response on the lever.

In substitution procedures, stable baselines of self-administration of a model compound such as cocaine, codeine, or pentobarbital is first established, and then a novel compound is substituted for the model compound. After substitution, experiments are conducted to determine if self-administration continues at the same rate or if responding is eventually extinguished. Because the rate of self-administration of different doses of the novel compound can be evaluated in reference to a standard drug, very precise estimates of a drug's relative ability to support self-administration can be determined.

There are a number of other behavioral methods useful in the evaluation of the abuse liability of drugs and chemicals including place preference procedures, or drug discrimination techniques which use behavioral methods used to study the development of tolerance to drug effects during the course of exposure. Together, they provide ample proof of the usefulness of behavioral methods in evaluating the abuse liability of different classes of drugs and chemical compounds.

5   Summary

One of the factors which limits the development of any scientific discipline is the availability of methods for studying particular problems. This has been particularly true for the scientific discipline of behavioral toxicology. It is not enough to just want to know whether or not a chemical can cause learning impairment, memory loss, deafness, or incoordination, one must be able to measure these behavioral processes as well. Behavioral measurement methods form the empirical basis for the scientific discipline of behavioral toxicology, and over the last 10 years there have been significant advances in our understanding of chemical effects on behavior and how to go about measuring them.

The present chapter reviews some of the techniques which are currently being used to study the adverse effects of chemicals and drugs on different aspects of behavioral functioning. The chapter begins with a discussion of what behavioral scientists mean when they use the term *behavior*; the distinctions which can be drawn between different classes of behavior, i.e., respondent and operant, based on the functional relationships which control their occurrence; and the role of conditioning or learning in modifying behaviors from these two general categories. In Section 2 of the chapter, the concept of "tiered" testing in animals is introduced and screening methods comprising the first step (tier 1) in identifying potential neurotoxins including standardized observational techniques and motor activity assessment are presented.

Quantitative techniques for characterizing different types of neurotoxicant-induced behavioral effects (tier 2) are also discussed. For quantifying clinical motor signs, a number of different methods for assessing different types of motor impairment including neuromuscular weakness, incoordination, and tremor are presented as well as the advantages and limitations of different approaches. In order to examine the effects of chemicals on sensory function in animals, basically two different types of behavioral paradigms are being used — those based on operant conditioning (e.g., active avoidance paradigms and psychophysical operant discrimination methodologies) and those using respondent techniques (e.g., reflex modification). Such techniques can be used not only for identifying specific sensory changes, but can also be used to track the progression of sensory impairments during the course of exposure. With respect to cognitive changes, a large number of paradigms are available for studying different aspects of learning, memory, and other cognitive behaviors. In order to assess changes in performance of learned behavior, both intermittent schedules of reinforcement as well as discrete-trial operant discrimination techniques can be used to engender very reliable baselines in order to evaluate toxicant-induced changes in the performance of previously acquired behavior. Such baselines of learned performance are proving very sensitive to the effects of a variety of chemicals including metals and organic solvents. Passive and active avoidance learning paradigms have been extensively used in the past to measure memory and learning. However, such paradigms are not suitable for tracking the development and progression of learning and memory impairment. Different paradigms, using both mazes and operant chambers, have been recently described which may prove more valuable as techniques for measuring changes in short-term working memory and learning capacity.

The advances made in the development of methods at the animal level have been accompanied by considerable advances in the study of neurotoxicant effects at the human level as well and these methods are discussed in Section 3 of the chapter. In clinical cases of frank poisoning, a comprehensive neurologic examination and clinical neuropsychological testing are required for establishing a differential diagnosis. Based on the use of such techniques in the clinic, a general diagnostic scheme has been devised for classifying different types of CNS toxic disorders. Although clinical methods can be effectively used in patients with documented overexposure to neurotoxins, an equally important concern is the possibility that insidious neurobehavioral changes may be occurring in exposed workers or in the general population at low exposure levels which only manifest themselves years later. Thus, in order to study workers or other segments of the population exposed to neurotoxins, neurobehavioral techniques suitable for use in epidemiological studies have also been devised. The World Health Organization, for example, has been instrumental in supporting the development of a shortened test battery derived from the more extensive clinical tests which does not require complicated equipment and can be administered virtually anywhere it is needed. With the advent of low-cost computer technology, however, it is becoming increasingly possible to use automated methods for administering tests and recording a person's response accurately and objectively. One such test battery, the Neurobehavioral Evaluation System, consists of subtests designed to measure attention, verbal and visuospatial memory, perceptual coding, reaction time and hand-eye coordination. Such automated tests are sensitive to both acute and chronic effects and are suitable for both field studies and experimental exposure studies. In addition, specialized behavioral tests are also being employed to monitor sensory changes in exposed workers.

Finally, the last section of the chapter surveys some of the techniques that are currently being used to study a very specific neurotoxic effect which is rarely discussed in toxicology textbooks, namely the potential of some drugs and chemicals to produce addiction. Different paradigms for assessing physical dependence including direct dependence studies, substitution studies, and precipitation withdrawal studies are presented. In addition, because drugs of abuse can function as reinforcers of drug-taking behavior, operant technology has gained a prominent place in evaluating the abuse potential of drugs and chemicals.

In conclusion, there is a growing acceptance in toxicology of the need for evaluating the effects of chemicals on behavior. Neurobehavioral methods suitable for neurotoxicity screening and characterization at the animal level and for monitoring exposed populations at the human level provide a much needed *instrumentarium* for evaluating the health risks associated with neurotoxic exposures. As our understanding of neurotoxic effects continues to grow, behavioral methods can be expected to play an increasingly important role in assessing the significance of neurotoxic exposures to human health and well-being.

References

1. Aceto, M.D. (1990) Assessment of physical dependence techniques for the evaluation of abused drugs. In: Adler, M.W. and Cowan, A. (Eds.) *Testing and Evaluation of Drugs of Abuse. Modern Methods in Pharmacology*, vol. 6. Wiley-Liss, New York, p 67.

2. Anger, W.K. (1984) Neurobehavioral testing of chemicals: impact on recommended standards. *Neurobehav. Toxicol. Teratol.* **6**: p 147.

3. Anger, W.K. (1990) Worksite behavioral research: results, sensitive methods, test batteries, and the transition from laboratory data to human health. *Neurotoxicology* **11**: p 629.

4. Balster, R.L. (1991) Drug abuse potential evaluation in animals. *Br. J. Addict.* **86**: p 1549.

5. Cory-Slechta, D.A. (1994) Neurotoxicant-induced changes in schedule-controlled behavior. In: Chang, L.W. (Ed.) *Principles of Neurotoxicology*. Marcel Dekker, New York, Chapter 11.

6. Crofton, K.M. and Sheets, L.P. (1989) Evaluation of sensory system function using reflex modification of the startle response. *J. Am. Coll. Toxicol.* **8**: p 199.

7. Dodgson, O.L. (1976) *The Complete Works of Lewis Carroll*. Vintage Books, Random House, New York.

8. Eckerman, D.A. and Bushnell, P.J. (1992) The neurotoxicology of cognition: attention, learning and memory. In: Tilson, H. and Mitchell, C. (Eds.) *Neurotoxicology*. Raven Press, New York, p 213.

9. Hänninen, H. (1971) Psychological picture of manifest and latent carbon disulphide poisoning. *Br. J. Ind. Med.* **28**: p 374.

10. Hartman, D.E. (1988) *Neuropsychological Toxicology: Identification and Assessment of Human Neurotoxic Syndromes*. Pergamon Press, New York.

11. Hogstedt, C., Andersson, K., Hane, M.A. (1984) Questionnaire approach to the monitoring of early disturbances in central nervous functions. In: Aito, R., Riihimaki, K., Vaino, H. (Eds.) *Biological Monitoring and Surveillance of Workers Exposed to Chemicals.* Hemisphere Publishing, Washington.

12. Hooisma, J., Hänninen, H., Emmen, H.H., Kulig, B.M. (1994) Symptoms indicative of the effects of organic solvent exposure in Dutch painters. *Neurotoxicol. Teratol.* **16**: p 613.

13. Horan, J.M., Kurt, T.L., Landrigan, P.J., Melius, J.M., Singal, M. (1985) Neurologic dysfunction from exposure to 2-t-butylazo-2-hydroxy-5-methylhexane (BHMH): a new occupational neuropathy. *Am. J. Pub. Health* **75**: p 513.

14. Hunter, D. (1974) *The Diseases of Occupations*, 5th edition. English Universities Press, London.

15. Jaspers, R.M.A., Muijser, H., Lammers, J.H.C.M., Kulig, B.M. (1993) Mid-frequency hearing loss and reduction of acoustic startle responding in rats following trichloroethylene exposure. *Neurotoxicol. Teratol.* **15**: p 407.

16. Johnson, B.L. (1987) *Prevention of Neurotoxic Illness in Working Populations.* John Wiley & Sons, New York.

17. Kulig, B.M. (1990) Methods and issues in evaluating the effects of organic solvents. In: Russell, R.W., Flattau, P.E., Pope, A.M. (Eds.) *Behavioral Measures of Neurotoxicity.* National Academy of Sciences, Washington, D.C., p 159.

18. Kulig, B.M. and Lammers, J.H.C.M. (1992) Assessment of neurotoxicant-induced effects on motor function. In: Tilson, H. and Mitchell, C. (Eds.) *Neurotoxicology.* Raven Press, New York, p 147.

19. McMillan, D. E. (1987) Risk assessment for neurobehavioral toxicity. *Environ. Health Perspect.* **76**: p 155.

20. Maurissen, J.P.J. (1979) Effects of toxicants on the somatosensory system. *Neurobehav. Toxicol.* **1** [Suppl. 1]: p 23.

21. Maurissen, J.P.J., Weiss, B., Davis, H.Y. (1983) Somatosensory thresholds in monkeys exposed to acrylamide. *Toxicol. Appl. Pharmacol.* **71**: p 266.

22. McKim, W.A. (1986) *Drugs and Behavior.* Prentice-Hall, New York.

23. Merigan, W.H., Barkdoll, E., Maurissen, J.P.J., Eskin, T.A., Lapham, L.W. (1985) Acrylamide effects on the Macaque visual system. I. Psychophysics and electrophysiology. *Invest. Opthalmol. Vis. Sci.* **26**: p 309.

24. Moser, V.C. (1989) Screening approaches to neurotoxicity: a functional observational battery. *J. Am. Coll. Toxicol.* **8**: p 885.

25. Office of Technology Assessment (1990) *Neurotoxicity: Identifying and Controlling Poisons of the Nervous System.* U.S. Government Printing Office, Washington, D.C.

26. Pryor, G.T., Uyeno, E.T., Tilson, H.A., Mitchell, C.L. (1983) Assessment of chemicals using a battery of neurobehavioral tests: a comparative study. *Neurobehav. Toxicol. Teratol.* **5**: pp 91.

27. Rice, D.C. and Gilbert, S.G. (1982) Early chronic low-level methylmercury in monkeys impairs spatial vision. *Science* **216**: p 759.

28. Sanger, D.J. (1991) Screening for abuse and dependence liabilities. In: Willner, P. (Ed.) *Behavioural Models In Psychopharmacology.* Cambridge University Press, Cambridge, Chapter 18.

29. Schaumberg, H.H. and Spencer, P.S. (1980) Selected outbreaks of neurotoxic disease. In: Spencer, P.S. and Schaumberg, H.H. (Eds.) *Experimental and Clinical Neurotoxicology.* Williams & Wilkins, Baltimore, p 883.

30. Spencer, P.S., Beaubernard, C.M., Bischoff-Fenton, M.C., Kurt, T.L. (1985) Clinical and experimental neurotoxicity of 2-t-butylazo-2-hydroxy-5-methylhexane. *Ann. Neurol.* **17**: p 28.

31. Silverman, P. (1978) *Animal Behavior in the Laboratory.* Chapman & Hall, London.

32. Tilson, H.A. and Harry, G.J. (1982) Behavioral principles for use in behavioral toxicology and pharmacology. In: Mitchell, C.L. (Ed.) *Nervous System Toxicology.* Raven Press, New York, Chapter 1.

33. Weiss, B. (1990) Neurotoxic risks in the workplace. *Appl. Occup. Environ. Hyg.* **5**: p 587.

34. White, R.F., Feldman, R.G., Proctor, S.P. (1992) Neurobehavioral effects of toxic exposures. In: White, R.F. (Ed.) *Clinical Syndromes in Adult Neuropsychology.* Elsevier Science Publishers, Chapter 1.

35. Wood, R.W. (1979) Reinforcing properties of inhaled substances. *Neurobehav. Toxicol.* Suppl. 1, pp 67–72.

36. World Health Organization (1986) *Principles and Methods for the Assessment of Neurotoxicity Associated With Exposure to Chemicals.* Environmental Health Criteria Document 60, Geneva.

Contents Chapter 4

Basic principles of disturbed CNS and PNS functions

0-8493-7802-8/99/$0.00+$.50
© 1999 by CRC Press LLC

Chapter 4

Basic principles of disturbed CNS and PNS functions

Gaylia Jean Harry

1 Basic structural components of the nervous system

It is the intention of this chapter to convey an impression of the overall organization and structure of the nervous system as far it is important to neurobehavioral toxicology. The treatment of the subject will not be exhaustive, and only a few major substructures will be outlined particularly where these are important for discussion in later chapters. The reader is referred to the general references at the end of this chapter for more detailed information on neuroanatomy, neurophysiology, and neuropathology.

TABLE 4.1

Major divisions of the central nervous system and its major structures
Di-, mes-, met- and meylencephalon together form the brainstem.

Forebrain	Telecephalon	Olfactory bulb	
		Cerebrum	Olfactory cortex
			Neocortex
		Limbic system	Amygdala
			Hippocampus
			Septum
	Diencephalon	Corpus striatum	
		Globus pallidus	
		Thalamus	
		Hypothalamus	
		Pituitary (hypophysis)	
		Geniculate nuclei	
		Ventral posterior nucleus	
Midbrain	Mesencephalon	Tegmentum	
		Colliculi	
Hindbrain	Metencephalon	Cerebellum	
		Pons	
Spinal cord	Myelencephalon	Medulla oblongata	

The nervous system comprises two components: the central nervous system (CNS) which is composed of the brain and the spinal cord, and the peripheral nervous system (PNS) which is composed of ganglia and the peripheral nerves which lie outside of the brain and spinal cord.

The vertebrate nervous system is considered to have five major divisions: the telencephalon, the diencephalon, mesenscephalon, the metencephalon, and the myelencephalon. These five parts reflect the changes that the brain undergoes during its embryological development. In the developing embryo, the tissue that will eventually develop into the CNS is first noticeable as a simple fluid-filled tube, the neural tube. The first indications of the developing brain are three swellings that occur at the rostral end of the neural tube. These are the antecedents of the forebrain, the midbrain (mesencephalon), and the hindbrain. Table 4.1 and Figure 4.1 cover the major structures of the mammalian brain.

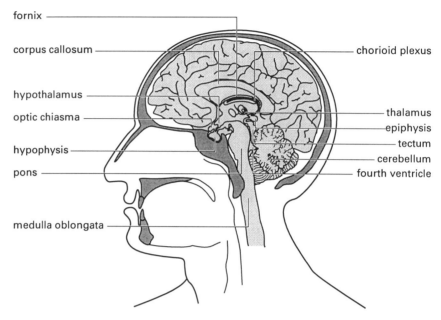

fornix
corpus callosum
hypothalamus
optic chiasma
hypophysis
pons
medulla oblongata

chorioid plexus
thalamus
epiphysis
tectum
cerebellum
fourth ventricle

FIGURE 4.1

Schematic representation of the major anatomical structures of the brain

Two equal numbered general types of cells are predominant in the nervous system: the nerve cells (neurons) and the neuroglial cells (oligodendroglial, astroglia, microglia, Schwann cells). The human nervous system contains at least 10 billion (10^9) neurons and even more neuroglial cells.

1.1   NEURONS

The neurons are similar to all other cells of the body in general structure, yet they have additional anatomical features, axons and dendrites, which allow nerve impulses to travel and communicate with other neural cells. Depending on the number of processes that originate from the cell body a neuron can be classified as unipolar, bipolar, or multipolar. The neuronal cell body is responsible for the synthesis of the molecular components necessary for the cell's survival. The neurons are responsible for reception, integration, transmission, and storage of information. Afferent or sensory neurons carry information into the nervous system, efferent or motor neurons carry commands to muscles and glands, and interneurons process local information or transfer information within the nervous system. Specialization is such that there are cells which respond only to a particular stimulus. For example, the chemoreceptors in the mouth and nose are responsible for gathering information about taste and smell, then sending this information to the brain. The skin contains cutaneous receptors that are involved in the sensation of heat, cold, and/or touch. Specific neurons in the retina respond to light, the rods and cones. Other specific neurons are designated to respond to painful stimuli.

During development, postmitotic neurons can be influenced in their phenotypic choices by environmental signals and the identity of a neuron's target can influence transmitter and peptide choices. As a process of normal development, many of these choices are reversible, and the transmitter and peptide phenotype can be changed as the environment changes for the growing neurons.

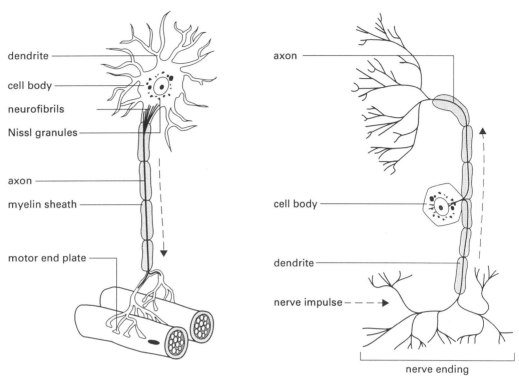

FIGURE 4.2

Schematic representation of a motor (left) and a sensory neuron (right)

The nerve cells of the PNS are generally found in aggregates called ganglia and the neurons in the CNS are segregated into functionally related aggregates called nuclei. Although anatomically distinct, these systems are functionally interconnected and interactive. Functionally related nuclei form higher organizational levels called systems, e.g., motor, visual, and associative systems. The vast array of processes are responsible for the coordination of activities, maintaining the balance of all other organ systems, and mediating communication and interaction with surrounding environment. For example, the neuro-endocrine system is formed around several small nuclei in the hypothalamus that secrete chemical messengers and stimulate the pituitary gland (hypophysis) to release chemical messengers into the general circulation to regulate other glands. Proper integrative functioning is responsible for communication and interaction with the surrounding external environment by speech, emotion, learning and memory, complex cognitive functions, the coordinated motion of the various body parts, and the functioning of peripheral organs. This massive communication between neuronal populations is made possible by structural features of the dendritic and axonal projections. In general, the dendrites are responsible for receiving information from the external environmental and the axon is responsible for transmitting information to the external environment. In most neurons, the axon is easily identified and conveys electrical impulses from the nerve cell body toward a target site.

In the CNS, the axons are relatively small and represent only a fraction of the cytoplasmic volume of the nerve cell body, however, in the larger motor and sensory neurons contributing to the sciatic nerve, the axons are long and large, and the axonal volume may be 100- to 1,000-fold of the perikaryon. The axon arises at the specialized region of the cell body, the "axon hilloc", and extends to the final synaptic ending at the target site. Axons rely heavily on the protein synthesis capacity of the cell body since they themselves are unable to synthesize any substantial amounts of protein due to the lack of intra-axonal ribosomes. Structurally, the axon consists of longitudinally oriented microtubules and neurofilaments, the intermediate filaments of neurons which are interlaced by crosslinkers. Biochemically, the axon is an extremely dynamic structure as will be covered in the following section on axonal transport.

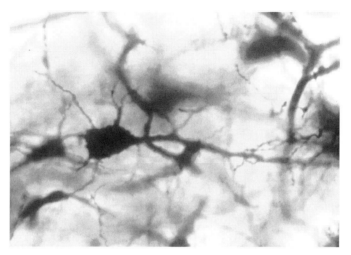

FIGURE 4.3
Photomicrograph of a Golgi stained CNS neuron

1.2   NEUROGLIA

The various glial cells in the nervous system play many different roles and have been grouped as follows:

1. True glial cells (macroglia) such as astrocytes and oligodendrocytes of ectodermal origin
2. Microglia of mesodermal origin
3. Ependymal cells, of ectodermal origin

Unlike the neuron, the glial cells have no synaptic contacts so they retain the ability to divide, particularly in response to injury. There is a critical interdependency between neurons and astroglia in both normal and injured states of the CNS. Cell-cell contacts between the two cells regulate both neuronal and glial differentiation and are critical for the glial-guided migration of neurons. Glia provide both trophic (nutritional) and tropic (turning toward) support for neurite extension, e.g., growth factors and cell adhesion systems. *In vitro* studies suggest that the levels of many glial proteins, including the integral structural protein of astrocytes, glial fibrillary acidic protein (GFAP), ion channels and neurotransmitters, extracellular matrix, and adhesion ligands are regulated by neuron-glia interactions. Glial cells have a very dynamic nature and provide critical processes necessary to maintain the normal

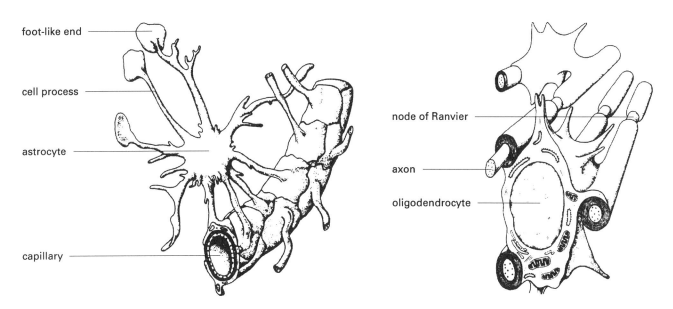

FIGURE 4.4

Schematic representation of different astroglial cells (an astrocyte and an oligodendrocyte)

functioning of the nervous system, e.g., regulation of local pH levels and ionic balances. The microglial cell is derived from mesodermal origin and is seen in the normal brain in a resting state. However, during disease or injury states, it can become mobile and assume macrophage-like characteristics. An immunologic role for microglia has been proposed based upon the ability to express cytokines and class II MHC antigens in the absence of a T cell response.

The *myelinating cells* in the nervous system, e.g., the Schwann cells and the oligodendrocytes, are responsible for forming a sheath around the axon, and in doing so the relatively small soma produces and supports many more times its own volume of membrane and cytoplasm. This sheath may be multiple

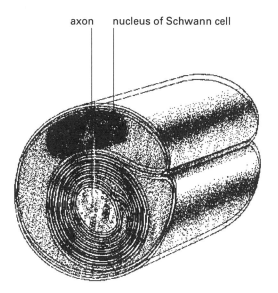

FIGURE 4.5

Schematic representation of the way in which Schwann cells form sheaths around an axon

119

wrappings of myelin lamella or simply a cytoplasmic sheath around each axon. The myelin sheath allows for efficient conducting of signaling, and many axons are wrapped with a myelin sheath. In the peripheral nervous system, the Schwann cells line up along the axon during development with intervals between them that will become the nodes of Ranvier. Then the plasmalemma (external cell membrane) of each Schwann cell surrounds a single axon membrane structure called the mesaxon, which elongates and spirals around the axon in concentric layers. As the wrappings of myelin sheath are extended the cytoplasm of the Schwann cell is extruded to create the compact lamellae of the mature myelin sheath[20]. The thickness of the sheath is proportional to the axonal diameter[27]. It has been estimated that the internodal distance (distance from one node of Ranvier to the next) is approximately 1–1.5 mm in the primary sensory axon resulting in 300–500 nodes of Ranvier along a primary afferent fiber between the thigh muscle and the cell body in the dorsal root ganglion. Since each internodal segment is formed by a single Schwann cell, it requires as many as 500 Schwann cells to myelinate a single peripheral sensory axon. In the central nervous system the cell responsible for myelination is the oligodendrocyte. While in the PNS there is a one-to-one relationship between the axon and the myelinating Schwann cell[29]. In the CNS one oligodendrocyte can wrap numerous axons[26], yet form only one specific internode on each axon thus creating the nodes of Ranvier. Signaling interactions between the axonal membrane and the myelinating cells appear to play a critical role in maintaining both the functional and structural integrity of the system. The presence of axons turn on the genes in Schwann cells that encode myelin while expression of the genes in oligodendrocytes that encode myelin are influenced by the presence of astrocytes, the other major glial cell type in the CNS. The presence of a myelin sheath can influence the distribution of ion channels along the axon; for example, potassium channels are localized in the myelinated region of the axon while sodium channels are clustered at the non-myelinated nodes of Ranvier.

In mature peripheral myelin, Po is the major protein and spans the plasmalemma of the Schwann cell. It is thought to contribute to the compaction of myelin through interaction with identical immunoglobulin domains on the surface of the apposed membrane. The CNS lacks Po but contains a proteolipid protein (PLP) which constitutes more than half of the total protein in the central myelin. Both central and peripheral myelin contain a group of proteins called myelin basic protein (MBP) which comprises a major portion of the central myelin and a lesser amount of the peripheral myelin. These proteins (MBP) are located at the major dense line of the myelin lamella (formed by fusing of the inner leaflets of the spiral myelinating cell process) and can undergo post-translational modification. Although a minor protein in the mature nervous system, myelin associated glycoprotein (MAG) is located at the margin of the mature myelin sheath and the axon, and may play a critical role during development and regeneration as a cell adhesion factor , and as a mediator of axon-glia interactions.

1.3   RESPONSE OF NEURONS AND GLIAL CELLS TO INJURY

The classification of a chemical as to the type of neuropathy produced following disease or chemical exposure has been based primarily on the structural site of morphological effects. A neuronopathy is characterized by an initial degenerative effect in the neuronal cell body, an axonopathy by an initial effect in the axon, a myelinopathy as a direct effect on the myelin sheath, and a Schwannopathy by an initial effect on the Schwann cell. This section will briefly describe some the pathological changes seen in different neural cells. However, it must be emphasized that a chemical may not produce its neurotoxicity by a direct pathological change in the cellular structure, but instead interfere with the vast array of cellular processes necessary for the functioning of the neural cell.

The reactions of neurons to injury vary dramatically. Neurons can degenerate following a direct action on the perikaryon or loss of synaptic target site influences and deprivation of trophic factors. Structural alterations of the nervous system associated with neurotoxicity can involve changes in the general morphology of the cell and its subcellular structures. In many cases a toxicant-induced neuropathology can resemble naturally occurring neurodegenerative disorders in humans (Table 4.2).

TABLE 4.2

Examples of known neuropathic agents

Target site	Pathology	Toxicant	Disease
Neuronal cell body	Neuronopathy	Methylmercury A.E.T.T. Quinolinic acid 3-acetylpyridine Aluminium Trimethyltin Tetraethyllead Ibotenic acid 3-acetylpyridine (3-AP)	Minamata disease Ceroid lipofuscinoses Huntington's disease Cerebellar ataxia
Nerve terminal	Neurodegeneration	MPTP	Parkinson's disease
Schwann-cell	Schwannopathy	Tellurium	
Schwann-cell myelin	Myelinopathy	Lead Buckthorn toxin	Neuropathy of matachromatic leukostrophy
Central-Peripheral Distal axon	Distal axonopathy	Acrylamide Hexacarbons Carbon disulfide	Vitamin deficiency
Central axons	Central axonopathy	Clinoquinol	Subacute myelooptico Neuropathy
Proximal axons	Proximal axonopathy	B,B'-Iminodi propionitrile	Motor neuron disease
Oligodendroxyte myelin (CNS)	Myelinopathy	Inorganic lead Triethyltin Lysolecithin Ethylnitrosourea Hypocholest-eremic agents Ethidium bromide Diptheria toxin Pyrithiamine biscyclohexanone (Cuprizone)	

The type of neuronal degeneration rarely identifies the damaging agent. It more commonly reflects the severity and duration of the injury and the acuteness or chronicity of exposure. The various types include the accumulation, proliferation, or rearrangement of structural elements or organelles. The degenerative process of the nerve cell can be either relatively quick or a slow prolonged process depending on the underlying mechanism responsible for the degeneration. Early morphological emphasis was placed on changes in the Nissl substance of the neuron however, Nissl substance is easily damaged and its disappearance is probably not significant unless the cell and its nucleus also show changes. Some classifications generally used in evaluating neuronal degeneration include the following.

In *central chromatolysis*, Nissl granules disappear around the nucleus, the cytoplasm swells, and the nucleus swells and becomes eccentric. This pattern of damage was initially described in a cell whose axon had been severed. The severity of the cell damage and its survival of the cell depend partially on the distance between the cell body and the damage site.

Peripheral chromatolysis defines the lack of Nissl substance near the cell membrane and may be an indication of direct damage to the nerve cell body or the recovery from central chromatolysis.

Simple chromatolysis is also known as Nissl's acute swelling where there is a diffuse loss of Nissl granules, the cytoplasm and axons swell, and the cell stains palely. The change may be reversible or it may progress to further loss of stainable Nissl substance and cytoplasmic vacuolization.

Other neuronal responses to injury include severe cell changes that are associated with neuronal degeneration. In such cases the cell nucleus is hyperchromatic, shrunken, and pyknotic, and the cytoplasm is swollen and eosinophilic. Neurofibrils and Nissl granules are lost. Ischemic changes in the neuron are usually irreversible and are characterized by a loss of affinity for stains, the nucleus becomes elongated and pyknotic with chromatin clumps, and the nuclear membrane may disappear. A common chronic response of the neuron is a shrinkage of the cell body and nucleus with both staining deeply. The intact internal structures are crowded and the cell border takes on a triangular shape. Neurons may survive in this state for an extended period of time; however, the changes are usually irreversible and eventually lead to cell sclerosis where the cell constituents have clumped or disintegrated.

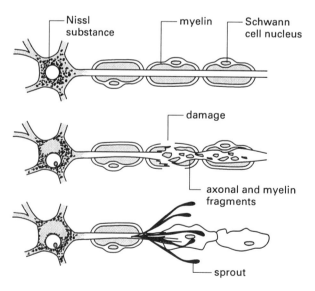

FIGURE 4.6

Sequence of changes following an axonal injury. The injury induces a neuronal response, the chromatolysis, and an axonal degeneration, the Wallerian degeneration. Axonal sprouting from the axonal end may occur when the neuron survives.

Neurofibrils are more resistant to injury than Nissl granules yet they disappear in cell atrophy and necrosis. During the course of neurofibrillary degeneration, the neurofibrils become coarse, thickened, and form a dense spiral within and around the neurons. The changes are nonspecific but occur in Alzheimer's and Pick's presenile dementias, senile dementia, and in amyloid angiopathy. Experimentally, neurofibrillary degeneration can be produced by aluminum and inhibitors of cellular mitosis such as cholchicine and vincristine.

Many of the changes in the neuronal cell body can be the result of an injury in the axonal projection of the neuron. Most of the understanding of the process of axonal degeneration has come from both clinical and experimental examination of the peripheral nervous system. Physical interruption of the axon by transection, crush, or freezing results in degeneration of the axon distal to the injury, known as Wallerian degeneration. In the PNS, axoplasm in the distal

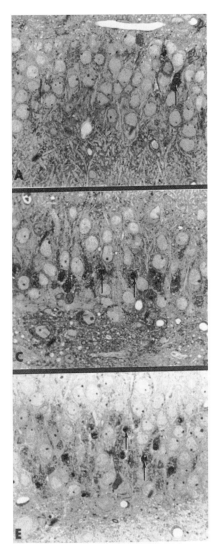

FIGURE 4.7

Photomicrograph of granule cell neurons in the dentate gyrus of the hippocampus following 4 mg trimethyltin per kg body weight per day for 3 days.

A dentate gyrus 1 day following 3rd dose of TMT illustrating rare pyknotic dead granule cell (arrow). C 3 days TMT illustrating a line of pyknotic dead neurons (arrow).E dentate gyrus 6 days post-TMT illustrating more extensive neuronal degeneration (arrow).

stump (the site distal to the injury) is reduced to granular and amorphous debris within 24 hours while in the CNS the axoplasm can persist for several days. The degeneration is an active process possibly associated with a rise in intra-axonal calcium along the distal stump and with the infiltration of phagocytic cells.

The heterogeneity of neurons is thought to contribute to the differential susceptibility of specific neurons to toxic degeneration. A number of chemicals appear to have distinct cellular specificity for neuronal populations. Although a pattern of neuronal degeneration has been established for various chemicals, elucidation of underlying mechanisms of the neurotoxicity requires an integrative strategy involving subcellular, cellular, and systems approaches (Table 4.2). One example, is the selective toxicity of MPTP in dopaminergic neurons. Research has focused on the metabolism of MPTP and its relationship to specific neurochemical features of these dopaminergic cells.

The selective neurotoxicity caused by 1-methyl-4-phenyl-1,2,3,6-tetrahydropyridine (MPTP) for dopaminergic neurons has received considerable attention. The strongly neurotoxic properties of this substance were first noticed in 1983 in a number of heroin addicts in the US. Their own attempts to synthesize meperidine (a centrally acting analgesic) yielded MPTP as a byproduct, which they unwittingly injected. Intravenous administration of MPTP causes akinesia and tremor — a picture similar to that of Parkinson's disease. Further research revealed that dopamine depletion in the substantia nigra was the likely cause of toxicity. MPTP is an uncharged molecule at physiological pH and thus easily crosses the blood-brain barrier. Once in the brain, it diffuses into astrocytes and other nerve cells. In the astrocytes, MPTP is now known to be oxidized by monoamine oxidase (MAO) to MPP$^+$ via a dihydropyridinium intermediate (see Figure 4.8).

MPP$^+$ appears to be selectively taken up into the substantia nigra via the neuronal absorption mechanisms for dopamine. As an inhibitor of the oxidative phosphorylation in the cell, MPP+ acts as a general cellular poison. Systemic injection of MPP+ does not cause brain damage, because MPP+ itself does not cross the blood-brain barrier. It is not known why humans and monkeys are much more sensitive to the effects of MPP+ than rats and mice are.

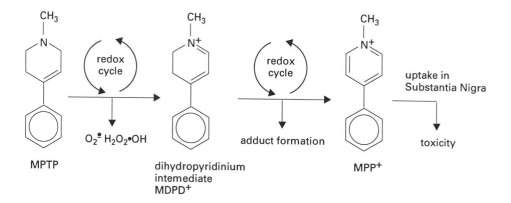

FIGURE 4.8

Conversion of methyl-phenyl-tetrahydropyridine (MPTP) to the methyl-pyridinium ion (MPP+). In the astrocytes MAO-B catalyzes the reaction in which MPTP is oxidized into MDPD+. MPP+ enters the dopaminergic cells in the substantia nigra via the dopamine uptake system.

124

Chemicals, such as MPTP, which work on a neuronal population based upon a specific neurochemical feature can be identified within pharmacological agents; however, they are relatively few in the area of environmental agents. Environmental chemicals such as metals, pesticides, and solvents may produce distinct morphological cellular changes in the nervous system; however, due to a multiplicity of toxic effects identifying the underlying mechanisms is a more complicated process and has been less successful. Attempts to identify site and mechanism of action for various chemicals are presented in Tables 4.2 and 4.3.

In addition to primary effects on a neuronal population, neuronal damage can be of a secondary nature, for example *transneuronal degeneration*.

TABLE 4.3

Neurotoxicants with known neurological mechanisms

Ionic balance		
a	Inhibit sodium entry	Tetrodotoxin
b	Block closing of sodium channel	p,p'-DDT, Pyrethroids
c	Increase permeability to sodium	Batrachotoxin
d	Increase intracellular ions	Chlordecone
e	Inhibit chloride movement	Cyclodiene organochloride insecticides
Cytotoxicants - depend on uptake into nerve terminal		MPTP
Uptake blockers		Hemicholinium
Metabolic poisons		Cyanide
Hyperactivation of receptors		Monosodium glutamate Domoic acid Kainic acid
Block of neurotransmitter release (ACh)		Botulinum toxin
Inhibition of transmitter degradation (ACh)		Organophosphates carbamates nerve gas
Microtubule disruptors		Vincristine Cholchicine

If a population of neurons is deafferentated, *anterograde transneuronal degeneration* occurs and is characterized by the degeneration of the target neuron due to loss of synaptic innervation. In an attempt to sustain the life of such post-synaptic target neurons, the appropriate surrounding neurons will undergo what is known as "reactive synaptogenesis" in order to replace the lost synaptic innervation. As a result of a loss of target population neurons, *retrograde transneuronal degeneration* occurs in the neurons that originally innervate the deafferented populations of neurons. It still remains a question as to if this retrograde transneuronal degeneration is initiated by the presence of a new signal or by the loss of signal from the innervated neuron. Such neuronal degeneration is secondary, not the result of a direct action of a toxic substance, and can serve to confound efforts to identify the primary site of action in the nervous system. It therefore must be emphasized that a chemical may not directly produce its neurotoxicity by a histological change in the cellular structure, but may instead interfere with the vast array of cellular processes necessary for the functioning of the neural cell. Ibotenic acid and 3-acetylpyridine (3-AP) affect the soma of the neuron and result in degeneration of both the soma and the axons. The neurons of the substantia nigra are damaged by 3-AP with a subsequent perturbation to the axons projecting to the dorsal and lateral aspects of the caudato-putamen. Consistent with the anatomical findings, neurochemical studies showed a slight decrease in dopamine in the striatum following 3-AP with a 40% in the dorsal and lateral regions. DSP-4, a specific noradrenergic neurotoxin, spares the noradrenergic soma in the locus coeruleus; however, it destroys the noradrenergic terminals originating from this area. Other neurotoxicants like trimethyltin (TMT) seem to produce stages of degeneration suggesting a role for multiple factors[3]. Transneuronal sites following TMT exposure include the central auditory system, the amygdala, and the dorsal division of the lateral septal nucleus. Several of the psychotropic amphetamine derivatives have been shown to release serotonin (5-HT) from nerve terminals. This acute depletion is followed by a progressive degeneration of 5-HT terminals, swollen fragmented nerve fibers, and axonal degeneration. If the cell bodies do not die, they sprout new axonal projections which appear structurally thick and abnormal.

The myelin membrane is highly vulnerable to numerous substances, toxic agents, and *demyelinating* autoimmune diseases such as multiple sclerosis and Guillain-Barré which specifically target and breakdown the myelin sheath. Toxicants can also directly result in the loss of the myelin sheath (demyelination) and/or produce an altered myelin sheath without a loss of myelin. In order to aid in identifying the underlying mechanisms involved in the maintenance of the sheath, this process has been recently referred to as

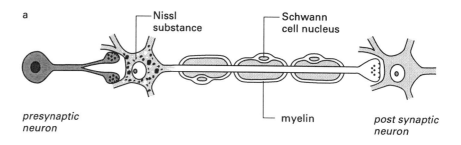

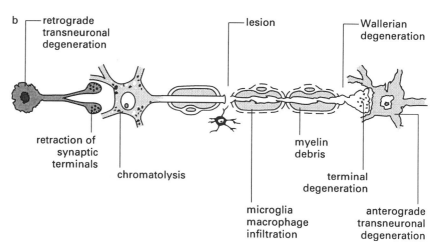

FIGURE 4.9

Axonal injury induces degenerative changes in the injured neuron as well as in the neurons with which it has synaptic connections

Part (a) shows the normal anatomical relationship of pre- and postsynaptic neurons. Part (b) shows the many reactive changes that take place in the injured neuron as well as in the presynaptic and postsynaptic neuron.

dysmyelination rather than demyelination. Any disturbances in myelin metabolism or assembly during development would result in less myelin (hypomyelination) as characterized by fewer myelin wrappings or thinner lamella. Demyelination can occur following a direct perturbation to the myelinating cell, its myelin sheath, or as a response to axonal degeneration (Table 4.2). In many cases of demyelination, effects are not uniform along the axon but rather selective effects can be seen at the larger internodes. This may be due to the increased metabolic demands of a larger plasmolemma in the vulnerability of larger myelinating cells. In the PNS, the process of demyelination is a sequence of the Schwann cell differentiating to autophagocyse its own plasmalemma, proliferating and remyelinating the damaged region of the axon with smaller internodes. Various models of remyelination result in an abundance of shorter internodes at the site of injury and can result in long term alterations in nerve conduction velocities. Usually the compound action potentials recorded from a diffuse chemical induced demyelinated nerve are delayed, dispersed, and reduced in amplitude. During the demyelinated stage, nerve conduction can be blocked at the focal site and the compound action potential lost.

Example

One interesting developmental model of toxicant-induced demyelination is rapidly generated in the PNS following early post-weaning exposure to tellurium. In this model the primary effect is on the Schwann cells resulting in a preferential loss of the large myelin internodes. The demyelination is associated with a specific metabolic block in the biosynthesis of cholesterol with the inhibition of squalene epoxidase[17]. This effect is seen primarily in the young animal and may be associated with an absence of sufficient cholesterol for the structural formation of myelin. The affected Schwann cells differentiate, proliferate, and autophagocyze their myelin sheath. The highly synchronized pattern of demyelination is rapidly followed by remyelination.

Following injury to the CNS, the astrocyte and the microglia both display a profound response[2,3]. The major characteristic of the *astrocyte response* to brain injury is an increase in the expression of an astrocyte specific protein, glial fibrillary acidic protein (GFAP) at both the protein and mRNA levels possibly due to its promotor with multiple domains sensitive to a number of factors like cAMP. This increase in protein is usually accompanied by an increase in either the number of astrocytes responding or in a morphological change in the responsive astrocytes. In some types of injury, e.g., physical injury, the astroglial cells proliferate; however, this is not the case for all perturbations. This process of reactive gliosis has been proposed as an early marker of damage to the nervous system. The sequence of the process can be rapid and transit or can continue over an extended period of time. In many cases such as physical injury to the brain, the astrocytes hypertrophy and can form what is called an "astrocytic scar". The presence of such a scar has been proposed to impede axonal regeneration either by presenting a physical barrier or by altering cell adhesion molecules to produce a more inhibitory site for the axons to regenerate. Attempts to intervene with anti-inflammatory agents and endogenous substances such as gangliosides to prevent the astrocyte response have offered some success in the process of regeneration following spinal cord injury.

In many cases of brain injury, there is also an increase in the "immune type" microglia cells both resident and infiltrating cells from the blood supply. Microglia have more in common with peripheral macrophages than with other cells in the nervous system, both in molecules they express and in their behavior when activated. Resting microglia can be described as "immunoallert"; they are highly ramified between neurons in the gray matter and parallel to axons in the white matter, but make no direct contact with each other. Resting microglia cells are so sensitive that almost any manipulation of the nervous system activates them. Following any challenge to the integrity of the nervous system, microglia respond within hours with a change in morphology, and an upregulation of molecules characteristic of macrophages, including the major histocompatibility (MHC) class I and class II antigens. Once activated, microglia draw in their branches, proliferate, become amoeboid and motile, and move to the site of injury or surround affected neurons. The majority of studies examining the activation of microglia have used models of physical or viral induced injury, both of which perturb the blood-brain barrier and allow infiltration of circulating factors and monocytes. Activation is graded, depending on the severity of the insult: similar to the macrophage in the PNS it is assumed that only when neuronal degradation is present do they become cytotoxic phagocytes. However, just as this has been shown not always to be the case in the PNS, the likelihood that the microglia may play a more active phagocytic role in the early stages of CNS injury must be considered.

In the PNS, the macrophage assumes a phagocytic role following nerve injury. The resulting debris is removed, so no glial scar is formed, and the axons are allowed to regenerate to appropriate target sites. However, the overzealous macrophage, both the resident macrophage in the nerve as well as the infiltrating macrophage from the circulating blood, can attack not only debris but also healthy neural tissue. Following traumatic insult, the Schwann cell undergoes an active phase of mitosis, can migrate locally, and with demyelination can assume phagocytic properties. In many cases of demyelination, the macrophage is responsible for infiltrating the nerve and stripping the myelin away. In the previously mentioned case of tellurium-induced demyelination,

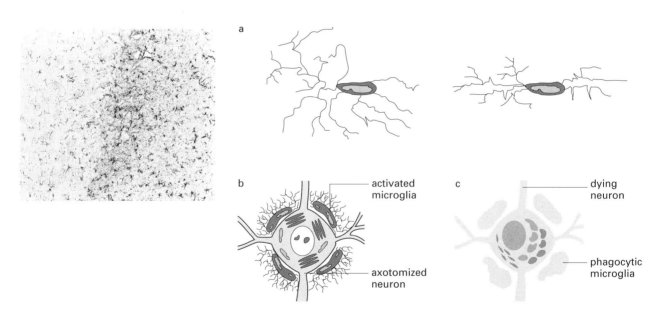

FIGURE 4.10

Schematic drawing of the proces of activation of microglia surrounding an axotomized neuron (b) and of phagocytic microglia surrounding a dying neuron (c). Part (a) shows resting microgliacells.

The photograph shows an area (densely packed) of activated microglia in the CNS.

128

the myelin is altered first, followed by the macrophages infiltrating to remove the debris. In this case, the macrophages remain in the nerve for an extended period of time and have been shown to play a role in the remyelination process. Reinnervation in the PNS is critically dependent upon the presence and integrity of the Schwann cell basal lamina with the formation of tubes, Bungner bands, to serve as channels along which regenerating axons can grow. The absence of such "guide tubes" has been proposed as one factor influencing the failure of axonal regeneration in the CNS.

1.4   DEVELOPMENTAL ORIGIN AND TIMING OF CELL TYPES

The embryonic ectoderm derives the neural crest from which all neurons and glial cells of the nervous system originate. Cells emerge from the neural crest and migrate to specific sites in the periphery according to characteristic migration patterns determined by the local environment. Cell adhesion and extracellular matrix molecules have a profound regulatory effect on neural crest cell motility and morphology. Some neural crest derivatives retain developmental plasticity and the commitment of terminal differentiation is delayed. These cells can undergo transdifferentiation without cell divisions in response to changes of environmental conditions. Once the neural tube has formed, neuroepithelial cells proliferate and four *basic embryonic zones* are formed: the ventricular, marginal, intermediate, and subventricular zones. The cerebral cortex is formed from the migration of postmitotic neuroblasts from the ventricular zones of neuroepithelial cells toward the surface. The cells that arrive in the cortex early remain in the deep layers and become macroneurons. Primitive cells from the germinal matrix proliferate and migrate toward the cortex, and then pass over the deep neuronal layers. Neuroglia originate from the primitive cells of the germinal matrix. They migrate into the intermediate or mantle zone, into fiber tracts, and continue to proliferate. The astrocytes extend their processes and guide the outward migration of neurons from the germinal matrix zones to the cerebral cortex. Several features are common to all regions of the brain during the period of histogenesis:

1. There is a separate origin for different cytoarchitecture
2. A spatial and temporal gradient exist for cellular proliferation migration
3. Neurons precede glial cells
4. Orderly sequence from large to small neurons
5. The phylogenetically older parts of the brain develop earliest

The development of the nervous system begins in the fetus, and for humans is not complete until approximately the time of puberty. Organogenesis for the nervous system occurs during the period from implantation through midgestation with synaptogenesis and myelination predominent during late gestation and early neonatal period (see Figure 2.1, page 31). The development of the mammalian nervous system is a highly complex process with very specialized morphological and biochemical patterns of ontogenesis that continue as a critically timed multistage process guided by chemical messengers. The basic framework of the nervous system is laid down in a sequential process with each step dependent upon the proper completion of the previous one. During the temporal and spatial organization of the developmental process, the result of a relatively minor disturbance can be severely detrimental to the nervous system. The developing nervous system has been reported to be selectively vulnerable to particular types of damage[1,10,18,25]. During this period of active growth, intricate cellular networks are formed. The axon is the first process to develop from the neuron and it always grows from the cell body along a specific pathway to form a synaptic connection with a specific target. At the growing tip of each axon is a growth cone which responds to environmental cues and "guides" the axon to a final terminal connection. This is followed by the outgrowth of the dendrites.

The dendrites mature in a ventrodorsal sequence, thus the dendrites of the deep layers of the cortex matures before the superficial layers. The timing between dendrites of different neurons is also established to follow the sequence of the dendrites of neurons with long axons (Golgi type I) maturing before those of short axons (Golgi type II neurons). During the same ontological time period, the synapse formation and the outgrowth of dendrites are ongoing. Since there is no turnover of neurons in the brain, the natural redundancy of the system results in an excess of necessary elements that requires a selection process to insure establishment of the proper final connections. This final profile of the synaptic network and connections is determined by competition and cell death resulting in a "pruning" of the excess. This allows this complex system to perform at a maximal level of efficiency.

Given the complexity of the nervous system and the critical role that timing plays in the final product, even minor alterations in either a cell type or process at a critical time point have the potential to produce long-term perturbations of overall functioning of the system. Any perturbation in a timing process critical

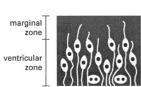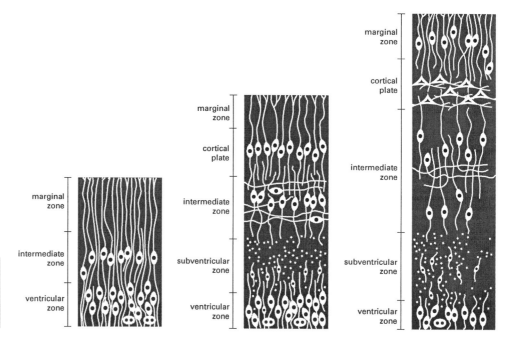

FIGURE 4.11

Process of cellular migration in the cerebral cortex

At the early stage, the wall consists only of epithelium. Cells that lose their capacity for synthesizing DNA withdraw from the mitotic cycle and form the intermediate zone. Cells that pass this zone in the forebrain aggregate and form the cortical plate. The cortical plate is the region in which the various layers of the cortex develop. Cells in the deeper layers of the cortical plate develop first (later IV). The cells in the superficial layer must migrate through the older layers to reach their appropriate position. The subventricular zone is a second proliferative region in which many glial cells and some neurons in the forebrain are generated.

to the maturation of the nervous system can interrupt the developmental interactions between selective cells resulting in effects ranging from altered growth and gross structural abnormalities, to death. Early gestational exposure to toxic substances has been reported to produce neural tube defects. Exposure to ionizing radiation produces microencephaly and mental retardation. Exposure to heavy alcohol produces craniofacial abnormalities and mental retardation while moderate levels consumed during gestation are reported to delay motor development. Low levels of lead can perturb cognitive functioning. In most cases, teratogens do not grossly affect the developing brain but instead alterations in the last trimester of gestation result in functional changes with little or no evidence of a morphological lesion or damage. Such insults may interfere with cell proliferation, neuronal migration, and differentiation.

During the process of development of the nervous system, the cellular events include:

1. Neurogenesis, migration to final target destinations
2. Aggregation of neurons to form organized nuclear groups
3. Differentiation and maturation of neurons with branching and elongation of axons and dendrites and myelinization of axons
4. Synaptogenesis, including elaboration of axonal pathways

A major factor determining the pattern of damage is the ontogenic stage at the time of exposure. The differential vulnerability of the developing nervous system may involve deprivation of rapidly differentiating cells from adequate metabolic supplies during critical periods when metabolic demand is high and a nonspecific insult to the metabolism of the brain may cause different patterns of damage for different ontological stages. The developing nervous system is critically dependent upon the synthesis of lipids for newly formed membranes; therefore, lipophilic agents or chemicals that directly perturb synthesis of lipids and cholesterol (e.g., tellurium) can be highly detrimental to the developing nervous system, especially as compared to the effects seen on the mature system. During development, the process of myelination is particularly vulnerable to a number of perturbations, including genetic disorders, viral infections, substances of abuse, environmental conditions, toxins, nutrition, etc., leading to the hypothesis that myelination is a "vulnerable" feature of the developing brain[11]. An early deficit in myelin is usually irreversible; the critical period associated with a permanent hypomyelination is prior to the onset of rapid myelin membrane synthesis from 8–14 days of age in the rat[31]. Perturbations during development may qualitatively differ from those seen with adult exposure. For example, rather than cellular necrosis changes may be seen in tissue volume, misplaced or misoriented neurons and processes, or by a delay or acceleration of functional or structural maturational landmarks and endpoints. Neurons appear to be sensitive to certain signals during their development, specifically those signals that induce neuronal differentiation, growth, migration, and synaptogenesis (see Figures 2.1 and 2.2, pages 31 and 32), thus alterations in the developmental signals may result in nervous system dysfunction that persists long after neuronal maturation. A number of environmental conditions may cause differentiating cells to depart from the normal plan of development. Such conditions include reduced oxygen and circulatory supply, nutrient deficiency, metabolic deficiency, toxicity, abnormal temperatures, etc. In many cases, perturbations induced during development of the nervous system may not be evident until the organism begins to mature or even age. An alteration may not become evident until an environmental condition, internal or external, imposes a challenge upon the organism.

1.5   INTRACELLULAR ORGANELLES

All cells of the nervous system contain all features associated with mammalian cells. The biological membrane is a lipid bilayer, the cell body contains both smooth and rough endoplasmic reticulum, Golgi apparatus, lysosomes, and mitochondria. The shape of the neuron and its processes is the responsibility of the fibrillar proteins of which there are three main components of the neuronal cytoskeleton. These include microtubules, neurofilaments, and microfilaments which consist of a core of acidic backbone proteins. The helical filaments are regulated by calcium and phosphorylation, and specific protein kinases can influence interactions with themselves or other related proteins. The proteins that constitute the cytoskeleton mediate the movement of organelles within the cell and anchor membrane constituents and receptors at appropriate surface sites. The cytoskeleton forms a highly cross-linked gel that fills the entire volume of the cell, thus determining the shape of the neuron. Each axon and dendrite contains long filaments that are cross-linked to each other. In the axon they are oriented longitudinally with polarity always in the same direction presumably associated with the direction of axonal transport. Although the mature cytoskeleton is very stable it is able to undergo transition by the polymeric filaments, actin and tubulin, to a dynamic state in regeneration. Actin and tubulin exist in a controlled balance of monomeric and polymeric forms. This balance can quickly shift between assembly and disassembly in response to physiological cues such as growth cones, extracellular matrix, disease, age, or environmental factors. Cytoskeletal elements of the nervous system are target sites of neurotoxic compounds. In addition to chemicals, neurofilaments are altered by the naturally occurring compounds cytochalasins which inhibit the rate of growth of the actin filament, and by phalloidin which accelerates the rate of polymerization and prevents breakdown in destabilizing conditions. Alterations in neurofilaments appear in many cases of peripheral neuropathies.

2   Role of the blood-brain and blood-nerve-barriers

The nervous system is dependent upon the extensive system of blood vessels and capillaries for both the delivery of oxygen and nutrients and the removal of toxic waste products. The barrier which separates the nervous system from the blood is a specialized structure composed of two continuous layers, endothelial cells of the blood vessels and the basement membrane. The adjoining endothelial cells physically merge tight junctions[7] preventing most substances from entering the brain.

The gap junctions negate oncotic and osmotic forces that normally control blood-tissue exchange. Each capillary is surrounded by astrocyte foot processes which exert modulatory effects on the capillary. These tight junctions and cellular interactions form the barrier and prevent the free passage of the majority of blood-borne substances producing a finely controlled extracellular environment for the nerve cells. This isolation from the transient changes in the composition of the blood is critical for maintaining homeostasis of the nervous system. The blood-cerebral spinal fluid barrier exists between the brain and the CNS compartments that contain the cerebrospinal fluid (CSF). The ependyma and adnexa of the choroid plexi are responsible for the actively secreted CSF. A barrier also exists between the brain and the interstitial fluid of the brain that is formed by the ependyma and the subjacent glial cells.

A number of factors determine the ability of a substance to cross the blood-brain barrier. This is largely determined by the lipid solubility of a substance with lipophilic substances easily able to cross. A series of specific carrier-mediated transport systems or facilitated diffusion mechanisms exist through which required nutrients, hormones, amino acids, proteins, peptides, fatty acids, etc. are able to reach the brain. These carriers are distributed either symetrically or asymetrically on the luminal and abluminal membranes of the endothelial cells, depending upon the homeostatic needs of the system[5]. For example, in order to maintain a low concentration of glycine in the brain, the carrier for glycine is located on the abluminal membrane. The abluminal membrane which also has more of the enzyme ATPase presumably as an aid to keeping the concentration of potassium in the extracellular fluid of the CNS low. The carriers for the essential amino acids are located on both the luminal and abluminal membranes to allow for these required substances to move freely across the membrane.

Although the barrier is finely tuned to control the entry of normal physiological substances into the nervous system, it is incapable of preventing the exchange of substances from the blood to the brain when the transport is dictated by the same physiological properties of lipid solubility or active transport carrier mechanisms. Metallic mercury can pass through membranes in both directions, however, it can be fixed in the CNS by the last oxidation reaction binding to -SH containing ligands. Thus, the enhanced brain accumulation of mercury in the CNS subsequent to exposure to the elemental form is related both to its lipid solubility and its rapid conversion to the nondiffusible divalent form. Increased exposure to methyl mercury can result in a disruption of the blood-brain barrier. Many toxic chemicals can pass through the blood-brain barrier and damage neural cells, yet a chemical disruption of the barrier can result in a cascade of effects due not only to the exogenous agent but also to the endogeneous substances that are normally excluded. Alterations in the barrier can possibly be induced by an excessive amount of an endogeneous substance, for example, the cytokines. In these cases, permeability of the barrier can be altered by both direct and indirect means.

Certain regions of the nervous system appear to be devoid of a blood barrier thus allowing direct exposure to all blood-borne substances. These areas include the *area postrema, hypophysis, pineal body, hypothalamic regions, subfornical organ, supraoptic crest,* and *choroid plexi.* A barrier to the blood supply is also absent in the dorsal root ganglion of the peripheral nervous system, and the

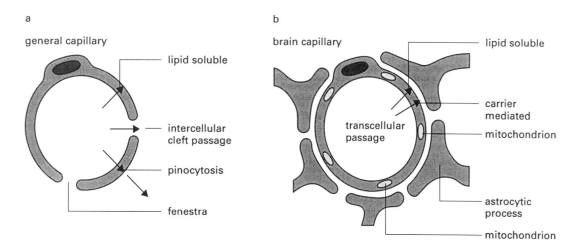

FIGURE 4.12

Ultrastructural features of general capillaries (a) and the capillary endothelial cells of the brain (b), forming the blood-brain barrier

olfactory nerve which may allow chemicals to penetrate directly from the nasal region to the frontal cortex. During development the barriers that protect much of the adult brain, spinal cord, and peripheral nerves are yet to be formed. Damage to the fragile capillaries of the immature brain can lead to swelling or rupturing of vessels that can result in encephalopathy. Such a pattern can result from cadmium, thallium, mercury, lead, and triethyltin. The differential susceptibility of the developing brain to environmental factors has been linked to the development of the blood-brain barrier. The pattern of susceptibility of neurons, petechial hemorrhages, and endothelial vacuolation following postnatal exposure to cadmium is thought to parallel the maturation of capillaries in the brain parenchyma[30]. Adult exposure to cadmium chloride produces massive hemorrhages in the spinal and trigeminal ganglia with no evidence of CNS perturbation. It must also be considered, that during the process of maturation not only are the various organ systems forming with their protective mechanisms but the ability of the organism to metabolize and detoxify toxic substances is also in the process of maturing.

3   Metabolic processes of the nervous system

The brain is metabolically one of the most active organs of the body, consuming approximately one-fifth of the total oxygen utilized by the body. Within the brain, oxygen is utilized almost entirely for the oxidation of carbohydrates. The enormous energy demand of the brain is constant, therefore dependent on continuously uninterrupted oxidative metabolism. This continuous supply is dependent upon replenishment of oxygen by the circulation since the oxygen stored in the brain is extremely small and would be consumed within seconds. In order to provide the supply of oxygen, the blood flowing through the brain is a considerable proportion, approximately 15%, of the total cardiac output, although the weight of the brain is only 2.5% of the body. This demand does not change during the course of the day. For example, during sleep cerebral oxygen consumption is only slightly decreased and in fact may even increase during rapid eye movement (REM) sleep. Energy metabolism is not the only demand for oxygen in the brain. A part of oxygen consumption is contributed toward the formation of various oxidases and hydroxylases critical in the synthesis and metabolism of a number of neurotransmitters. In the brain, the blood constituent that is most likely to be the main substance oxidized is glucose. Unlike the muscles of the body, the brain has only a limited ability to oxidize substances other than glucose. Therefore, a significant fall in blood glucose is followed by clinical alteration in cerebral function depending on the severity and duration of the hypoglycemia.

The quantitative 2-[^{14}C]deoxyglucose (2-DG) method measures glucose utilization and allows for the investigation of functional events in the brain related to various physiological, pharmacological, and behavioral activities. A differential level of energy demand can be seen in distinct brain regions and interference with glucose metabolism in the mature organism affects the cerebral cortex, cerebellar folia, and the hippocampus. The functional activation of glucose utilization occurs mainly in nerve terminals rather than in cell bodies; therefore, any alterations in energy metabolism of a specific brain region are due to the afferent inputs to that structure and not the cell bodies within the structure. However, distinct cellular vulnerability does exist with small cell bodies, with many processes showing vulnerability to anoxia while motor cells with long axons and fewer processes are less sensitive. Alterations in brain glucose utilization have been examined following administration of cocaine and suggest activation in the dopaminergic mesocorticolimbic system which is consistent with the regional neurochemical data on cocaine's actions.

A significant proportion of brain energy is required for the process of excitation, conduction, and subsequent neurotransmission. For such needs mitochondria and glycolytic enzymes are concentrated in dendritic and axonal terminals. Most of the ATP produced in the brain is required to restore the ionic gradients of the membrane constantly altered by synaptic transmission. Direct stimulation of specific neuronal sites in the brain have shown a relationship between glucose utilization and electrical activity. In humans, a correlation exists between cerebral metabolic rate and mental activity, and has been demonstrated in a variety of pathological states of altered consciousness. Irrespective of the origin, a reduction in cerebral oxygen consumption is accompanied by a reduction in the degree of mental alertness. However, increased demands for mental effort such as problem solving have shown no increased energy utilization. Also, disorders that alter the quality of cerebral functioning but not the level of consciousness (e.g., functional neuroses, psychoses, and psychotomimetic states) have no apparent effect on the average blood flow and oxygen consumption of the brain. This does not rule out the possibility of altered energy demands in focal regions of the brain. For example, drugs which "activate" the system by increased synaptic activity, (e.g., cocaine, amphetamine) can produce focal demands for energy, and by doing so possibly result in a depletion of the energy stores available with subsequent damage to the neural tissue.

The energy demands of the brain and required blood flow vary as a function of age, disease, or drug state of the organism. Animal data suggests that cerebral oxygen consumption is low at birth, rises during the period of cerebral growth and development, and reaches a maximal level at maturation which varies for different regions of the brain. During the process of aging, cerebral oxygen consumption and blood flow is not less *per se*; however, changes may be seen as a secondary effect of a disease state such as arteriosclerosis or other tissue pathology. In Alzheimer's disease, positron emission tomography (PET) scanning has shown that glucose metabolism is decreased in the temporoparietal lobe of the brain prior to manifestations of cognitive deficits. Postmortem analysis of brain tissue has shown significant reductions in the activities of enzymes important for energy metabolism, e.g., pyruvate dehydrogenase, a-ketoglutarate dehydrogenase, and mitochondrial cytochrome-c oxidase. In Parkinson's disease and in the chemical model of the disease, 1-methyl-4-phenylpyridinium (MPP+), the substantia nigra shows a reduced activity of complex I of the mitochondrial electron transport chain. PET studies have shown that in Huntington's disease both glucose and oxygen metabolism are decreased in the basal ganglia and cerebral cortex. Ultrastructural studies have supported this observation with the presence of abnormal mitochondria. One hypothesis for neurodegeneration due to altered energy functioning is based upon the premise that the excitotoxic mechanisms of the brain, e.g., altered membrane ion flux, formation of free radicals, are triggered by energy failure.

4   Transport processes

All types of cells are required to transport proteins and other molecular components from their site of synthesis near the nucleus to the various other sites of usage in the cell. Neurons have the unique distinction of requiring the cell body to maintain the functions normally associated with its own support, and to provide continuous support to its various processes. Essentially all newly synthesized membraneous organelles within axons and dendrites are exported to the axon from the cell body. The terminal is dependent on the cell body for all of the macromolecular components needed for transmission —

biosynthetic and degradative enzymes, proteins of the synaptic vesicles (both those in vesicle membranes and those that may be contained within the vesicle in soluble form), and most of the lipid. This requires transport of materials over distances that can be very long, and represents a process that is critical to the functioning and maintainence of the neuronal processes while being vulnerable to interruption by toxic chemicals. At the nerve terminals, the vesicle membranes are recycled many times through exocytosis for reuse in synaptic transmission. The membrane is constantly replaced by new components arriving from the cell body, and existing membrane components are returned from nerve terminals to the cell body where they are either degraded or reused. A continual supply of certain trophic factors as well as membraneous material is necessary for cell functioning; therefore, these factors of transport play a significant role in the normal growth and maintainence of the neural cells.

4.1   FAST ANTEROGRADE TRANSPORT

Fast transport is the process by which the neuron provides the proteins and lipids necessary to maintain the axonal and nerve terminal membranes. Fast axonal transport moves material from the cell body down the axon to the nerve endings at a rate of 400 mm per day. This transported material has been shown by electron microscopy to be associated with the axonal smooth endoplasmic reticulum and associated vesiculotubular structures[13]. The vesiculotubular structures are involved in the active movement of material, while the axonal smooth endoplasmic reticulum (SER) is stationary and possibly associated with the exchange of newly synthesized material to the axonal membrane. It depends on one or more of the filaments that make up the neuron's cytoskeleton and on microtubules that provide an essentially stationary track on which specific organelles move in a saltatory fashion driven by a microtubule associated ATP-ase, kinesin. Once material enters the rapid transport vector, the continuation of rapid transport is independent of the cell body and the electrical activity of the axon. It is however, critically dependent on oxidative metabolism within the axon for a continued supply of ATP and on extracellular Ca^{2+}.

4.2   SLOW ANTEROGRADE TRANSPORT

Cytoskeletal elements and soluble proteins are transported down the axon by slow axoplasmic flow. Slow transport involves two groups of proteins termed "slow components a and b"[19] or "IV and V"[32]. The slower components travel at rate of 0.2-1.0 mm per day and carry the proteins used to make up the fibrillar elements of the cytoskeleton, the subunits that make up neurofilaments, and the alpha and beta tubulin subunits that make up the microtubule. The second component travels at a rate of 2.0–5.0 mm per day and contains a more complex protein content including actin, myosin, clathrin, enzymes of intermediary metabolism, and calmodulin. Slow axonal transport requires a considerable metabolic stability since in larger mammals it may take many months for slowly transported molecules to reach the terminal ends of long peripheral axons. On reaching the nerve terminals, much of the slow transported material is rapidly degraded presumably by specific calcium-activated proteases.

4.3   RETROGRADE TRANSPORT

Material returns from the terminals to the cell body either for degradation or recycle at a rate approximately 50–70% of that of fast anterograde transport[6]. Particles move along microtubules driven by a motor molecule in the form of dynein which is a microtubule-associated ATPase (MAP-1C). The composition of these materials is similar to that of the anterograde fast component and the materials to be transported are packaged in large membrane-bound organelles that are part of the lysosomal systems. It is assumed that much of this material is the result of a reversal of the rapid anterograde stream, turnaround, at the nerve ending. Retrograde transport is dependent upon an axonal supply of ATP and Ca^{2+}, is independent of the cell body, and its rate can be altered by variations in body temperature. Extracellular components transported in the retrograde direction are taken up at the terminals by endocytosis. For example, nerve growth factor, a peptide synthesized by the target cell that stimulates growth of certain neurons, is transported from the nerve endings to the cell body. This may play a critical role in informing the cell body of the events occurring at the nerve ending and be a critical factor during the development of the nervous system and during regeneration processes. Some neurotropic viruses and toxins can reach the central nervous system by ascending from the peripheral nerve terminals to cell bodies by fast retrograde transport. Herpes simplex, rabies, polio viruses, and tetanus toxin can all reach the neuron by retrograde transport. It has also been reported that methyl mercury can reach the neurons by an uptake mechanism at the nerve terminal and subsequent axonal transport.

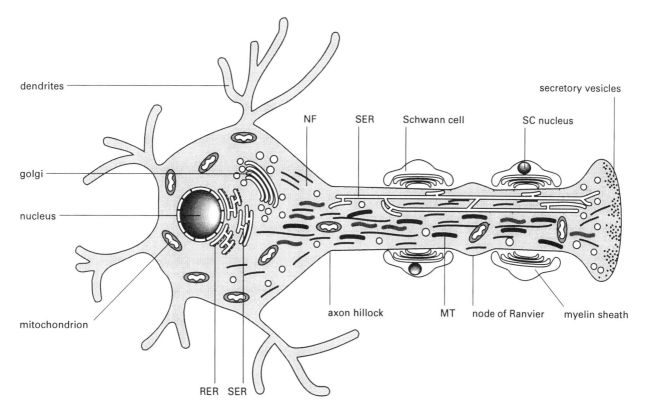

FIGURE 4.13

Schematic of a myelinated nerve indicating several general features of axonal transport

All protein and lipid synthesis necessary to maintain the axolemma is conducted in the cell body within the Golgi apparatus and endoplasmatic reticulum (SER). Organelles are assembled in membraneous and cytoskeletal compartments (NF: neurofilaments; MT: microtubules) and exported for transport to destination sites along the axon. Synaptic vesicles and other membraneous organelles involved in synaptic transmission are transported to the nerve terminal and transmitter is released by exocytosis. Degraded material is returned to the cell body by retrograde transport for reutilization.

A number of toxicants can produce morphological alterations in the axons of peripheral nerves and alter various aspects of axonal transport. An agent may interact with a specific step in the mechanism of axonal transport, the movement of a specific transported component may be altered, or in a general biochemical process, e.g., energy metabolism. Although axonal transport is altered by various environmental agents and pharmaceuticals it is still unclear if an axonopathy can be produced by a direct effect on axonal transport. Neurofilaments are a major target site for a number of agents, and since they are a structural component intrinsic to the axonal transport system an alteration in axonal transport can be seen. Focal accumulations of neurofilaments within neurons and their processes have been observed following exposure to certain chemicals[2] (Table 4.4).

TABLE 4.4

Examples of chemicals that produce axonal transport perturbations in neurofilaments

Model	Proximal axon	Distal axon	Neurofilament transport
IDPN	massive assumulations of neurofilaments	atrophy	decreased
2,5-HD	atrophy	multifocal accumulation of neurofilaments	increased
Aluminium	neurofilament swellings	atrophy	decreased
Acrylamide	atrophy	some neurofilament swellings; degeneration	decreased
Axotomy	atrophy	Wallerian degeneration	decreased

It has been proposed that toxicants which produce focal accumulations of neurofilaments do so by covalently modifying the neurofilament proteins leading to a destabilized cytoskeletal framework, uncoupling microtubule and neurofilament transport[28]. This hypothesis is supported by hexacarbons producing the crosslinking of neurofilaments[12,22]; however, it does not support the alterations in rapid anterograde and retrograde transport. This may be due to a general alteration in energy metabolism supporting the proximal axon at the expense of the distal portion, resulting in what is usually called a "distal-dying back axonopathy" since the distal area of each internodal region shows the first signs of degeneration. Slow axonal transport is the main transport component altered by neurofilamentous accumulations.

Axonal swellings can also occur from an accumulation of vesiculotubular membrane structures presumably derived from the smooth endoplasmic reticulum. Such accumulations occur following exposure to several neurotoxic agents such as *p-bromophenylacetylurea*, *zinc pyridinethione*, and certain organophosphorous compounds. Chemicals that produce an accumulation of membraneous structures appear to have little or no effect on slow axonal transport until the axonopathy is severe and the axons undergo degeneration.

Example

Following exposure to acrylamide, peripheral nerves display a multifocal distal axonopathy characterized by distal axonal swellings containing intermediate filaments and other organelles localized at the nodes of Ranvier. Various reports suggest alterations in fast bidirectional transport and slow transport. The conflicting data concerning alterations in axonal transport following acrylamide exposure emphasize the need to control for various experimental factors such as, extent of morphological changes in the axon at time of study, body and nerve temperature, as well as the technique used to examine axonal transport. Biological factors such as the heterogenicity of axons that comprise any nerve also need to be considered when interpreting transport data. For example, it has been demonstrated that early acrylamide-induced alterations in fast anterograde transport of glycoproteins is localized in the myelinated axons while transport in unmyelinated axons is unaltered[16]. The myelinated and unmyelinated axons contribute equally to the sciatic nerve and an effect on the overall transport within the sciatic nerve would miss such a biologically significant effect. This dramatic and preferential impairment of axonal transport of glycoproteins in myelinated axons prior to any observed axonal pathology provides additional support for the hypothesis of differential cell body responses as well as a direct involvement of axonal transport abnormalities in the pathogenesis of acrylamide-induced distal axonal degeneration.

Toxicants can alter axonal transport by directly affecting the initiation of transport from the neuronal cell. A direct effect on the cell body is characterized as a neuronopathy. Methyl mercury is a specific toxicant that alters somal synthesis and processing, thus resulting in changes in axonal transport. The critical role of the cell body makes it essential in determining the initiation, processing, and subsequent commitment to transport in the cell body when examining alterations in axonal transport. Many neurotoxicants and disease processes appear to have a direct pathological effect on axons; however, the question remains as to whether any of these agents produce a direct effect upon axonal transport prior to the appearance of morphological alterations or clinical signs. The disruption of the transport process remains a critical part of the cellular pathology. If alterations in axonal transport are a cause or an effect of peripheral neuropathy remains a critical question; however, the information obtained about the process is an important factor in the further understanding of the pathogenesis and progression of toxicant-induced neuropathies.

5   Neural signalling

5.1   ION CHANNELS

There exists a number of processes by which neurons interact with other neurons and non-neuronal tissue. The major focus of such research has mostly been on those processes by which the nerve transmits and receives information and the resulting mechanisms by which these signals are transduced to produce cellular responses. All aspects of neuronal regulation are influenced by the neuronal milieu consisting of various ions, input from other neurons, and a variety of growth and hormonal factors. Neurons interact with their extracellular environment through membrane and intracellular receptors and membrane ion channels. The excitable membrane is particularly sensitive to changes in intra- and extracellular ion concentrations. This requires a tight regulation of intraneuronal ion concentrations accomplished by a relatively impermeable membrane, various ion pumps, intracellular ion binding sites, and a variety of specific ion channels. These membrane ion channels may be voltage sensitive, directly associated with, membrane receptors, or operationally linked with or controlled by, a cascade of intracellular signals (i.e. second messengers). Membrane receptors modulate certain ion channels and are linked to a variety of intraneuronal processes that play roles in neuronal growth, differentiation, plasticity, and survival. One of the more critical processes is the maintenance of calcium homeostasis.

Intracytoplasmic calcium homeostasis is maintained at concentrations below 0.1 mM by Ca^{2+} binding proteins, voltage-sensitive plasma membrane Ca^{2+} channels, plasmalemma Ca^{2+}-ATPases pumping Ca^{2+} across the membrane, plasma membrane Na^+/Ca^{2+} exchanges, and intracellular Ca^{2+} storage organelles, e.g., mitochondria and endoplasmic reticulum. Alterations in any one of these systems can result in altered calcium homeostasis and produce cytotoxicity in the neuron. For example, the neuronal toxicity induced by glutamate and other excitatory amino acids is characterized by acute swelling of the dendrites and cell body subsequent to a slower Ca^{2+}-dependent neuronal degeneration[8].

Neuronal signaling is dependent on rapid changes in the electrical potential differences across nerve cell membranes. These rapid changes are possible by ion channels, a class of integral glycoproteins that traverse the cell membrane. The channels conduct ions, recognize and select among specific ions, and they open and close in response to specific electrical, mechanical, or chemical signals. Ion channels are needed in the nervous system since ions are not easily able to cross the phospholipid bilayer membranes. The gated channels are useful for rapid neuronal signaling since they open and close in response to various stimuli. The non-gated channels, however, are always open and contribute to the resting potential. The use of each channel for signalling comes

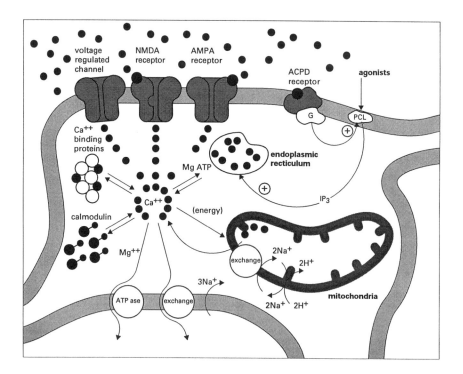

FIGURE 4.14

Calcium homeostasis in a neuron.

The figure reflects the balance between ligand- and voltage-regulated influx, receptor mediated release from the endoplasmic reticulum, the buffering by cytoplasmic proteins, the sequestration by mitochondria and the export from the cell by a calcium pump or in exchange for sodium. (Adapted from *Science and Medicine* **1**(1): p 25, 1995.)

from their ion specificity and susceptibility to regulation. In nerve and muscle cells, ion channels are important for controlling the rapid changes in membrane potential associated with the action potential and postsynaptic potentials. Distinct from ligand-gated channels are the voltage-regulated ion channels. The sodium channel plays a role in the propagation of the action potential, potassium channels control excitability and shape signals, and calcium channels regulate intracellular calcium levels. The membrane potential of the resting nerve cell is determined by the potassium ion channels while the sodium channels are activated during the action potential. Sodium is actively pumped out of the cell by the Na^+/K^+ transporting ATPase and helps set up the resting membrane potential of $-80mV$. The cytoplasmic Ca^{2+} concentration is maintained at a level some 1000x less than that outside the cell by means of

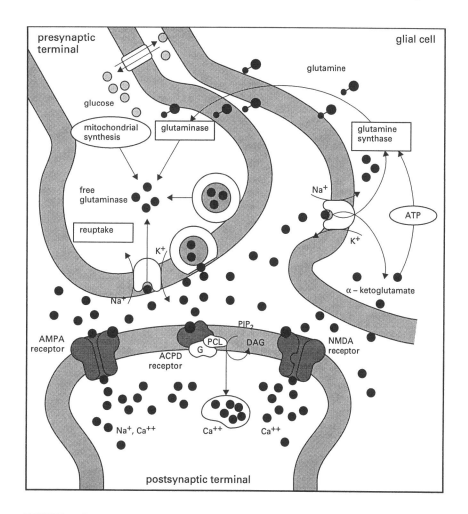

FIGURE 4.15

The glutamate synapse. Glutamate receptors include the N-methyl-D-aspartate (NMDA), the alpha-amino-3-hydroxy-5-methyl-4-isoxazole-proprionic acid (AMPA) and kainate receptors. These are receptor-channel complexes that regulate transmembrane conductane of Na^+, K^+, and Ca^{2+} ions.

The metabolic receptors such as the aminocyclopentyl dicarboxyl acid (ACPD) receptor, are coupled to G proteins and function through intracellular second messengers including inositol triphosphate (IP3), cyclic adenosine monophosphate (cAMP), and calcium. The figure shows vesicular release into the synaptic space, activation of postsynaptic receptor systems, reuptake into the presynaptic terminal and surrounding glial cells, translocation of glutamine back to the presynaptic terminals reconversion to glutamate and de novo synthesis of glutamate from glucose. (Adapted from *Science and Medicine* **1**: p 23, 1995).

ATPases that actively move Ca^{2+} out of the cell or into intracellular storage organelles. The calcium influx controlled by these channels can alter many metabolic processes within cells, resulting in activation of various enzymes and proteins and also act as a trigger for the release of neurotransmitters. Each channel can exist in one of three states; closed, open (activated), and desensitized (inactivated). They differ in ion specificity, time and voltage dependence of activation and inactivation, single-channel conductance, and pharmacological susceptibility.

Ion selectivity is achieved through physical-chemical interaction between the ion and various amino acid residues that line the walls of the channel pore. Gating involves a conformational change of the channel in response to various external stimuli, including voltage, ligands, and stretch or pressure. Neurotransmitters, acetylcholine (ACh), g-amino butyric acid (GABA), glycine, and glutamate, regulate the opening and closing of an ion pore by binding to a site on the channel protein and mediate rapid postsynaptic responses. The nicotinic ACh receptor, located in the postsynaptic membrane at the vertebrate neuromuscular junction, is the most widely characterized ligand-gated channel. Glutamate regulates ion channels via three types of glutamate receptors: the kainate, quisqualate/AMPA (alpha-amino-3-hydroxy-5-methyl-4-isoxazole-propionic acid) and N-methyl-D-aspartate (NMDA) receptors. Internal ligands, e.g., cGMP and cAMP, can regulate other channels while cytoplasmic messengers (phosphoinositol biphosphate (IP3), arachidonic acid metabolites, and calcium) modulate other channels.

The activity of ion channels can be modified by cellular metabolic reactions including protein phosphorylation, various ions that act as blockers, and by toxins, poisons, and drugs. If the flow of ions across the cell membrane is changed, the transmission of information between nerve cells will be altered. A substance may directly interfere with the ionic balance of a neuron.

Example

Organophosphate and carbamate insecticides produce autonomic dysfunction while organochlorine insecticides increase sensorimotor sensitivity and produce tremors which can result in seizures and convulsions. This process can be seen following exposure to lindane, dichlorodiphenyltrichloroethane (DDT), pyrethroids, and the organometals like trimethyltin and trimethyllead. The reverse is true for solvents in that they can act to raise the threshold for eliciting seizures or may act to reduce the severity or duration of an elicited convulsion. Local anesthetic drugs and cocaine interfere with the uptake of sodium through voltage-regulated sodium channels. By this process of inhibition of sodium flux, cocaine interacts with the the downstream processes triggered by inositide hydrolysis generated second messengers. Ethanol acts on several ligand-gated and non-ligated-gated ion channels. At low doses ethanol has effects on the N-methyl-D-aspartate (NMDA), GABAa, 5-hydroxytyptamine (5HT) receptors, and the L-type calcium channel. At higher doses, the action of serotonin is potentiated at the 5-HT_3 receptors and g-amino butyric acid at some GABAa receptors. The 5-HT_2 receptor is a G protein-linked receptor that mobilizes intracellular calcium and can be stimulated by various hallucinogenic drugs. Cannabinoids act as agonists on the cannabinoid receptor present in the hippocampus, cerebellum, basal ganglia, and the ventromedial nucleus of the hypothalamus and linked to a G protein-coupled receptor and acts to inhibit the activation of adenylate cyclase. These specific locations may be responsible for the effects of cannabinoids on perception and memory, motor incoordination, and glucose craving respectively.

Membrane ion channels are also important targets in various diseases. For example, autoimmune neurological disorders such as myasthenia gravis and the Lambert-Eaton syndrome are thought to result from the actions of specific antibodies interfering with channel function. Cystic fibrosis is thought to involve a genetic defect in the control of a certain type of chloride channel.

5.2   NERVE CONDUCTION

Information is transferred between neurons by two types of synaptic transmission, electrical and chemical.

Electrical transmission is mediated by the direct flow of current from the presynaptic to the postsynaptic neuron through gap junctions. Electrical synapses can be rectifiying or non-rectifiying and occur by means of current flow through gap junction channels that directly connect the cytoplasm of both cells. Transient electrical signals are important for transferring information rapidly and over long distances. These electrical signals are produced by changes in the current flow into and out of the cell that drives the electrical potential across the cell membrane away from its resting state. The passive electrical properties of neurons do not change during signaling and are determined by the conductance of non-gated ion channels, the membrane capacitance, and the conductance of the cytoplasm. They affect the time course of synaptic potentials and conductance from site of origin to trigger zone. These properties contribute to synaptic integration, the process by which a nerve cell adds up all incoming signals and determines whether or not it will generate an action potential. During signaling, the rate of change in the membrane potential is critically dependent on membrane capacitance. The ion channel plays a critical role in this process which is important in determining the rate of information transfer within a neuron.

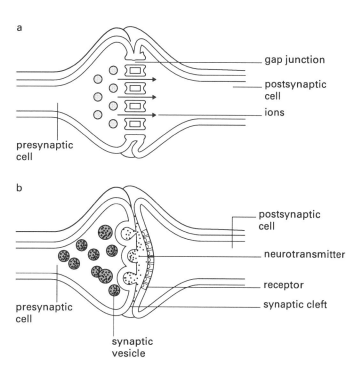

FIGURE 4.16

Schematic drawing of two types of synapses, the electric (a) and the chemical synapse (b).

143

5.3   SYNAPTIC TRANSMISSION

The classic form of "chemical" synaptic transmission is the action of acetylcholine (ACh) at the neuromuscular junction. In response to an action potential in the presynaptic motor neuron, ACh is released from the terminals of the motor neuron, diffuses across the synaptic cleft, and activates nicotinic ACh receptor-channels. Binding of ACh to the receptor leads to the opening of a channel that is an integral part of the receptor protein. This channel is permeable to cations (Na^+, K^+, and Ca^{2+}) and its opening leads to a net influx of sodium ions, producing a depolarizing synaptic potential called the end-plate potential (EEP). Calcium, entering through voltage-dependent channels in the presynaptic terminal, is essential for synaptic transmission. The synaptic delay, the time between the onset of the action potential and the release of transmitter, reflects the time it takes for incoming calcium to diffuse to its site of action and trigger the discharge of transmitter from the synaptic vesicle. The development of the neurotransmitter systems begins during the late fetal period of the rodent. Although the morphological development of the system is complete within 2 weeks after birth, the rate of synthesis and the concentration of some neurotransmitters take up to 2 months in rodents and 3–4 years in humans to reach mature levels.

Chemical synaptic transmission in the central nervous system can be either excitatory or inhibitory with glutamate being the major excitatory transmitter in the brain and spinal cord. Receptors for neurotransmitters can be divided into two major families according to how the receptor and effector functions are coupled.

1. The first family consists of receptors that gate ion channels directly and the two functions, receptor and effector, are carried out by different domains of a single macromolecule. It includes the nicotinic acetylcholine, the g- aminobutyric acid, the glycine, the AMPA (kainate-quisqualate) and the N-methyl-D-aspartate (NMDA) class of glutamate receptors.

2. In the second family, the receptors gate channels indirectly of G-protein and the coupled receptor-recognition of the transmitter and activation of effectors are carried out by separate molecules.

For a and b adrenergic, serotonin, dopamine, and muscarinic ACh receptors, and receptors for neuropeptides and rhodopsin, the receptor molecule is coupled to its effector molecule by an intermediary guanosine nucleotide-binding protein (G protein). Activation of the effector component requires the coordinated effort of several distinct proteins. Usually, the effector is an enzyme that produces a diffusible second messenger, such as cyclic adenosine monophosphate (cAMP), diacylglycerol, or an inositol polyphosphate. These second messengers trigger a biochemical cascade either activating specific protein kinases that phosphorylate a variety of the cell's proteins or mobilizing calcium ions from intracellular stores. This leads to the reactions that change the cell's biochemical state. The G protein, second messenger (cAMP, cGMP, or metabolites of arachidonic acid) can act directly on an ion channel. Second messenger kinases not only produce covalent modification of preexisting proteins, but also induce the synthesis of new proteins by altering gene expression. This can lead to other long-term changes such as neuronal growth.

In many cases, long term alterations in the presynaptic cell will result in a modification of the postsynaptic receptor. For example, following long term dopamine antagonist therapy there is an up-regulation of dopamine receptors and a resulting tardive dyskinesia. Many substances can alter the synthesis and release of specific neurotransmitters and activate their receptors in specific neuronal pathways. Perturbation may occur by an overstimulation of receptors, a block in transmitter release and/or inhibit transmitter degradation, or a block reuptake of neurotransmitter precursors. Altered neurotransmission can indicate the inability of various target organs to compensate for the altered signal as is the case for acetylcholinesterase inhibitors.

In many cases pharmacological agents have been designed to act at distinct receptor sites allowing for a specificity of action. For example, nicotine acts on the nicotinic cholinergic receptor with an initial stimulation in the peripheral sympathetic and parasympathetic ganglia, the adrenal gland, and the brain, followed by a subsequent inhibition. Amphetamines block monoamine reuptake and release catecholamines from intraneuronal stores. Methamphetamine can also produce cytotoxicity in neurons in the striatum. The serotonergic system is altered by 3,4-methylene dioxyamphetamine (MDA) and 3,4-methylene dioxymethamphetamine (MDMA; also called XTC or Ecstasy) producing long-lasting decreases in serotonin levels, loss of 5-HT reuptake sites, and eventual neuronal degeneration of serotonin neurons. In most of the amphetamine–like effects of cocaine, the dopaminergic system plays a major role. Its different pharmacological actions in the central nervous system include an inhibition of dopamine uptake, indirect-acting dopamine receptor agonist, and a local anesthetic.

6   Basic neuroanatomy of selected nervous system regions

The CNS is comprised of the brain which is the enlarged anterior portion enclosed within the cranial cavity of the skull, and the long axonal processes within the spinal cord protectively encased within the spinal vertabral column. Each is protected by three membranes known as meninges. The general divisions of the brain are the: (a) prosencephalon, (b) mesencephalon, and (c) rhombencephalon which are then subdivided respectively based upon divisions of the neural tube to the: (a) telencephalon, diencephalon,

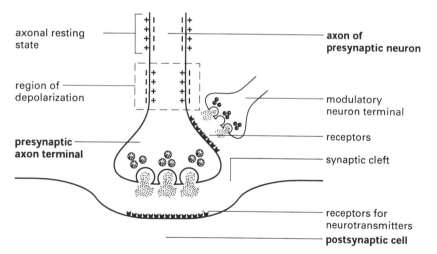

FIGURE 4.17

General structures of chemical synapse

The synaptic cleft separates the plasma membranes of the presynaptic and postsynaptic cells. The depolarization of the axonal membrane results in the transmission of electrical impulses by the release of a neurotransmitter from the synaptic vesicles, its diffusion across the synaptic cleft, and its binding at postsynaptic receptors. The postsynaptic cell can be a dendrite, cell body, or axon of a neuron, muscle, or gland cell. The depolarization state of the neuron is influenced by receptor activation by modulatory neuron terminals.

(b) mesencephalon, (c) metencephalon and myelencephalon. Selected regions of the central nervous system have been chosen for discussion: the cerebral cortex, hippocampus, hypothalamus, and spinal cord. Interaction with the rest of the body is conducted through the axonal projections within the peripheral nervous system. With few exceptions, once the nerve fibers leave the spinal cord, no matter if the motor fibers from neurons in the spinal cord, or if they are sensory fibers from the cell bodies in the dorsal root ganglion, features of the peripheral nervous system are assumed.

6.1   CORTEX

The cortex is gray matter comprised of cell bodies of neurons, neuroglia, nerve fibers, and blood vessels. In the cortex of a fully developed mammal there are, apart from the neuroglia, two neuronal classes. The two neuronal types are called pyramidal and non-pyramidal . Among the latter are several subtypes. The names of the cells reflect their morphology as seen with stains such as Golgi stain, that reveal the morphology of the whole cell. The lamination of the cortex is, at the routine histological level seen with the Nissl stain, a lamination of cell bodies.

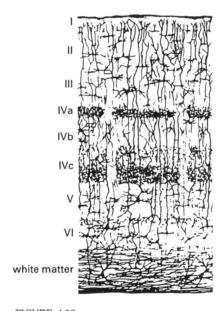

FIGURE 4.18

Cellular layering pattern of the cerebral cortex. Layers II, III, V and VI contain pyramidal cell bodies.

The cortical layers I-III (molecular, external granular, and external pyramidal) have numerous stellate cells and play a role in association and higher functions such as memory, interpretation of sensory input, and certain discriminative functions. Layer IV, the internal granular layer is mainly a receptive layer for the various other regions of the brain. Layers V and VI, the ganglionic and multiform layers, are primarily efferent layers that contain nerve cell bodies that project into the corticospinal tract. Five type of cells are found in the cerebral cortex: pyramidal cells, granule cells, fusiform cells, horizontal cells, and the cells of Martinotti (the ascending axon cells). Some of the general topographic areas of the cortex are known to have specialized functions. The primary motor area controls fine, highly skilled, voluntary movements of the body.

146

The premotor area is located in the frontal region of the cortex and is concerned with the development of motor skills. The occipital eye field is located in the occipital lobe and is associated with involuntary eye movements that may be induced by visual stimuli while another eye field is concerned with voluntary eye movements. The prefrontal cortex receives input from the thalamus and all lobes of the cerebrum. In the human this area is responsible for abstract thinking, mature judgement, foresight, tactfulness, and self-control.

Just as there are defined motor control regions of the cortex, there are sensory functions associated with specific regions of the cortex. The primary sensory cortex allows the body to recognize pain, temperature, and pressure, and allows for proprioception and kinesthetic sense. The second somesthetic cortex receives input from several sensory modalities with pain being predominant. In humans, the other main sensory area of the brain is the somesthetic association cortex, and damage to this area can result in agnosia. Both the visual system and the auditory system have specialized regions in the cortex. The visual cortex sends projections to the pons, midbrain, and to the oculomotor complex, and it receives input from the lateral geniculate body. The auditory cortex receives the majority of its input from the medial geniculate nucleus which receives input from the organ of Corti.

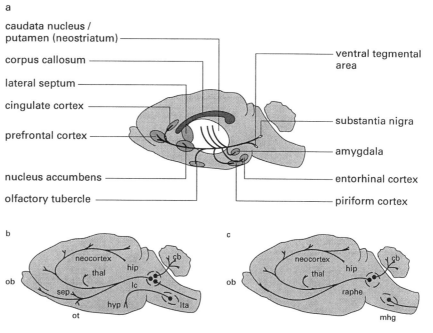

FIGURE 4.19

Dopamine (a), norepinephrine (b) and 5-hydroxytryptamine (c) neuronal pathways in the rat brain. The principal long pathways are shown; for clarity, the shorter neuronal pathways are omitted.
CB = cerebellum, OB = olfactory bulb, SEP = septum, THAL = thalamus, HIP = hippocampus, HYP = hypothalamus, LC = locus coeruleus, LTA = lateral tegmental area, OT = olfactory tubercle, MHG = medullary 5-HT group.

The entire cortex is innervated by norepinephrine fibers arising from the locus ceruleus in the brainstem and contains both a- and b-adrenergic receptors. The action of norepinephrine on cortical neurons has generated a great deal of interest in neurotoxicology due to its role in cortical information processing, neuronal plasticity, and synapse formation during development. The cortex contains a relatively low concentration of nicotinic receptors varying in density throughout the cortical regions. Muscarinic receptor sites in the cortex are concentrated in layer IV of the rat parietal cortex. Serotonin receptors are present in high concentration in the cortex and are associated with a serotonin-sensitive adenylate cyclase. Serotonin denervation resulting from administration of the neurotoxin 5,7-dihydroxytryptamine or lesions of the midbrain raphe nuclei produces an increase in the number of high-affinity binding sites. The number of serotonin receptor binding sites is decreased following repeated treatment with either *d*-fenfluramine, which releases serotonin, or with monoamine oxidase inhibitors. In contrast to the widespread innervation by noradrenergic and serotonergic fibers, dopaminergic innervation is limited to the frontal and limbic areas of the cortex. Both GABA and glutamate decarboxylase activity, the enzyme that decarboxylates glutamic acid, have been found in the cortex. A calcium dependent release of GABA in the cortex has been shown during spreading depression, potassium stimulation, electrical stimulation of the contralateral cortex, and stimulation with calcium ionophores. Cortical synaptosomes possess a high affinity, sodium-dependent uptake mechanism for glutamate and aspartate. Iontophoretic application of either glutamate or aspartate to cortical neurons has been shown to produce a rapid and strong excitation. A number of peptides have also been localized in the cortex: somatostatin, vasoactive interstinal polypeptide, cholecystokinin, avian pancreatic polypeptide. Very low levels have been found of neurotensin, substance P, or met-enkephalin.

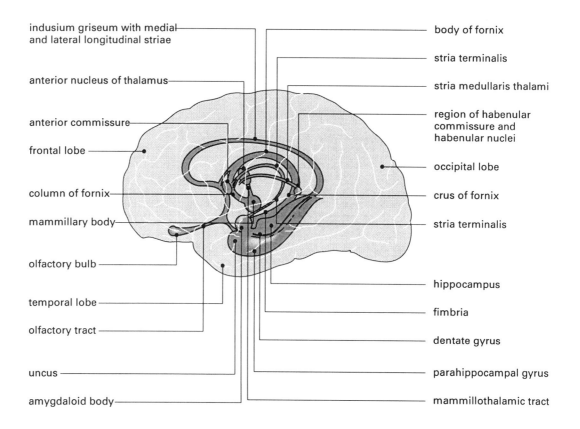

FIGURE 4.20

Medial aspect of the right cerebral hemisphere, showing the structures that form the limbic system.

6.2   LIMBIC SYSTEM

The limbic system of the CNS consists of a large number of neural components involved in the regulation of autonomic and somatic behaviors and has widespread connections with many regions of the forebrain and brainstem. The limbic system includes the following cortical structures: the olfactory cortex, hippocampal formation, cingulate gyrus, and subcallosal gyrus; as well as subcortical regions: the amygdala, septum, hypothalamus, epithalamus, anterior thalamic nuclei, and parts of the basal ganglia.

The hippocampus is a major component of the limbic system and plays a role in arousal, motivational, attentional, and emotional behaviors, and in learning and memory processing. It is involved in the intergration and classification of incoming stimuli from the environment, both internal and external. In the rodent brain, the hippocampal formation is located on the posteromedial border of the hemispheres. It borders the roof of the lateral ventricle and extends from the rostromedially located septum to the ventrolaterally located *amygdaloid area.*

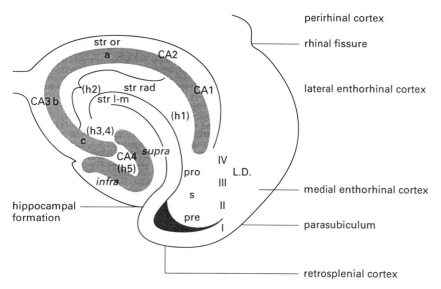

FIGURE 4.21

Schematic drawing of a cross section through the hippocampal formation.
SUPRA = suprapyramidal blade of the dentate gyrus.
INFRA = infrapyramidal blade of the dentate gyrus.
str l-m = stratum lacunosum moleculare.
str or = stratum oriens.
PRO = prosubiculum.
S = subiculum.
PRE = presubiculum.
Hip = hippocampal fissure.
Roman numerals I through IV refer to successively deeper cortical layers.

The most obvious internal features of the hippocampus are rows of densely packed neurons that are readily apparent in histological sections with Nissl dyes. The cells in these rows have a characteristic appearance. The areas that contain the distinctive rows of densely packed cells are defined as the hippocampus proper, or Ammon's horn, and the closely associated dentate gyrus. The term hippocampal formation is used more broadly to include the hippocampus proper, the dentate gyrus, and, in addition, the cortical regions that are interposed between the hippocampus and the neocortical surface of the brain in mammals. Many attempts have been made to provide a useful nomenclature for the areas of the hippocampus and dentate gyrus. In mammals, the granule cells of the dentate gyrus appear as a V formation at the end of the band of hippocampal pyramidal cells. Overall, the layer of hippocampal pyramidal cells appear as a curved line that ends in the middle, open part of the V. The simplest subdivision of the hippocampus is that of a superior and an inferior region. A further subdivision of the hippocampal cell system has been made into four subfields designated cornu ammonis (CA) 1–4 (CA_1–CA_4). CA4 is also known as the hilus of the dentate gyrus. The two main cell types are pyramidal neurons of the hippocampus proper and granule cells of the dentate gyrus. The molecular layer of the hippocampus actually is a continuation of the molecular layer (layer I) of the cerebral cortex. Both in the molecular layer and close to the granule cells many interneurons exist, some using GABA as a transmitter. The dentate granule cells form a dense layer beneath the molecular layer of the pyramidal cells and it is their dendrites that offer the major entrance to the hippocampus. The cells propagate the impulses from the perforant path, a massive excitatory fiber tract originating outside of the hippocampus in the cerebral cortex region, in the *entorhinal cortex*, through their excitatory axons, the mossy fibers, to the CA4 pyramidal cells in the *hilus*. Some of these cells then send excitatory projections back to the *dentate*. The mossy fibers also make contact with the pyramidal cells in the CA_3 region of the hippocampus. Activity in the hippocampal formation is modulated by local and projection-type inhibitory neurons from the medial septal area, by monoaminergic fibers from the brainstem, and by the cholinergic pathways from the ventral forebrain.

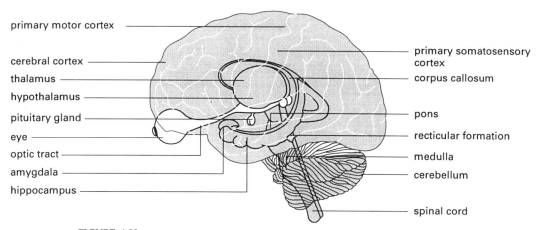

FIGURE 4.22

Schematic drawing of the hippocampus in relation to the rest of the brain

The complex interactions between various neurons and the monoamine modulation of cellular activity influence the processing and gating of incoming stimuli in the hippocampus. The hippocampal formation contains acetylcholine, catecholamines, norepinephrine, dopamine, serotonin, GABA, L-glutamate, and L-aspartate. Not only have these chemicals been found in the hippocampus but they have also been shown to influence the excitability of hippocampal neurons. The majority of the cholinergic innervation to the hippocampus is from the septum[23]. The hippocampus influences other brain regions by means of extensive cortical and subcortical projection. Most of the hippocampal efferents are excitatory and may use glutamate as their transmitter.

The hippocampal formation displays a form of plasticity as demonstrated by long-term potentiation which has been reported to represent a major substrate of learning and memory, and is involved in the formation of cognitive spatial maps[24]. Behaviors that detect alterations in the development, maturation or functioning of the hippocampus include cued Morris watermaze, conditional spatial discrimination and short-term memory, and some scheduled controlled behaviors. The vulnerability of the hippocampus to environmental manipulation is well-documented. As mentioned earlier, it is a target site for organometals like trimethyltin and triethyllead. It is also highly vulnerable to excitotoxins and to conditions of anoxia and ischemia, with the pyramidal cells of the CA1 appearing differentially affected. In many cases, a distinction must be made with regard to primary and secondary effects of a toxicant-induced injury to the hippocampus in that any alterations in oxygen supply or in generalized induced seizures can perturb neurons in the hippocampus.

6.3   CEREBELLUM

The cerebellum occupies most of the posterian cranial fossa. It lies dorsal to the brainstem and is attached to the pons, medulla, and the mesencephalon by the cerebellar peduncles (three pairs of thick fiber bundles). The cerebellum is composed of the cerebellar hemispheres and a midline portion, the vermis, each involved in various aspects of motor movement and coordination. The cortex is a layer of gray matter covering the surface, and the medullary center is an internal core of white matter fibers from the cortex to the cerebellar nuclei, incoming fibers, and interconnecting fibers. The internal structure consists of five types of neurons: granule cells, Purkinje cells, basket cells, stellate cells, and Golgi cells. The molecular layer, the outer layer, receives input from the granular layer and deep cerebellar nuclei, as well as outside the cerebellum. It consists of nerve fibers and some basket and stellate cells. The Purkinje cell layer is the middle layer and it is the output layer comprised of the Purkinje cell bodies. The granular layer is the innermost layer containing granule cells and Golgi cell bodies, and it receives input from outside the cerebellum. The mossy fibers terminate in synaptic contact with granule cells, and climbing fibers enter the molecular layers.

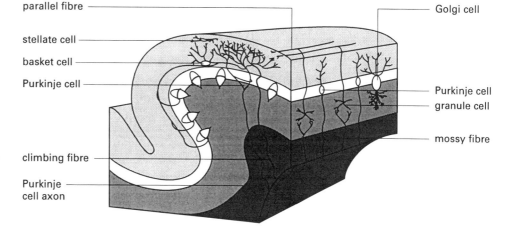

FIGURE 4.23

Drawing of a section through a part of the cerebellar cortex showing different layers and cell types

Studies have indicated that acetylcholine, acetylcholinesterase, and choline acetyltransferase are active in the cerebellum. Receptors for both aspartate and glutamate have been characterized and may be associated with the excitatory actions of the climbing fibers synapses onto the Purkinje cells. Both norepinephrine and serotonin innervate the cerebellar cortex. The basket, stellate, and Golgi cells have inhibitory functions in the cerebellum and use GABA as a transmitter. The Purkinje cells contain calbindin, a major calcium binding protein, and are also inhibitory using GABA as its transmitter to the deep cerebellar nuclei and the vestibular complex. Cyclic GMP is highly enriched in the cerebellum and its levels are altered by a variety of neuroactive agents such as ACh, amino acids, apomorphine, harmaline, and even temperature extremes. Only trace amounts of the various neuroactive peptides are found in the cerebellum, e.g., Substance P, somatostatin, and enkephalin. The cerebellum develops predominantly during the postnatal period and is a target site for various neurotoxicants: lead, methyl mercury, cadmium.

6.4   HYPOTHALAMUS

The hypothalamus is a control center for the visceral system and intergrates information for the visceral, limbic, and endocrine systems. Its activities are diverse and include influences on the secretory activity of the anterior lobe of the pituitary, regulatory influences on the autonomic nervous system, and has major connections with centers in the brainstem associated with autonomic regulation of respiration and cardiovascular activity. Interruption of the descending sympathetic pathway in humans causes ipsilateral Horner's syndrome: miosis (small pupil), ptosis (drooping eyelid), and an apparent enophthalmos (recession of the eyeball). The thermoreceptor-sensitive neurons in the hypothalamus maintain body temperature by vasodilation, vasoconstriction, sweating, shivering, etc. The anterior hypothalamus regulates heat loss and the posterior portion regulates heat conservation. The hypothalamus also regulates food and water intake, sleep, and influences sexual functions and emotional responses.

The neurons of the hypothalamus have been defined in clusters of systems. In the *magnocellular neurosecretory system*, the neurons contain neurohormones and their neurophysins, i.e. oxytocin, vasopressin, dynorphin, glucagon, cholecystokinin (CCK-8), and angiotensin II. Histamine and histamine receptors are very high in the supraoptic nucleus. The *parvocellular neurosecretory system* distributes releasing/releasing-inhibiting hormones e.g., luteinizing hormone-releasing hormone (LHRH), thyrotropin-releasing hormone (TRH), and somatostatin or growth hormone release-inhibiting hormone (SRIF). The hypothalamus also contains a distribution of neurons that concentrate peripheral hormones such as steroid hormones and thyroid hormones which have access to the brain and bind to intracellular receptor molecules. Gonadal steroids play a major role in the expression of certain behaviors and in the development of sexually dimorphic characteristics of the central nervous system[15]. In the hypothalamus, the arcuate nucleus (AN) surrounds the third ventricle of the brain and is one of the few areas that has a limited, if any, blood-brain barrier. The neurotransmitters associated with this region include dopamine, serotonin, acetylcholine. Peptides found in the AN are similar to the hormones in the pituitary and include adrenocorticotrophic hormone, a-melanoctye-stimulating hormone, b-lipotropic hormone, and b-endorphin. Given the neuroendocrine functions localized in the hypothalamus, alterations in this region are of major concern when dealing with environmental agents that may have estrogenic like activities, e.g. dioxin. Concern has been focused on the young and the long term consequenes of hypothalamic damage especially during the hormonally challenging time of puberty.

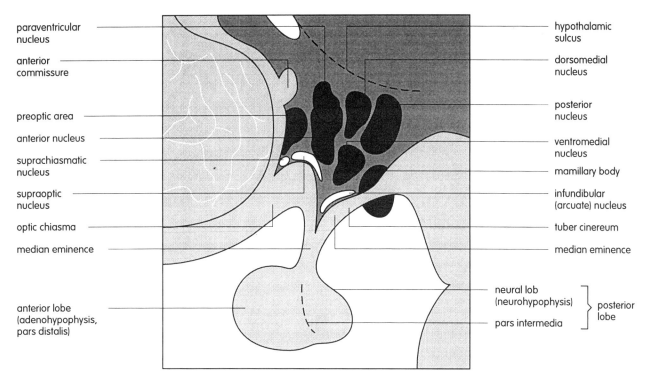

paraventricular nucleus

anterior commissure

preoptic area

anterior nucleus

suprachiasmatic nucleus

supraoptic nucleus

optic chiasma

median eminence

anterior lobe (adenohypophysis, pars distalis)

hypothalamic sulcus

dorsomedial nucleus

posterior nucleus

ventromedial nucleus

mamillary body

infundibular (arcuate) nucleus

tuber cinereum

median eminence

neural lob (neurohypophysis)

pars intermedia

posterior lobe

FIGURE 4.24

Midsagittal section showing the nuclei of the hypothalamus

6.5   BASAL GANGLIA

The term basal ganglia is used to refer to a group of closely connected structures including the caudate nucleus and putamen, together called corpus striatum, the globus pallidus, the substantia nigra, and the subthalamic nucleus. Together, the globus pallidus, and putamen form the lenticular nucleus. The caudate nucleus, globus pallidus and putamen are three large subcortical gray masses, derived from the telencephalon, anatomically interrupted by massive fiber bundles of the internal capsule.

The globus pallidus is the part of the basal ganglia considered to be concerned with somatic motor functions. The neurons have individual distinct morphologies classified as medium sized neurons that are either spiny or aspiny and large neurons that are also either spiny or aspiny. They are thought to contain acetylcholine (ACh), g-aminobutyric acid (GABA), taurine, aspartate, glutamate, and neuropeptides. The majority of the ascending fiber systems to the striatum join the medial forebrain bundle. The cerebral cortex projects onto the striatum in a highly ordered arrangement with fibers sent to the caudoputamen. The olfactory bulb and hippocampal formation send fibers to the ventral striatum. Subcortical afferents to the caudate nucleus and putamen come from the thalamus. Both the cortical and the hippocampal projections utilize the excitatory amino acid, glutamate and possibly aspartate, as neurotransmitters. The afferent pathways into the striatum contain biogenic amines (dopamine, serotonin, histamine, norepinephrine), amino acids (glutamate, aspartate, GABA), AChE, ACh, and neuropeptides. The main outputs of the striatum are the striatopallidal pathways to the globus pallidus, and the striatonigral pathway to the pars reticulata of the substantia nigra. The neurotransmitter, GABA, and the neuropeptide, enkephalin, are associated with each pathway. The basal ganglia control system modulates motor

153

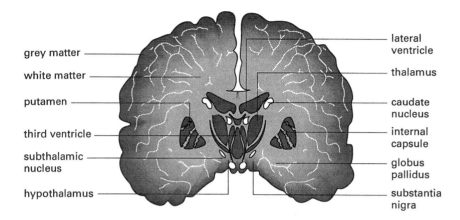

FIGURE 4.25

Section through the brain, showing the basal nuclei and thalamus
Putamen and globus pallidus together form the lentiform nucleus; the lentiform nucleus, the claustrum and caudate nucleus together form the basal nuclei.

activities and includes the putamen, globus pallidus, caudate nucleus, ventral anterior nucleus of the thalamus, subthalamic nucleus, substantia nigra, and their connections with the cerebral hemispheres and the diencephalon. Disruption to this control system can result in major alterations in motor function such as hypokinesia and hyperkinesia. Such human disorders are seen in Parkinson's disease, Huntington's chorea, Sydenham's chorea, and tardive dyskinesia.

6.6   SPINAL CORD AND PERIPHERAL NERVES

The spinal cord is that part of the CNS that is continuous with the medulla oblongata of the brainstem. It is located within the vertebral canal and extends from the foramen magnum to the lumbar region.

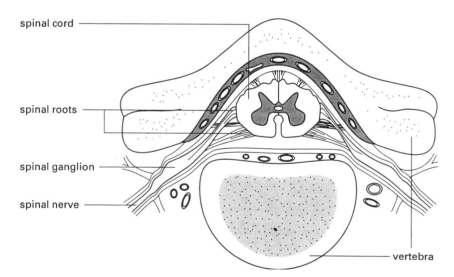

FIGURE 4.26

Cross section showing the relationship of the spinal cord and spinal nerve roots to a vertebra

TABLE 4.5
Diseases of the spinal cord

Disease of injury	Pathology
Physical injury spinal cord transection spinal cord hemisection	paraplegia unilaterial transverse lesion- Brown-Sequard syndrome
Syphilitic infection	degeneration in dorsal root ganglia and dorsal roots secondary degeneration in dorsal column
Syringomyelia	softening and cavitation of around central canal ofcervical spinal cord
Poliomyelitis	lesions in anterior gray horn lesions in ventral nerve roots
Amyotrophic lateral sclerosis	degeneration of motor cells in brain stem, cerebral cortex, spinal cord secondary degeneration of peripheral nerve axons

Although it is an unsegmented structure, areas that receive or give dorsal and ventral roots to a pair of spinal nerves are called segments. There are two basic divisions of the spinal cord: the gray matter in the central part which contains glia cells and cell bodies of efferent and internucial neurons; and the outer layer of white matter containing nerve fibers and glia. Serotonin, norepinephrine, epinephrine, and dopamine containing neurons project onto the spinal cord. Motor neurons of the ventral horn use acetylcholine as a transmitter. Peptides are localized within the unmyelinated primary afferents and interneuronal systems, and are not seen in the large myelinated axons. Lesions to the spinal cord can occur by physical injury, viruses, or can be chemically induced (Table 4.5). During development the neural groove undergoes a closure process. If this process is perturbed a number of malformations in the spinal cord and brain can be seen e.g. spina bifida.

The peripheral nervous system is made up of two divisions, somatic and autonomic. The somatic division is composed of sensory neurons of the dorsal root and cranial ganglia that innervate the skin, muscles, and joints, providing sensory information to the CNS about the muscle and limb position and about the environment outside the body. The axons of somatic motor neurons innervate skeletal muscles and project to the periphery. The autonomic division is composed of the motor system for the viscera, the smooth muscles of the body, and the endocrine glands. Within the autonomic division, the sympathetic system is responsible for the response of the body to stress, the parasympathetic systems conserve the body's resources and restore homeostasis, and the enteric nervous system controls the function of smooth muscle of the gut. It conveys its influence from the cerebral cortex, basal ganglia, and subcortical nuclei indirectly to the spinal cord via multisynaptic connections. The typical spinal nerve is formed by a dorsal and a ventral root coming together inside the vertebral canal and contains both sensory and motor fibers.

7   Processing of neural input

7.1   SENSORY SYSTEMS

There are four distinct somatic modalities: touch, proprioceptive sensations, pain, thermal sensations. In addition to these four main modalities, there are many submodalities and compound sensations. Mechanoreceptors and proprioception receptors are encapsulated and have axons with fast conduction velocities. Pain and temperature receptors are bare nerve endings and have axons with slower conduction velocities. Within each sensory receptor type, individual receptors are tuned to different levels of the stimuli. Even though each somatosensory modality is mediated by a separate class of receptors, all sensory information from the body is conveyed by the dorsal root ganglion neurons.

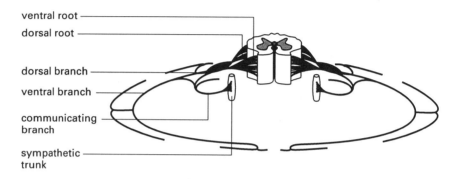

FIGURE 4.27

Schematic drawing of a spinal cord with a pair of spinal nerves

Sensory information conveyed to the brain is first relayed through the spinal cord and the different classes of primary afferent fibers take specific routes ending in different regions of the spinal cord. Sensory information is conveyed by two major ascending systems in the spinal cord: the dorsal column-medial lemniscal system relays tactile sensation and limb proprioception information, and the anterolateral system relays information concerning pain and temperature. Tactile and proprioception information traverses through a brainstem pathway to the thalamus and then to the anterior parietal cortex. Most of the axons of the anterolateral system terminate in the reticular formation of the pons and medulla, the midbrain, and the thalamus. From here the information is relayed to either the posterior parietal cortex or the association cortex. The input from the various submodalities interacts only once it reaches the somatic sensory areas of the cortex.

7.2   MOTOR SYSTEMS

The motor unit is composed of the cell body of the motor neuron, the axon of the motor neuron that projects in the peripheral nerve, the neuro-muscular junction, and the muscle fibers innervated by that neuron. Information concerning reflex or forced muscle stretching is conveyed by afferent (sensory) neurons to the CNS. The sensory information acts directly on motor neurons in the spinal cord or indirectly through interneurons to inhibit motor neurons that contract antagonist muscles and relay information to higher regions of the brain. The thalamus mediates motor function by conveying information from the cerebellum and basal ganglia to the motor regions of the frontal lobe — the primary motor cortex and higher-order motor areas. Axons from layer 5 of the primary motor cortex project directly to motor neurons in the spinal cord via the corticospinal tract. Like the ascending sensory system, the descending corticospinal tract on each side of the brainstem crosses to the opposite side of the spinal cord. These axons terminate on groups of motor neurons in the spinal cord that innervate specific limb muscles and on interneurons associated with the motor neurons. This tract of fibers is primarily concerned with controlling distal muscles responsible for precise movements. Other motor pathways originating in the brainstem nuclei mediate postural adjustments during movement. Ongoing behavior and movement is critically dependent upon motor integration, which is the role of the cerebellum and the basal ganglia. They both receive sensory input and the cerebellum is involved in the initiation and timing of movements, while the corpus striatum of the basal ganglia is involved in regulating the speed of movement. The motor cortex controls movement directly through projections to motor neurons, while the control mechanisms of the cerebellum and basal ganglia are mediated by brainstem and thalamic motor nuclei.

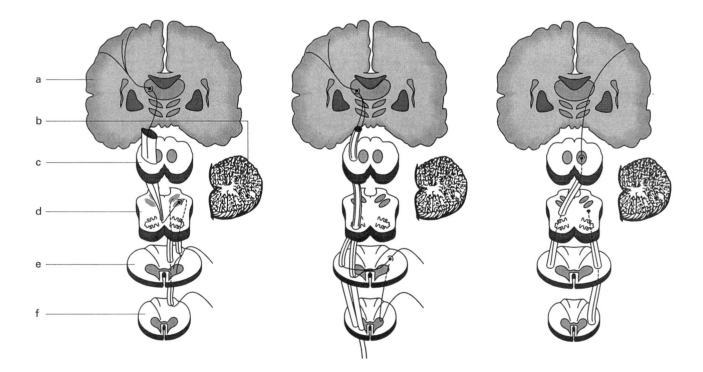

FIGURE 4.28
Schematic drawing of the sensory (left) and motor (middle-primary; right = secondary) systems in the spinal tract
a = cerebral hemisphere.
b = cerebellum.
c = mesencephalon.
d = medulla oblongata.
e + f = spinal cord.

8   Nervous system control of physiological functions

8.1   HOMEOSTASIS

Homeostatic processes include temperature regulation, feeding, and thirst, and correspond to motivational states. For each of these processes there is a control system that maintains each variable within a certain biological range and the adjustment is accomplished by the interactive functioning of inhibitory and excitatory mechanisms. The regulation of body temperature requires integration of autonomic, endocrine, and skeletomotor responses localized in the hypothalamus. Information is received from peripheral temperature receptors located throughout the body and central receptors concentrated in the hypothalamus. Although the limits of body temperature are usually maintained within a close range, this set point can be altered in pathogenic states such as is seen by fever produced by the action of pyrogens. The hypothalamus also plays a role in feeding behavior. Lesions to the ventromedial hypothalamic nuclei produce hyperphagia and obesity, whereas, lesions to the lateral hypothalmic nuclei produce aphagia. These effects on feeding are due to the alteration of several factors including sensory information, level of set point, hormonal balance, and the fibers of passage through these distinct regions. Water balance is also regulated by the hypothalamus with the two main physiological variables of tissue osmolality and vascular volume being major controlling factors. The osmoreceptor (sodium-level receptor) in the hypothalamus can be directly acted upon. Angiotensin and baroreceptor afferents play a major role in regulating drinking with angiotensin acting on the subfornical organ.

8.2.   MOTIVATION

The motivational system regulates the initiation of a voluntary movement by acting on the somatic motor system in the brain and the autonomic nervous system. The hypothalamus serves as the main control center for the autonomic nervous system, and as such it regulates autonomic output and endocrine function, and is responsive to varied behavioral stimuli. The limbic system is also a major component of the motivational system. Most behavioral acts require the integration of all three major functional systems of the brain, the sensory, motor, and motivational systems. Each of these systems contains synaptic relays and is composed of various subdivisions. Stimulation of various brain sites has been found to be reinforcing, possibly due to evoking a drive state and activation of systems normally activated by a reinforcing stimulus. Hypothalamic sites have been found to be particularly effective for stimulus reinforcement. It has been proposed that addictive drugs like cocaine may induce euphoria via the dopaminergic system in the nucleus accumbens. Much of the existing evidence suggests a critical role for dopamine in determining motivation state; however, any complex behavioral function is unlikely to rely on any one neurotransmitter system.

8.3   LEARNING AND MEMORY

The biological basis and anatomical localization of learning and memory have been the driving force behind immense amount of research for a number of years. Some of the basic general information has come from examining elementary forms of reflexive learning: habituation, sensitization, and classical conditioning, and has indicated that most of these modifications involve changes in the effectiveness of specific synaptic connections. In habituation an animal learns about the properties of a stimulus following repeated presentation and involves depression of synaptic transmission. Sensitization requires the animal to learn about the properties of a noxious stimulus and to remember to respond more effectively to other stimuli. This form of learning has been associated with an enhancement of synaptic transmission. Classical conditioning requires that associations between two stimuli are learned and responded to appropriately. In all learning situations there is a need for maintaining memory for either a short time (short-term memory) or over a long period (long-term memory) depending on the continuing demands of the stimulus presentation. It is thought that long-term memory requires the synthesis of new proteins and the establishment of new synaptic connections. In the mammalian system, the hippocampus has been shown to be important for memory storage. Also, the neurons in the hippocampus display synaptic plasticity that would be required for associative learning. The hippocampus has now been intensively studied as a target site responsible for learning and memory. A brief high-frequency train of stimuli delivered to the afferent pathways to the hippocampus will facilitate the excitatory synaptic potential in postsynaptic hippocampal neurons that can last for weeks. This phenomenon of long-term potentiation has been used as a model of learning and memory to examine the associated structural plasticity and biochemical alterations in the hippocampus. In addition to the establishment of new synaptic connections, the signal transduction pathways currently appear to play a major role in the over-all process of learning and memory. Any toxicant or pharmacological agent that alters the normal functioning of the limbic system or the plasticity of the nervous system can have detrimental effects on the ability of the animal to learn new tasks and to remember responses critical for normal interactions with the external environment.

9   Neuroanatomy and aging

Not all subgroups of the population respond similarly to any environmental or pharmacological insult. Not only is the system vulnerable during its development but it is also at high risk when in any compromised state such as age, nutritional status, drug status, mental state, and occupational environment. For example, neurotoxic substances may exacerbate existing neurological or psychiatric disorders in a population. With aging, the level of risk for a number of health-related factors increases as well as the risk for toxic responses to environmental substances. Even though the nervous system is unable to replace lost or damaged cells following neurogenesis, the system is able to display considerable responsiveness or plasticity. To adjust for this inability to regenerate, many of the cells of the nervous system are present in excess to allow for the existence of a buffer in the face of injury. During development, the nervous system uses plasticity and redundancy to insure proper neural network connections. With the process of aging, comes a decreased ability of the nervous system to respond to adverse events or to compensate for biological, physical, or toxic effects. Damage to the nervous system may evoke an alteration of connectivity between the surviving neurons allowing for a limited amount of functional adjustment in order to compensate for the damage. A response of this nature changes the normal connectivity of the brain and can have profound consequences for neurological, behavioral, and related body functioning. The aging process results in nerve cell loss, neurofibrillary tangles (abnormal accumulation of certain filamentous proteins), and neuritic plaques (abnormal clusters of proteins and other substances near synapses).

One of the major age-related diseases of concern today is Alzheimer's disease. This disease is characterized by a well-defined pathogenetic cascade that may be the result of distinct gene defects or unknown environmental factors. In this disease there is a loss of neurons, medium and large-sized pyramidal cells, and the presence of intraneuronal neurofibrillary tangles, extracellular deposits of amyloid filaments surrounded by altered neuritic processes, and glia (senile plaques). These pathological changes are seen in the hippocampus, amygdala, and other limbic structures. Alzheimer's induced dysfunction and loss of basal forebrain cholinergic neurons and their cortical projections is seen as a marked decline in choline acetyltransferase and acetylcholinesterase activities. Other sites of injury include noradrenergic and serotonergic cells in the brainstem, cortical cells producing somatostatin or corticotropin-releasing factor, and glutamate, GABA, substance P, or neuropeptide Y containing neurons.

Not only does susceptibility of the nervous system to perturbation increase with aging, but manifestations of perturbations induced earlier may become evident. The clinical manifestation of neurodegenerative disorders may have a contributing component of past exposures to environmental chemicals. By depleting neuronal reserves at an earlier age, various agents may contribute to Alzheimer's disease, Parkinson's disease, or amyotrophic lateral sclerosis. It has been suggested that the incidence of Parkinson's disease has been correlated with exposure to pesticides, especially pyridines such as the herbicide paraquat[4]. Individuals occupationally exposed to organic solvents or metal vapors have reported symptoms of Alzheimer-type syndromes[14].

10   Summary

The nervous system plays a major role in coordinating and executing the activities of the entire organism. Given the complexity of the nervous system in its chemical, metabolic, structural, and physiological basis any alterations, however subtle, induced by environmental or pharmacological chemicals can have long-term detrimental effects on the integrative functioning of the organism. Although much is known about the basic structural components and pathways in the brain, the mapping of the brain by structure, chemical, and function is a continuous process. The integrity of the system is dependent upon maintaining ionic balance across membranes, cell-cell contact and communication between both neurons and glial cells, neurotransmitter synthesis and release, and the integrity of the blood-brain and blood-nerve barriers. Identification and elucidation of mechanisms associated with neuroactivity and neurotoxicity of pharmaceutical agents is usually dependent upon the known therapeutic site of action of a drug. Identifying such sites of action for environmental agents is complicated with the non-specificity and toxicity of various chemicals. For both drugs and environmental chemicals, the concern exists for identifying susceptible populations, e.g., the developing organism, the aged, the mentally compromised, or the population with other ongoing disease processes. Each of these populations may be differentially vulnerable to chemically induced neurotoxicity while the "average population" may not demonstrate signs of adverse reactions. Although examples of environmental chemicals and drugs of abuse are given to illustrate how various processes are altered and the resulting neurotoxicity, it must be remembered that the current amount of information available is limited. Technical advancement, identification, and further understanding of basic nervous system processes will allow for further examination of the mechanisms associated with neurotoxicity. The following chapters will address specific aspects of neurotoxicity: structural, neurochemical, and behavioral. In all cases the complex, integrative, and interdependent nature of the nervous systems needs to be taken into consideration in interpreting the consequences of chemical exposure.

References

1. Annau, Z. and Eccles, C.U. (1986) Prenatal exposure. In: Annau, A. (Ed.) *Neurobehavioral Toxicology*, Johns Hopkins University Press, Baltimore, 1986, pp 153-169.
2. Amaducci, L., Forno, K.I., Eng, L.F. (1981) Glial fibrillary acidic protein in cryogenic lesions of the rat brain. *Neurosci. Lett.* **21**: pp 27-32.
3. Balaban, C.D., O'Callaghan, J.P., Billingsley, M.L. (1988) Trimethyltin-induced neuronal damage in the rat brain: comparative studies using silver degeneration stains, immunocytochemistry and immunoassay for neuronotypic and gliotypic proteins, *Neuroscience* **26**: pp 337-361.
4. Barbeau, A., Roy, M., Bernier, G., Gampanelia, G., Paris, S. (1987) Ecogenetics of Parkinson's disease; prevalence and environmental aspects in rural areas, *Can. J. Neurol. Sci.* **14**: pp 36-41.
5. Betz, A.L. and Goldstein, G.A. (1978) Polarity of the blood-brain barrier: neutral amino acid transport into isolated brain capillaries, *Science* **202**: pp 225-227.
6. Bisby, M.A. (1980) Retrograde axonal transport. *Adv. Cell Neurobiol.* **1**: pp 69-117.
7. Brightman, M.W. and Reese, T.S. (1969) Junctions between intimately apposed cell membranes in the vertebral brain, *J. Cell Biol.* **40**: pp 648-677.
8. Choi, D.W. (1988) Glutamate neurotoxicity and diseases of the nervous system, *Neuron* **1**: pp 623-634.
9. Cowan, W.M. (1970) Anterograde and retrograde transneuronal degeneration in the central and peripheral nervous system. In: Nauta, W.J.H. and Ebbesson, S.O.E. (Eds.) *Contemporary Research Methods in Neuroanatomy*. Springer, Berlin, pp 217-251.
10. Cushner, I.I.M. (1981) Maternal behavior and perinatal risks: alcohol, smoking, and drugs, *Annu. Rev. Pub. Health* **2**: pp 201-218.
11. Davison, A.N. and Dobbing, J. (1966) Myelination as a vulnerable period in brain development, *Br. Med. Bull.* **22**: p 40.
12. DeCaprio, A.P., Strominger, N.L., Weber, P. (1983) Neurotoxicity and protein binding of 2,5-hexandione in the hen. *Toxicol. Appl. Pharmacol.* **68**: pp 297-307.
13. Ellisman, M.H. and Lindsey, J.D. (1983) The axoplasmic reticulum within myelinated axons is not transported rapidly. *J. Neurocytol.* **12**: pp 393-411.
14. Freed, D.M., Kande, I.E. (1988) Long-term occupational exposure and the diagnosis of dementia. *Neurotoxicology* 9: pp 391-400.
15. Gorski, R.A. (1980) Sexual differentiation of the brain. In: Krieger, D.T., Hughes and H.J.C. (Eds.) *Neuroendocrinology*. Sinauer, Sunderland, MA, pp 215-222.
16. Harry, G.J. (1992) Acrylamide-induced alterations in axonal transport: biochemical and autoradiographic studies. *Mol. Neurobiol.* **6**: pp 203-216.
17. Harry, G.J., Goodrum, J.F., Bouldin, T.W., Toews, A.D., Morell, P. (1989) Tellurium-induced neuropathy: metabolic alterations associated with demyelination and remyelination in rat sciatic nerve. *J. Neurochem.* **52**: pp 938-945.
18. Hill, R.M. and Tennyson, L.M. (1986) Maternal drug therapy: effect on fetal and neonatal growth and neurobehavior. *Neurotoxicology* 7: pp 121-140.
19. Hoffman, P.N. and Lasek, R. (1975) The slow component of axonal transport identification of major structural polypeptides of the axon and their generality among mammalian neurons. *J. Cell Biol.* **66**: pp 351-366.
20. Jacobson, S. (1963) Sequence of myelination in the brain of the albino rat. *J. Comp. Neurol.* **121**: p 5.
21. Kosik, K.S. and Salkoe, D.J. (1983) Experimental models of neurofilamentous pathology. In: Marotta, C.A. (Ed.) *Neurofilaments*. University of Minnesota Press, Minneapolis, pp 155-195.
22. Lapadula, D.M., Sywita, E., Abou-Donia, M.B. (1988) Evidence for multiple mechanisms responsible for 25-hexandione-induced neuropathy. *Brain Res.* **458**: pp 123-131.
23. Lewis, P.R. and Shute, C.C.D. (1967) The cholinergic limbic system: Projections to the hippocampal formation medial cortex nuclei of the ascending cholinergic reticular system and the subfornical organ and supraoptic crest. *Brain* **90**: pp 521-540.
24. O'Keefe, J. and Madel, L. (1978) *The Hippocampus as a Cognitive Map*. Oxford University Press, New York.

25. Pearson, D.T. and Dietrich, K.N. (1985) The behavioral toxicology and teratology of childhood: models methods and implications for intervention *Neurotoxicology* **6**: pp 165–182.

26. Peters, A., Palay, S.L., Webster, H. deF. (1970) *The Fine Structure of the Nervous System.* Harper Row, New York.

27. Raine, C.S. (1984) Morphology of myelin and myelination. In: Morell, P. (Ed.) *Myelin*, 2nd edition. Plenum Press, New York.

28. Sayre, L.M., Autilio-Gambietti, L., Gambietti, P. (1985)Pathogenesis of experimental giant neurofilamentous axonopathies: a unified hypothesis based on chemicalmodification of neurofilaments. *Brain Res.* **10**: pp 69–83.

29. Webster, H. deF. (1975) Peripheral nerve structure. In: Hubbard, J.I. (Ed.) *The Peripheral Nervous System.* Plenum Press, New York.

30. Webster, W.S. and Valois, A.A. (1981) The toxic effects of cadmium on the neonatal mouse CNS. *J. Neuropathol. Exp. Neurol.* **40**: p 247.

31. Wiggins, R.C. and Fuller, G.N. (1978) Early postnatal starvation causes lasting brain hypomyelination. *J. Neurochem.* **30**: 1231.

32. Willard, M., Cowan, W.M., Vagelos, P.R. (1974) The polypeptide composition of intra-axonally transported proteins evidence for four transport velocities. *Proc. Natl. Acad. Sci. USA* **71**: pp 2183-2187.

Additional reading

Aschner, M. and Kimelberg, H.K. (1995) *The Role of Glia in Neurotoxicity.* CRC Press, Boca Raton, FL.

Harry, G.J. (1991) *Developmental* Neurotoxicology. CRC Press, Boca Raton, FL.

Jacobson, M. (1991) *Developmental Neurobiology*, 3rd edition. Plenum Press, New York.

Kandel, E.R., Schwartz, J.H., Jessel, T.M. (1991) *Principles of Neural Science*, 3rd edition. Elsevier, New York.

Morell, P. (1984) *Myelin*, 2nd edition. Plenum Press, New York.

Siegel, G., Agranoff, B., Albers, R.W., Molinoff, P. (1994) *Basic Neurochemistry*, 5th edition. Raven Press, New York.

Spencer, P.S. and Schaumburg, H.H. (1980) *Experimental and Clinical Neurotoxicology.* Williams & Wilkins, Baltimore, MD.

Tilson, H.A., Mitchell, C.L. (1992) *Neurotoxicology.* Target Organ Toxicology series, Hayes, A.W., Thomas, J.A., Gardner, D.E. (Eds.) Raven Press, New York.

Contents Chapter 5

Behavioral toxicological effects and pathology

Chapter 5

Behavioral toxicological effects and pathology

Michel Panisset and Donna Mergler

1   Introduction

Knowledge of the mind-altering capabilities of certain chemicals goes as far back as recorded history. Indeed, natural, and more recently, synthetic chemicals have been used by humans to induce mind states of euphoria, hallucinations, and apparent well-being. Many environmental and occupational neurotoxic substances have properties similar to those of, alcohol and recreational drugs, initially producing excitatory effects; as the dose increases in quantity or in time, these same neurotoxic substances depress the nervous system. Children are familiar with Lewis Carroll's febrile "mad-hatter" in Alice in Wonderland, who was exposed to mercury vapors in the manufacture of felt hats. Ramazinni in "De morbis artificum diatriba", published in 1700, notes that the artists of Modena, Italy, where he lived and worked, manifested similar behavior and nervous system deficits that he attributed to their exposure to a wide variety of chemical agents[86].

Modifications in nervous system functioning brought about by exposure to neurotoxic substances are often reflected by changes in behavior. These changes can take the form of an immediate, acute reaction to intoxication, or appear insidiously and gradually when exposure occurs, even at low levels, over a long period of time. Continued interference with molecular and cellular processes causes neurophysiological and psychological functions to undergo slow alterations, which in the early stages may go unseen due to nervous system plasticity, compensatory adaptive processes and learning. For most neurotoxic chemicals, the level at which no adverse effect will occur over long term exposure is still unknown.

The severity of acute or chronic effects depends on exposure dose, which includes both the quantity and duration of exposure, as well as factors such as genetic susceptibility, age, health status, and lifestyle. Behavioral changes associated with exposure to neurotoxic substances can be described on a continuum (Figure 5.1).

Initial alterations in the nervous system are not necessarily accompanied by functional disorders and may be reversible. Early symptoms are often non-specific, including mood lability, memory and concentration difficulties, headaches, blurred vision, excessive tiredness, dizziness, slowness, tingling sensation in hands or feet, and loss of libido. With increasing dosage, changes are observed in motor, sensory, and cognitive functions in the absence of clinical evidence of abnormality. As nervous system damage progresses, symptoms and signs become more apparent so individuals may seek medical attention. Finally, impairment may become so severe that a clear clinical syndrome is present.

Different types of studies have provided an overall portrait of nervous system changes along this continuum. In experimental set-ups, healthy human volunteers are placed in air-tight chambers, where they are exposed to specific levels of toxicants; these controlled human chamber studies have served to document acute neuropsychological and psychophysical alterations in healthy subjects. Surveys of occupationally and environmentally exposed active populations have provided information on early behavioral changes associated with long term, low exposures, while clinical reports and epidemiological studies have described the more lasting, severe changes associated with intoxication.

Detection of early nervous system changes in humans resulting from exposure to neurotoxic substances, and the screening and diagnosis of neurotoxic substance-related illnesses require different, but complementary approaches. In the initial stages, when clinical signs and symptoms are absent, indications of subtle nervous system alterations may be detected by examining exposed populations using sensitive measurements of dysfunction. Although these early changes are not necessarily predictors of eventual illness, they often reflect diminished physical, mental, or emotional well-being. Their consequences are important not only on an individual level, but for society. For example, the well-documented association between blood lead levels in early life and reduced intellectual ability (see Chapter 12) is not only meaningful for the particular person, but also for the exposed group, within which will be a higher level of retarded children and fewer very bright children. Since exposure is often hand-in-hand with poverty[22], it makes recognition doubly difficult. Surveillance of early indicators of nervous system alterations in populations with potential exposure to neurotoxic substances may prove particularly useful for initiating pro-active preventive intervention, as well as reducing the risk of eventual illness.

Clinical diagnosis relies on a constellation of signs and symptoms, coupled to the medical and exposure history for an individual; etiologies other than exposure must be systematically ruled out. For the surveillance of early dysfunction, the group portrait of dysfunction is important in establishing a relationship with exposure. Most often, the overall pattern of dysfunction observed for the exposed group as a whole will be similar to the clinical pattern of impairment observed in the disease. It is somewhat like summing up mild alterations among all of the individuals who make up the group to produce a global picture of what is happening to the nervous system. The pattern or profile of the overall early response provides an indication of the specificity and the type of action of the particular neurotoxic substance or mixture.

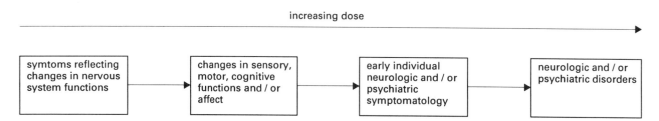

FIGURE 5.1

Progressive deterioration of nervous system function with increasing dose, encompassing both concentration and duration of exposure

Initial reactions are insidious and manifest as non-specific symptoms, followed by changes in affect sensory, motor, and cognitive performance, which can be measured in exposed populations, using sensitive neurofunctional test batteries. With increasing severity of nervous system damage, subclinical signs and symptoms appear in affected individuals, with the final stage presenting as a full-blown neurological or psychiatric syndrome.

This chapter describes behavioral manifestations along the continuum from well-being to pathology. In the first part, manifestations and assessment of early changes associated with low dose exposure are discussed. Then, factors which may influence susceptibility to neurotoxic substances are reviewed, and finally, specific neurotoxic syndromes are examined.

2   Early indicators of nervous system dysfunction

As seen in the previous chapter, the nervous system is complex and while certain areas may be affected by specific chemicals, other areas are sensitive to the action of a large number of toxic agents. Animal and *in vitro* studies have demonstrated that neuronal components, such as the membrane channels, neurotransmitter production, release and receptor mechanisms, axonal terminals, the myelin sheath, sensory receptors, and the neuromuscular junction constitute potentially vulnerable and often multiple targets for different substances. Recent studies indicate that astrocytes and other glial cells may also be affected by toxic chemicals. Neurotoxic damage may also occur secondarily to toxic effects on blood vessels or to hypoxia. The initial behavioral manifestations, which result from the action of low-levels of toxic substances on one or more neural nervous system targets, are usually widespread and non-specific, and can be detected by measuring subtle changes ir performance and affect.

2.1   PSYCHOMOTOR AND COGNITIVE FUNCTIONS

Over the past 30 years research efforts have focused on measuring the subtle changes in nervous system functions that are present in the early stages of intoxication. The first neurobehavioral test battery for use in worksite studies was developed by Helena Hanninen[52], a pioneer in the field of neurobehavioral deficits associated with toxic exposure. Her studies of workers who had no obvious clinical abnormalities but were exposed over a long period to carbon disulfide demonstrated deficits in psychomotor and cognitive performance when compared to unexposed control groups [53]. Today, there an extensive literature on early neurobehavioral deficits associated with environmental and occupational exposure to a wide variety of neurotoxic substances, and a number of textbooks, covering different aspects of the subject have been published (see Further Reading).

Controlled human chamber experiments, described above, and field studies comparing performance between exposed and non-exposed persons, and analyzing neuro-outcomes with respect to the degree of exposure have shown that certain tasks are particularly sensitive to one or several neurotoxic agents. These include reaction time, hand-eye coordination, short-term memory, visual and auditory memory, attention and vigilance, manual dexterity, vocabulary, switching attention, motor speed, and hand steadiness. Table 5.1, adapted from Anger, lists the type of early cognitive and psychomotor deficits that have been consistently observed with some of the most common neurotoxic substances[6]. Coding, intelligence, memory, spatial relations, coordination, and speed appear to be the behavioral functions most frequently affected by chemical exposures.

Toxic interference in the neuronal circuitry required to perform these tasks translates most often into depression of nervous system functions and a slowing down of integrative processes. However, in some cases, exposure may initially speed up or enhance processes. With increasing dose, over time and/or concentration, the opposite effect occurs and nervous system functions progressively diminish. This is probably the case for most recreational drugs in which enhanced sensory sensations or feelings of euphoria are sought. Toluene intoxication, which can occur through glue-sniffing or in industrial settings, is a good example of this type of response: initial reaction to toluene is excitatory, but over time, or with higher concentrations, the nervous system is depressed[13]. Other neurotoxic substances that may produce this inverted U-shape response include alcohol, nicotine, volatile organic solvents, lead, and methyl mercury[36]. Manganese, which is an essential element, is an interesting substance in this respect: insufficient manganese intake results in deficiency; optimum performance is obtained with a certain amount of manganese, too much manganese is toxic, associated with reduced performance; while very high manganese exposure can produce a Parkinson-like syndrome, called manganism[75, 77].

TABLE 5.1

Consistent behavioral effects of worksite exposures to some leading neurotoxic substances

(Adapted from Anger, W.K. (1990) *Neurotoxicology* **11**: p 629.)

Multiple solvents	Carbon disulfide	Styrene	Organo-phosphates	Lead	Mercury
acquisition categori-zation coding concept shifting	coding			acquisition coding	coding
				distractibility	
intelligence memory spatial relations vigilance	intelligence memory spatial relations vigilance	memory	intelligence memory	intelligence memory spatial relations vigilance vocabulary	intelligence memory
coordination speed	coordination speed	coordination speed		coordination speed	coordination speed

2.2 SENSORY FUNCTIONS

Sensory systems include a receptor organ, afferent pathways, and central integrative networks and neurotoxic substances may affect one or all of three of them. Moreover, some receptor organs are in contact with neurotoxic substances. Olfactory receptors are the most evident since airborne pollutants pass directly over the receptor cells. Somatosensory receptors may be affected by percutaneous chemicals, which cross the skin's protective barrier and seep into the underlying tissues. Even the retina may be directly affected by substances absorbed through the mucous membranes of the eye. Quantitative measures of sensory functions have been used in a systematic fashion to further examine early nervous system dysfunction associated with exposure to toxic substances[76].

The visual system is composed of a complex network of interconnecting neurons, whose receptor organ, the retina, is one of the most intricate sensory organs; changes in retinal transmission, or in the optic pathways, can affect visual information processing. Subtle alterations in visual functions may prove particularly useful for the surveillance of early neurotoxic impairment and in determining etiology. In population studies visual deficits have been associated with exposure to various neurotoxic substances in the absence of visual acuity loss[74]. These include acquired dyschromatopsia, reduced contrast sensitivity threshold at intermediate spatial frequencies, and restricted visual fields.

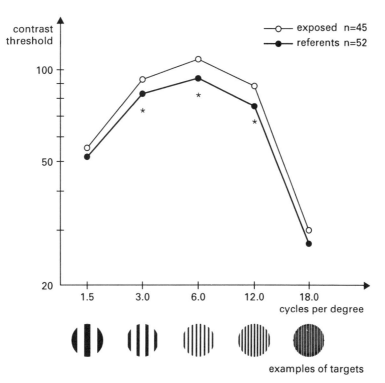

FIGURE 5.2

Mean Group Near Visual Contrast Sensitivity profile, assessed with Vistech 6000 cards, for microelectronics workers with a history of exposure to neurotoxic substances and a reference group of similar age, educational level, socio-economic status, and geographical location

Asterisks indicate significant differences between the two groups (Student's *t*-test; * $p < 0.05$). The circles below, are representative of the targets with their corresponding spatial frequencies. The threshold level indicates the level at which the subject cannot distinguish between the darker and lighter bars.

Although the underlying neuropathological action of toxic substances on visual pathways has not been entirely elucidated, different hypotheses have been advanced including changes in retinal dopamine, a neuromodulator of the light response.

Figure 5.2 illustrates how visual functions can provide information on the effects of exposure to neurotoxic substances. The detection of lighter and darker contrasts is an important element in visual coding. Measurement of the contrast sensitivity threshold to large objects (low spatial frequencies) and to small objects (high spatial frequencies) helps to assess the integrity of neuro-optic pathways. In contrast sensitivity tests, objects such as the sinusoidal grating bars illustrated in Figure 5.2 are presented with diminishing contrast between the object and the surround. The threshold of contrast is determined and plotted on a near visual contrast sensitivity profile. Visual contrast sensitivity loss has been observed in the absence of loss of visual acuity among patients who have cerebral lesions, optic nerve dysfunction caused by glaucoma and demyelinating, as well as retinal disease. Visual contrast sensitivity profiles, coupled to other ophthalmological data, can be used for individual diagnosis. However, for detecting the effects of neurotoxic exposure, group mean profiles are useful.

The mean near visual contrast sensitivity profiles presented in Figure 5.2 are from a study of microelectronics assembly workers (exposed group)[46], who have a history of exposure to a mixture of a large number of neurotoxic

169

chemicals, including such substances as 1-1-1 trichloroethane, toluene, and trichloroethylene, compared to the mean profile of persons of similar age, educational status, gender, and geographical location, with no history of exposure (reference group); all had normal visual acuity. It can be seen that while the mean differences between the two groups for the lowest spatial frequency (1.2 cycle per degree) and the highest (18 cycles per degree) are not statistically significant, the mean values for the two groups differ significantly at the intermediate spatial frequencies (3, 6 and 12 cycles per degree). These results are consistent with those reported by Broadwell et al. for another group of microelectronics workers with similar exposures[20].

Loss of somatosensory functions is an important sign of peripheral neuropathy and has been used for a long time in clinical diagnosis. The "glove and sock" pattern of paresthesia is a common presentation in severe acute or chronic intoxication. In the search for quantitative indicators of early nervous system alterations associated with exposure, a certain number of instruments have been developed to measure vibrotactile perception threshold[76]. These instruments are basically of similar design and include a vibrating stimulus of varying amplitude; most provide vibrations at a fixed frequency, although in more sophisticated devices the frequency of stimulus can be varied. Elevated vibrotactile perception thresholds, in the absence of manifest illness, have been reported in a number of studies of exposure to neurotoxic substances, including mercury[3], mixed solvents, and pesticides[18, 38, 70].

TABLE 5.2

Sensory effects of worksite exposures to some leading neurotoxic substances

Multiple solvents	Carbon disulfide	Styrene	Organo-phosphates	Lead	Mercury
chromatic discrimination contrast sensitivity	chromatic discrimination	chromatic discrimination			chromatic discrimination contrast sensitivity visual fields
vibrotactile perception threshold odor perception threshold odor identification			vibrotactile perception threshold	odor identification	

As mentioned above, the functioning of the olfactory system can also be very important as an early indicator for nervous system dysfunction. Although as many as 120 substances have been reported to produce some form of temporary or permanent olfactory dysfunction, it is relatively recently that valid and reproducible psychophysical tests, designed to assess odor identification and olfactory perception threshold, are being used to assess early neurotoxic alterations[76]. Tests of odor identification[39] and olfactory perception threshold are the most commonly used to quantify early changes in olfactory functioning[45]. The particular vulnerability of the olfactory system to airborne toxins results from the anatomical location of receptor cells and possibly from the highly active transneuronal transport mechanisms within the olfactory nerve cells[54]. Smell hyposensitivity has been associated with a number of neurotoxic substances[5]. There are also reports of cacosmia (altered smell) and odor hypersensitivity among workers with long-term exposure to manganese [77, 90]. Although chemically induced, olfactory loss may be secondary to upper respiratory tract blockage. It can also result from neurotoxic injury to olfactory mucosa and neural receptors, and/or central lesions. For example, manganese-related hyperosmia may be related to the olfactory bulb's affinity to this metal; Bonilla and co-workers showed that the olfactory bulb has one of the largest concentrations of manganese in the human brain[17].

Table 5.2 summarizes results from studies of visual, somatosensory, and olfactory alterations associated with certain neurotoxic exposures; there is also evidence that deficits in hearing[66], temperature sensitivity, and vestibular functions may also be important indicators of early neurotoxicity[7,15].

2.3   AFFECT AND PERSONALITY

Altered mood states are often the earliest indicator of chemically induced changes in nervous system functioning. Euphoria, lightheadedness, irritability, hyperactivity, sudden mood changes, excessive tiredness, feelings of hostility, anxiousness, depression, and tension are among the mood states most often associated with neurotoxic exposures. Although in the early stages these symptoms are usually not sufficiently severe to interfere with work, they do reflect diminished well-being and affect one's capacity to fully enjoy family and social relations.

Population based studies on neurotoxic effects of occupational and environmental pollutants usually include measures of affect or personality disturbance, in the form of symptoms questionnaires, mood scales, or personality indices. Anger lists 32 different tests of affect or personality that have been used in a total of 63 studies on a wide variety of workplace chemical exposures, including mercury, lead, specific organic solvents and mixtures, as well as pesticides[6].

In a study of mood among active workers exposed to styrene in reinforced plastics manufacture, Sassine et al. showed significant correlations between the level of end-shift urinary mandelic acid (a metabolite of styrene that is used as a biological indicator of exposure) and the scale scores for aggressiveness and fatigue on the Profile of Mood States (POMS) Test[91]. The POMS test provides quantitative measures for six scales: Tension, Depression, Anger, Vigor, Fatigue, and Confusion[73]. Figure 5.3 shows the mean scores for the profile of POMS scale scores for the styrene exposed workers from the reinforced plastics plants that have with high levels of end-shift mandelic acid (> 0.1 mmol/mmol creatinine) as compared to those with lower levels (≤ 0.1 mmol/mmol creatinine). It can be seen that although the standard deviation overlaps between those with the higher and lower levels of the biological indicator of exposure, on a group basis, the mean scores are significantly higher on the scales of Tension, Anger, Fatigue, and Confusion.

Animal experiments have shown that exposure to styrene, an organic compound often used in the manufacture of plastics, is associated with changes in brain catecholamines[2,79,80]; however, it is not clear whether or not this type of action would explain the mood states of the more highly exposed workers.

Although the mood states of the more highly exposed styrene workers were not necessarily in the clinical range (Figure 5.3), these changes in temper are a reflection of the loss of emotional well-being. It has been our experience that persons exposed to neurotoxic substances are often ostracized by their colleagues or employers due to their erratic and irritable emotional state.

3   From early indicators to pathology

From a public health point of view, detection of early indicators of nervous system dysfunction among groups exposed to toxic substances in workplaces, from environmental pollution, or from recreational drugs can lead to preventive intervention. If exposure increases, in quantity or over time, certain individuals will begin to show signs and symptoms that require clinical care. Factors such as age, genetic susceptibility, health status, and lifestyle can influence the prognosis by speeding up or slowing down the process of nervous system deterioration along this continuum.

3.1   AGE

In studies of the effects of neurotoxic agents on the nervous system, age is an important factor to take into account. Toxic substance exposure may be manifested by different syndromes depending on if the exposed person is in childhood or in adulthood. An example of this is the effect of lead, which gives mainly a neuropathy in the adult, whereas it gives an encephalopathy in children. Age may also be a factor of susceptibility to a specific agent. An example of this is MPTP (*N*-methyl-4-phenyl-1,2,3,6-tetrahydropyridine), a toxic agent that has been discovered to cause a syndrome similar to Parkinson's disease (PD) in humans, and has been extensively used thereafter in producing a model to study PD in animals. When MPTP is injected intravenously in to young monkeys, neurons of the median part of the substantia nigra are selectively destroyed and dopamine subsequently shows a decrease in the striatum. In older animals the same treatment induces a further destruction of

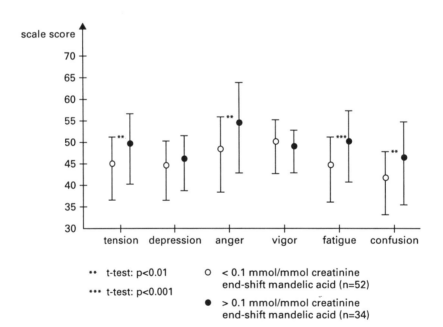

FIGURE 5.3

Scale scores (mean and standard deviation) of mood assessed with Profile of Mood State (POMS) for workers from reinforced plastics plants, with respect to levels of exposure, estimated by end-shift urinary mandelic acid. Workers did the same job throughout the week and had similar exposures from day to day. Higher scores indicate greater tension, depression, anger, fatigue, and confusion, as well as greater vigor.

The asterisks indicate significant differences between the two groups. The group of workers with higher end-shift urinary mandelic acid report significantly higher levels of tension, anger, fatigue, and confusion.

172

neurons of the locus ceruleus. Of the monkeys 15 years of age and older, 82% had neuronal loss in both the substantia nigra and at the locus ceruleus, whereas only 27% of the monkeys aged 5 years or less had similar lesions[44]. Moreover, the amount of Lewy bodies found in the brain appears to vary with age. Lewy bodies are the pathological markers of Parkinson's disease and consist of intraneuronal cytoplasmic inclusion bodies. Similar bodies can be induced in the MPTP-treated animal PD model. These "pre-Lewy bodies" as they are often called are found in more locations in the older animals compared to the younger ones, including: on top of the ventro-medial substantia nigra, the locus ceruleus, the raphe nuclei near the dorsal motor nucleus of the vagus, the nucleus basalis of Meynert, and the periamygdalian cortex. Irwin et al. showed that young and old mouse brains were relatively resistant to the effects of MPTP, whereas brains of young adults were more sensitive, giving an age versus MPTP effect inverted U curve[59]. One possible explanation for this observation is that the activity of monoamine oxidase-B (MAO-B) shows a similar inverted U curve in relation to age. MAO-B has been shown to be responsible for the transformation of MPTP (the inactive molecule) into the active compound, 1-methyl-4-phenylpropionoxypiperidine (MPPP, see Section 4.1.1).

In population studies of early indicators of neurotoxicity, age must thus be controlled. This can either be done by using adequate epidemiological design involving, for example, comparing groups with similar socio-demographic status, or by using statistical methods of adjustment. There are, however, certain pitfalls that should be avoided. Older persons may have longer histories of exposure, making it difficult to separate cumulative exposure from aging.

3.2   GENETIC PREDISPOSITION/SUSCEPTIBILITY

Genetic susceptibility to neurotoxic agents can account at least in part for the expression of particular syndromes. For example, individuals with the antigen HLA B40 (Human Leukocyte Antigen) have an increased predisposition to aluminum encephalopathy[47]. Proponents of the aluminum (Al) hypothesis for Alzheimer's disease (AD) have also suggested that a genetic susceptibility factor may explain why only selected individuals develop Alzheimer's disease, since Al is an ubiquitous element in the environment, being the most abundant element on Earth after oxygen and silicon (8% of the Earth's outer crust). Moreover, it has been shown that Alzheimer's disease patients may inherit a reduced capacity of sulfoxidation and a reduced activity of the enzyme thiolmethyltransferase when compared to normal controls These abnormalities in the metabolism of xenobiotics would confer upon Alzheimer's disease patients a higher susceptibility to neurotoxic agents[96]. It needs to be mentioned that at this time no toxic agent has been well established in association with the development of AD.

It has been suggested that Parkinson's disease (PD) could result in a hereditary predisposition to environmental risk factors[55]. It has also been suggested that debrisoquine is related to the development of PD[12]. The poor metabolizer genotype for debrisoquine is over-represented in patients with Parkinson's disease when compared to normal control subjects. Also, patients with PD would have an increased potential to transform non-methylated pyridines into N-methylated pyridines, which are substances related to MPTP, thus predisposing them to neurotoxic damage[104]. Other obvious indications for a link between genetics and neurotoxicity come from studies on drug metabolism. In humans, genetic determination of the rate of drug metabolism has been shown. The acetylation of isoniazid in the human population exhibits a bimodal distribution, with half of the population being fast acetylators and

173

half slow acetylators. Slow acetylators tend to develop isoniazid (INH) toxicity (e.g. peripheral neuritis) more frequently, and slow acetylation is known to be inherited as an autosomal recessive trait[42]. The following example illustrates the individual heterogeneity in the susceptibility to neurotoxic agents.

A patient came to our clinic because of ataxia and memory difficulty. The syndrome had started insidiously 2 years ago and had slowly progressed. The patient is a 60–year–old man who had been working all of his life as a machinist and welder in different companies. His neurotoxicologic questionnaire revealed significant exposure to various neurotoxic agents including organic solvents used in the daily cleaning of machined pieces, welding fumes, and metal fumes. His family history was negative for similar syndromes and for any degenerative disorder. On exam, he showed mild memory problems and frontal lobe impairment. He had significant limb and axial ataxia. Imaging techniques revealed cerebellar and cerebral cortex atrophy. Repetitive investigations for an underlying cancer in the context of a possible paraneoplastic syndrome were negative. We thought this represented the manifestations of exposure to a variety of neurotoxic substances, but it is still unclear why he and not any of his coworkers developed this syndrome, though some cases of leukemia had been discovered in his colleagues.

While different genotypes may explain some of the individual heterogeneity, other factors (dose and duration of exposure, for example) need to be considered to explain the variance of syndromes in individuals with the same genes. The next chapter (Chapter 6) will go further into the matter of genetic predisposition and neurotoxicity.

4   Clinical syndromes

In this section the different syndromes which are most often encountered in the practice of a behavioral neurology practician and differential diagnosis will be reviewed. Some agents will be given more emphasis because of their actual importance in our understanding of certain disorders such as aluminum and Alzheimer's disease (AD) or MPTP and Parkinson's disease (PD). Table 5.3 provides examples of neurotoxic substances associated with major neurological clinical syndromes. In the cases of agents that give central, peripheral nervous system, and systemic manifestations, only those related to the brain will be emphasized here.

We limit ourselves here to environmental exposures. Nevertheless, it needs to be mentioned that endogenous substances can also be considered as neurotoxic. For example, glutamate causes excitatory neuronal death when in higher concentration. Also, many reactions of the organism may lead to the formation of reactive oxygen species (ROS) (H_2O_2, O_2^-, OH) and free radicals. Such a hypothesis is supported by many experts in the field of Parkinson's disease but still is not widely accepted[26]. Interaction between ROS, glutathione peroxidase, and superoxide dismutase copper-zinc (SOD 1), which are enzymes involved in the detoxification of free radicals and ROS, have been explored for a better understanding of the disease.

4.1   PARKINSONISM AND MOTOR SYSTEM DISORDERS

4.1.1   *Parkinsonism*

Parkinsonism is characterized by bradykinesia, tremor, and muscular rigidity. The tremor may involve the limbs and the jaw, and happens at rest, disappearing during movement or sleep. Rigidity is characterized by an increased tone that can be shown on examination by mobilizing the limbs

TABLE 5.3

Major neurologic syndromes produced by various neurotoxic substances
(Adapted from Rosenberg, N.L. (1994) Toxic encephalopathies: with particular reference to the psycho-organic (solvent) syndrome). In: *Occupational Neurology and Neurotoxicology*, Syllabus #132. American Academy of Neurology Annual Meeting, pp 29–44.)

	Solvents	Metals	Pesticides	Pesticides
Seizures	methanol ethylene glycol benzene carbon tetrachloride	lead mercury thallium tin aluminum arsenic	organophosphates organochlorines carbamates	hydrogen sulfide carbon monoxide methyl chloride
Encephalopathy	all	aluminum antimony arsenic bismuth lead manganese mercury silicon tin lithium	organophosphates organochlorines carbamates	hydrogen sulfide carbon monoxide methyl chloride carbon disulfide ethylene oxide nitrous oxide
Cerebellar dysfunction	all	aluminum bismuth manganese mercury thallium zinc tin lithium	organophosphates organochlorines	carbon monoxide methyl chloride
Peripheral neuropathy	n-hexane methyl butyl ketone	lead mercury arsenic gold thallium organotin lithium	organophosphates organochlorines carbamates (?)	nitrous oxide carbon disulfide ethylene oxide carbon monoxide
Parkinsonism and other movement disorders	methanol toluene trichloroethane carbon tetrachloride	lead mercury bismuth manganese thallium bromides zinc aluminum lithium	organophosphates organochlorines carbamates	carbon monoxide carbon disulfide hydrogen sulfide methyl chloride

Cranial neuropathy	methanol
	ethylene glycol
	trichloroethane
	n-hexane
	methyl butyl
	   ketone
	benzene
	toluene
	carbon
	   tetrachloride
		lead
		mercury
		arsenic
		thallium
		gold
		bromides
		bismuth
			organophosphates
			organochlorines
			carbamates
				nitrous oxide
				carbon disulfide

passively which often give an impression of a cogwheel. There is a generalized paucity of movement (bradykinesia), especially automatic movements, and this can account for the "mask-like" facies. Many conditions manifest parkinsonism, including idiopathic Parkinson's disease (paralysis agitans), other degenerative disorders of the central nervous system (like progressive supra-nuclear palsy), rare cases of encephalitis, and side effects from medications, in particular the anti-psychotic neuroleptics.

Parkinson's disease is a disease of later life with a mean age of onset of about 60 years. It affects 0.15% of the total population, but the prevalence of those over 50 increases to 1–2%. The disease is progressive, with a gradual exacerbation of symptoms. Pathophysiologically Parkinson's disease is characterized primarily by a degeneration of the nigrostriatal pathway. Other cerebral structures like the raphe nuclei, the locus ceruleus, and the motor nucleus of the vagus are usually involved. The pathological marker of Parkinson's disease is the Lewy body, which is an intraneuronal cytoplasmic inclusion body. A significant number of patients with Parkinson's disease have cognitive deficits and some of them appear to have concomitant Alzheimer's disease.

Some of the risk factors for PD identified in epidemiological studies include: professional exposure to manganese, iron, or aluminum; exposure to pesticide products and metals; living near mines or industries; rural living; farming; private well water drinking; and use of pesticides[25,27,85,107]. A specific agent still cannot be precisely identified. As mentioned above, in at least one study a history of familial parkinsonism was one of the factors needed to predict the occurrence of the disease[25]. An inverse relationship has been found with cigarette smoking[97].

4.1.2   MPTP and parkinsonism

N-Methyl-4-phenyl-1,2,3,6-tetrahydropyridine (MPTP) causes selective damage to the neurons of the substantia nigra (SN) in humans and different animals. The discovery of the neurotoxicity of MPTP dates back to 1971 when a student developed parkinsonism after using illicit drugs containing 1-methyl-4-phenylpropionoxypiperidine (MPPP), a potent meperidine analog purported to be "synthetic heroin". MPPP is metabolized into various molecules, one of them being MPTP. This student continued to abuse drugs and finally died of an overdose. His brain showed neuronal loss and gliosis of the zona compacta of the substantia nigra.

Many more people have used MPPP. Some of them have developed parkinsonism, while an asymptomatic loss of dopaminergic neurons have been shown in others[28]. An industrial chemist exposed to MPTP by inhalation or skin contact developed parkinsonism. The parkinsonian syndrome induced by MPTP seems to be identical to idiopathic PD and to show similar responses to L-dopa[10]. Nevertheless, fluctuations would appear earlier than in PD and the response to dopaminergic cell transplantation would be better than in the PD patients[103]. Stern et al. have shown that the neuropsychological features of MPTP parkinsonism is similar to the profile of PD and mainly consists in executive and visuo-spatial dysfunctions[95].

MPTP has become very useful in producing animal models of PD. It has been shown that monkeys given MPTP intravenously have selective destruction of the pigmented cells in the SN and loss of dopamine (DA) in the striatum. Nevertheless, some discrepancies exist between the animal models and the human pathology. First, the multiplicity of the Lewy bodies (LB) is not as marked in the animal as it is in PD. Second, the "LB" of the MPTP monkeys lack certain features of the LB of PD. Finally, factors such as age, time between MPTP exposure and death, route of administration, and species all seem to induce variations in the manifestations of MPTP. MPTP is a substrate to monoamine oxidase-B (MAO-B), an extraneuronal enzyme present in glial cells, and is transformed in 1-methyl-4-phenylpyridinium (MPP$^+$) which is more toxic than MPTP. MPP$^+$ accumulates in dopaminergic terminals because it is taken up by the DA reuptake system. It is also accumulated by mitochondria and is a reversible inhibitor of mitochondrial oxidative phosphorylation at the level of NADH dehydrogenase (complex I). This inhibition is responsible for a decrease in the capacity to generate ATP which induces changes in the ability to maintain membrane potentials and calcium homeostasis. Free radical formation would ensue giving rise to neuronal degeneration[99]. The antipsychotic haloperidol is known to cause parkinsonism as a side effect. On top of acting as a dopamine antagonist, haloperidol (Figure 16.4) has a similar structure to MPTP (Figure 4.8) and was found to also depress mitochondrial complex I activity[23].

There is evidence that substances similar to MPTP can be found in the environment. The herbicide paraquat has a similar structure and similar effects as MPP^{+}[12]. Diquat, another herbicide, can also produce parkinsonism[93]. Tetrahydroisoquinoline, present in cheese, wines, and a variety of foods, can induce parkinsonism in the monkey at high doses[106]. This compound is also of particular interest since 2-methyl-6,7-dihydroxy-1,2,3,4-tetrahydroisoquinoline has recently been found in the cerebrospinal fluid of four out of nine PD patients, whereas it was not present in any of the control subjects (Scholtz et al., personal communication, 1993).

4.1.3 Manganese and parkinsonism

Manganese (Mn) is an essential ingredient in steel, and is also used in electric battery factories. Methylcyclopentadienyl manganese tricarboxyl (MMT) is now being used instead of lead in gasoline as an anti-knock agent. Canada is the only country where MMT has totally replaced lead in the gasoline. Intoxication has mainly been described in miners in Cuba, Chile, and Morocco. The long term effects of the increased concentration of Mn oxides (from the combustion of MMT) in the environment is largely unknown.

Chronic manganese exposure produces a syndrome similar to PD (including resting tremor, rigidity, freezing episodes, akinesia, hypersialorrhea) but is also characterized by dystonia, adiadochokinesia, mental status changes, and psychiatric disturbances[16,77]. Recent studies have shown that compared to normal control subjects, workers exposed to Mn showed less manual dexterity and deficits on tests highly dependent on attention processes (or working memory)[14,56,58,77]. Parkinsonian exposed workers had qualitatively similar but more severe deficits than non-parkinsonian workers, and were much like PD patients[56].

Some studies have shown that the parkinsonian syndrome induced by chronic Mn exposure improves if the exposure is stopped sufficiently early, whereas others have shown that it may progress long after the cessation of high exposure[57]. L-Dopa has been successful in the treatment of manganese -induced parkinsonism, but the beneficial effects faded away after 2–3 years. No fluctuations occurred in hypokinetic patients when given a regimen of L-dopa that would produce similar abnormal movements in PD.

In Mn parkinsonism, the bulk of the neuropathology is located at the striatum and pallidum with little change at the substantia nigra. Lesions are also found throughout the cerebrum, the brainstem, and cerebellum. In animals given Mn, there is a decrease in dopamine, in homovanillic acid (HVA), and an even more pronounced decrease in norepinephrine in the striatum[12a].

4.1.4 Motor neuron system

Diseases of the motor neurons can have either an acute or chronic character. Motor neuron diseases do not affect the sensory neurons. The best known disorder is amyotrophic lateral sclerosis (ALS), also called Lou Gehrig's Disease.

Amyotrophy refers to the neurogenic atrophy of muscle and lateral sclerosis to the hardness felt when the spinal cord is examined at autopsy. ALS is a disease marked by the progressive degeneration of the neurons that give rise to the corticospinal tract and of the motor cells of the brainstem and spinal cord, resulting in a deficit of upper and lower motor neurons. Usually the disease ends fatally within 2–3 years.

Foci of increased incidence of ALS sometimes associated to parkinsonian features and dementia (ALS-PD or lytico-bodig) has been described on certain islands of the West Pacific which include the Chamorros on some of the Mariana Islands (Guam and Rota), the Auyo and Jakai people of west New Guinea, and Japanese in certain districts of the Kii peninsula of Honshu Island (see map).

The ALS of these islands is clinically similar, but is distinct pathologically. There is an additional and apparently specific feature of an excess of neurofibrillary tangles and intracytoplasmic (granulovacuolar) bodies widely distributed in the brain and occasionally in motor neurons of the spinal cord[69].

It was initially thought that this disease is due to the use of the seed of the indigenous cycad (Cycas circinalis) as food and medicine[48,94]. It was found that cycad seeds contain two potential neurotoxins, β-L-methyl-amino-alanine (BMAA), which causes chromatolysis of giant Betz's cells and smaller pyramidal cells in the cynomolgus monkey, and cycasin. BMAA is chemically and neuropharmacologically related to BOAA (β-N-oxalyl-amino-L-alanine), the plant-derived amino acid found in chickling peas (Lathyrus sativa) and related species, causing lathyrism (spastic paraparesis).

BMAA's acute toxic action is susbstantially reduced by N-methyl-D-aspartate (NMDA) receptor antagonists. Controversies concerning these hypotheses stem from the demonstration that both BMAA and cycasin do not produce significant neurotoxicity in animal models, except for one study in the monkey[94]. Moreover, more than 80% of BMAA is lost from seeds during the thorough washing given before it is ground to make flour[40]. A recent epidemiological study showed that sporadic ALS was weakly related only to exposure to lead, suggesting the possibility of a multifactorial etiology[8]. Furthermore, conjugal cases have been described suggesting environmental factors[33].

Exposure to mercury, especially to organic mercury, can cause various neurological syndromes depending upon the degree of exposure. For example, in Minamata Japan, an important exposure of fishermen to methylmercury lead to intellectual and neurobehavioral changes such as emotional lability, depression and insomnia, cataracts, constriction of the visual fields up to complete blindness, nystagmus, metal taste, deafness, dysarthria, ataxia, postural tremor, hypesthesia involving the limbs, perioral area and tongue, proximal and distal weakness, hyperreflexia with Babinski signs, primitive signs such as the snout, hypersialorrhea. These signs can take the form of a polyneuropathy or of a syndrome similar to amyotrophic lateral sclerosis[3,31,60,71,87]. Clonic seizures, myoclonus and epileptiform changes on the electroencephalogram have been reported[19].

With cerebral imaging and neuropathology, a diffuse or focal atrophy can be appreciated at the level of the cerebellum, the calcarine fissure, the pre- and post-central gyri and the temporal cortex[60,100]. Microscopically, there is loss of the granular cells of the cerebellum, degeneration of posterior columns, ganglia and axons of posterior radicles. Similar lesions using methylmercury have been reproduced in experimental animals[60].

Consequences of chronic and smaller levels of exposure (10-60 µg per g in hair which is where mercury concentrates in humans) are still not known. Projects to study the effects of long-term exposure to mercury are being

BMAA

$$CH_2 - CH - COO$$
$$NH_2^+ - NH_2$$
$$CH_3$$

BOAA

$$CH_2 - CH - COO$$
$$NH \quad {}^+NH_3$$
$$CO$$
$$COO$$

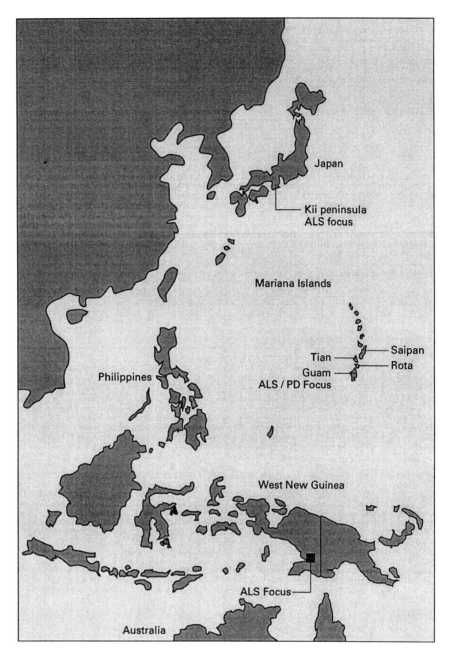

FIGURE 5.4

developed in Northern Quebec where large populations of Cree Indians are exposed, and in the Brazilian Amazon region, where mercury used in gold-mining is contaminating the foodchain.

Methyl mercury from fish or topical exposure to phenyl mercuric nitrate, used as a preservative agent for pilocarpine, are associated to the formation of cataracts before the appearance of neurological signs[1].

4.2   MEMORY LOSS AND DEMENTIA

It is known that lesions in distinct brain regions disturb memory in different ways, although our knowledge of how the brain mediates the various aspects of memory, and how distinct lesions disrupt separate components of memory is far from complete. Many strategies have been employed to study the physiological bases of learning and memory in the brain. Much of what we know about the brain and memory in humans comes from patient studies. Amnesia is the failure of memory processes. Retrograde amnesia is the loss of memory for events learned before the particular amnesia-inducing trauma (e.g. a blow to the head), whereas anterograde amnesia is the failure to learn and store experiences encountered after the trauma.

The amnestic syndrome is characterized by the impairment of memory, both retrograde and anterograde amnesia, occurring in a normal state of consciousness. This syndrome may be caused by thiamin deficiency, but also results from any pathological process causing bilateral damage to certain structures in the medial temporal lobe and diencephalon (e.g. the hippocampal formation). Causes include degenerative disorders like Alzheimer's disease, nutritinal deficiencies like thiamine deficiencies often encountered associated to chronic alcoholism, head trauma, brain tumors, infarction, cerebral hypoxia, and carbon monoxide poisoning.

Drug- or toxicant-induced memory loss has also been reported. Impairment of memory has been described following amphetamine use or consumption of short-acting benzodiazepines[78], and a certain loss of short-term memory is associated with tetrahydrocannabinol (THC) intoxication. There are a few examples of neurotoxins inducing acute, reversible amnesia, e.g. domoic acid and *Datura fastuosa*.

4.2.1 *Domoic acid*

The amnestic shellfish poisoning syndrome was first described in Canada when 107 people in 1987 became ill and 3 died after eating cultivated blue mussels. The symptoms included vomiting, diarrhea, abdominal cramps, headache and loss of short-term memory. The causative agent was identified as an algal toxin, domoic acid[83]. Domoic acid could selectively activate kainate receptors primarily found in the hippocampus causing an intracellular overload of calcium which would disrupt the mitochodrial functioning and activate various proteases leading to cell destruction (excitotoxicity)[32].

4.2.2 *Ketjubung poisoning*

Ketjubung is the common name for *Datura fastuosa*, a common shrub in Indonesia. The seeds, roots, leaves and flowers of Datura species contain the belladonna alkaloids atropine and scopolamine. *Datura fastuosa* is used as a folk medicine in Indonesia, for instance as a remedy for rheumatic pains and asthma. Ketjubung poisoning can arise unintentionally if, through unfamiliarity with the plant, too many of its seeds or leaves are eaten. But the poisoning may also be the result of evil intent, for instance, to poison a person one intends to rob, or to take one's revenge, or by way of a joke. The aim is easily achieved, for example by mixing in the seeds with the victim's food or drink. The case history of an Indonesian man who, because of his odd behavior, was brought to a missionary hospital in Jogyakarta on a November evening in 1939 illustrates some of the effects of ketjubung poisoning. A man had consumed a ketjubung seed infusion instead of tea. Shortly afterwards, he began to behave strangely. The police were notified, and he was brought to the hospital. The patient, 30 years of age, had consumed one glass of the "tea". Contact with him was impossible; he was pacing up and down the treatment room in an excited state. He squatted under a laboratory table for lengthy periods, purposefully reaching out in all directions trying to grasp something. This odd behavior evoked much laughter from the spectators. He was not aggressive, but did resist physical examination. His pulse rate was 132, his skin was red and hot, and his tongue was dry. His pupils were fully dilated and unresponsive to light. Insertion of a stomach pump failed due to the patient's resistance. Phenobarbital 100 mg was administered subcutaneously, after which he allowed himself to be escorted to a ward. His body temperature was 38.5 °C; his leukocyte count was 12 400 mm^{-3}. Chloral hydrate 0.5 g was administered rectally.

The following day, the physician on duty visited the patient at 7 a.m. To his surprise, he found a quiet, cooperative man, who did not remember the incident. He no longer had fever, and his pulse rate was about 70; the pupils were still dilated but now showed some response to light. Physical examination did not reveal any abnormalities. The man was discharged that same day[102].

4.2.3   *Alzheimer's disease*

Alzheimer's disease is the most common form of dementia in the Western world. It is characterized by an insidious onset of memory deficits associated to other intellectual problems such as aphasia or agnosia, and it slowly progresses over time. The onset of the symptoms may sometimes start in the 50s, but the prevalence of the disease increases steadily with age. Women outnumber men with AD for reasons that are still not completely understood. Education, inflammatory response, and an abnormality in the cholesterol transport by apolipoprotein E have been identified as factors to the development of Alzheimer's disease. Pathologically, AD is characterized by neuronal loss, senile plaques, and neurofibrillary tangles. The senile plaques consist of the accumulation of degenerating neurons and amyloi,d and the neurofibrillary tangles are abnormal intraneuronal filaments. These changes are mainly evident in the medial temporal lobes but also in the neocortex. The nucleus basalis of Meynert is particularly involved which explains the profound cholinergic deficiency. No good animal models of AD are available at the present time. The cholinergic model of AD is very controversial. The best correlation with the cognitive deficits is found with the amount of cell loss and the synaptic disruption.

Widely accepted risk factors for Alzheimer's disease at this point do not include environmental factors, but some data suggest a possible role in the expression of the disease. In the US the prevalence of Alzheimer's disease is about the same in the white and in the black populations, whereas in Africa, Alzheimer's disease seems to be very rare. The heterogeneity in the age of onset of homozygote twins with Alzheimer's disease suggests that environmental agents may influence the evolution of the disease. Alzheimer's disease patients have been reported to have an impaired metabolism of xenobiotics[96]. Exposure to aluminum (Al) has been suggested as a risk factor and smoking could have a protective role. Though some argue for a pharmacological action of nicotine, it may also be that cigarette smoking causes a population selection bias, i.e. smoking predisposes for vascular events and thus to vascular dementia.

Aluminum (Al) containing perspirants were initially associated with the development of Alzheimer's disease. Abnormally high concentrations of insoluble Al salicylates have been found in Alzheimer's disease neurons with neurofibrillary tangles (NFT)[29]. Al has also been found in senile plaques of Alzheimer's disease, Down's syndrome, dementia pugilistica, and in the ALS-parkinsonism-dementia syndrome of Guam[84]. Miners submitted to an Al based treatment aimed at preventing pulmonary silicosis have experienced cognitive problems, and high concentrations of Al were found in their brains. Desferroxamine, a chelator that extracts Al from blood and brain, had been of some benefit in these miners and was thought to be protective in Alzheimer's disease patients[68].

In contrast, recent studies found that chronic ingestion of Al-containing antacids were not associated with Alzheimer's disease[21,51]. Also, Al salts injected in cerebrospinal fluid caused dogs, cats and rabbits to develop NFT, but these NFT are different from those found in Alzheimer's disease in that they are composed of straight instead of double helicoidal neurofilaments[34,105]. Al is associated to the amyloid of the SP but not to the amyloid of the congophilic angiopathy, and the B-amyloid is the same in SP and in the angiopathy in Alzheimer's disease. Furthermore, Al is not increased in all Alzheimer's disease patients brain, and encephalopathy related to Al exposure does not induce NFT[72]. Methodological problems could explain the relationship between Al in drinking water and Alzheimer's disease, or Al in NFT and amyloid plaques[92]. Furthermore, Foncin and El Hachimi reported the case of a man into whose brain a metallic aluminum device was accidentally inserted[43]. The patient was asymptomatic for 15 years and then developed a seizure disorder and died in status epilepticus. At autopsy, there were no NFT and no senile plaques. It is now accepted that even if Al is not an etiological factor for Alzheimer's disease, it can be considered as an aggravating factor.

A very recent hypothesis suggests that Zinc (Zn) is able to accelerate amyloid formation. BA4 binds strongly to Zn, and if concentrations of Zn are increased, amyloid solubility is reduced and tends to aggregate. This reaction may be speeded up by the addition of Al. But, at this point, there is no evidence linking environmental exposure to Zn and Alzheimer's disease.

4.3   ENCEPHALOPATHIES

Almost any toxic agent if in sufficient amount is likely to cause an impairment in consciousness and delirium. Exposure to acetone, benzene, toluene, xylene, chloroform, or ether are associated to feelings of drunkenness, headache, memory loss, and concentration problems. Long-term effects have also been reported. Exposure to low concentrations, sometimes even lower than permissible limits, often cause health damage that does not have obvious clinical manifestations. Depending on the agent of exposure and the subject, different symptoms may be manifested. Early toxic encephalopathy is characterized by mood changes, neurotic behavior, and a deficit in cognitive functions.

Pesticides and parathion are acetylcholinesterase inhibitors and are associated to memory loss and visuo-spatial problems[89]. Below, lead and aluminum exposure are discussed as examples. Other agents and their main characteristics are given in Table 5.3 and inthe following chapters of this book.

4.3.1   *Lead*

Classically, inorganic lead has been known to cause encephalopathy in children and pure motor neuropathy in adults. Encephalopathy has also been described in the adult, but it is more insidious. In children, intoxication most often occurs from ingestion of lead-containing paints, particularly by small children with pica. The exact effect of environmental contamination by lead on children's intelligence is an important concern. Exhaust fumes from vehicles using petrol with added lead as an anti-knock agent are blamed for increasing environmental exposure.

The manifestations of encephalopathy in children are characterized by irritability and anorexia which give way to ataxia, vomiting, seizures, and then coma. There is evidence of increased intracranial pressure and anemia: X-rays show lead lines at the end of the long bones and increased density of the metacarpals, phalanges, and ribs. Treatment consists of controlling seizures and removal of lead by chelation therapy. Mapou and Kaplan have reported an attentional deficit in a stained-glass artist chronically exposed to lead[67]. These deficits improved after chelation therapy. Chronic exposure to low lead levels has been associated to fatigue, irritability, depression, decreased appetite, decreased libido, insomnia, abdominal and muscle pain, headache, polyneuropathy, and various cognitive skills[9,35].

At autopsy, acute encephalopathy is characterized by brain edema, hyperemic meninges, and petechial hemorrhages of the white matter. A perivascular protein rich transudate is responsible for the brain edema and gives gray and white matter a spongy appearance. Microscopically, there are degenerating neurons, proliferating glia, and fat-laden cells. These changes involve all regions of the brain but the cerebellum is the most severely affected. Changes of chronic encephalopathy are similar to those of acute encephalopathy, but mixed with atrophic changes[81]. Experimental evidence suggests that lead's action primarily involves the developing vessels[52,98].

Organic lead (tetra-ethyl-lead) has been added to petrol as an anti-knock agent since 1923. Dangerous exposures have occurred in workers cleaning out petrol storage tanks. Encephalopathy has also been reported with petrol sniffing. This encephalopathy is characterized by nightmares, sleep disturbances, excitement leading to maniacal psychosis (delusions, hallucinations, anxiety). Treatment consists of sedation and usually no sequelae ensue. Autopsy studies have shown mild cerebral cortical atrophy, mild ventriculomegaly as well as neuronal loss in certain sectors of Ammon's horn in the hippocampus, in the cerebellum (Purkinje cells) and in the dentate nuclei[61,101]. Tetra-ethyl-lead is transformed in tri-ethyl-lead by the liver and it is in this latter form that is neurotoxic.

4.3.2 *Dialysis encephalopathy*

Toxicity from aluminum (Al) occupational exposure is extremely rare[72]. Daily intake of aluminum for people in the general population has been reported to range from 9 mg/day to 36 mg/day, with an average of about 20 mg/day. Aluminum encephalopathy has occurred in some patients undergoing chronic renal dialysis, due to high concentrations of aluminum in the water used for dialysis. Around 1976, progressive encephalopathy became one of the most common causes of death in renal dialysis units — Alfrey et al. found that brain Al was increased[4]. This was found to be due to the high aluminum content in the water used in dialysis[41,82]. The disease has practically disappeared following control of the water aluminum concentration[37]. Nevertheless, it has been recently shown that even using water containing 0.08 g Al per liter and using desferoxamine, hemodialysed patients, especially those with lower premorbid intelligence, had decreased visual memory, and a decline in attention/concentration, frontal lobe functions in relation to the blood aluminum levels.

Dialysis encephalopathy is characterized by progressive dementia with speech difficulty, myoclonus, focal epilepsy, focal neurological signs. A milder syndrome is characterized by uncoordination, poor memory, impairment in abstract reasoning, and depression. Burks et al. reported six autopsied cases of dialysis encephalopathy[24]. They found no neuronal loss, but many neurons were shrunken, gliosis and cortical spongiform changed and no NFT.

4.4 PERSONALITY CHANGES

Many neurological disorders appear to have associated changes in personality: Parkinson's patients may be depressed; Huntington's chorea may produce depression, bipolar-like symptoms, or delusionary-hallucinatory state; presenile dementias may produce numerous adverse personality changes; temporal lobe and psychomotor epileptic patients may experience personality changes after several years of seizures; moderate to severe head injuries cause major personality changes. The changes may include emotional dulling, disinhibition, reduced anxiety, euphoria, decreased social sensitivity, depression, hypersensitivity, hyper-/hypo-sexuality, irritability, restlessness, low frustration tolerance and apathy.

Numerous attempts have been made to find a relationship between personality and drug use. Practical reasons for investigating such a relationship are: 1. the ability to identify those who are likely to experience difficulties in managing their drug use, 2. to be able to match therapy to the individual, and 3. to identify those individuals most likely to relapse into drug use after a drug free period.

Cocaine and other illicit drug intoxication can be associated with memory loss, confusional states, and hallucinations. Chronic neuropsychological effects of LSD and heroin have never been demonstrated, but chronic intoxication with cannabis is associated to prolonged effects[30]. A cause and effect relationship is often impossible because multiple drugs are often used in conjunction so adulterants may be responsible for some manifestations. This cocktail of drugs and adulterants may give rise to a wide variety of neurological and other medical syndromes, some of which include myelopathy, neuropathy, myopathy, and infectious complications. There is an incidence of strokes by vasculitis associated to the use of cocaine[62-64]. Amphetamines have also been associated with vasculitis.

Barbiturism or chronic barbiturate intoxication/addiction is similar to alcohol addiction, and is characterized by bradyphrenia, emotional lability, untidiness, dysarthria, nystagmus and ataxia, while striking tolerance develops. Withdrawal symptoms are similar to other withdrawal states, but convulsions and delirium tremers may be encountered more frequently. Barbiturate withdrawal seizure will respond only to barbiturate treatment and to no other antiepileptic medication. Barbiturates are found in the streets and are used by addicts as opiates. Barbiturism is also frequent in those who have ready access to barbiturates.

5   Conclusion

Early behavioral disturbances may be a sign of neurotoxic exposure. These indicators serve to identify substances and circumstances that may become a health hazard. If these exposures are not corrected, they may lead to behavioral, neurologic, and systemic syndromes in particularly in susceptible individuals. Among the factors that influence this course of events are age, genetic background, and lifestyle. A better knowledge of these risk factors may help identify individuals at risk or dangerous situations and can provide guidelines for correction of the environment or the withdrawal of individuals.

6   Summary

The present chapter discussed the behavioral manifestations of neurotoxic exposure along a progressive course model. The effects of substances depend on an equation taking into account exposure dose, including both quantity and duration, as well as individual susceptibility. Among the factors that affect the predisposition of individuals are age, genetics, and lifestyle. Examples of methods used to detect early signs of impairment related to exposures are given. Finally, behavioral syndromes associated to specific substances are discussed.

References

1. Abrams, J.D. and Majzoub, U. (1970) Mercury content of the human lens. *Br. J. Ophthalmol.* **54**: pp 59-61.
2. Agrawal, A.K., Srivastava, S..P, Seth, P.K. (1982) Effect of styrene on dopamine receptors, *Bull. Environ. Contamin. Toxicol.* **29**: p 400.
3. Albers, J.W., Kallenbach, L.R., Fine, L.J. et al. (1988) Neurological abnormalities associated with remote occupational mercury exposure. *Ann. Neurol.* **24**: pp 651–659.
4. Alfrey, A.C., LeGendre, G.R., Kaehny, W.D. (1976) The dialysis encephalopathy syndrome. Possible aluminum intoxication. *N. Engl. J. Med.* **294**: pp 184–188.
5. Amoore, J. (1986) Effects of chemical exposure on olfaction in humans. In: Barrow, C.S. (Ed.) *Toxicology of the Nasal Passage.* Hemisphere Publishing, Washington, D.C.
6. Anger, W.K. (1990) Worksite neurobehavioral research: results, sensitive methods, test batteries and the trensition from laboratory data to human health, *Neurotoxicology* **11**: p 629.
7. Antti-Poika, M., Ojala, M., Matikainen, E., Vaheri, E., Juntunen, J. (1989) Occupational exposure to solvents and cerebellar, brainstem and vestibular functions. *Int. Arch. Occup. Environ. Health* **61**: p 397.
8. Armon, C., Kurland, L.T., Daube, J.R., O'Brien, P.C. (1991) Epidemiologic correlates of sporadic ALS. *Neurology* **41**: pp 1077–1084.
9. Baker, E.L., White, R.F., Murawski, B.J. (1985) Clinical evaluation of neurobehavioral efffects of occupational exposure to organic solvents and lead. *Int. J. Ment. Health* **14**: pp 135–158.
10. Ballard, P.A., Tetrud, J.W., Langston, J.W. (1985) Permanent human parkinsonism due to 1-methyl-4-phenyl-1,2,3,6-tetahydropyridine (MPTP): seven cases. *Neurology* **35**: pp 949–956.
11. Barbeau, A., Roy, M., Cloutier, T., Plasse, L., Paris, S. (1986) Environmental and genetic factors in the etiology of Parkinson's disease. *Adv. Neurol.* **45**: pp 299–306.
12. Barbeau, A., Cloutier, T., Roy, M., Plasse, L., Paris, S., Poirier, J. (1985) Ecogenetics of Parkinson's disease; 4-hydroxylation of debrisoquine. *Lancet* ii: pp 1213–1216.
12a. Barbeau, A., Inoue, N., Cloutier, T. (1976) Role of manganese in dystonia. In: Eldridge, R. and Fahn, S. (Eds.) *Advances in Neurology*, vol. 14. Raven Press, New York, pp 339–352.
13. Benignus, V.A. (1981) Neurobehavioral effects of toluene: a review. *Neurobehav. Toxicol. Terartol.* **3**: p 407
14. Beuter, A., Mergler, D., de Geoffroy, A., et al. (1995) Diadochokinesimetry: a study of patients with Parkinson's disease and manganese exposed workers. *Neurotoxicology,* **13**: p 655.
15. Bleeker, M.L. (1985) Quantifying sensory loss in peripheral neuropathies. *Neurobehav. Toxicol. Teratol.* **7**: p 305.
16. Bleeker, M.L. (1988) Parkinsonism: a clinical marker of exposure to neurotoxins. *Neurotoxicol. Teratol.* **18**: pp 475–478.
17. Bonilla, E., Salazar, E., Villasmil, J.J., Villalobos, R. (1982) The regional distribution of manganese in the hormonal human brain, *Neurochem. Res.* **7**, p 221.
18. Bove, F., Litwak, M.S., Arezzo, J.C., Baker, E.L. (1986) Quantitative sensory testing in occupational medecine, *Semin. Occup. Med.* **1**: p 185.
19. Brenner, R.P. and Snyder, R.D. (1980) Late EEG findings and clinical status after organic mercury poisoning. *Arch. Neurol.* **37**: pp 282–284.
20. Broadwell, D.K., Darcey, D.J., Hudnell, H.K., Otto, D.A., Boyes, W.K. (1995) Work-site clinical and neurobehavioral assessment of solvent exposed microelectronics workers. *Am. J. Ind. Med.* **27**: pp 677.
21. Broe, G.A., Henderson, A.S., Creasey, H., McCusker, E., Korten, A.E., Jorm, A.F., Longley, W., Anthony, J.C. (1990) A case control study of Alzheimer's disease in Australia. *Neurology* **40**: pp 1698–1707.
22. Bullard, R.D. (Ed.) (1993) *Confronting Environmental Racism Voices from the Grass Roots.* South End Press, Boston.
23. Burkhardt, C., Kelly, J.P., Lim, Y.H., Filley, C.M., Parker, W.D. (1993) Neuroleptic dedications inhibit complex I of the electron transport chain. *Ann. Neurol.* **33**: pp 512–517.
24. Burks, J.S., Alfrey, A.C., Huddlestone, J., Nortenberg, M.D., Lewin, E. (1976) A fatal encephalopathy in chronic hemodialysis patients. *Lancet* **1**: pp 764–768.

25. Butterfield, P.G., Valanis, B.G., Spencer, P.S., Lindeman, C.A., Nutt, J.G. (1993) Environmental antecedants of young-onset Parkinson's disease. *Neurology* **43**: pp 1150–1158.

26. Calne, D.B. (1992) The free radical hypothesis in idiopathic parkinsonism: evidence against it. *Ann. Neurol.* **32**: pp 799–803.

27. Calne, S., Schoenberg, B., Martin, W., Uitti, R.J., Spencer, P., Calne, D.B. (1987) Familial Parkinson's disease: possible role of environmental factors. *Can. J. Neurol. Sci.*, **14**: pp 303–305.

28. Calne, D.B., Langston, J.W., Martin, W.R.W. et al. (1985) Observations relating to the cause of PD: PET after MPTP. *Nature* **317**: pp 246–248.

29. Candy, J.M., Oakley, A.E., Klinowski, J., Carpenter ,T.A., Pery, R.H., Atack, J.R. et al. (1986) Aluminosilicates and senile plaque formation in Alzheimer's disease. *Lancet* **1**: pp 354–357.

30. Carlin, A.S. and Trupin, E. (1977) The effect of long-term chronic cannabis use on neuropsychologic functioning *Int. J. Addict.* **12**: pp 617–624

31. Chapman, L.J., Sauter, S.L., Henning, R.A., Dodson, V.N., Reddan, W.G., Matthews, C.G. (1990) Differences in frequency of finger tremor in otherwise asymptomatic mercury workers. *Br. J. Indust. Med.* **47**: pp 838–843.

32. Choi, D.W. (1985) Glutamate neurotoxicity in cortical cell culture is calcium dependent. *Neurosci. Lett.* **58**: pp 293–297.

33. Cornblath, D.R., Kurland, L.T., Boylan, K.B., Morrison, L., Radhakrishnan, K., Montgomery, M. (1993) Conjugal amyotrophic lateral sclerosis: report of a young married couple. *Neurology* **43**: pp 2378–2380.

34. Crapper, G., McLachlan, D.R., De Boni, U. (1980) Aluminum in human brain disease — an overview. *Neurotoxicology* **1**: pp 3–16.

35. Cullen, M.R., Robins, J.M., Eskenazi, B. (1983) Adult inorganic lead intoxication: presentation of 31 new cases and a review of recent advances in the literature. *Medicine* **62**: pp 221–247.

36. Davis, J.M. and Svensgaard, D.J. (1990) U-shaped dose-response curves: their occurence and implications for risk assessment, *J. Toxicol. Environ. Health* **30**: p 71.

37. Davison, A.M., Walker, G.S., Oli, H., Lewins, A.M. (1982) Water suppply aluminum concentration, dialysis dementia, and effect of reverse osmosis water treatment. *Lancet* **2**: pp 785–787.

38. Demers, R.Y., Markell, B.L., Wabeke, R. (1991) Peripheral vibratory sense deficits in solvent-exposed painters, *J. Occup. Med.* **33**: p 1051.

39. Doty, R.L., Shaman, P., Dann, M. (1984) Development of the University of Pennsylvania smell identification test: a standardized microencapsulated test of olfactory function, *Physiol. Behav.* **32**: p 489.

40. Duncan, M.W., Kopin, I.J., Garruto, R.M., Lavine, L., Markey, S.P. (1988) 2-Amino-3 (methylamino)-propionic acid in cycad-derived foods is an unlikely cause of amyotrophic lateral sclerosis/parkinsonism. *Lancet* **2**: pp 2631–2632.

41. Elliot, H.L., Dryburgh, F., Fell, G.S., Sabet, S., MacDougall, A.I. (1978) Aluminum toxicity during regular hemodialysis. *Br.Med. J.* **1**: pp 1101–1103.

42. Evans, D.A.P., Manley, K.A., McKusick, V.A. (1960) Genetic control of isoniazid metabolism in man. *Br.Med. J.* **2**: pp 485–491.

43. Foncin, J.F., El Hachimi, K.H. (1986) Neurofibrillary degeneration in Alzheimer's disease: a discussion with a contribution to aluminum pathology. In: Bes, A., Cahn, J., Cahn, R., Hoyer, S., Marc-Vergnes, J.P., Wisniewski, H.M. (Eds.) *Senile Dementias: Early Detection*. John Libbey Eurotext, London-Paris, pp 191–200.

44. Forno, L.S., DeLanney, L.E., Irwin, I., Langston, J.W. (1993) Similarities and differences between MPTP-induced parkinsonism and Parkinson's disease. Neuropathological considerations. In: Narabayashi, H., Nagatsu, T., Yanagisawa, N., Mizuno, Y. (Eds.) *Advances in Neurology*, vol. 6. Raven Press, , New York, pp 600–608.

45. Fortier, I., Ferraris, J., Mergler, D. (1991) Measurement precision of an olfactory perception threshold test for use in field studies, *Am. J. Ind. Med.* **20**: p 495.

46. Frenette, B., Mergler, D., Bowler, R. (1991) Contrast sensitivity loss in a group of former microelectronics workers with normal visual acuity. *Optometry Vision Sci.* **68**, p 556.

47. Garrett, P., Spencer, S., Muleahy, D., Hanly, P., O'Hare, J.A., Carmody, O., O'Dwyer, W.F. (1985) *Proc EDTA-ERA* **22**: pp 360–362.

48. Garruto, R.M., Yase, Y. (1986) Neurodegenerative disorders of the Western Pacific: the search for mechanisms of pathogenesis. *Trends Neurol. Sci.* **9**: pp 368–374.

49. Garruto, R.M. (1991) Pacific paradigms of environmentally-induced neurological disorders: clinical, epidemiological and molecular perspectives. *Neurotoxicology* **12**: pp 347–377.

50. Goldstein, G.W., Asbury, A.K., Diamond, I. (1974) Pathogenesis of lead encephalopathy. Uptake of lead and reaction of brain capillaries. *Arch. Neurol.* **31**: pp 382–389.

51. Graves, A.B., White, E., Koepsell, T.D., Reifler, B.V., Van Belle, G., Larson, E.B. (1990) The association between aluminum-containing products and AD. *J. Clin. Epidemiol.* **43**: pp 35–44.

52. Hanninen, H., Lindstrom, K. (1971) *Neurobehavioral Test Battery of the Institute of Occupational Health.* Helsinski, Institute of Occupational Health.

53. Hanninen, H. (1971) Psychological picture of manifest and latent carbon disulphide poisoning. *Br. J. Ind. Med.* **28**: p 374.

54. Hastings, L., Evans, J.E. (1991) Olfactory primary neurons as a route of entry for toxic agents into the CNS, *Neurotoxicology*, **12**: p 707.

55. Hornykiewicz, O. (1993) Parkinson's disease and the adaptive capacity of the nigrostriatal dopamine system: possible neurochemical mechanisms. *Adv. Neurol.* **60**: pp 140–147.

56. Hua, M.S., Huang, C.C. (1991) Chronic occupational exposure to manganese and neurobehavioral function. *J. Clin. Exp. Neuropsychol.* **13**: pp 495–507.

57. Huang, C.C., Lu, C.S., Chu, N.S. et al., Progression after chronic manganese exposure. *Neurology* **43**: pp 1479–1483.

58. Iregen, A. (1990) Psychological test performance in foundry workers exposed to low levels of manganese. *Neurotoxicol. Teratol.* **12**: pp 673–675.

59. Irwin, I., DeLanney, L.E., Langston, J.W. (1993) MPTP and aging. Studies in the C57BL/6 mouse. *Adv. Neuro.* **60**: pp 197–206.

60. Jacobs, J.M. and Le Quesne, P.M. (1992) Toxic disorders. In: *Greenfield's Neuropathology*, 5th edition, Oxford University Press, Oxford, pp 881–987.

61. Kaelan, C., Harper, C., Vieira, B.I. (1986) Acute encephalopathy and death due to petrol sniffing: neuropathological findings. *Aust. N.Z. J. Med.* **16**: pp 804–807.

62. Kaye, B.R. and Fainstat, M. (1987) Cerebral casculitis associated with cocain abuse. *JAMA* **258**: pp 2104–2106.

63. Klonoff, D.C., Andrews, B.T., Obana, W.G. (1989) Stroke associated with cocain use *Arch. Neurol.* **46**: pp 989–992.

64. Krendel, D.A., Ditter, S.M., Frankel, M.R., Ross, W.K. (1990) Biopsy-proven cerebral vasculitis associated with cocaine abuse. *Neurology* **40**: pp 1092–1094.

65. Kurland, T. (1988) Amyotrophic lateral sclerosis and Parkinson's disease complex on Guam linked to an environment neurotoxin. *Trends Neurol. Sci.* **11**: pp 51–54.

66. Manninen, O. (1988) Changes in hearing, cardiovascular functions, hemodynamics, upright body sway, urinary catecholamines and their correlates after prolonged successive exposures to complex environmental condition, *Int. Arch. Occup. Environ. Health* **60**: p 249.

67. Mapou, R.L. and Kaplan, E. (1991) Neuropsychological improvement from chelation after long-term exposure to lead: case study. *NNBN* **4**: pp 224–237.

68. Martyn, C.N., Barker, L.J.P., Osmond, C., Harris, E.C., Edwardson, J.A., Lacy, R.F. (1989) Geographical relationship between Alzheimer's disease and aluminum in drinking water. *Lancet* **1**: pp 59–62.

69. Matsumoto, S., Hirano, A., Goto, S. (1990) Spinal cord neurofibrillary tangles of guamanian amyotrophic lateral sclerosis and parkinsonism-dementia complex: an immunohistochemical study. *Neurology* **40**: pp 975–979.

70. McConnell, R., Keifer, M., Rosenstock, L. (1994) Elevated quantitative vibrotactile threshold among workers previously poisoned with methamidophos and other organophosphate pesticides, *Am. J. Ind. Med.* **25**: p 325.

71. McKeaon-Eyssen, G.E. and Ruedy, J. (1983) Prevalence of neurological abnormality in Cree Indians exposed to methylmercury in Northern Quebec. *Clin. Invest. Med.* **6**: pp 161–169.

72. McLaughlin, A.I.G., Kazantzis, G., King, E., Teare, D., Porter, R.J., Owen, R. (1962) Pulmonary fibrosis and encephalopathy associated with the inhalation of aluminum dust. *Br. J. Ind. Med.* **19**: pp 253–263.

73. McNair, D.M., Loor, M., Droppleman, L. (Eds.) (1981) *Manual: Profile of Mood States*. San Diego, CA.

74. Mergler D (1994) Neurotoxicology of the visual system. Part I. Early indications of visual dysfunction. In: Bleecker, M.L. (Ed.) *Occupational Neurology and Clinical Neurotoxicology*. Williams & Wilkins, Baltimore, MD, Chapter 7.

75. Mergler, D. (1994) An update of early manifestations of manganese neurotoxicity in humans. *5th International Symposium on Neurobehavioral Effects and Methods in Occupational and Environmental Health*, December, Cairo.

76. Mergler, D. (1995) Behavioral neurophysiology: quantitative measures of sensory toxicity. In: Chang, L. and Siller (Eds.) *Neurotoxicology: Approaches and Methods*, Academic Press, New York, pp 727–736.

77. Mergler, D., Huel, G., Bowler, R., et al. (1995) Nervous system dysfunction among workers with long-term exposure to manganese. *Environ. Res.*, in press.

78. Morris, H.H. and Estes, M.L. (1987) Traveller's transient global amnesia secondary to triazolam. *JAMA* **258**: p 945.

79. Mutti, A., Falzoi, M., Romanelli, A., Lucertini, S., Franchini, I. (1984) Regional alterations of brain catecholamines by styrene exposure in rabbits. *Arch. Toxicol.* **55**: p 173.

80. Mutti, A., Romanelli, A., Falzoi, M., Lucertini, S., Franchini, I. (1985) Styrene metabolism and striatal dopamine depletion in rabbits. *Arch. Toxicol.* [suppl 8]: p 447.

81. Osetowska, E. (1971) Metals. In: Minckler, J. (Ed.) *Pathology of the Nervous System*, vol. 2. McGraw-Hill, New York, pp 1644–1651.

82. Parkinson, I.S., Ward, M.K., Feest, T.G., Fawcett, R.W.P., Kerr, D.N.S. (1979) Fracturing dialysis osteodystrophy and dialysis encephalopathy. An epidemiological survey. *Lancet* **1**: pp 406–409.

83. Perl, T.M., Bédard, L., Kosatsky, T., Hockin, J.C, Todd, E.C.D., Remis, R.S. (1990) An outbreak of toxic encephalopathy caused by eating mussels contaminated with domoic acid. *N. Engl. J. Med.* **322**: pp 1775–1780.

84. Perl, D.P., Gajdusek, D.C., Yanagihara, R.T., Gibbs, C.J. (1982) Intraneural aluminum accumulation in amyotrophic lateral sclerosis and parkinsonism-dementia of Guam. *Science* **217**: pp 1053–1055.

85. Rajput, A.H., Uitti, R.J., Stern, W. et al. (1987) Geography, drinking water, industry, pesticides and herbicides and the etiology of Parkinson's disease. *J. Neurol. Sci.* **14**: pp 414–418.

86. Ramazzini, B. (1990) *De Morbis Artificum Diatriba-1700*, translated by de Foucray, A. (Ed.) AlexitÅre Editiona, Paris.

87. Rice, Gilbert (1990) Effects of developmental exposure to methylmercury on spatial and temporal visual function in monkeys. *Toxicol. Appl. Pharmacol.* **102**: pp 151–163.

88. Rosenberg, N.L. (1994) Toxic encephalopathies. In: *Occupational Neurology and Neurotoxicology*, Syllabus #132. American Academy of Neurology Annual Meeting, pp 29–44.

89. Rosenstock, L., Keifer, M., Daniell, W.E., McConnell, R., Claypoole, K. (1991) The Pesticide Health Effects Study Group. Chronic central nervous system effects of acute organophosphate pesticide intoxication. *Lancet* **338**: pp 223–227.

90. Ryan, C.M., Morrow, L.A., Hodgson, M. (1988) Cacosmia and neurobehavioral dysfunction associated with occupational exposure to mixtures of organic solvents, *Am. J. Psychiatry* **145**: p 1442.

91. Sassine, M.P., Mergler, D., Bélanger, S., Larribe, F. (1995) Détérioration de la santé mentale chez des travailleurs exposés au styrène, *Rev. d'épidémiologie et santé publique*, in press.

92. Schupf, N., Silverman, W., Zigman, W.B., Moretz, R.C., Wisniewski, H.M. (1989) Aluminum and Alzheimer's disease. *Lancet* **1**: p 267.

93. Sechi, G.P., Agnetti, V., Piredda, M., Canu, M., Deserra, F., Omar, H.A., Rosati, G. (1992) Acute and persistent parkinsonism after use of diquat. *Neurology* **42**: pp 261–263.

94. Spencer, P.S., Nunn, P.B., Hugon, J., Ludolph, A.C., Ross, S.M., Roy, D.N., Robertson, R.C. (1987) Guam amyotrophic lateral sclerosis-parkinsonism-dementia linked to a plant excitant neurotoxin. *Science* **237**: pp 517–522.

95. Stern, Y., Langston, J.W. (1985) Intellectual changes in patiens with MPTP-induced parkinsonism. *Neurology* **35**: pp 1506–1509.

96. Steventon, G.B., Heafield, M.T.E., Sturman, S., Waring, R.H., Williams, A.C. (1990) Xenobiotic metabolism in Alzheimer's disease. *Neurology* **40**: pp 1095–1098.

97. Sunchuck, K.M., Love, E.J., Lee, R.G. (1993) Parkinson's disease: a test of the multifactorial etiologic hypothesis. *Neurology* **43**: pp 1173–1180.

98. Thomas, J.A., Dallenbach, F.D., Thomas, I.M. (1971) Considerations of the development of experimental lead encephalopathy. *Virch.ow's Arch fur Pathol. Anat. Physiol.* **352**: pp 61–74.

99. Tipton, K.F., McCrodden, J.M., Sullivan, J.P. (1993) Metabolic aspects of the behavior of MPTP and some analogues. *Adv. Neurol.* **60**: pp 186–193.

100. Tokuomi, H., Uchino, M., Imamura, S., Yamanaga, H., Nakanishi, R., Ideta, T. (1982) Minamato disease (organic mercury poisoning): neuroradiologic and electrophysiologic studies. *Neurology* **32**: pp 1369–1375.

101. Valpey, R., Sumi, M., Copass, M.K., Goble, G.J. (1978) Acute and progressive encephalopathy due to gasoline sniffing. *Neurology* **28**: pp 507–510.

102. Vergiftiging door Datura fastuosa (ketjubung) (1985). *Ned. Tijdschr. Geneeskd.* **129**: nr 29.

103. Widner, H., Tetrud, J., Rehncrona, S., et al. (1992) Bilateral fetal mesencephalic grafting in two patiens with parkinsonism induced by 1-methyl-4-phenyl-1,2,3,6-tetrahydropyridine (MPTP). *N. Engl. J. Med.* **327**: pp 1556–1563.

104. Williams, A.C., Pall, H.S., Green, S., Buttrum, S., Molloy, H., Waring, R.H. (1993) N-methylation of pyridines and Parkinson's disease. In: Narabayashi, H., Nagatsu, T., Yanagisawa, N., Mizuno, Y. (Eds.) *Advances in Neurology*, vol. 60. Raven Press, New York, pp 194–196.

105. Wisniewski, H., Terry, R.D., Hirano, A. (1970) Neurofibrillary pathology. *J. Neuropathol. Exp. Neurol.* **29**: pp 163–176.

106. Yoshida, M., Ogawa, M., Suzuki, K., Nagatsu, T. (1993) Parkinsonism produced by tetrahydroisoquinoline (TIQ) or the analogues. *Adv. Neurol* **60**: pp 207–211.

107. Zayed, J., Ducic, S., Campanella, G. et al. (1990) Facteurs environmentaux dans l'étiologie de la maladie de Parkinson. *Can. J. Neurol. Sci.* **17**: pp 286–291.

Additional references

Arlien-Søberg, P. (1992) *Solvent Neurotoxicity*. CRC Press Boca Raton, FL.

Bleecker, M.L. (Ed.) (1994) *Occupational Neurology and Clinical Neurotoxicology*. Williams & Wilkins, Baltimore, MD.

Change, L.W. and Slikker, Jr., W. (Ed.) (1995) *Neurotoxicology: Approaches and Methods*. Academic Press, San Diego.

Hart, D.E. (1988) *Neuropsychological Toxicology: Identification and Assessment of Human Neurotoxic Syndromes*. Pergamom Press, New York.

Johnson, B. (Ed.) (1987) *Prevention of Neurotoxic Illness in Working Populations*. World Health Organization, John Wiley & Sons, New York.

Johnson, B. (Ed.) (1990) *Advances in Neurobehavioral Toxicology*. Lewis Publishers, Ann Arbor, MI.

Valciukas, J. (1991) *Foundations of Environment and Occupational Neurotoxicology*. Van Nostrand Reinhold, New York.

Contents Chapter 6

Genetic aspects of behavioral toxicity

Chapter 6

Genetic aspects of neurobehavioral toxicology

Frank R. George

1 Introduction

As biological organisms, we are sensitive to agents in the environment which disrupt physiological processes. For centuries, humankind has been aware of a variety of substances which can produce toxic effects. The range of toxic effects and their specific mechanisms of action are substantial. This chapter begins with a brief discussion of toxic agents and several of the ways in which they produce their adverse effects. A second section describes some important genetic methods used in the analysis of genetic risk for susceptibility to toxic agents. Subsequently, individual differences in responsivity to some of these agents are described. The chapter ends with a discussion of the importance of interactions between genes and environments, and some of the scientific and ethical questions raised by our growing understanding of genetic risk factors in toxicology.

1.1 TOXIC AGENTS ARE PART OF OUR ENVIRONMENT

The effects of toxic agents on biological organisms can range from acute alterations in physiology and behavior such as depression, stimulation, coma, or seizures, to chronic effects such as increased risk for cancer and other diseases, decreased learning capabilities, or motor disorders. While some toxic agents produce relatively specific effects at certain receptor or tissue sites, many toxicants produce a number of effects based upon a broader, multiple tissue toxicity. Individual differences in response to exposure suggest differences in the degree of inherited risk for susceptibility to these environmental toxicants.

When thinking about toxicity, many people think of poisons, radiation, and other severely harmful agents which make up part of the air we breathe or the materials we work with in our industrial/technological society. It is important to realize that most toxic substances begin as a natural part of our world and have been in existence throughout the course of human evolution. However, the form in which some of these toxic agents now exist may have been devised only recently. Toxicity can be in the form of an acute episode, such as ingestion of a poison, or a build-up of effect over a period of days, months or years, such as the result of chronic tobacco smoking or exposure to certain pesticides. While many substances can be toxic even in small concentrations, just about any substance has the potential to be toxic, given the appropriate circumstances.

Recent work has shown that individuals can differ in their biological responsivity to toxic agents. Some persons, by reason of their genetic heritage, are less prone towards developing skin cancers, and thus ultraviolet radiation has less impact upon them, at least with regards to this effect. Inheritance also appears to play an important role in determining the degree of toxicity seen with regards to behavioral changes resulting from exposure to environmental toxicants, such as drugs and certain metals.

Why are there individual differences in response to toxic agents? What is it about certain persons or animals which causes these differences? This chapter will explore these questions and help to answer them by introducing the reader to the study of a newly emerging field known as toxicogenetics.

1.2 TECHNIQUES AND TERMS USED IN TOXICOGENETICS

Many terms important to understanding neurobehavioral toxicology have been presented in the preceding chapters. However, there are some additional terms which are important in learning about the genetic aspects of toxicology: some of these are presented here.

1.2.1 *Inbred strains*

Inbreeding is a reduction in genetic variance within a population so that individuals become more genetically alike with subsequent generations. Inbreeding results from the mating of individuals who are closely related, such as siblings. A strain maintained by brother-sister matings for at least 20 generations is considered to be inbred, and use of the word "strain" denotes a history of inbreeding.

Most of us are aware of the deleterious effects of inbreeding on survival. For example, virtually all cultures have incest taboos which limit the mating of close relatives. We all carry a few deleterious genes in our chromosomes. However, most of us are fortunate in that out of each gene pair only one is a deleterious, or potentially lethal, gene, while the other in the pair is normal. However, if two closely related individuals mate, they may each contribute their lethal gene to their offspring, with the result that the offspring receives a pair of the deleterious gene and does not survive. An example of this deleterious effect of inbreeding is the high rate of blood disorders, such as hemophilia, within certain European families. In addition, animals which are highly inbred appear to be at greater risk for showing deleterious responses to many types of toxic substances, such as cancers caused by radiation, or impaired behaviors following exposure to certain drugs.

Inbreeding can, however, be of significant value. Because members of an inbred strain are genetically identical, use of a single strain controls for genetic variation, resulting in controlled populations over time and across laboratories. This eliminates much of the need for repeated replications of basic findings. Thus, if inbred strains can be developed which survive the perils of inbreeding, these animals can provide a powerful and economical approach to research.

1.2.2 *Selective breeding*

Selective breeding is a program of specific matings over a number of generations with the intention of changing particular traits. Selection is often conducted in a bidirectional manner to produce maximally distinct populations, for example high tumor susceptibility versus low tumor susceptibility. There are a number of reasons for conducting a selective breeding program. One is that if the selection is successful, this is a strong indication that genetic factors influence the trait being measured. Another reason is that once these different populations are created, they may be studied to help understand the genetic and biological bases of the trait for which they were bred.

Several programs have been conducted in which mice have been selectively bred for toxic responses to ethanol. For example, mice will experience alcohol withdrawal when, after they have been made physically dependent on ethanol by chronic exposure to this drug, the ethanol is no longer provided. One of the common symptoms of alcohol withdrawal is a seizure. Based upon this effect, mice have been bred to be either prone to having withdrawal seizures (withdrawal seizure-prone mice (WSP)) or resistant to withdrawal seizures (withdrawal seizure-resistant (WSR) mice.)

1.2.3   *Genetic correlation*

A powerful genetic tool is a methodological approach known as genetic correlation. Genotypes which differ for a given trait can be used to test relationships between variables by determining correlations between traits hypothesized to be causally related. A lack of correlation indicates that the measures studied are not mechanistically related. A strong positive correlation, especially a perfect rank order correlation across genotypes, enables one to reasonably conclude that the measures are causally related. The incorporation of several inbred strains into an experiment can provide important insights into the genetic and biochemical mechanisms of the toxic effect under study.

1.2.4   *Recombinant inbred strains*

Another behavioral genetic method, one which is gaining rapidly in popularity and utility, is the use of recombinant inbred strains. Recent advances in molecular biology and genetics are part of the reason for the increasing interest in this approach. To understand this method, one need only look at its' name. When using the recombinant inbred strain method, one simply takes two inbred strains and "recombines" them to produce new genetic strains. By recombine we mean that the two strains, and thus their genes, are cross-mated to produce new combinations, or recombinations, of genes. Following this initial recombination, the new animals which have been produced are then systematically inbred as described above to produce several new inbred strains. An important feature of these new inbred strains is that they are all derived from the same two parent strains.

Recombinant inbred (RI) mice were originally developed to identify characteristics mediated by a single gene controlling a primary physiological effect. However, recently RI mice have begun to be used to identify genes responsible for less major physiological effects, such as a gene important in determining whether alcohol has a slightly more positively or negatively perceived taste. These genes associated with measurable (i.e., quantitative) but minor aspects of various traits are known as quantitative trait loci (QTL), referring to the specific chromosomal location, or locus, where the DNA coding for the trait resides. QTL represent genes that contribute in detectable but minor ways to traits such as behavioral responses to toxic agents where there are probably several genes involved. Thus, the use of QTL analyses has greatly expanded the geneticist's tool kit for understanding the mechanisms of behaviors and other traits. Instead of being limited to studying those traits governed by a single major gene, now researchers can study gene loci which contribute to as little as 15 or 20% of the variation in the trait being measured. For example, the genetic basis for Huntington's disease was described recently, and we now know that this neural degenerative disease is due to a single major gene. However, other neural degenerative disorders, such as Alzheimer's disease, also appear to have a significant genetic basis, but are controlled by more than one gene. Here, the use of QTL analyses to associate several gene loci, each of which contributes in a small but significant manner to the development of the disease, is allowing researchers to uncover the complex genetic and biochemical basis for this disorder, as discussed later in this chapter.

The utilization of quantitative trait loci in toxicogenetics has several advantages. These include prospects of generating animal models of toxic effects, studying the dynamics of gene interaction relevant to toxic responses, and providing clues in the search for similar, or homologous, genes in humans. Taken together, several QTL may account for a substantial proportion of the variability seen in response to a toxicant, providing researchers with the prospect of being able fully to describe the genetic mechanisms responsible for determining biological risk to various environmental agents.

1.2.5   *Transgenic mouse strains*

The rapid expansion of knowledge and techniques in the area of molecular genetics has provided the basis for developing new biological organisms which could not occur through the normal biological processes of sexual reproduction and natural selection. One example of such organisms which are proving to be useful in furthering our understanding of many disease processes are transgenic mice.

Transgenic mice are animals which have had a new gene inserted into their DNA, so that the animals incorporate the new gene into their physiology and pass that gene on to their offspring. In this way, the effects of a single specific gene can be studied in detail. While the creation of a transgenic animal could theoretically occur in any species, mice are popular for this approach because they are small, much is known about their genetic makeup, and they are mammals with physiological systems similar to those in humans.

Thus, there are many tools which toxicogeneticists have at their disposal. Some of these tools have been used in research for many years, while others have been developed recently. But, regardless of which methods are used, these genetic approaches are providing important insights into the mechanisms by which substances produce toxic effects as well as the specific genetic differences that affect an individual's susceptibility to various forms of toxicity.

2   Genetic factors in systemic effects of toxicants

2.1   IMPORTANCE OF PERIPHERAL TOXICITY TO CHANGES IN BEHAVIOR AND BRAIN FUNCTION

The focus of this chapter is on genetic factors that may regulate behavioral responses to toxic agents. However, one important manner in which behavioral response to a toxicant may be altered is through alterations in the course or process by which the toxicant moves through a person's body and its associated physiological compartments and organs.

One important way in which we can be protected against many toxic compounds is through the physiological processes associated with metabolism. Most metabolism, or transformation, of substances occurs in the liver, which is an organ critical for the process of detoxification. The liver is where many substances are found in high concentration since that is where they must accumulate in order to be metabolized. Thus, the liver can be highly susceptible to the toxic effects of these agents. If the liver is damaged through exposure to a toxicant, an effect known as hepatotoxicity, the liver may become impaired in its ability to break down not just toxic substances taken in from the environment, but also chemicals which occur naturally within our bodies. The results of this impairment may have serious negative consequences on many physiological processes, including many behaviors. As we will see below, sometimes our bodies' own chemicals can be toxic if they are not removed appropriately, and this can significantly damage the nervous system and impair behavior.

2.2   EFFECTS OF TOXICANTS ON METABOLISM AND ORGAN FUNCTION

Since our blood circulates chemicals through the liver where they can be metabolized for eventual elimination from the body, highly toxic compounds can affect the ability of the liver to function properly, with adverse consequences for the liver as well as many other tissues and organs. Thus, there have been many studies investigating the effects of various substances on liver function. Recently, there has been increased interest in genetic vulnerability and individual differences in hepatotoxicity.

One group of compounds for which the liver plays a vital role in metabolism and detoxification is drugs of abuse, such as cocaine, alcohol, nicotine, and heroin. When most people think about the effects of these drugs, they tend to think mostly about their euphoric or addicting effects. However, it is important to understand that drugs almost always have more than one effect, and often drug use can have serious toxic consequences. Because alcohol and cocaine have widespread use and much is known about their chemistry and toxicity, our discussion on peripheral organ toxicity will focus on susceptibility to the toxic effects of these two addictive substances.

2.2.1   *Susceptibility to cocaine hepatotoxicity*

Both gender and inbred strain differences have been reported for hepatotoxicity in response to cocaine. Even a single exposure to cocaine can produce substantial changes in liver function, including the ability of certain enzymes to function properly as well as direct lethal effects of cocaine on certain liver cells. The genetic differences in the effects of cocaine on liver function may be related to differences in responses to cocaine. For example, individuals who are more able to metabolize cocaine will have a decreased risk for cocaine overdose and symptoms such as seizures or cardiac arrest. Changes produced in the liver by chronic exposure to cocaine may be related to long term changes in response to cocaine. For example, it appears that some individuals become more susceptible to certain toxic effects of cocaine, such as seizures, following repeated exposure to this drug. There also appear to be genetic differences with regards to the extent of this sensitizing effect.[19]

2.2.2   *Susceptibility to alcohol hepatotoxicity*

Advanced alcoholism is the primary cause of a liver disease known as cirrhosis. Cirrhosis of the liver is a chronic disease characterized by the replacement of normal tissue with fibrous tissue and the loss of functional liver cells. In one study[14] of nearly 16,000 middle-age male pairs of twins, concordances for liver cirrhosis were 15% for identical twins and 5% for fraternal twins. Thus, if one twin acquired this disease, there was a 15% probability that the co-twin would have the disease if the twins were identical, and a 5% probability if the co-twin was a non-identical, that is fraternal, twin. The fact that even a 5% probability of acquiring this disease is much higher than that seen in the overall population, plus the fact that the concordance rate was higher for identical twins than for fraternal twins, is strong evidence for a genetic risk factor. Since twins have been the subjects for many studies on genetic risk factors in alcohol and drug abuse, the use of concordance rate studies with twins will be discussed more fully in Chapter 18.

Ethanol can be broken down, or metabolized via a number of enzyme pathways in the body. However, the primary pathway for ethanol metabolism is via the enzymes alcohol dehydrogenase (ADH) and aldehyde dehydrogenase (ALDH), as illustrated in Figure 6.1. Liver ADH and ALDH are polymorphic in human beings. This means that different individuals may possess different versions of these enzymes. The different versions exist because of slight genetic differences with regards to the DNA codes for these enzymes. The different genetic codes result in the production of enzymes which have slightly different structural configurations, and hence, somewhat different, although still highly similar, properties.

Genetically determined differences in the enzymes of alcohol metabolism have been shown to be important in contributing to susceptibility to liver damage from alcohol consumption. In one study[19] of alcohol misusers versus healthy subjects, two different versions of the liver ADH enzyme were found. These two ADH polymorphisms were termed version "A" and version "B". As shown in Figure 6.2, 85% of the control subjects had version "A" of the ADH enzyme, while only 37% of the alcohol misusing subjects had the "A" enzyme. In contrast, while only 15% of the control subjects had the "B" enzyme, 63% of the alcohol misusers had this version of liver ADH. Thus, in this study, the "B" liver ADH allele was significantly associated with liver damage, alcohol abuse, and family history of alcohol abuse. These findings suggest that genetic differences in the enzymes of alcohol metabolism may contribute to both susceptibility to alcohol induced liver damage as well as susceptibility to alcohol abuse.

Other studies have substantiated these findings.[4] Overall, it appears that genetically determined differences in alcohol metabolism may, in part, explain predisposition to alcohol-related cirrhosis, which itself is a consequence of long term alcohol misuse.

Many persons of Asian descent have genes which code for the production of relatively inactive forms of ADH or ALDH. Because these persons can become quite ill after consuming modest amounts of alcohol, the rates of alcohol-related

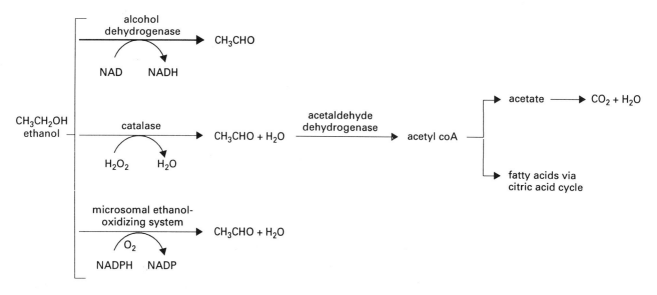

FIGURE 6.1

Schematic representation of ethanol metabolism pathways
Bold lines and print represent primary metabolic path.

196

toxicities, such as liver disease, are fairly low in these populations. However, these diseases still exist, and it is of interest to determine the relationship between the presence or absence of active ADH and ALDH enzymes and the development of alcoholism and associated liver toxicities.

Polymorphisms of alcohol and aldehyde dehydrogenase genes have been found which are associated with alcoholic cirrhosis in Chinese patients.[4] The data strongly suggest that genetic variation in both ADH and ALDH influences drinking behavior and risk for alcoholism.

2.2.3   *Susceptibility to alcohol pulmonary toxicity*

ADH and ALDH enzymes also exist in other tissues, such as the lungs. If individuals carry the inactive form of ALDH they will have an impaired ability to break down acetaldehyde and eliminate this highly toxic substance from their tissues. If acetaldehyde builds up in the lungs it may produce severe tissue damage. Fortunately, most individuals appear to have active forms of the necessary enzyme. However, within certain Chinese and other Asian populations there appears to be a significant portion of persons who have the inactive form of the ALDH enzyme.

For example, one study of Chinese patients found that 51% lacked ALDH enzyme activity in the lungs[27]. These findings suggest that individuals with the inactive form of this enzyme would likely end up with acetaldehyde accumulation if they consume alcohol, rendering them susceptible to lung tissue injury caused by this toxic metabolite of alcohol. However, it is not yet clear what the relationship is between the presence of the slow form of ALDH, acetaldehyde accumulation, and the onset of pulmunary disorders.

3   Genetic factors in effects of drugs on behavior

3.1   ACUTE TOXIC EFFECTS OF DRUGS ON BEHAVIOR

3.1.1   *Cocaine*

Cocaine and amphetamine are effective reinforcers and have become popular drugs of abuse. Both drugs have traditionally been viewed as belonging to the same pharmacological class since they produce numerous physiological, behavioral, and subjective effects in common, primary among them being marked physiological responses (e.g., changes in heart rate and blood pressure) as well as behavioral effects including euphoria and central nervous system (CNS) stimulation.

In laboratory animals, the CNS effects of low cocaine doses have been observed as decreases in locomotor activity, while moderate doses typically increase locomotor activity.[11] However, large doses of cocaine can induce stereotypy, disrupt behavior, and produce severe toxic responses including seizures and death.[9,10] Similarly, low to moderate doses of amphetamine produce increases in locomotor activity, while higher doses produce stereotypy and death.[11] Unfortunately, with the recent increase in self-administration of crack, a highly potent and rapidly absorbed form of cocaine, the incidence of toxic responses to cocaine, especially seizures and death, has risen significantly.[16]

Inherited factors play an important role in determining the magnitude of behavioral responses to cocaine, amphetamine, and other stimulant drugs. One way in which these effects have been studied in animals is by measuring the rate and amount of movement, or locomotion, an animal, such as a rat or mouse, makes within a specified period of time following a dose of the drug

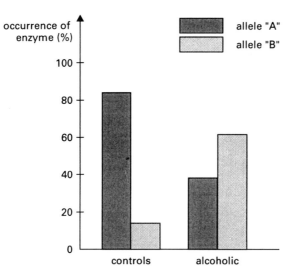

FIGURE 6.2

Percent of occurrence of two versions of liver ADH in control and alcoholic subjects
(Data from Sherman, D.L. et al.[19])

(Figure 6.3). For example, mice from some inbred strains, such as the C57BL/6J, are substantially more activated by cocaine than mice from other inbred strains, such as the A/J (Figure 6.4).[21] There are also substantial genetic differences in the effects of cocaine on heart and respiration rates.[18] In addition, substantial genetic differences have been found in susceptibility to seizures induced by cocaine[9], as well as the lethal effects of this drug.[10]

Reports of death attributable to cocaine appeared before the beginning of the twentieth century, and the first report of convulsions associated with

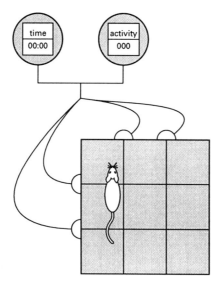

FIGURE 6.3

A representation of a locomotor activity monitoring device for use with rodents
An electronic grid is constructed through the use of photocell beams and detectors. When an animal interrupts one of the beams, it is the electronic equivalent of crossing over a line. Interrupting any of the beams causes an activity counter to increment. The device is turned on and off by an electronic timer. In this way, the amounts of movement the animal makes during a specified time period can be accurately assessed. Some highly sophisticated monitors can record not only the amount of activity, but the patterns of behavior as well.

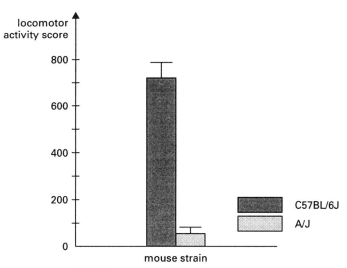

FIGURE 6.4

A bar graph representing averaged locomotor activity for mice from two different inbred strains

After the same 20 mg/kg dose of cocaine, mice from the C57BL/6J strain become more activated than do mice from the A/J strain. The vertical and horizontal lines at the top of each bar represent the standard error of the mean for each strain, a measure of variability in response within each group. (Data from Shuster, L. et al.[21])

cocaine use appeared shortly thereafter.[16] More recently, it has been demonstrated that there are important genetic differences in the convulsant and lethal responses to toxic doses of cocaine. Large differences in sensitivity to cocaine-induced seizures have been found between various strains of rats and mice, such as the SS and LS mice (Figure 6.5).[9] This is consistent with the apparent wide variation in susceptibility to toxic effects of cocaine in humans.

Genetic differences with regard to the lethal effects of cocaine have also been studied in rodents. In one experiment,[10] four rat strains, the ACI, F344, LEW and NBR, showed significant genetic differences in cocaine-induced

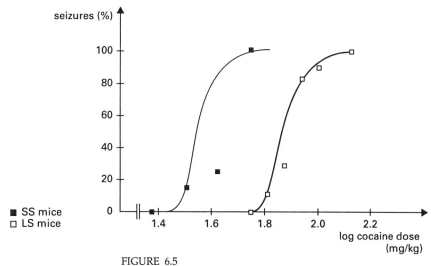

FIGURE 6.5

Percent seizures as a function of the logarithm of cocaine dose in SS/Ibg and LS/Ibg male mice

Filled squares denote SS mice. Open squares denote LS mice. 1.4 = log of 25 mg/kg cocaine. 2.0 = log of 100 mg/kg cocaine. (Adapted from George, F.R. (1991) *Psychopharmacology* **104**: p 307.)

199

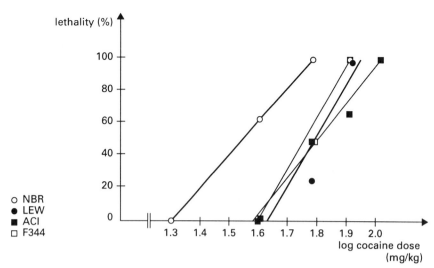

FIGURE 6.6

Percent lethality as a function of the logarithm of cocaine dose in ACI, F344, LEW and NBR male rats

As the dose of cocaine is increased, the lethal effect increases accordingly. 1.4 = log of 25 mg/kg cocaine. 2.0 = log of 100 mg/kg cocaine. (Adapted from George, F.R. (1991) *Pharmacol. Biochem. Behav.* **38**: p 893.)

lethality, with a nearly two-fold difference in sensitivity to cocaine toxicity seen between the most sensitive (NBR) and least sensitive (ACI) strains (Figure 6.6). Interestingly, sensitivity to the lethal effects of cocaine was not correlated with sensitivity to the stimulant effects of this drug. These findings suggest that the stimulant and lethal effects of cocaine are not highly related. This may help to explain the sudden death syndrome associated with cocaine use: some individuals may be more sensitive to the lethal effects of cocaine and so succumb to toxicity at doses that, for them, produce marginal euphoric or stimulant effects.

3.1.2 *Amphetamine*

Several reports have also shown that genetic differences exist in response to amphetamine, and these differences have been linked to differences in dopamine pathways and function in the brain. For example, the effects of amphetamine on locomotor activity have been studied in rats from the ACI, F344, LEW, and NBR inbred strains.[11] Dose-dependent increases in locomotor activity were found in all strains. However, large potency and efficacy differences were found. Rats from the ACI and Lewis strains were much more sensitive to the stimulant effects of amphetamine, while NBR rats were much less sensitive (Figure 6.7). In addition, although F344 rats were relatively sensitive to amphetamine, the maximum effect of the drug, that is, its efficacy in producing a stimulant effect, was limited in these animals. In contrast, while NBR rats were relatively insensitive to amphetamine, once a sufficient dose was achieved these rats showed high levels of stimulant response.

Interestingly, a similar study of cocaine locomotor activity effects in rats from these same four strains showed that the strain rank order for stimulant response to the two drugs was not identical.[11] In other words, rats which were highly responsive to amphetamine were not necessarily as responsive to cocaine, and vice versa. Thus, while both drugs produce robust stimulant as well as other toxic effects, these findings support the conclusion that these two drugs produce their effects through different sites of action.

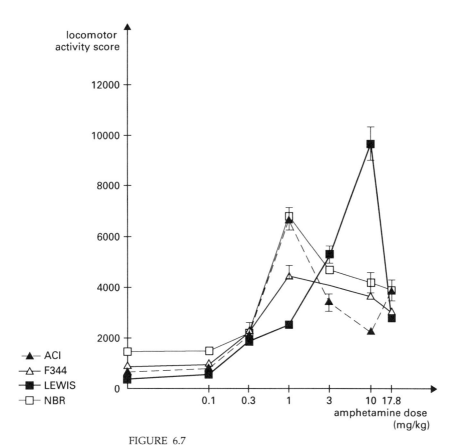

FIGURE 6.7

Sixty-minute locomotor activity scores as a function of amphetamine dose in four inbred rat strains

Points and bars represent mean ± SEM. (Adapted from George, F.R. (1991) Cocaine toxicity: genetic differences in cocaine-induced lethality in rats. *Pharmacol. Biochem. Behav.* **38**: p 893.)

3.2 TOXIC EFFECTS ON BEHAVIOR FOLLOWING CHRONIC EXPOSURE TO DRUGS

3.2.1 *Enhanced responsivity to cocaine*

A series of studies has indicated that genotype influences sensitization to cocaine effects following repeated exposure to the drug. Shuster et. al.[21] reported that greater sensitization to the locomotor effects of cocaine was observed in A/J mice relative to C57BL/6J mice. Furthermore, the results from studies of recombinant inbred strain mice suggest that the level of initial loco-motor stimulant response to cocaine and sensitization to cocaine following repeated exposure to this drug are mediated by separate genetic determinants[21].

3.2.2 *Neurodegeneration caused by methamphetamine and related drugs*

Methamphetamine (METH) is another stimulant-type drug which, in addition to its stimulant and addictive properties, is known to produce severe damage to certain neuronal pathways in the brain. This damage produces notable changes in behavior ranging from alterations in locomotor activity and patterns of locomotion to changes in the ability to control limb movements. In fact, these neurotoxic effects are so robust that METH is commonly used as an experimental agent in studies on brain mechanisms of behavior. However, it is also known that there are substantial species and individual differences in response to METH. Some species apparently show little or no toxic response to METH, and reports of human brain damage related to METH abuse suggest wide variation in the levels of METH used and the length of time during which the use occurred.

In terms of its actual effects on neurons, administration of METH causes damage to and loss of neurons in brain pathways which are responsive to the neurotransmitter dopamine. However, the mechanism by which METH causes these neurotoxic effects remains unknown. One hypothesis is that METH somehow causes the production of molecules known as oxygen radicals. These molecules are highly corrosive and are known to cause damage to cells similar to that seen following administration of METH.

Using genetic models is a useful approach to determining if differences seen in response to METH may be genetic in origin, as well as for testing hypotheses concerning the biological mechanisms of METH neurotoxicity, such as the oxygen radical hypothesis. To test the oxygen radical hypothesis in METH-induced neurotoxicity, a transgenic mouse model has been developed.[2] In this study, transgenic mice were created which express the human gene for an enzyme known as superoxide dismutase (SOD). This enzyme is important to human cellular function as it is responsible for the neutralization of harmful oxygen radicals. Thus, if METH is producing neurotoxicity by causing the formation of oxygen radicals, then the SOD transgenic mice should show less neurotoxicity than the control non-SOD mice since the SOD transgenic mice containing the human SOD gene should have an enhanced capability for neutralizing the oxygen radical molecules before they cause damage to neurons.

In non-SOD transgenic mice, METH administration caused significant decreases in levels of dopamine. In contrast, there were no significant decreases in dopamine in the SOD transgenic mice. Thus, these results suggest that METH-induced neurotoxicity may in fact result from an increased production of oxygen radicals in dopaminergic neurons. The use of a transgenic mouse model in this study[2] illustrates the elegant manner in which genetic approaches may be used to test hypotheses concerning neurotoxic mechanisms.

3.2.3 *Neurodegeneration caused by MDMA*

Over the past several years another derivative of amphetamine, 3,4-methylenedioxymethamphetamine, commonly known as MDMA or "ecstasy", has gained popularity as a recreational drug. MDMA produces a complex array of psychoactive effects, including euphoria and hallucinations, and it appears to be for these reasons that the drug is used in non-therapeutic settings. Recently, however, it has been shown that MDMA is capable of producing serious neurotoxic effects which result in alterations in behavior, affecting such things as locomotor activity and limb movement. Interestingly, the actual sites of neurotoxicity resulting from exposure to MDMA appear to differ from the neurotoxicity sites associated with METH.

As discussed above, METH exposure results in depletion of dopamine pathways in the brain. However, MDMA exposure causes the degeneration and depletion of serotonin pathways in the brain. No direct comparisons of MDMA neurotoxicity have been made across strains within a single species such as mice. However, studies[1,22] have shown large differences in the effects of MDMA across different species, suggesting that a significant genetic component regulates responsivity to this drug.

3.2.4   *Neurodegeneration caused by MPTP*

MPTP is a compound which has been discussed extensively in chapters 4 and 5. MPTP appears to be the cause of serious motor behavior effects which have been described in a number of drug abusers. These effects resemble the symptoms of Parkinson's disease, a neurodegenerative disorder which will be described below. While several persons known to use MPTP have been stricken with a parkinsonian-type disorder, others using this drug have suffered little or no lasting consequences in terms of motor behavior and performance. These individual differences in response to MPTP raise the possibility that there may be genetically conveyed differences in the degree of risk for this neurotoxic effect on behavior. A number of studies[13,22] have established that there are in fact significant genetic differences in the neurotoxic response to MPTP and that these genetic differences may play a role in determining the level of susceptibility to the toxic effects of this drug.

Thus, a number of psychoactive drugs which are used primarily for their stimulant and euphoric effects are also capable of producing serious neurotoxic effects. These effects are seen as profound changes in performance related to motor skills such as locomotion and limb movement and control. The use of genetic models in this area has provided substantial evidence that genetic differences exist in the degree of risk for these neurotoxic effects, and they have contributed to an improved understanding of the specific neuronal mechanisms through which these drugs produce their toxic effects.

3.3   GENETIC FACTORS IN SENSITIVITY TO METAL-INDUCED NEURO- AND BEHAVIORAL TOXICITY

It is known that exposure to high concentrations of certain metals can produce substantial damage to the nervous system and impair behavior, as discussed in detail in Chapter 12. In part because of its wide use in most parts of the world today, the potential toxic effects of aluminum have been studied more that than those of most other metals, with the possible exception of lead. Aluminum neurotoxicity has been suggested to play a role in the development of senile dementia related to Alzheimer's disease, as described in Chapter 5. However, the specific role and extent of aluminum exposure in generating senile dementia is unclear, since not all exposure to aluminum results in neurotoxicity and the development of Alzheimer's disease.

Concentrated aluminum produces accumulations of neurofilaments in certain nerve cells, or neurons. This accumulation of filaments, which are similar to very fine strands or tiny tubular lengths of proteins, results in disruptions of neuronal cell function. What causes the production of these abnormalities is unclear. However, the ability of the genes which control the production of the filament proteins to function properly is changed following exposure to aluminum.

It may be that the less than clear relationship between aluminum exposure and the development of behavioral disorders is due to different degrees of genetic susceptibility to the effects of aluminum. The results of studies using inbred strains of mice suggest that this may be the case. For example, in one study,[7] mice from five different inbred strains were divided into two groups each. For each inbred strain, one group, the control group, was fed a normal diet, while the second group was fed the same diet plus an additional amount of aluminum. After receiving the diets for 4 weeks, brains, livers, and bones were analyzed for their aluminum content. The results showed that there were substantial differences between the different genetic strains with regards to how much aluminum had accumulated in the different tissues. In addition, while in some strains, such as the DBA/2J strain, the aluminum-fed group showed higher concentrations of this metal in their tissues, other strains, such as the A/J, the aluminum-fed group was not different from the control group in terms of aluminum accumulation. Thus, variability in aluminum toxicity may be, in part, due to genetic differences in the degree to which individuals accumulate toxic concentrations of this metal in their tissues.

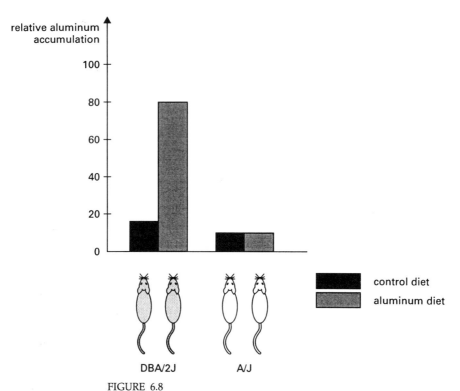

FIGURE 6.8

Genetic differences in accumulation of aluminum in mice

Following chronic dietary exposure to high levels of aluminum, mice of the DBA/2J strain show a substantial accumulation of aluminum in various body tissues, whereas mice of the A/J strain do not show such an accumulation. (Data from Fosmire, G.J. et al. (1993) *Biol. Trace Elements Res.* **37**: p 115.)

Another potent neurotoxin is the organotin compound trimethyltin (TMT). TMT is primarily a central nervous system neurotoxin which affects neurons within the hippocampus.[5] TMT exposure results in a variety of toxic effects on behavior, including tremors, spontaneous seizures, vocalizations, hyper-reactivity to sensory stimuli, and heightened aggressive responses. Impairments in learning and memory processes are also apparent following exposure, and these effects may persist for several years.

In laboratory animals, studies[3,25,26] have shown that there are genetic differences in responsivity to the effects of trimethyltin on a number of behavioral measures. When trained to press a small lever to obtain milk, C57BL/6N mice show a greater sensitivity to the disruptive effects of trimethyltin on this pattern of behavior relative to BALB/c mice even following just a single dose of trimethyltin. The behavioral deficits were closely paralleled by the presence or absence of significant neuropathology. C57BL/6N mice are also more sensitive to the lethal effects of this metal, as well as the locomotor stimulant effects. In addition, C57BL/6N mice show more severe tremors and are more hypersensitive to external stimuli following administration of trimethyltin relative to mice of the BALB/c strain.

More work is needed in the area of metal neurotoxicity in order to determine if there are individuals at greater relative risk for deleterious effects on neural function and behavior following exposure to the many metals and related products that exist in our current technological society.

4   Genetic susceptibility to behavioral disorders associated with neurotoxic diseases

4.1   GENETIC FACTORS IN NEURODEGENERATIVE TOXICITIES

For many years it has been known that certain behavioral disorders, such as Parkinson's disease, Alzheimer's disease, and several others, involve degeneration and destruction of neurons in the brain. However, the specific mechanisms of neuronal degeneration, why degeneration occurs, and why this appears to only happen in some individuals, has remained unknown. Recently, researchers have gained new insights on the mechanisms of these neurodegenerative disorders, and it appears that a significant portion of the reason for the substantial individual differences in acquiring these disorders is linked to differences in our genes.

One intriguing area of study which has provided important clues towards understanding neurodegenerative toxicities is the study of endogenous neurotoxins, substances which occur naturally in our bodies but which can be highly toxic under certain conditions. As has been discussed above, metabolism and breakdown of chemicals is an important process for detoxification, and in the absence of appropriate enzymes, toxic levels of compounds can build up and damage tissues and organs. As we learned above, the lack of efficient breakdown of acetaldehyde is one example of a genetically based susceptibility to toxicity. However, sometimes the body's own chemicals can be toxic if they are not removed appropriately, and this can also significantly damage the nervous system and impair behavior. An example is the neurotransmitter glutamate.

Glutamate is the primary excitatory neurotransmitter in the mammalian brain. Indeed, there appear to be glutamate receptors on every neuron in the central nervous system. Glutamate is important in the processes of neurotransmission and in establishing certain aspects of memory. It has been established that, in addition to its normal function as a neurotransmitter in the healthy brain, glutamate can also damage and kill neurons when present in high concentrations or for extended periods of time. Therefore, glutamate is a "two-edged sword" which, under certain unknown conditions, undergoes a transition from neurotransmitter to neurotoxin. Glutamate-related toxicity has been implicated in damage to neurons in the neurodegenerative disorders such as Alzheimer's and Parkinson's disease.[12]

In the laboratory, *in vitro* exposure of neurons to glutamate significantly reduces cell viability relative to control neurons. However, treating the cells with a substance called nerve growth factor (NGF) significantly reduced the glutamate-related neuronal damage.[20] These findings suggest that NGF can protect neurons against glutamate-induced neurotoxicity. Since it is likely that there are different versions of the gene which codes for NGF, this could result in different levels of innate protection against the toxic effects of glutamate, and make some individuals more prone to developing neurodegenerative toxic disorders such as Alzheimer's or Parkinson's disease.

4.2   ALZHEIMER'S DISEASE

Over the past several years, scientists have learned much about the factors which contribute to the development of Alzheimer's disease, and the biological and genetic mechanisms which initiate and maintain this behavioral disorder. In Alzheimer's disease, certain regions of the brain show degeneration. These neurodegenerative changes are characterized by the appearance of abnormal protein aggregates. The most common form of protein found in these aggregates is called b-amyloid protein. This protein and its chemical relatives appear to play an important role in the development of Alzheimer's disease.

Human genetic studies have shown that a mutation exists in the gene which codes for a beta-amyloid-related protein in certain forms of Alzheimer's disease, especially those forms which show a high familial incidence. In the laboratory, inducing overexpression of beta-amyloid-related proteins results in the degeneration of neurons, suggesting that beta-amyloid protein and its derivatives are neurotoxic. What remains unclear is why the beta-amyloid protein appears in the first place. It is possible that it is linked to the production of glutamate, and results from the initiation of neuronal necrosis during glutamate-induced excitotoxicity.

In the laboratory, however, transgenic mice which bear the defective beta-amyloid-related gene have not developed convincing neuronal changes which parallel those observed in patients with Alzheimer's disease.[8] This interesting discrepancy between the clinic and the laboratory suggests that there are other agents occurring in the brain which can serve to suppress beta-amyloid protein-related neurotoxicity. Substances such as NGF, described above, could be playing an important role in suppressing neurotoxicity, and genetic differences in production of NGF could contribute to differences in risk for developing neurodegenerative disorders.

4.3   PARKINSON'S DISEASE

Parkinson's disease is a serious motor disorder which is associated with degeneration of certain dopamine-related pathways in the central nervous system. Suggestions of a genetic link in the development of Parkinson's disease first appeared nearly 50 years ago.[4] Recent studies suggest that several genes may have a role in determining individual susceptibility to this disease.[17] For example, it has been shown that Parkinson's patients have a greater frequency, relative to control patients, of a particular version of the enzyme monoamine oxidase (MAO),[13] which is important in the metabolism of the neurotransmitter dopamine. Since degeneration of dopamine pathways is a primary mechanism in Parkinson's disease, differences in the ability to metabolize dopamine may be related to the sensitivity of these neurons to toxicants such as glutamate.

Oxidants such as oxygen radicals, discussed above, are plentiful in our environment and could also play an important role in neurodegenerative diseases such as Parkinson's disease. All cells contain several antioxidant enzymes such as superoxide dismutase (SOD). It has been suggested that changes in the activity of SOD might be related to neurotoxic damage such as that seen in glutamate-induced neurotoxicity as discussed above.[2] Interestingly, SOD activity in certain brain regions involved in Alzheimer's disease and Parkinson's disease is apparently different in organisms more susceptible to the types of brain damage seen in these disorders.

One additional set of findings related to genetic risk for Parkinson's and Alzheimer's disease and their neurotoxic mechanisms come from epidemiological studies which have examined the times, places, and populations related to the occurrence of these disorders.[24] As described above, Parkinson's and Alzheimer's diseases are degenerative disorders that share common clinical, neuropathological, and biochemical aspects. In addition, these diseases show an unusually similar distribution across places and times[24], and there are great similarities among the populations who have been identified as being at higher risk for developing these disorders. Thus, it may be that these disorders, as well as other neurodegenerative disorders, share common genetic and biochemical mechanisms, and persons at higher risk for neurodegenerative diseases may share some form of common genetic makeup which conveys greater responsivity to such events as glutamate-induced neurotoxicity or oxygen radical formation.

An area of great potential which is developing from these genetic studies is the use of gene therapy for treating disorders such as Alzheimer's and Parkinson's disease.[23] Numerous advances have been made in the ability to incorporate new genes into an organism's existing genetic code, and there is now a better understanding of the specific mechanisms and target sites for such genes. Hopefully, it will eventually be possible to replace the defective gene or add an additional gene which is able to correctly regulate or metabolize the toxic products associated with neurodegenerative disorders. Of course, our ability to replace genes carries with it serious ethical and moral considerations. Some of these are discussed below.

While we want to avoid the spread of agents which can be toxic at any concentration in the air or water, it is important to know the range of variation in toxic response to these substances, so that we may gain a better understanding of not only whether or not a substance has toxic potential, but the conditions which promote toxicity and the extent of these effects.

5   Considerations for the future

As we learn more about the specific nature of neurotoxins and the role of certain genes in determining susceptibility to these effects, we are encountering an increasing number of ethical, legal, and economic issues. How we face these issues will influence many aspects of our lives as we enter the twenty-first century.

For example, should we screen students for sensitivity to certain chemicals before allowing them to enroll in courses where they may be exposed to these agents? There is at least one legal case in the United States where a student brought suit against a university for allowing the student to be exposed to volatile chemicals in a course. Where does protection end and rights begin? Related to this issue, if we can show that certain persons are more genetically sensitive to industrial toxicants, should those persons not be allowed to work in jobs where they may be exposed to such agents? Should a person's insurance company be given information about the individual's genetic makeup when determining insurability?

If we can replace genes and, in so doing, relieve the suffering of Parkinson's disease, how do we decide which individuals are good candidates for such treatment? Since these are genes which affect aspects of behavior, where do we draw the line, or we, in terms of allowing genes to be added, removed or replaced?

These are issues which can only be dealt with through education and informed opinions. Hopefully, this introduction to these topics will aid the student in understanding some of the terms and concepts in toxicogenetics and in forming independent and thoughtful positions on these important issues.

References

1. Battaglia, G., Yeh, S.Y., De Souza, E.B. (1988) MDMA-induced neurotoxicity: parameters of degeneration and recovery of brain serotonin neurons. *Pharmacol. Biochem. Behav.* **29**: p 269.

2. Cadet, J.L., Sheng, P., Ali, S., Rothman, R., Carlson, E., Epstein, C. (1994) Attenuation of methamphetamine-induced neurotoxicity in copper/zinc superoxide dismutase transgenic mice. *J. Neurochem.* **62**: p 380.

3. Chang, L.W., Wenger, G.R., McMillan D.E., Dyer, R.S. (1983) Species and strain comparison of acute neurotoxic effects of trimethyltin in mice and rats. *Neurobehav. Toxicol. Teratol.* **5**: p 337.

4. Chao, Y.C., Liou, S.R., Chung, Y.Y., Tang, H.S., Hsu, C.T., Li, T.K., Yin, S.J., (1994) Polymorphism of alcohol and aldehyde dehydrogenase genes and alcoholic cirrhosis in Chinese patients. *Hepatology* **19**: p 360.

5. Duvoisin, R.C. and Johnson, W.G. (1992) Hereditary Lewy-body parkinsonism and evidence for a genetic etiology of Parkinson's disease. *Brain Pathol.* **2**: p 309.

6. Feldman, R.G., White, R.F., Eriator, I.I. (1993) Trimethyltin encephalopathy. *Arch. Neurol.* **50**: p 1320.

7. Fosmire, G.J., Focht, S.J., McClearn, G.E. (1993) Genetic influences on tissue deposition of aluminum in mice. *Biol. Trace Elements Res.* **37**: p 115.

8. Fukuchi, K., Ogburn, C.E., Smith, A.C., Kunkel, D.D., Furlong, C.E., Deeb, S.S., Nochlin, D., Sumi, S.M., Martin, G.M. (1993) Transgenic animal models for Alzheimer's disease. *Ann. NY. Acad. Sci.* **695**: p 217.

9. George, F.R. (1991) Cocaine toxicity: genetic evidence suggests different mechanisms for cocaine-induced seizures and lethality. *Psychopharmacology* **104**: p 307.

10. George, F.R. (1991) Cocaine toxicity: genetic differences in cocaine-induced lethality in rats. *Pharmacol Biochem Behav* **38**: p 893.

11. George, F.R., Porrino, L.J, Ritz, M.C., Goldberg, S.R. (1991) Inbred rat strain comparisons indicate different sites of action for cocaine and amphetamine locomotor stimulant effects. *Psychopharmacology* 104: p 457.

12. Henneberry, R.C. (1992) Cloning of the genes for excitatory amino acid receptors. *Bioessays* 14: p 465.

13. Hotamisligil, G.S., Girmen, A.S., Fink, J.S., Tivol, E., Shalish, C., Trofatter, J., Baenziger, J., Diamond, S., Markham, C., Sullivan, J. (1994) Hereditary variations in monoamine oxidase as a risk factor for Parkinson's disease. *Move. Disord.* **9**: p 305.

14. Hrubec, Z. and Omenn, G.S. (1981) Evidence of genetic predisposition to alcoholic cirrhosis and psychosis: twin concordances for alcoholism and its biological end points by zygosity among male veterans. *Alcohol Clin. Exp. Res.* **5**: p 207.

15. Jarvis, M.F., Wagner, G.C. (1985) Neurochemical and functional consequences following 1-methyl-4-phenyl-1,2,5,6-tetrahydropyridine (MPTP) and methamphetamine. *Life Sci.* **36**: p 249.

16. Karch, S.B. (1989) The history of cocaine toxicity. *Hum. Pathol.* **20**: p 1037.

17. Mullan, Crawford, F. (1993) Genetic and molecular advances in Alzheimer's disease. *Trends Neurosci.* **16**: p 398.

18. Ruth, J.A., Ullman, E.A., Collins, A.C. (1988) An analysis of cocaine effects on locomotor activities and heart rate in four inbred mouse strains. *Pharmacol. Biochem. Behav.* **29**: p 157.

19. Sherman, D.I., Ward, R.J., Warren-Perry, M., Williams, R., Peters, T.J. (1993) Association of restriction fragment length polymorphism in alcohol dehydrogenase 2 gene with alcohol induced liver damage. *Br. Med. J.* **307**: p 1388.

20. Shimohama, S., Ogawa, N., Tamura, Y., Akaike, A., Tsukahara, T., Iwata, H., Kimura, J. (1993) Protective effect of nerve growth factor against glutamate-induced neurotoxicity in cultured cortical neurons. *Brain Res.* **632**: p 296.

21. Shuster, L., Yu, G., Bates, A. (1977) Sensitization to cocaine stimulation in mice. *Psychopharmacology* **52**: p 185.

22. Sonsalla, P.K., Heikkila, R. E. (1988) Neurotoxic effects of 1-methyl-4-phenyl-1,2,3,6-tetrahydropyridine (MPTP) and methamphetamine in several strains of mice. *Prog. Neuropsychopharmacol. Biol. Psychiatry* **12**: p 345.

23. Suhr, S.T., Gage, F.H. (1993) Gene therapy for neurologic disease. *Arch. Neurol.* **50**: p 1252.

24. Treves, T.A., Chandra, V., Korczyn, A.D. (1993) Parkinson's and Alzheimer's diseases: epidemiological comparison. 2. Persons at risk, *Neuroepidemiology* **12**: p 345.

25. Wenger, G.R., McMillan, D.E., Chang, L.W. (1984) Behavioral effects of trimethyltin in two strains of mice. I. Spontaneous motor activity. *Toxicol. Appl. Pharmacol.* **73**: p 78.

26. Wenger, G.R., McMillan, D.E., Chang, L.W. (1984) Behavioral effects of trimethyltin in two strains of mice. II. Multiple fixed ratio, fixed interval, *Toxicol. Appl. Pharmacol.* **73**: p 89.

27. Yin, S.J., Liao, C.S., Chen, C.M., Fan, F.T., Lee, S.C. (1992) Genetic polymorphism and activities of human lung alcohol and aldehyde dehydrogenases: implications for ethanol metabolism and cytotoxicity. *Biochem. Genet.* **30**: p 203.

Suggestions for further reading

Henneberry, R.C. (1992) Cloning of the genes for excitatory amino acid receptors, *Bioessays* **14**: p 465.

Plomin, R., DeFries, J.C., McClearn, G.E. (1990) *Behavioral Genetics: A Primer*, 2nd edition. W.H. Freeman, New York.

Suhr, S.T. and Gage, F.H. (1993) Gene therapy for neurologic disease. *Arch. Neurol.* **50**: p 1252.

Part II

Factors in Food

Contents Chapter 7

Assessing neurotoxic hazards of food-related chemicals

Chapter 7

Assessing neurotoxic hazards
of food-related chemicals

Thomas J. Sobotka

1   Introduction

The safety of foods and food-related products for human consumption represents an unusually complex issue. On the one hand, food is the primary source of nutrients needed by humans to sustain growth, effect cellular repair, and maintain vital biological processes. Yet, on the other hand, food is also one of the most common avenues of human exposure to varieties of chemicals, many of which singly and in combination have diverse biological effects, some of which may be quite toxic.

The reduction of health risks, particularly neurotoxicity, associated with food intake is the focus of the present chapter. The initial sections of this chapter identify some potential sources of exposure to neurotoxic food components and describe reported episodes of human poisoning. Attention is then turned to the role of toxicological testing in the process of risk assessment followed by a brief discussion of the determinants of toxicological hazard. The remainder of this chapter contains a comprehensive account of how food-related chemicals are assessed, or should be assessed, for neurotoxicity. To exemplify this process, the neurotoxicological testing strategy proposed by the U.S. Food and Drug Administration (FDA) in the safety assessment of food additives is described.

2   Categories of food-related chemicals

Among the array of chemicals contributing to the multivariate nature of food are nutrients, naturally occurring food constituents, additives, pesticide and animal drug residues, and industrial and natural environmental contaminants (Table 7.1)[4,5,7,16,23].

The chemical mix or composition of any particular food will vary considerably, even among similar foods, depending upon multiple environmental, genetic, agricultural, manufacturing, processing, and cultural factors. Across the categories of food chemicals there are numerous examples of substances which have neurobiological properties and under certain conditions may be neurotoxic[2,4,5,10-12,16-18,23,24,27].

In this context neurotoxicity may be defined as any adverse effect of exposure to a chemical on the structure or functional integrity of the adult or developing nervous system[21] (Table 7.2). Neurotoxic effects may involve a spectrum of neurochemical, morphological, behavioral, and physiological changes, and may result from a direct action of a toxicant/metabolite on the nervous system (primary or direct neurotoxicity) or from an indirect action of a

TABLE 7.1

General categories of food-related chemicals

Nutrients	– Macronutrients: proteins, carbohydrates, fats – Micronutrients: vitamins, mineral, trace elements – Fiber
Natural constituents	– Plant-derived – Meat/poultry/seafood – Dairy products
Additives	– Directs – Indirects – GRAS
Chemical residues	– Pesticides – Animal drugs
Environmental contaminants	– Industrial chemicals – Natural environmental chemicals

toxicant/metabolite on non-neuronal systems (secondary, indirect or, "proneurotoxicity"). Changes which significantly compromise an organism's ability to function appropriately in its environment are considered adverse.

Nutrients

Nutrients in proper balance are essential for normal neurobiological function. If the bioavailability of the various nutrients is altered, for example, by dietary changes or dysfunctional homeostatic regulation, nutrient balance is disrupted and profound effects may result[2,12,17,24,27]. The adverse consequences of nutrient deficiency states on the development and functional integrity of the nervous system are well established and there is growing appreciation of the potential neurotoxicity associated with excess levels of various nutrients, e.g., pyridoxine (B6), tryptophan, iron, manganese, and vitamin A[24]. In addition to the extreme states of nutrient deficiency and excess, research attention has also begun to focus increasingly on the biological effects associated with marginal nutrient imbalances (deficiency and excess). Another area of growing importance is the reciprocal interactions between nutrients/nutritional status and other exogenous compounds (including drugs and toxicants) involving, for example, effects on the processes controlling bioavailability and metabolism, diet mediated changes in threshold of toxic agents, and secondary mechanisms of neurotoxicity.

Naturally occurring food constituents

Natural food constituents include plants, meats, poultry, seafood, and dairy products. Among these, plants represent a particularly interesting source of a rich variety of chemical substances, many of which may affect the nervous system[8,10]. Among the main groups of plant-derived neuroactive chemicals are the *alkaloids,* such as morphine, caffeine, harmine, strychnine, and mescaline; *glycoalkaloids,* such as the solanines in potatoes which are examples of plant stress metabolites; *cyanogenic glycosides,* such as linamarin in lima beans and cassava (a staple food in many parts of the world); *amino acids,* such as β-methylaminoalanine, β-N-oxalylaminoalanine, 2,4-diaminobutyric acid, and 3-nitropropionate; and *monoterpenes and terpenoids,* such as cannabinoids, camphor, thujone, myristicin, capsaicin, and chicory. Many plant-derived xenobiotics have served as models in the development of therapeutic agents. Typically, however, the varieties of plants and plant-derived chemicals used in

TABLE 7.2

Definition and characteristics of neurotoxicity

– Any adverse effect of exposure to a chemical on the structure or functional integrity of the adult or developing nervous system

– May involve a spectrum of neurochemical, morphological, behavioral and physiological changes

– Changes which significantly compromise an organism's ability to function appropriately in its environment are considered *adverse*

– Adverse effects may stem from direct action of toxicant/metabolite on the nervous system *(primary or direct neurotoxicity)* or from indirect action of toxicant/metabolite on non-neuronal systems *(secondary, indirect or pro-neurotoxicity)*

foods and beverages, e.g., as flavors, spices, and herbs, do not include those with potent neurotoxic chemicals or may contain such chemicals only in minute quantities which, when consumed in typical quantities, represent little hazard.

Additives

Certain chemicals are approved for use in food as direct or indirect additives to provide a variety of technical effects, including anticaking, emulsifier, antioxidant, non-nutritive sweetener, lubricant, packaging, and thickener. Typically, the process for approving the use of chemicals as food additives includes an assessment of safety based on the development of toxicological data[7,22]. Historically, efforts specifically to assess the neurotoxic potential of these types of chemicals have been limited[19,21,22]. However, on the basis of information that is available, few of the additives currently in use appear to have any significant potential for adversely affecting the nervous system, and then usually only under certain conditions[7]. For example, the flavor enhancer, monosodium glutamate, has been shown to affect the hypothalamus in experimental animals but apparently only when administered in large doses during a brief window shortly after birth. As will be detailed below, increasing efforts are being made by regulatory agencies, such as the U.S. Food and Drug Administration (FDA), to focus more specific attention on the neurotoxic potential of proposed food additives as part of the safety assessment process[21].

Pesticide and animal drug residues

Residues of pesticides (e.g., insecticides, herbicides and fungicides) and of animal drugs (e.g., antibiotics, sulfonamides, anthelmintics, and growth promotants) are an inevitable consequence of the widespread use of such chemicals in agriculture. Examples of chemical residues with neurotoxic potential include DDT, chlordecone, endosulfan, the organophosphates, and the carbamates[10,11,16,19,20]. Continued efforts are being focused on ways to minimize the presence and levels of such residues in foods and food products, particularly in the diets of infants and children.

Industrial and natural environmental contaminants

As a group, industrial and natural environmental chemicals include some of the most potent types of neurotoxicants known[4,5,10,11,19,20]. Industrial chemicals have become widely dispersed in the environment and their residues have been detected as contaminants in agricultural, biological, and environmental specimens from around the world. Examples of neurotoxic industrial contaminants found at low levels in food include the toxic elements such as lead, mercury, arsenic, and selenium, and the polyhalogenated biphenyls, specifically the polychlorinated biphenyls (PCBs) and the polybrominated biphenyls (PBBs), all of which can have dramatic effects on the nervous system, particularly in the immature developing organism (see Chapters 9–13). Many naturally occurring neurotoxic chemicals, which may be present in certain foods as environmental contaminants, are produced by micro-organisms associated with those foods.

These chemicals may include *fungal toxins*, such as the tremorgenic mycotoxins (e.g., lolitrem-B, verruculogen, and penitrem-A), the ergot alkaloids, and fumonisin (associated with equine leukoencephalomalacia); *bacterial toxins*, such as botulinum; and certain *seafood toxins*, such as saxitoxin (associated with paralytic shellfish poisoning), brevetoxin, and domoic acid (associated with amnesic shellfish poisoning). The extent of food contamination by natural neurotoxicants is determined by a number of factors including the occurrence of environmental conditions conducive to the growth of the particular micro-organisms responsible for their production, including conditions related to the methods of processing and preparation of the food.

3 Episodes of human poisoning

There have been a number of documented instances of human poisoning resulting specifically from the ingestion of neurotoxicants in food[15,19]. These include the chemical contamination of cooking oil by tri-*o*-cresyl-phosphate in 1937 and again in 1959 and 1988; the contamination of barley with thallium in 1932; the ingestion of foods contaminated with methylmercury in the 1950's, 1960, 1964, 1969 and 1971; melons contaminated with aldicarb in 1965; the presence of domoic acid in mussels in 1987; and the ingestion of *l*-tryptophan products resulting in eosinophilia-myalgia syndrome. Virtually all of these episodes involved acute exposure to toxic levels of chemical substances and none were associated with exposure to the normal background levels of food-related chemicals. Such recorded instances of poisoning entailed unusual circumstances such as the consumption of quantities of toxin-containing foods not normally used, the presence of higher than normal levels of neurotoxic chemicals and the accidental presence of neurotoxic contaminants not normally found in food. Indeed, it would appear that under normal conditions the presence of potentially neurotoxic chemicals in human food is limited so that the risk of overt neurotoxicity from exposure to biologically potent quantities of toxicants in food is minimal. Good manufacturing practices, careful ingredient selection, proper food preparation and, particularly, government regulatory programs, when in place and effectively implemented, all contribute to minimizing the presence of hazardous chemicals in food. One of the primary missions of government regulatory agencies, such as the United States Food and Drug Administration or the Commission of the European Union, is to ensure that foods are safe for human consumption. In carrying out this charge, such agencies play a substantial role in minimizing the risk of toxicological hazard associated with exposure to food in part by regulating the use and/or presence of certain chemicals, including the additives, residues, and contaminants, in human foods and food products[19,20,22]. Such regulatory control is based on a number of considerations including the safety of those chemicals. In the countries of the European Union food safety evaluation is formally carried out by the Commission of the European Union (see Figure 7.1).

Proposals made by these working groups for the safe use of food additives and for maximum residue limits are, if adopted by Regulatory Committees, enforced by the Council of Ministers. Enforced proposals are published in the Official Journal of the European Union and are, from that time on, imperative for the regulatory authorities in the member countries. As in the U.S. (see Sections 7.6 and 7.7 of this chapter), the neurotoxic potential of a chemical is considered as an integral component of the toxicological profile. Also, a committee in the European Union is presently working to give greater emphasis to studies of, among others, nuerotoxicity.

4   Toxicological hazard in safety assessment

In discussing chemical safety, a fundamental principle of toxicology established in the 16th century by Paracelsus, is that "All substances are poisons: There is none which is not a poison, the right dose differentiates a poison from a remedy"[6]. While some level of toxicity may be an inherent property of every chemical, the mere presence of a chemical does not necessarily make it hazardous (with the possible exception of certain types of carcinogens)[6,22]. Any substance (noncarcinogenic) can adversely affect a biological organism but only at a high enough dosage (threshold) and consequently has some definable level of hazard associated with certain levels of exposure. The safety of a particular chemical is not equated with the absolute absence of adverse effects, but rather is considered more in terms of minimal risk or reasonable certainty of no harm at the levels of exposure associated with the conditions of proposed or intended use[7]. In this context the primary objective of toxicological testing is to identify and quantitate the potential dose-related hazards/toxicity associated with exposure to chemical substances[7,15,19,20,22]. The information about hazard/toxicity would then be used to assess potential human risk of adverse effects from exposure to chemicals added to food or present in food as contaminants or residues, and to help define the conditions of use, including acceptable levels of exposure, which would minimize this risk. The process of risk assessment, as applied to food safety, is typically geared toward the assessment of individual chemicals. Given the numbers of chemicals in foods, this is an awesome and difficult task

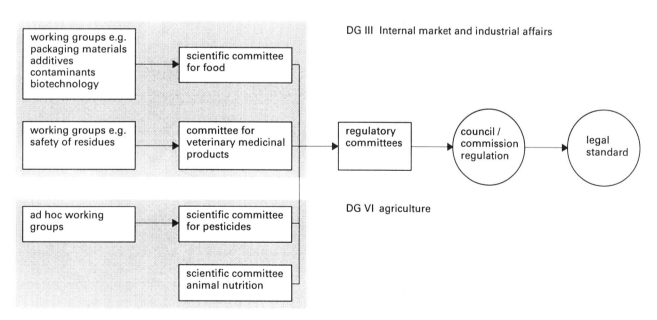

FIGURE 7.1

Scientific working groups involved in food safety evaluation within the European Union

in itself. However, it is becoming increasingly clear that, in addition to concerns about the inherent toxicity of single chemical substances, regulatory attention should begin to address the even more difficult questions regarding possible interactive effects of combinations of food chemicals. Of course, one of more difficult aspects of the latter is the almost limitless permutations and combinations of food chemicals.

5   Determinants of toxicological hazard

The potential toxicological hazard of a chemical substance is a function of several interdependent factors including the nature of the biological effects, the dose-response characteristics, and the modulators of sensitivity[20]:

Nature of effects
Across a range of doses or exposures a toxic substance may produce a variety of adverse effects involving multiple organ systems acting through common or different mechanisms. A toxicological profile may include effects such as carcinogenicity, teratogenicity, immunotoxicity, cardiovascular dysfunction, and neurotoxicity. In the evaluation of safety, all adverse effects are of importance and should be included in an integrated assessment of risk [19-21,25,26]. At the present time, empirical testing represents the only reliable means of determining the spectrum of toxic effects for any particular chemical and in particular of identifying which chemicals may be potentially neurotoxic. Structure activity databases have not yet been sufficiently developed to be of significant value in predicting toxicity; this is particularly true for neurotoxicity. To provide a reasonable level of assurance that the full spectrum of potential hazardous effects of a chemical are being adequately determined, a broad-based profile of toxicological information is needed. Some toxicological data can be derived from humans, for example from cases of food-related poisoning, epidemiological studies, and clinical investigations. However, the nature and extent of available human toxicological data are too incomplete to serve as the basis for an adequate assessment of potential health hazards. Consequently, for most toxicological assessments it is necessary to rely on information derived from valid animal models. A series of short-term and long-term dietary or oral exposure studies which mimic the mode of human exposure are typically used to provide a comprehensive assessment of the toxicological profile of chemical substances including general toxicity, carcinogenicity, developmental and reproductive toxicity, immunotoxicity and neurotoxicity.

Dose-response characteristics
The relationship between dose/exposure and the production of adverse effects represents one of the most critical elements in assessing the potential hazard of a chemical substance[1,6,7,16,20,25]. In terms of individual responses, dose determines the magnitude and severity of an adverse effect. In terms of population, dose determines the probability that a given effect will occur in a certain percentage of the population. For neurotoxicity and other non-carcinogenic effects, the generally accepted assumption is that a threshold level of exposure exists for the induction of an adverse effect by a chemical agent below which there are no biologically significant increases in the frequency or severity of adverse effects between the exposed population and its appropriate control. The assessment of safety is based on the estimation or approximation of this threshold or "no observed adverse effect level" (NOAEL) relative to the intended or expected uses of the chemical agent in food[6,7,19,20,22,25]. In the experimental assessment of potential neurotoxic hazard, factors which impact on the pharmacokinetics (i.e. absorption, distribution and excretion) of the

toxicant are very important determinants of effect. In this regard, route of administration is particularly critical. Depending upon the physico-chemical properties of certain substances, the gastrointestinal system and enterohepatic circulation may play important roles in regulating their bioavailability following oral administration. Bypassing this "barrier" by systemic administration enables a more rapid and concentrated exposure to such substances than might otherwise occur resulting in atypical mass action effects not seen with oral exposure. For example, the lowest effective dose of glutamate to produce hypothalamic neuropathology in the weanling mouse is 0.7 g/kg when given subcutaneously and 2 g/kg by oral intubation, whereas a dose of even 35 g/kg administered in the diet produces no neuronal damage[7]. Systemic administration, through this differential effect on the neurotoxicity of chemical substances, may serve a useful purpose as an optional mode of exposure in the early stages of hazard identification to detect potential neurotoxic chemicals. Systemic dosing may also provide information about potential neurotoxic susceptibility of particular subpopulations, such as those with a compromised gastrointestinal barrier. In the routine determination of NOAELs, however, as a basis for predicting the risk of potential human neurotoxic hazards and for making regulatory decisions, it is important that the experimental design mimics the human mode of exposure as closely as possible. For food related chemicals this would typically involve dietary exposure or, possibly, oral intubation[22].

Modulators of sensitivity

As with other manifestations of toxicity, there are a number of factors which may modulate the sensitivity to a particular neurotoxic chemical and may reflect certain characteristics of vulnerable subpopulations[1,2,10-12,15,16,20,21,25,26]. Species or genetic variation, gender, age/developmental stage, and nutritional status may each uniquely affect the nature or extent to which the nervous system is affected by a particular toxicant.

One of the most notable factors in determining vulnerability to neurotoxicity is the age of the organism[3,16]. In many cases, the development and/or expression of a chemically induced neurotoxicity is dependent upon the maturational stage of the organism. Although toxicology has traditionally focussed on the mature adult, it has become widely appreciated that the immature developing organism, as well as the aged individual, may be particularly vulnerable to noxious chemicals. For many chemicals adverse effects may result from exposure during the prenatal and postnatal periods of ontogenetic development to dose levels lower than would be necessary to elicit toxicity in adults. Certainly, for many other chemicals adults may be more sensitive than the immature organism, or there may be no age-related differences. During the developmental period the nervous system is actively growing and establishing intricate cellular connection. The protective blood-brain and blood-nerve barriers, which in the adult help control passage of substances into and out of the nervous system, are not fully developed in the immature organism and may be less effective than in the adult in preventing the entrance of certain neurotoxicants into the nervous system. Also, the various protective excretory and metabolic mechanisms by which the organism deals with toxic substances, such as the hepatic glucuronide conjugation and the detoxification systems, are not fully developed in the immature organism.

With advanced age, the level of risk for a number of health-related factors increases as well as the vulnerability for toxic perturbations to the nervous system. The adaptive capacity of the aged nervous system gradually becomes compromised as reflected by nerve cell loss, abnormal accumulations of proteins and other substances, altered neurotransmitter concentrations and related enzymatic activity, and slowed conduction of nerve impulses stemming from the loss of myelin. The aged nervous system exhibits decreased ability to respond appropriately to stimuli or to compensate for biological, physical, or toxic effects.

Compounding the influence of age on the susceptibility to neurotoxicity is the role that nutrition plays in the neurotoxic response of an organism to chemical exposure. Both general nutritional status (for example, caloric restriction) and specific nutritional deficiencies or excesses (involving, for example, protein, thiamine, iron, and calcium) can significantly influence the response to a neurotoxic substance, possibly via changes in its bioavailability or metabolism, or by altering the neurotoxic threshold level through changes in receptor mediated processes. To the extent possible, the empirical assessment of neurotoxic hazards should control for or otherwise consider those factors which may significantly modulate the sensitivity of the nervous system to toxic insult, for example, by the appropriate choice of animal models, by the inclusion of paradigms involving multiple-age exposure, by the use of male and female animals, and by the use of standardized dietary conditions.

6   Neurotoxicity evaluation as an integral component in the hazard assessment of food-related chemicals

In the evaluation of food chemical safety, the neurotoxic potential of a chemical substance should be considered an integral component of the toxicological profile. As such, the assessment of neurotoxicity should be routinely included in the experimental studies designed to determine that toxicological profile.

Neurotoxicity can be described at multiple levels of organization of the nervous system, including chemical, anatomical, physiological, and behavioral[1,13,15,19,20,25,26]. Neurotoxic chemicals invariably initiate their effects at the molecular level, altering neurochemical processes involved, for example, in energy production, ionic transport, protein synthesis, neurotransmitter synthesis and metabolism, receptor functions, and maintenance of intraneuronal cytoarchitecture. These neurochemical changes may be expressed by or associated with the following: dysfunctions of physiological processes under neuronal control (e.g., thermoregulation, cardiovascular function, respiratory control, etc.); behavioral abnormalities in motor, sensory, or cognitive function; and neuropathology involving variable effects on neurons, axons, glia, and myelin sheaths.

Among the various approaches that can be used for assessing neurotoxicity, behavioral testing represents a practical means of obtaining a relatively comprehensive assessment of the functional development and integrity of the nervous system as part of standardized toxicity studies[3,13,15,19-21,25,26]. Behavior is an adaptive response of an organism, orchestrated by the nervous system, to some set of internal and external stimuli. A behavioral response represents the integrated endproduct of multiple neuronal subsystems including sensory, motor, cognitive, attentional, and integrative components, as well as an array of physiological functions. As such, behavior can serve as a measurable index of the status of multiple components of the nervous system. Behavioral testing typically is non-invasive and can be used repeatedly for longitudinal assessment of the neurotoxicity of a test compound, including the occurrence of persistent or delayed treatment-related effects.

Behavioral testing has been established as a reliable toxicological index in safety assessment[3,13,20,25,26]. Considerable progress has been made in the standardization and validation of neurobehavioral and neurodevelopmental testing procedures. As a result, a variety of behavioral methodologies is available for use in determining the potential of chemical substances to

TABLE 7.3

Toxicological profile needed for safety assessment of direct food additives
(Based on FDA 1982 Redbook)

General toxicity	– Morbidity – Mortality – Cage-side clinical observations • Signs of toxicological/pharmacological effects • Signs of abnormal behavior – Food/water intake – Body weight growth – Hematology/clinical chemistry – General pathology (gross and micro)
Reproductive performance/ developmental effect	– Mating – Parturition – Offspring growth and survival – Teratogenicity (structural)
Carcinogenic potential	– Mutagenesis/DNA repair – Carcinogenicity (bioassay)
Pharmacokinetics	– Absorption – Distribution – Metabolism – Excretion

adversely affect the functional integrity of the nervous system in adult and developing organisms. Behavioral testing can be readily incorporated into toxicity testing protocols and can improve and expand current approaches to assess neurotoxic hazard. Although behavioral endpoints may be sensitive indicators of neurotoxicity, they can be influenced by other factors, such as the status of other organ systems within the body. The interpretation of functional endpoints of neurotoxicity can be facilitated if integrated with other measures of nervous system toxicity (neurochemical and neuropathological), as well as indices of systemic toxicity.

7 Testing strategy for assessing neurotoxic potential

To exemplify one strategy for assessing neurotoxic potential of food-related chemicals, the approach currently being proposed by the FDA will be presented[21]. In 1982, the FDA published a document entitled "Toxicological Principles for the Safety Assessment of Direct Food Additives and Color Additives Used in Food"[22]. This document, referred to as the Redbook, describes the nature and extent of the toxicological information that is needed to assess the safety of food and color additives (Table 7.3) and recommends a variety of conventional toxicological studies to obtain that information (Table 7.4).

Although neurotoxicity is neither explicitly discussed nor defined in that document, certain indices are included in the general toxicity evaluation to screen for the presence of adverse effects to the nervous system (Figure 7.2). These consist only of a general non-systematic cage-side observation for clinical signs of toxicity and a general pathological evaluation of sections of the brain, spinal cord, and peripheral nerves in adult animals but notably not in immature developing animals. If significant signs of nervous system toxicity are found, the conduct of special neurotoxicity studies may be recommended on a case-by-case basis. Beyond mentioning these screening elements as part of the general toxicity evaluation, the 1982 Redbook provides little specific guidance as to the manner in which this information is to be developed, recorded, or reported. Based on this limited guidance, the type of information developed enables little

TABLE 7.4

Varieties of toxicological protocols recommended for safety assessment of direct food additives

Short-term toxicity
Subchronic (rodent)
Subchronic (non-rodent)
Long-term (non-rodent)
Chronic with *in utero* exposure (rodent)
Reproduction/multigeneration
Teratology (phase of reproduction study)
Teratology (gavage)*
Oncogenicity (rat)
Oncogenicity (other rodent)
Short-term oncogenicity (mutagenicity, DNA)
Special studies*

* May be indicated by results of other studies or mode of exposure

more than the detection of evident nervous system toxicity associated with general neuropathology and overt neurological dysfunction. Little consistent information is provided about other possibly less severe, but equally important, neurotoxicity including, for example, behavioral and physiological dysfunction, developmental disorders, and discrete neuropathological changes. Numerous publications and reports from expert committees and regulatory agencies have stressed the need to evaluate neurotoxicity as a routine and specific component of safety assessment, and have emphasized the importance of including measures with which to assess more effectively the full spectrum of neurotoxic hazard. In concert with this concern, the FDA has proposed a draft revision of

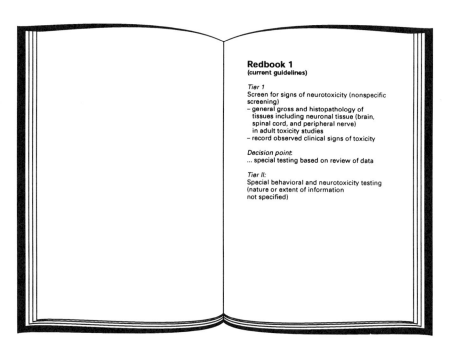

Redbook 1
(current guidelines)

Tier 1
Screen for signs of neurotoxicity (nonspecific screening)
– general gross and histopathology of tissues including neuronal tissue (brain, spinal cord, and peripheral nerve) in adult toxicity studies
– record observed clinical signs of toxicity

Decision point:
... special testing based on review of data

Tier II:
Special behavioral and neurotoxicity testing (nature or extent of information not specified)

FIGURE 7.2

Toxicological principles for the safety assessment of direct food additives and color additives used in food as defined by the U.S. Food and Drug Administration.

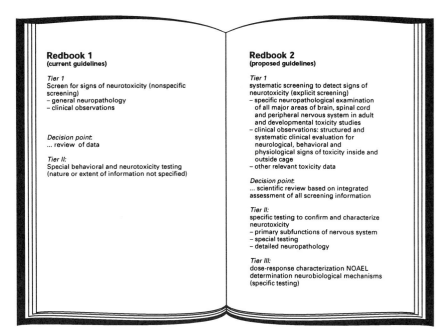

Redbook 1
(current guidelines)

Tier 1
Screen for signs of neurotoxicity (nonspecific screening)
– general neuropathology
– clinical observations

Decision point:
... review of data

Tier II:
Special behavioral and neurotoxicity testing (nature or extent of information not specified)

Redbook 2
(proposed guidelines)

Tier 1
systematic screening to detect signs of neurotoxicity (explicit screening)
– specific neuropathological examination of all major areas of brain, spinal cord and peripheral nervous system in adult and developmental toxicity studies
– clinical observations: structured and systematic clinical evaluation for neurological, behavioral and physiological signs of toxicity inside and outside cage
– other relevant toxicity data

Decision point:
... scientific review based on integrated assessment of all screening information

Tier II:
specific testing to confirm and characterize neurotoxicity
– primary subfunctions of nervous system
– special testing
– detailed neuropathology

Tier III:
dose-response characterization NOAEL determination neurobiological mechanisms (specific testing)

FIGURE 7.3

Tiered testing approach as adopted by the US FDA for use in assessing neurotoxicity

its toxicological guidelines (revised Redbook) to include the assessment of neurotoxicity (structural and functional aspects) as an integral part of the toxicological profile needed for the safety assessment of food-related chemicals[21]. A draft of this revised Redbook was released in 1993 for general review by the national and international communities. Comments have been received and are addressed in a finalized version of the revised guidelines, which is currently under agency review.

Consistent with approaches recommended for use in assessing neurotoxicity[13,19,20], the general strategy adopted by the FDA involves a process of tiered testing (Figure 7.3). Briefly, testing would commence with an initial stage of *screening* (Tier I) in which chemicals are tested across a range of dose/exposure levels for any clinical or pathological signs of adverse effects, including those to the nervous system. Chemicals identified as exhibiting a potential for adversely affecting the nervous system (presumptively identified as neurotoxic chemicals) are then considered candidates for subsequent *specific neurotoxicity testing* (Tier II and III) to confirm and characterize further the nature and extent of the effects on the nervous system. This subsequent testing would utilize more specific neurotoxicological test methods/procedures and be designed to include characterization of dose-response relationships, determining no-observed-adverse-effect levels (NOAELs), compiling pharmacokinetic data and, as appropriate, developing mechanistic information about the neurobiological processes underlying the primary neurotoxic effects.

7.1   SCREENING

A tiered approach to neurotoxicity testing and evaluation allows for multiple decision points at which scientifically based decisions can be made about the adequacy of available information and the need for additional testing. In this process screening plays a particularly important role. The primary objective of

223

TABLE 7.5

Examples of clinical observations/examinations used to screen for potential neurotoxicity

---

I Irwin's observational assessment of behavioral, neurological and autonomic state (Irwin, S. 1968, *Psychopharmacologia* 13:222-257)

1. *Neurologic*
   - Posture
   - Muscle tone and grip strength
   - Equilibrium/gait, e.g. ataxia, righting, gait
   - CNS excitation, e.g. tremors, twitches, convulsion

2. *Behavioral*
   - Spontaneous activity, e.g. sleep, locomotion, bizarre behavior
   - Motor/affective response, e.g. struggle, escape, biting, vocalization, irritability
   - Sensori-motor reflexes/responses, e.g. visual placing, pain response, corneal and pinna response, startle, orientation (to visual, olfactory, tactile, auditory stimuli)

3. *Autonomic*
   - Eyes, e.g. pupil size, ptosis, exophthalmus
   - Secretions/excretions, e.g. salivation, lacrimation, diarrhea
   - General physiological signs, e.g. hypothermia, piloerection, skin color, respiration

---

II Behavioral endpoints used in a functional observational battery (Adapted from Moser, V. 1989 *J. Am. Coll. Toxicol.* 8:85-94)

Home cage and open field	*Manipulative*
Posture	Ease of removal
Convulsions and tremors	Ease of handling
Palpebral closure	Approach response
Lacrimation	Auditory startle
Piloerection	Response to aversive stimulus
Salivation	Righting reflex
Vocalizations	Landing foot splay
Rearing	Grip strength
Urination and defecation	Pupil reflex
Gait	
Arousal	
Mobility	
Stereotypy	
Bizarre behavior	

---

screening is detection, enabling the presumptive identification of those chemicals that may represent a neurotoxic hazard to the adult or developing organism that should be considered for specific neurotoxicity testing. The critical importance of detection in safety assessment resides in the fact that chemicals not identified during screening as potentially neurotoxic are typically considered to have no significant adverse effects on the nervous system; for such chemicals, the regulatory decisions about the conditions for their safe use in food would be based on endpoints other than neurotoxicity. The selection and conduct of screening tests should not only minimize the likelihood of missing potentially neurotoxic chemicals (alpha or false negative error), but also reduce the error of misidentifying a chemical as potentially neurotoxic when in fact it is not (beta or false positive error).

To provide some reasonable assurance that potential adult and developmental neurotoxicity is being identified, the FDA guidelines recommend that experimental data needed to screen for potential neurotoxicity should be routinely obtained as part of those studies commonly recommended for the toxicological testing of proposed food chemicals. Specifically, neurotoxicity screening information could be developed most appropriately in *short-term (e.g., 14- to 28-day rodent) studies* to screen adult animals exposed across a broad range of relatively high dose levels for brief periods of time, in *subchronic (e.g., 90-day rodent) and long-term (e.g. 1-year non-rodent) studies* to provide a reasonable means of screening adults for possible gradual onset, cumulative or progressive neurotoxic effects associated with repeated low dose exposure to the test chemical, and *reproduction/developmental studies* to screen for potential developmental neurotoxicity in perinatally exposed offspring. Neurotoxicity screening information could also be developed in other types of toxicity studies (e.g., chronic studies).

An adequate screen for neurotoxicity should include a sufficiently broad spectrum of endpoints to enable the detection of a variety of structural and functional changes to the central, peripheral, and autonomic segments of the nervous system. The FDA guidelines suggest that a relatively comprehensive and reliable screen would include the following elements: (1) a specific *neuropathological examination* of all major areas of the brain, spinal cord, and peripheral nervous system; (2) a structured and systematic *clinical examination* of experimental animals inside and outside of their cages using a clearly defined battery of clinical tests and observations to detect neurological, behavioral, and physiological signs of toxicity; and (3) *other relevant toxicological information*, since adverse effects to the nervous system should be evaluated within the context of all significant toxic effects of a test chemical. The clinical examination should be comprehensive yet relatively simple, inexpensive and rapidly carried out, and designed to detect relatively overt signs of neurotoxicity, such as seizure, tremor, paralysis, general changes in appearance, altered levels of arousal, motor incoordination or weakness, sensory dysfunction, abnormal reflex behaviors, physiological changes, changes in the developmental ontogeny of the nervous system, and other apparent signs of toxicity (Table 7.5).

In compiling reliable and useable neurotoxicity screening information, there are a number of factors which should be considered in the study design and in the conduct of testing :

1. *age-appropriate testing* – Since screening is intended to be a routine part of both adult and developmental toxicity studies, the specific composition of the screen and the endpoints recorded should be appropriate to the age (and species) of the animals tested. For example, when screening immature animals, it would be appropriate to include certain physical (e.g., body weight, vaginal opening, prepucial separation) and functional (e.g., reflex behaviors) milestones as indices of development.

2. *description of screen* – To help ensure the complete and consistent application of the neurotoxicity screen throughout a particular study, each protocol should include detailed information about the screen to be used in that study, including its composition, the test procedures to be followed, the neuronal structures to be examined, the endpoints to be used, and the methods for recording and analyzing the data.

3. *sufficient numbers of animals* – There should be a sufficient number of animals from both sexes to provide adequate statistical power to determine statistical significance, taking into consideration the variability of the endpoints.

4. *satellite groups* – While the same animals are typically used for neurotoxicity screening and for obtaining general toxicity information, the use of satellite groups of animals may be considered as an acceptable alternative in screening for neurotoxic effects.

5. *representative intervals of testing* – The clinical examinations should be carried out at representative intervals throughout the duration of the study to provide some information about the consistency or variability of the neurotoxic effects, and, if possible, about their onset, duration, and reversibility.

6. *appropriate control conditions* – The experimental design should incorporate appropriate measures or controls to minimize experimental bias and other potentially confounding variables. To the extent possible, standard operating procedures should be applied across treatment groups for elements such as random assignment to treatment, blinded assessment, housing, diet/nutritional status, circadian cycles, test-to-test interactions, handling, environmental conditions of temperature and humidity, etc.

7. *document all data* – All of the data developed in the neuro-toxicological screen, including positive and negative results, should be recorded, analyzed by the use of appropriate statistical procedures, and reported.

The information collected during screening is used to make the explicit determination of whether or not the test chemical represents a potential neurotoxic hazard and whether or not to recommend additional specific testing to confirm and characterize further the neurotoxicity, define NOAELs, and develop other necessary information. In making this evaluation, there are a number of considerations which should enter into the scientific interpretation of the neurotoxicity screening information . These include: (1) the adequacy and completeness of the screening assessment; (2) the nature, severity, and duration of effects detected; (3) consistency of effects across dose; (4) consistency of effects across testing intervals within a study; (5) replicability of effects across different types of toxicity studies, e.g., 28-day study versus 90-day study; (6) presence of other toxic effects, since it is important that the neurotoxicity screening information be evaluated together with other relevant toxicological data to enable an integrated assessment of all toxic effects in determining neurotoxic potential; and (7) the margin of difference between the doses producing neurotoxic effects and those producing other toxicity. The extent to which screening provides the information to address these issues adds to the level of confidence in identifying a chemical as a potential neurotoxic hazard and making the decision to proceed from screening to specific neurotoxicity testing.

7.2   SPECIFIC NEUROTOXICITY TESTING

Specific neurotoxicity testing focuses attention on determining the nature and extent to which the nervous system is affected by the test chemical. At this level the neurotoxic effects identified during screening are confirmed and further characterized, and additional efforts are made to determine whetheror not the test chemical has any other, possibly more subtle, effects on the functional and structural integrity of the nervous system in both the mature and developing organism. Functional testing would involve the use of a core battery of valid behavioral and physiological tests, selected from the varieties of procedures referenced in the literature[1,3,13,19,20,25,26] (see also Chapter 3) to assess the primary subfunctions of the nervous system, including sensory, motor, cognitive, reactivity, and physiological functions. As appropriate, this core battery may be supplemented with special tests to follow up significant observations of neurobehavioral dysfunctions made during screening. Neuropathology would include a detailed histopathological examination involving the use of special stains and, as appropriate, morphometric techniques to highlight relevant treatment-related effects. The characterization of the dose-response relationships and the determination of NOAELs are central to these studies. The NOAEL should be based upon an accurate and reliable determination of the dose-response and dose-time relationships derived from repeated exposure studies, e.g., intermittent and continuous exposure regimens, typically using the most relevant and sensitive endpoint(s). In concert with conventional toxicity testing protocols, routine neurotoxicity evaluation would generally be carried out using the rodent as the primary species of choice. However, as necessary, neurotoxicity studies using an alternative non-rodent species may be recommended, on a case-by-case basis, to develop information needed for more reliable cross-species extrapolation of data.

The availability of any additional information, for example, regarding the factors that may modulate the sensitivity of the organism to the neurotoxic compound, the presence of treatment-related neurochemical changes, or the pharmacokinetics of the test compound would help to identify characteristics of susceptible subpopulations and, in general, help to minimize the uncertainties involved in predicting human risk from neurotoxicity data derived from animal studies. To provide a more comprehensive assessment of neurotoxic hazard, efforts to develop such types of supplemental information is encouraged.

Increasing attention is being devoted to the development of *in vitro* systems for assessing the neurotoxicological impact of chemical agents[19,20]. *In vitro* methods would have practical advantages, such as minimizing the use of live animals, but validation studies remain to be done to correlate *in vitro* results with neuro-toxicological responses in whole animals. Such systems, once appropriately validated, may have particularly useful application in screening for potential neurotoxicity and in helping to elucidate mechanistic information.

8   Summary

In summary, food is a complex matrix of diverse chemical substances which may include some chemicals with a potential for neurotoxic hazard. Under most normal conditions, the presence of potentially neurotoxic chemicals in human food appears to be limited such that the risk of overt neurotoxicity from exposure to biologically potent quantities of toxicants in human food appears to be limited. The broader question of subtle or progressive neurotoxic effects from intermittent or chronic exposure to low levels of potentially toxic chemicals remains more difficult to address. However, the appropriate application of well-defined testing strategies now available can help identify those chemicals with a potential for even subtle forms of neurotoxicity, and can enable effective measures to be taken to minimize the risk of any neurotoxic hazards associated with exposure to those chemicals in food.

References

1.   Abou-Donia, M.M. (Ed.) (1992) *Neurotoxicology.* CRC Press, Boca Raton, FL.
2.   Abou-Donia, M.M. (1992) Nutrition and neurotoxicology. In: Abou-Donia, M.M. (Ed.) *Neurotoxicology.* CRC Press, Boca Raton, FL, pp 319–335.
3.   Buelke-Sam, J., Kimmel, C., Adams, J. (1985) Design considerations in screening for behavioral teratogens: results of the collaborative behavior teratology study. *Neurobehav. Toxicol. Teratol.* 7: pp 537–789.

4. Concon, J.M. (1988) *Food Toxicology. Part A: Principles and Concepts*. Marcel Dekker, New York, pp 1–676.

5. Concon, J.M. (1988) *Food Toxicology. Part B: Contaminants and Additives*. Marcel Dekker, New York, pp 677–1371.

6. Harvard School of Public Health, Center for Risk Analysis (March/1994) *A Historical Perspective on Risk Assessment in the Federal Government*. Harvard School of Public Health, Boston, MA.

7. Hattan, D.G., Henry, S.H., Montgomery, S.B., Bleiberg, M.J., Rulis, A.M., Bolger, P.M. (1983) Role of the Food and Drug Administration in regulation of neuroeffective food additives. In: Wurtman, R. and Wurtman, J. (Eds.) *Nutrition and the Brain. Vol. 6: Physiological and Behavioral Effects of Food Constituents*. Raven Press, New York, pp 31–99.

8. Huxtable, R.J. (1992) Neurotoxins in herbs and food plants. In: *The Vulnerable Brain and Environmental Risks. Vol. 1: Malnutrition and Hazard Assessment*. Plenum Press, New York, pp 77–108.

9. Irwin, S. (1968) Comprehensive observational assessment. Ia: A systematic, quantitative procedure for assessing the behavioral and physiologic state of the mouse. *Psychopharmacologia* **13**: pp 222–257.

10. Issaacson, R.L. and Jensen, K.F. (Eds.) (1992) *The Vulnerable Brain and Environmental Risks. Vol. 1: Malnutrition and Hazard Assessment*. Plenum Press, New York.

11. Issaacson, R.L. and Jensen, K.F. (Eds.) (1992) *The Vulnerable Brain and Environmental Risks. Vol. 2: Toxins in Food*. Plenum Press, New York.

12. Kotsonis, F., Mackey, M., Hjelle, J. (Eds.) (1994) *Nutritional Toxicology: Target Organ Toxicology Series*. Raven Press, New York.

13. Leukroth, Jr., R. (1987) Predicting neurotoxicity and behavioral dysfunction from preclinical toxicologic data. *Neurotoxicol. Teratol.* **9**: pp 395–471.

14. Moser, V.C. (1989) Screening approaches to neurotoxicity: a functional observational battery. *J. Am. Coll. Toxicol.* **8**: pp 85–93.

15. National Research Council (1992) *Environmental Neurotoxicology*. National Academy Press, Washington, DC.

16. National Research Council (1993) *Pesticides in the Diets of Infants and Children*. National Academy Press, Washington, DC.

17. Somogyi, J.C. and Hotzel, D. (Eds.) (1986) *Nutrition and Neurobiology*. Karger, New York.

18. Steyn, P.S. and Vlegaar, R. (1985) Tremorgenic mycotoxins. *Fortschr. Chem. Org. Naturst.*, **48**: pp 1–80.

19. U.S. Congress, Office of Technology Assessment (1990) *Neurotoxicity: Identifying and Controlling Poisons of the Nervous System*. OTA-BA-436. U.S. Government Printing Office, Washington, DC.

20. U.S. Environmental Protection Agency (1994) Final Report of Working Party on Neurotoxicology: Principles of Neurotoxicity Risk Assessment. *Federal Register*, **59** (#158): pp. 42360–42404.

21. U.S. Food and Drug Administration (1993) *Draft Revision: Toxicological Principles for the Safety Assessment of Direct Food Additives and Color Additives Used in Food (Redbook II)*. Food and Drug Administration, Center for Food Safety and Applied Nutrition, Washington, DC.

22. U.S. Food and Drug Administration (1982) *Toxicological Principles for the Safety Assessment of Direct Food Additives and Color Additives Used in Food (Redbook I)*. Food and Drug Administration, Bureau of Foods, Washington, D.C.

23. Watson, D.H. (Ed.) (1987) *Natural Toxicants in Food*. Ellis Horwood, Chichester, England.

24. Wenk, G.L. (1992) Dietary factors that influence the neural substrates of memory. In: *The Vulnerable Brain and Environmental Risks. Volume 1: Malnutrition and Hazard Assessment*. Plenum Press, New York, pp 67–75.

25. Weiss, B. and O'Donoghue, J. (Eds.) (1994) *Neurobehavioral Toxicology: Analysis and Interpretation*. Raven Press, New York.

26. World Health Organization (WHO) (1986) *Principles and Methods for the Assessment of Neurotoxicity Associated with Exposure to Chemicals*. Environmental Health Criteria Document 60. World Health Organization, Geneva.

27. Wurtman, R.J. and Wurtman, J.J. (Eds.) (1986) *Nutrition and the Brain. Vol. 7: Food Constituents Affecting Normal and Abnormal Behaviors*. Raven Press, New York.

Contents Chapter 8

Natural food factors

Chapter 9

Neurotoxic food contaminants: polychlorinated biphenyls (PCBs) and related compounds

Susan L. Schantz

1   Introduction

Food contaminants are substances that are unintentionally present in food. These chemicals originate from many sources, including human activity. It is difficult to give an unequivocal definition of the term food contaminant, and the substances involved may be either synthetic or natural. The previous chapter summarizes the naturally occurring neurotoxicants as they can be found in plants and animals used for human consumption. Unnatural contaminants, originating from production and technological applications can be divided in two subcategories: metals and organic substances. Neurotoxic metals will be dealt with in Chapter 12.

Table 9.1 lists some categories of neurotoxic food contaminants. The present chapter will focus on the neurotoxicology of the second group, the organic substances. For polyhalogenated biphenyls and related compounds, extensive experimental and epidemiological neurotoxic research has been carried out. Therefore the present chapter will focus on the neurotoxicity of these chemicals.

TABLE 9.1

Examples of classes of possible unnatural food contaminants associated with neurotoxicity

Metals	Pesticides	Organometals	Antibiotics	Antibacterial and antifungal agents	Halogenated aromatic hydro-carbons
Aluminium	Nicotine	Methylbromide	Doxorubicin	Pyridinethione	PCBs
Gold	Carbamates	Methylmercury		Hexachlorophene	TCDD
Lead	Organochlorines	Trimethyltin			PCDFs
Manganese	(e.g. chlordecone)				
Platinum	Organo-				
Thallium	phosphorus				
	compounds (e.g.				
	TOCP)				
	Pyrethroids (e.g.				
	deltametrin)				

1.1   CHEMICAL AND PHYSICAL PROPERTIES OF PCBS

The polychlorinated biphenyls (PCBs) and related halogenated aromatic hydrocarbons such as the chlorinated dibenzodioxins (PCDDs) and dibenzofurans (PCDFs) are a family of widely dispersed, environmentally persistent organic compounds. PCBs were synthesized for the first time in 1881, but were not produced commercially until 1929. In the United States, PCBs were produced and marketed by Monsanto under the trade name Aroclor. Other commercial PCBs include Kanechlor which was produced in Japan, Clophen which was produced in Germany, and Phenclor which was produced in Italy, among others. Production of PCBs peaked in the late 1960s and then declined precipitously in the 1970s after PCBs were discovered as environmental contaminants in numerous geographical locations throughout the world. Production ceased altogether in most countries by the late 1970s or early 1980s. Commercial PCBs were manufactured by reaction of gaseous chlorine with molten biphenyl. The resultant products were complex mixtures of chorobiphenyl congeners with different chlorine substitution patterns (Figure 9.1). Different manufacturing methods were used to produce PCB mixtures with different degrees of chlorination and, thus, different physical and chemical properties. For example, Monsanto produced Aroclor mixtures that ranged from 21-68% chlorine by weight (Aroclors 1221–1268; see Table 9.2). The first two digits of the four digit product code refer to the 12 carbons on the biphenyl ring. The second two digits refer to the % chlorine by weight (e.g., Aroclor 1242 is 42% chlorine). The only exception to this rule is Aroclor 1016 which is a distillation product of Aroclor 1242 and also has a chlorine content of about 42%. The lightly chlorinated PCB mixtures (e.g., Aroclors 1221–1248) are clear oils, while the more highly chlorinated mixtures are viscous resins.

TABLE 9.2

Chlorine content and properties of the aroclors

Aroclor	Chlorine (by weight) (%)	Description
1016	42	clear mobile oil
1221	21	clear mobile oil
1232	32	clear mobile oil
1242	42	clear mobile oil
1248	48	clear mobile oil
1254	54	yellow viscous liquid
1260	60	yellow sticky resin
1262	62	viscous resin
1268	68	white solid

1.2   INDUSTRIAL USES OF PCBS

The PCBs have several desirable physical and chemical properties that make them useful in a wide range of industrial applications. These include resistance to acids and bases, compatibility with organic materials, resistance to oxidation and reduction, thermal stability, and nonflammability. PCBs were used most extensively as dielectric fluids in capacitors and transformers, but they were also used as heat transfer fluids, hydraulic fluids, lubricants, sealants and microscope immersion oils, as additives in paints, plastics and dyes, and as extenders in pesticides. More than 95% of all capacitors produced before the early 1970s in the United States contained PCBs. These included the phase correction capacitors on power lines and the ballast capacitors for fluorescent lighting. Although older electrical equipment containing PCBs is slowly being phased out, many PCB-containing capacitors and transformers are still in use in the United States and elsewhere.

1.3   ENVIRONMENTAL OCCURRENCE OF PCBS

Unfortunately, the same properties that make PCBs attractive for industrial use cause them to persist and bioaccumulate in the environment. PCBs are readily detected in every component of the global ecosystem including air, soil, water, and sediments and are present in the tissues of wildlife, domestic animals, and humans worldwide.[40] They are a particular problem in aquatic ecosystems such as the Great Lakes[29] and the Baltic Sea.[17] Total worldwide production of PCBs from 1929 through the early 1980s has been estimated at about 1.2×10^9 kg. It is believed that about 30% of the total amount produced has been released into the global environment. Another 60% is either still in use in older electrical equipment or has been deposited in landfills and dumps for storage. Thus, even though PCBs are no longer produced, the potential for release into the environment continues to exist. Most experts agree that PCB contamination will continue to be a major environmental problem for at least several more decades.

The primary route of human exposure to PCBs is through consumption of contaminated food products, including fish from polluted waters. The average background level of PCBs in the blood of adults living in industrialized nations is about 2 ng/ml.[40] However, people in some geographic areas are more heavily exposed. For example, in Scandinavia the average serum PCB concentration is about 10 ng/ml.[74] In the United States, people who are heavy consumers of sport-caught Great Lakes fish average about 20 ng/ml.[26,28]

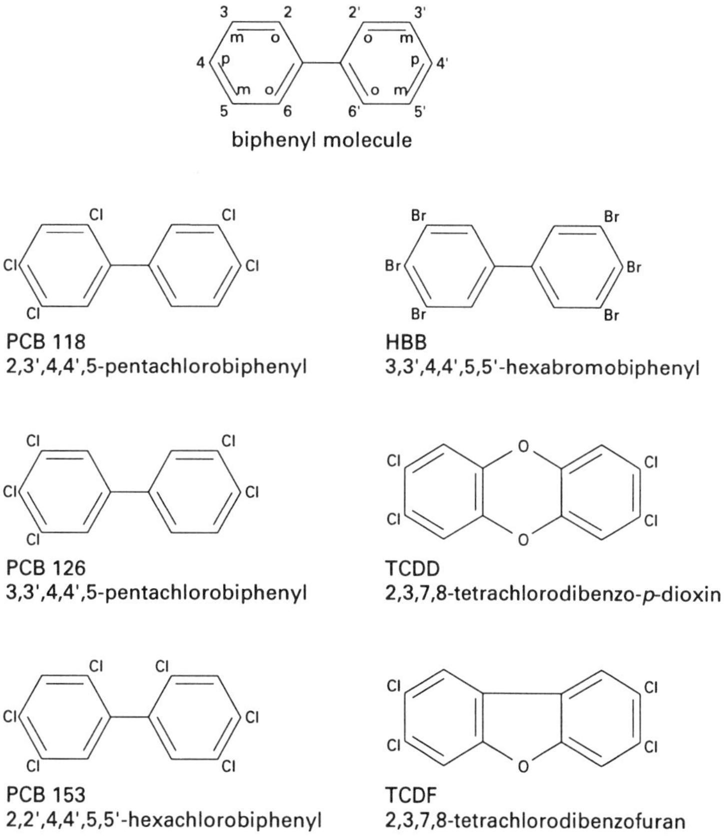

biphenyl molecule

PCB 118
2,3',4,4',5-pentachlorobiphenyl

HBB
3,3',4,4',5,5'-hexabromobiphenyl

PCB 126
3,3',4,4',5-pentachlorobiphenyl

TCDD
2,3,7,8-tetrachlorodibenzo-*p*-dioxin

PCB 153
2,2',4,4',5,5'-hexachlorobiphenyl

TCDF
2,3,7,8-tetrachlorodibenzofuran

FIGURE 9.1

Examples of chlorobiphenyl congeners with different chlorine substitution patterns

1.4 PROPERTIES AND OCCURRENCE OF RELATED COMPOUNDS

The polybrominated biphenyls (PBBs) are chemically similar to the PCBs (Figure 9.1), but only one PBB formulation, Firemaster BP-6, a product of the Michigan Chemical Company has been widely used in industry. Firemaster was used primarily as a flame retardant in plastics. PBBs are not widespread in the environment or in human tissue, except in the state of Michigan where a manufacturing mixup resulted in widespread contamination of the food supply (see Section 2.2, below). The chlorinated dibenzodioxins (PCDDs) and dibenzofurans (PCDFs) are industrial byproducts formed during the synthesis of halogenated aromatics such as PCBs, chlorinated phenols. and hexachlorobenzene, and are present at trace levels in these products. PCDDs and PCDFs are also formed during the bleaching of wood pulp with chlorine and are found in effluents, wastes and pulp from the paper and pulp industry. These compounds are also formed during combustion processes and are found in fly ash from municipal and hospital incinerators. Like PCBs, PCDDs and PCDFs have been identified as contaminants in nearly every component of the global ecosystem. They are found in the tissues of fish, wildlife, domestic animals, and humans worldwide. Trace amounts of PCDDs and PCDFs are found in many commercial foods including fish, meat, cow's milk, butter, chicken, and eggs. The sources and environmental occurrence of PCDDs and PCDFs have been reviewed.[2] Although these compounds are present at much lower concentrations than PCBs, they are still of concern because of their extreme toxicity. PCDDs and PCDFs are often found together with PCBs, which can complicate the determination of the causative agent(s) when human health effects are observed (see Section 5.1, below).

1.5 TOXIC EQUIVALENCY FACTORS

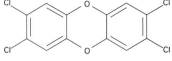

TCDD
tetrachlorodibenzo-p-dioxin

Many of the toxic responses observed following exposure to PCDDs, PCDFs, and PCBs are similar and appear to be mediated by the binding of these compounds to the aryl-hydrocarbon (Ah) receptor.[60] TCDD, which is the 2,3,7,8-substituted PCDD, binds with the highest affinity and is the most potent in eliciting toxic responses. The most potent of the PCBs are the coplanar, non-*ortho*-substituted congeners (Figure 9.1). Some relatively planar, mono-*ortho*-substituted PCBs also bind to the receptor, but have much lower affinities. Di-*ortho*-substituted PCBs cannot assume the planar configuration required for binding and have little or no affinity for the receptor. Because of the good correlation between Ah-receptor binding affinity and toxic potency, a toxic equivalency factor (TEF) approach has been suggested for monitoring levels of these chemicals in the environment and for determining the risks to humans from exposure.[2,61] In the TEF approach, TCDD is assigned a value of 1.0 and all other compounds are rated based on their relative potency in eliciting Ah-receptor mediated responses. For example, a TEF value of 0.1 has been proposed for the most toxic PCB congener, 3,3',4,4',5-pentachlorobiphenyl.[61] The TEF approach can be used to transform analytical data to TCDD toxic equivalents, or TEQs. This is done by multiplying the amount of each compound present in the sample by its TEF and summing across all compounds, yielding a number which represents the total amount of TCDD equivalents in the sample. It should be stressed that this approach is valid only if the effects of the individual compounds are truly additive. It should also be stressed that a growing number of toxicologists find the recent emphasis on this approach disturbing because *ortho*-substituted PCB congeners appear to have important toxicological effects, but do not bind to the Ah receptor and are

256

therefore not factored into the equation.[2,71] Prominent among these so-called, non-Ah-receptor mediated effects are changes in endocrine function and neurological function.[71] Thus, a policy that regulated PCBs using the TEF approach would, in all likelihood, underestimate the risk of nervous system effects.

2   Poisoning episodes

2.1   YUSHO AND YUCHENG

Two major outbreaks of human PCB poisoning have been recorded. The first occurred in Japan in 1968. In that incident, over 1000 people became ill after ingesting rice oil that had beencontaminated with PCBs during the manufacturing process. The PCBs were being used as heat transfer fluids during the decolorization process and leaked into the oil. The disease, which was best characterized by its dermal manifestations including acneform lesions, brown pigmentation of the skin, and ocular swelling became known as Yusho, or "rice oil disease". Many Yusho patients also complained of neurologic disorders including headache, memory loss and numbness, hypoesthesia and neuralgia of the limbs.[78] Detailed neurological exams of a subset of the exposed individuals did not reveal any evidence of central nervous system (CNS) damage. However, behavioral tests which might have revealed subtle functional impairments in CNS function were not administered.

Pregnant women who suffered from Yusho gave birth to babies that were small and had dark brown pigmentation of the skin.[80] Because of the pigmentation, these babies have been referred to in the literature as "cola-colored babies".[55] Biopsy of the pigmented skin revealed the presence of increased melanin and hyperkeratosis. The color and pattern of the pigmentation closely resembled that seen in patients with Addison's disease (failure of adrenocortical function), but no evidence of hypoadrenocorticism could be found.[80] Other abnormalities observed in the infants included hypersecretion of the meibomian glands, orbital edema, gingival hyperplasia, natal teeth, abnormal calcification of the skull, and rocker bottom heel.[80] The precocious dentition and abnormal calcification suggested a possible PCB-related alteration in calcium metabolism. However, the mechanism for such an effect remains unknown.

A follow-up study of a subset of the children reported a number of abnormalities including continued growth impairment, slowness, lack of endurance, hypotonia, jerkiness, clumsy movement, apathy, and IQs averaging around 70.[25] Nerve conduction velocities were apparently not assessed, but the reported disturbances in responsiveness and neuromuscular organization suggest peripheral nervous system damage. The impaired affective and intellectual functioning indicate CNS involvement.

In 1979, 11 years after the Yusho incident in Japan, a similar outbreak of PCB poisoning occurred in Taiwan. In Taiwan, where the disease was called YuCheng, over 2000 people became ill. Again, the disease was traced to rice oil contaminated during the manufacturing process.[27] The clinical manifestations seen in the PCB-poisoned Taiwanese were very similar to those reported a decade earlier in Japan. Dermal lesions and neurological complaints were the most common.[27] As in Japan, decreased sensory and motor nerve conduction velocities were observed, indicating the presence of peripheral nervous system damage. In contrast, EEGs and evoked potentials were normal. Dark brown pigmentation of the skin and intrauterine growth retardation were observed in infants born to mothers suffering from YuCheng.[27] The YuCheng children have been followed closely since their birth and a number of developmental abnormalities including lower body weight and height, continued hyperpigmentation of the skin, hypertrophy of the gums, deformities of the nails, increased frequency of bronchitis, and delays in neuropsychological development have been observed as they have matured. Effects of the exposure on neuropsychological functioning of the children are discussed in 5.1, below.

2.2 MICHIGAN PBB INCIDENT

PBB contamination of the food supply is not a widespread problem, except in the state of Michigan where a fire retardant, Firemaster BP-6, composed of PBBs was inadvertently mixed with cattle feed in place of a feed supplement, magnesium oxide.[9] The contaminated feed was shipped to farm co-ops and dispersed to farmers throughout the lower peninsula of Michigan. The problem was not discovered until a year later when livestock began to sicken and die and farm families began to complain of various health problems. At the time the problem was discovered, the FDA set "action guidelines" for levels of PBBs in milk, milk products, meat, and eggs. As a result, great numbers of cattle, swine, sheep, and chickens were destroyed, along with thousands of pounds of dairy products and millions of eggs. However, by that time PBBs had already entered the commercial food supply, and virtually every resident of the lower peninsula of Michigan had been exposed. Families residing on the affected farms or buying food directly from the affected farms had much higher exposure. Several studies were undertaken to ascertain the impact of the PBB exposure on the health status of farm families. Although heavily exposed individuals complained of various physical and neurological problems, the studies did not find any clear association between PBB body burden and physiological or neurological symptomatology in adults.[45] Neuropsychological development of a small group of children who were known to have been exposed to the PBBs *in utero* and/or through breast milk was also assessed and there was a suggestion of a relationship between PBB body burden and developmental abilities (see 5.3, below).[68,70]

3 Low level exposure from the food chain

3.1 CONTAMINATED FISH

Consumption of PCB-contaminated fish from polluted waters such as the Great Lakes is a major source of PCB exposure for humans.[29] The Great Lakes are one of the most important water resources on the planet. Together the Lakes contain over 20% of the surface freshwater on earth, and they support the largest freshwater fishery on earth. Commercial fishing on the Great Lakes has been severely curtailed because of high levels of various toxic contaminants, including PCBs, in the fish. However, sport fishing on the Great Lakes is not regulated and remains a multi-billion dollar industry. In 1985 more than 4 million people fished the Great Lakes. Over 1000 chemicals have been identified in the Great Lakes aquatic environment. Fortunately, only a few of these are detected in Great Lakes fish in greater than trace quantities. Most are organochlorine compounds or heavy metals. Of these, PCBs top the list of contaminants likely to reach a person's dinner plate. One of the major metabolites of dichlorodiphenyltrichloroethane (DDT), dichlorodiphenyl-dichloroethylene (DDE) is also present in relatively large amounts even though DDT has not been used in the United States since 1971. Long-lived, predatory fish such as trout, salmon, and walleye, and bottom dwelling fish such as carp are the most heavily contaminated. A study conducted in the early 1980s demonstrated that persons who ate one or more meals of sport-caught Great

Lakes fish per week had serum PCB levels that were three times higher than persons in a matched comparison group that did not eat Great Lakes fish.[28] The levels ranged from less than 3 ng/ml to over 200 ng/ml with a median of 21 ng/ml.[28] A follow-up assessment of a subset of the cohort a decade later indicated that PCB body burdens in these people were remaining steady, even though the PCB concentrations in the fish declined during the intervening 10 years.[26] In contrast, DDE levels declined substantially during the same period. Several large scale studies assessing the health status and neuropsychological functioning of Great Lakes fish eaters and their offspring are currently underway. The findings are discussed in sections 5.2 and 5.3, below.

The Baltic Sea is another heavily PCB-contaminated aquatic environment. The Baltic is the largest brackish water body in the world and it supports a large and commercially important fishery. Nine countries share the shoreline of the Baltic and 14 countries have territory within its drainage basin. A number of large cities are located within the Basin and the total population of the area is more than 70 million. High levels of PCBs and other organochlorine compounds have been found in fish and commercial fish products from the Baltic Sea. Seal populations in the Baltic region have elevated body burdens of PCBs, DDE, and other organochlorine compounds they and suffer from immune and reproductive disorders. Cod liver oil originating from the Baltic Sea contains high levels of organochlorine contaminants.[17] As in Great Lakes fish, PCBs are the dominant contaminant. DDE is also present at fairly high concentrations. A number of other compounds such as HCB, dieldrin, and chlordane are also detected, but they are present in much lower concentrations. In the recent past, the concentrations of PCBs and other organochlorines in cod liver oil from the Baltic have exceeded the tolerance levels of most countries, and it has been reported that cod liver of Baltic origin was unfit for human consumption.[17]

3.2   CONTAMINATED FARM SILOS

In some parts of the United States, dietary exposure to PCBs has occurred via consumption of contaminated farm products. In the 1940s and 1950s, concrete silos on many midwestern farms were coated on the inside with a sealant that contained the PCB mixture, Aroclor 1254. The coating material was used to protect the inside surfaces of the silos from the erosive actions of organic acids produced during silage fermentation. Over time, the sealant peeled off and became mixed with the silage, which was used as feed for beef and dairy cattle. Thus, families that regularly ate beef and dairy products from their own farms were unknowingly consuming PCB-contaminated food. A survey by the Michigan Department of Agriculture in the early 1980s identified approximately 400 Michigan farms with PCB contaminated silos. Approximately 900 farms with similarly contaminated silos are located in Indiana and Ohio. Additional contaminated silos are located on farms in Illinois, Kentucky, and Pennsylvania. The silos were constructed in the 1940s and 1950s and some were still in use as late as the mid-1980s, so contamination of farm food occurred over a period of more than 30 years. Because the total number of farms involved was relatively small, the domestic food supply was never adversely affected. However, for families who ate their own farm products on a daily basis, this represented a continuous exposure spanning many years.

The Michigan Department of Public Health has obtained blood samples from individuals residing on farms with contaminated silos and analyzed them for PCBs. As expected, most farm residents had serum PCB levels considerably above background.[28] The levels ranged from 5-300 ng/ml with a median value of 28 ng/ml. These figures are quite similar to those reported for heavy consumers of PCB-contaminated Lake Michigan fish.[28] The farmers were found to have a greater incidence of joint problems and numbness than a comparison group of Iowa farmers.[28] In Illinois, members of an extended family residing on a large dairy farm with four PCB-contaminated silos were reported to have serum PCB concentrations ranging from 50-166 ng/ml.[23] Family members complained of various health problems including joint pain, fatigue, irritability, persistent skin rashes, reduced libido and menstrual irregularities. Neuropsychological functioning of a group of children whose mothers resided on the Michigan farms with PCB-contaminated silos was evaluated and some behavioral changes similar to those reported in children of women who consumed PCB-contaminated Lake Michigan fish were observed (see Section 5.2, below).

4   Effects of PCBs and related compounds on CNS function: laboratory studies

4.1   PRIMATE STUDIES

Nearly all of the laboratory studies assessing neurobehavioral function following exposure to PCBs or related compounds have focussed on the effects of *in utero* and/or lactational exposure on later neurobehavioral function. One of the most extensive series of studies has been conducted in primates. In these studies, rhesus monkeys born to mothers with a history of chronic dietary exposure to commercial PCB mixtures (either Aroclor 1248 or Aroclor 1016) or to TCDD were followed longitudinally from weaning through adulthood.[4,6,63,66,71] The Aroclor 1248 and TCDD doses that were used were similar on a toxic equivalency basis,[71] so direct comparisons of the effects of this complex PCB mixture and TCDD could be made. Aroclor 1016 has little or no TCDD-like activity, thus, the doses could not be expressed in TEQs for comparison. Both cognitive functioning and locomotor activity were assessed periodically until the monkeys reached adulthood (Tables 9.3 and 9.4). Some of the groups of offspring were born concurrent with maternal exposure to the PCBs or TCDD, whereas other groups were born 1–3 years after the maternal exposure had ended.

Locomotor activity was assessed at yearly intervals. Locomotor hyperactivity was a consistent finding in both Aroclor 1248 — and Aroclor 1016 — exposed offspring.[4,63] The increased activity was present shortly after weaning and persisted for several years. Later monkeys that had been exposed to high levels of Aroclor 1248 (2.5 ppm in the maternal diet) became hypoactive.[4] In contrast, monkeys exposed to lower levels of Aroclor 1248 (0.5 or 1.0 ppm in the maternal diet), or to Aroclor 1016 (0.25 or 1.0 ppm in the maternal diet) eventually returned to control levels of activity.[63] Unlike the PCB exposed monkeys, monkeys exposed to TCDD did not show any consistent changes in locomotor activity.

The monkeys were also assessed on two learning tasks known as discrimination-reversal learning and delayed spatial alternation.[6,66,71] These tasks were selected because they have been used extensively in brain lesion research, and the brain areas and neural circuits critical for accurate performance are quite well understood.[46] Discrimination-reversal learning consisted of a series of four learning problems in which the monkey was required to make an initial discrimination between two stimulus objects. In the first two problems, the correct object occupied a particular spatial position (spatial discriminations), whereas in the last two problems the correct object was of a particular color or shape (object discriminations). On each problem, the monkey was required to learn the initial discrimination and then a series of reversals. Delayed spatial alternation was similar to spatial reversal learning except that the position of the reward was alternated on every trial. In other words, the monkey had to learn to alternate its responses back and forth between the two response sites rather than responding to one site for a number of trials and then switching to the other site when the original one was no longer rewarded. The spatial alternation task also had a memory component to it. This was achieved by placing delays of varying length between the trials. The monkey had to remember the position of the reward on the previous trial during the delay period in order to make a correct response to the opposite spatial position on the next trial. The monkeys were tested on discrimination-reversal learning when they were 1–2 years of age and on delayed spatial alternation when they were 4–6 years of age.

Monkey offspring whose mothers were fed Aroclor 1016 were impaired in their ability to learn the simple left-right spatial reversal that was the first problem in the discrimination-reversal learning series.[66] They required more trials than did their age-matched controls to learn the original discrimination and the first reversal, but they were not impaired on later reversals of spatial position. Offspring born to mothers with a prior history of exposure to Aroclor 1248 were tested concurrently with the Aroclor 1016-exposed offspring and were not impaired. However, a group of offspring born during maternal exposure to the Aroclor 1248 had been tested earlier and they were found to be profoundly impaired on spatial reversal learning.[6] Paradoxically, after training on a second spatial task which included irrelevant color and shape cues, both the Aroclor 1016- and Aroclor 1248-exposed offspring learned color and shape reversal problems in fewer trials than age-matched controls.[66]

The effects seen in TCDD-exposed monkeys tested on the same discrimination-reversal learning series were qualitatively different from those seen in the PCB-exposed monkeys. In fact, the pattern was just the reverse of that seen in the PCB-exposed monkeys. TCDD-exposed monkeys were facilitated on simple spatial reversal learning and impaired on object reversal learning.[71] The TCDD-related effects were present on original learning and the first reversal, but not on later reversals. In that sense, the pattern was similar to that for PCBs.

The pattern of the spatial reversal learning deficit observed in PCB-exposed monkeys was very similar to that seen in monkeys with lesions to an area of the brain known as the dorsolateral prefrontal cortex.[46] Like the PCB-exposed monkeys, lesioned monkeys have trouble learning the original discrimination and the early reversals, but are able to learn the later reversals as well as controls. In contrast, animals with damage to other brain areas such as the temporal lobe or the mammillary bodies show impaired learning across all of the reversals of spatial reversal learning problems.[46] The similarity of effects in the PCB-exposed and dorsolateral prefrontal cortex lesioned animals suggests that PCBs may alter the functioning of the dorsolateral prefrontal cortex or one of its input or output pathways in some way. The improved object reversal learning seen in the PCB-exposed monkeys would also be consistent with this interpretation because monkeys with dorsolateral prefrontal cortical lesions have been reported to show improved object reversal learning.[46] It is believed that because animals with damage to this brain region do not attend well to spatial cues, they are able to focus more exclusively on other dimensions such as color and shape, learning to discriminate them more readily.

Interestingly, brainstem dopamine input to the dorsolateral prefrontal cortex plays an important role in spatial learning and memory. Specific chemical lesions that disrupt dopamine input to the dorsolateral prefrontal cortex cause a spatial learning deficit nearly identical to that caused by surgical ablation of the

TABLE 9.3

Summary of Effects of Exposure to PCBs and Related Compounds on Locomotor Activity in Animals

Compound and species	% Chlorine	Maternal dose (mg/kg/day)	Exposure period	Age at test	Activity effect	Reference
Aroclor 1016 Monkeys	42	0.008 or 0.029	pregestation to weaning	24 months 36 months	↑ activity no effect	Schantz and Bowman, 1983
Aroclor 1248 Monkeys	48	0.084	pregestation to weaning	6 months 12 months 44 months	↑ activity ↑ activity ↓ activity	Bowman, 1982 Bowman and Heironimus, 1981
		0.084	ending 12 mo prior to gestation	12 months 36 months	↑ activity ↑ activity	Bowman, 1982 Schantz and Bowman, 1983
		0.014 or 0.030 (3 days/week)	pregestation to weaning	12 months 36 months	↑ activity no effect	Bowman, 1982 Schantz and Bowman, 1983
TCDD Monkeys	–	0.12 ng/kg/day (5 ppt)	pregestation to weaning	12 months 24 months 36 months	no effect	
Clophen A30 Rats	42	0.4 or 2.4	pregestation to weaning	22 days 120 days	↑ activity no effect	Lilienthal et al., 1990
Fenchlor 42 Rats	42	5.0 or 10.0	5 days pregestation	14 days 21 days	↓ activity ↓ activity	Panteleoni et al., 1988
		2.0 or 4.0	gestation days 6-15	14 days 21 days	no effect no effect	
		1.0 or 2.0	lactation days 1-21	14 days 21 days	↓ activity no effect	
Aroclor 1254 Mice	54	11 or 82 ppm	pregestation to weaning	27 days	↑ activity	Storm et al., 1981
PCB 77 Mice	–	32.0	gestation days 10-16	35 days 65 days 12 months	↑ activity ↑ activity ↑ activity	Tilson et al., 1979 Agrawal et al., 1981
PCB 77 Mice	–	0.41 or 41.0	postnatal day 10	120 days	↓ activity	Eriksson et al., 1991
Kanechlor 500 Mice	54	10.0 or 100.0	gestation days 8-14 or 15-21	84 days	no effect	Shiota, 1976

dorsolateral prefrontal region.[7,8] Seegal and colleagues have demonstrated that PCBs reduce dopamine levels and dopamine turnover in the brainstem of monkeys.[71] Thus, it is possible that PCBs cause deficits in spatial learning by altering dopamine input to the prefrontal cortex.

The ability of PCB-exposed offspring to learn the delayed spatial alternation task was also impaired.[66] The effect was most pronounced in offspring born to the Aroclor 1248-exposed mothers. Even after very prolonged testing, they were not able to reach control levels of performance. This profound deficit is particularly striking when one considers that the monkeys were 4–6 years of age at the time of the testing and had not been exposed to PCBs since they were weaned at 4 months of age. The deficit most likely represents a permanent consequence of early PCB exposure. It is also striking because deficits were present in monkey offspring born as much as 3.0 years after maternal PCB exposure ended. This suggests that offspring remain at risk for certain types of neurobehavioral dysfunction long after maternal PCB exposure has ended. Neurobehavioral disturbances have also been observed in YuCheng children born years after the poisoning episode occurred (see Section 5.1, below).

The pattern of the deficit across the different intertrial delay intervals was also striking. A large deficit in percent correct performance was present even when there was only a 5 s delay between trials. In fact, the deficit was just as pronounced after a 5 s delay as it was after a 10 or 20 s delay. This suggests that the deficit was probably cognitive or attentional rather than mnemonic. Interestingly, monkeys with lesions to the dorsolateral area of the prefrontal cortex show a very similar pattern of effects.[46] They are able to perform spatial alternation when immediate responding is allowed, but when as little as a 2 s delay is introduced they show a profound decrease in performance. Delayed spatial alternation deficits are also seen following specific chemical lesions to the mesocortical dopamine pathway which projects from the brainstem to the prefrontal cortex[7,8]. These results linking delayed spatial alternation deficits to altered dopamine function strengthen the hypothesis that PCBs may cause deficits in spatial learning and memory by altering brain dopamine content.

In contrast to the PCB results, TCDD did not seem to impair delayed spatial alternation. Instead, TCDD-exposed animals performed slightly, though not significantly, better than controls.[71] These results are consistent with the facilitated spatial reversal learning observed in the TCDD-exposed animals, and they further highlight the difference in behavioral effects in TCDD- and PCB-exposed monkeys. Given that the Aroclor 1248 and TCDD doses used in these studies were similar on a TEQ basis, these findings of severely impaired delayed spatial alternation performance in Aroclor 1248-exposed animals and no effect in TCDD-exposed animals suggest that the PCB-related deficit in behavioral function may be mediated by the non-dioxin-like, *ortho*-substituted PCB congeners present in the mixture. Interestingly, several investigators have published studies which suggest that the *ortho*-substituted PCB congeners are more neuroactive than the coplanar, dioxin-like congeners.[42,72]

4.2   RODENT STUDIES

The effects of *in utero* and lactational exposure to PCBs on motor activity and learning have also been studied in rodents (Tables 9.3 and 9.4). Like the primate studies summarized above, many of the rodent studies have focused on commercial PCB mixtures and have used exposure protocols that included both *in utero* and lactational exposure. However, a few studies have evaluated the effects of specific PCB congeners and/or have exposed subjectsduring specific periods of development.

Mice whose dams were exposed to a specific coplanar PCB congener, 3,3',4,4'-tetrachlorobiphenyl (PCB 77) during gestation showed increased latencies on a one way active avoidance learning task after weaning.[77] They also showed altered motor activity, which included circling behavior, head bobbing and generalized locomotor hyperactivity.[1,77] Decreased striatal dopamine levels and decreased striatal dopamine receptor binding were observed in the PCB 77-exposed mice.[1] Interestingly, administration of d-amphetamine to the hyperactive PCB 77-exposed mice decreased their activity level.[13] This paradoxical effect is similar to the effect observed in rats treated neonatally with 6-hydroxydopamine, which selectively destroys the brains dopamine pathways.[8,49] These results lend further support to the hypothesis that altered central dopamine function may play a role in mediating the behavioral effects observed following PCB exposure. However, they should be interpreted with some caution because the doses employed were overtly toxic to the pups. Both a decrease in the number of live births and a large increase in postnatal mortality were observed. Pups that survived to be tested behaviorally had reduced body weights relative to controls, and these reductions in body weight persisted into adulthood. Thus, it is difficult to determine whether the behavioral and neurochemical effects observed were direct effects of the PCB 77 or were secondary to other toxic processes occurring in the animals. Preliminary studies by Seegal and colleagues suggest that perinatal exposure of rats to lower doses of PCB 77 results in *increased* rather than decreased striatal dopamine concentrations.[71] At present, the reasons for these discrepant results remain unclear. Different toxic mechanisms could come into play at low versus high levels of exposure; alternatively, the discrepant findings may be explained by a species difference between mice and rats in response to PCBs.

A more recent study looked at the neurobehavioral effects of combined *in utero* and lactational exposure to another coplanar PCB congener, 3,3',4,4',5-pentachlorobiphenyl (PCB 126) in rats.[3] Interestingly, this study also employed overtly toxic doses, but in contrast to the earlier study, few neurobehavioral deficits were observed. There were numerous signs of developmental toxicity including reduced gestational weight gain, reduced litter size, and reduced pup body weights that extended until at least 18 weeks postpartum, yet early reflex development was normal and there were no changes in visual discrimination learning at 5–18 weeks of age. The authors did not observe any changes in activity level and no mention of circling behavior or head-bobbing was made. The only behavior effect they observed was a delay in neuromuscular development. No neurochemical endpoints were assessed in this study.

Eriksson and colleagues[16] exposed mice to a single dose of PCB 77 on postnatal day ten and measured the density of cholinergic receptors in the hippocampus and cerebral cortex. Their studies suggest that PCBs may impair cholinergic as well as dopaminergic function. They reported a significant decrease in the density of muscarinic receptors in the hippocampus. However, the change was moderate, amounting to only about a 16% decrease in receptor density. Furthermore, the two doses given differed by a factor of 100-fold (0.41 versus 41 mg/kg) yet had roughly the same magnitude effect. The authors do not speculate about possible reasons for the lack of a dose-effect relationship in the data. In contrast to the effects observed in the hippocampus, no changes in muscarinic cholinergic receptor density were observed in the cerebral cortex.

Eriksson and colleagues[16] also observed changes in locomotor activity in their postnatally exposed mice. However, the pattern of effects differed substantially from that reported in the earlier studies.[1,13] No circling behavior or head bobbing was reported. The postnatally exposed mice were hypoactive compared to control mice during the first 20 minutes of a 60 minute test session, but then failed to show a normal pattern of habituation to the test apparatus over time, so that by the last 20 minutes of the test session they were displaying greater amounts of locomotor activity than control mice. A similar pattern of effects was observed for rearing behavior.

Rodents exposed to Aroclor 1254 perinatally also show less habituation of locomotor activity[76] and changes in hippocampal cholinergic markers.[41] Juarez de Ku et al. measured the activity of choline acetyl transferase (ChAT), the biosynthetic enzyme for acetylcholine, in the hippocampus, basal forebrain and cerebral cortex of rats exposed to Aroclor 1254 perinatally. They found significant decreases in ChAT activity in both the hippocampus and basal forebrain of PCB-exposed rats. A dose-effect relationship was observed and the highest PCB dose, although it was not overtly toxic to the dams or pups, reduced ChAT activity in both the hippocampus and basal forebrain to about 50% of control levels. ChAT activity in the cerebral cortex was unchanged following PCB exposure. The authors point out that neonatal hypothyroidism, induced either surgically or via treatment with chemical goitrogens, causes a similar reduction in hippocampal and basal forebrain ChAT. Interestingly, they found that serum thyroxine (T4) concentrations were severely depressed in their PCB-exposed rat pups. Co-treatment of the PCB-exposed pups with T4 partially restored ChAT activity. This suggests that PCB effects on the cholinergic system may be mediated, at least in part, through PCB-induced changes in thyroid function.

Mice and rats exposed to commercial PCB mixtures *in utero* and/or via lactation have also been reported to show changes in locomotor activity[54,76] and deficits on various types of learning tasks including active avoidance learning (Table 9.4), visual discrimination learning, and spatial learning in a water-filled multiple T-maze. Two of these studies have investigated the relationship between timing of exposure and pattern of effects.[48,54]

Pantaleoni and colleagues exposed dams to Fenchlor 42 (42% chlorine by weight) prior to mating, during gestation, or during lactation. Motor reflexes, motor coordination, and activity level were assessed in the pups prior to weaning. After weaning, the pups were assessed on a one-way active avoidance learning task using a 0.5 mA shock. Some behaviors were affected in pups from all three exposure groups, whereas other behavioral effects seemed to depend on the timing of exposure. The comparisons between the pre- and postnatal exposure groups are the most interesting. For example, the development of cliff avoidance was markedly delayed in postnatally exposed pups, but was unchanged in prenatally exposed pups. Open field activity was also suppressed in postnatally exposed pups, but not in prenatally exposed pups. In contrast, the development of swimming behavior was markedly delayed in both pre-and postnatally-exposed pups, and both exposure groups learned the active avoidance task more slowly.

Lilienthal and Winneke[48] used a different approach to study the effect of timing of exposure. They exposed dams to either PCB-contaminated or control diets beginning 60 days prior to mating and continuing until weaning. Half of the pups from each group were cross fostered to dams of the other exposure condition at birth. This created four groups exposed to PCBs: not at all, prenatally only, postnatally only, or both pre- and postnatally. The PCB mixture they used was Clophen A30 (also 42% chlorine by weight). Male offspring were assessed on a two-way active avoidance learning task when they reached adulthood. These authors used a 1 mA shock for their active avoidance testing because earlier studies in their laboratory had shown that nonassociative factors like the animal's activity level and reactivity strongly influence active avoidance when lower shock intensities are used.[47] Both the prenatal only and the combined pre- and postnatal exposure groups showed impaired active avoidance learning. However, unlike Pantaleoni et al.'s findings, the postnatal only group did not differ from controls. There are several differences between the two studies that could explain this disparate finding. These include potential differences in the chemical composition of the two PCB mixtures that were used, differences in the rat strains that were used (Wistar versus Fischer 344), differences in the age at which behavioral data were collected (day 30 versus day 115), and differences in the methods that were used to collect behavioral data (one-way versus two-way active avoidance; 0.5 mA versus 1.0 mA shock).

TABLE 9.4

Effects of Exposure to PCBs and Related Compounds on Learning and Memory in Laboratory Animals

Compound and species	% Chlorine	Maternal dose (mg/kg/day)	Exposure period	Age at test	Activity effect	Reference
Aroclor 1016 Monkeys	42	0.008 or 0.029 (0.25 or 1.0 ppm)	pregestation to weaning	14 months	impaired spatial RL, improved object RL	Schantz et al., 1991
				4 years	high dose impaired DSA, low dose slightly facilitated DSA	
Aroclor 1248 Monkeys	48	0.084 (2.5 ppm)	pregestation to weaning	7 months	impaired spatial RL	Bowman et al., 1978
		0.084 (2.5 ppm)	ending 12 or 32 months prior to gestation	14 months	improved object RL	Schantz et al., 1991
				6 years	impaired DSA	
TCDD Monkeys		0.12 ng/kg/day (5.0 ppt)	pregestation to weaning	14 months	impaired object RL; slightly improved spatial RL	Seegal and Schantz, 1994
				2-6 years	slightly improved DSA	
Clophen A30 Rats	42	2.4	pregestation to weaning	115-180 days	impaired active avoidance; impaired retention of visual discrimination	Lilienthal and Winneke, 1991
			pregestation to birth	115-180 days	impaired active avoidance; impaired retention of visual discrimination	
			birth to weaning	115-180 days	no change on active avoidance or visual discrimination	
Fenchlor 42 Rats	42	5.0 or 10.0	5 days pregestation	30 days	no effect on active avoidance	Panteleoni et al., 1988
		2.0 or 4.0	gestation days 6-15	30 days	impaired active avoidance	
		1.0 or 2.0	lactation days 1-21	30 days	impaired active avoidance	
Kanechlor 500 Rats	54	20.0 or 100.0	gestation days 8-14 or 15-21	91 days	impaired maze learning	Shiota, 1976
PCB 126 Rats		0.001 or 0.002	gestation days 9-19 (every other day)	35-126 days	no effect on visual discrimination	Bernhoft et al., 1994
PCB 28 PCB 118 PCB 153 Rats		8.0 or 32.0 4.0 or 16.0 16.0 or 64.0	gestation days 10-16	90-200 days	no effect on radial arm maze working/ reference memory; impaired T-maze DSA in females	Schantz et al., 1995
PCB 77 Mice		32.0	gestation days 10-16	35 days	increased active avoidance latencies	Tilson et al., 1979
Aroclor 1254 Mice	54	11.0 or 82 ppm	pregestation to adult- hood	23 days	increased active avoidance latencies	Storm et al., 1981

Rats in Lilienthal and Winneke's study were also tested on a visual discrimination learning task. Long-term memory was assessed by retesting animals on the same visual discrimination again 4 weeks after the initial testing was completed. All groups learned the initial visual discrimination equally well, but the prenatal only and combined pre- and postnatal exposure groups showed poorer retention of the learned information. Again, the postnatal only group did not differ from controls. The authors conclude that prenatal exposure to PCBs places the offspring at greater risk of neurobehavioral dysfunction, even though the amounts of PCBs transferred during lactation are much greater. A similar conclusion has been reached by researchers conducting epidemiological studies in human populations (Table 9.5; see Section 5.2, below). However, Pantaleoni et al.'s findings suggest that lactational exposure may, indeed, have some adverse neurobehavioral effects. Further research will be needed to resolve this issue.

In our laboratory, we recently investigated the effects of combined *in utero* and lactational exposure to three *ortho*-substituted PCB congeners on spatial learning and memory in rats.[67] The congeners we selected for study were 2,4,4'-trichlorobiphenyl (PCB 28), 2,3',4,4'5-pentachlorobiphenyl (PCB 118), and 2,2',4,4',5,5'-hexachlorobiphenyl (PCB 153). These congeners were selected because they are prevalent in human tissue, including breast milk[62] and serum from breast-fed children[32], and because they represent a range of chlorine substitution patterns with different chemical and toxicological properties.[60] PCB 28 is a lightly chlorinated, *ortho*-substituted PCB which does not bind to the Ah receptor, PCB 118 is a mono-*ortho* coplanar PCB which does show some affinity for the Ah receptor, and PCB 153 is a highly chlorinated, extremely persistent PCB that has little or no affinity for the Ah receptor. The animals were tested on two spatial learning tasks, a working memory/reference memory task on a radial arm maze and a delayed spatial alternation task on a T-maze. The T-maze task was patterned after the delayed spatial alternation task we had used previously to assess perinatally PCB-exposed monkeys.[66]

The animals did not show any learning deficits on the radial arm maze task. However, all three congeners resulted in slower acquisition on the T-maze delayed spatial alternation task. Interestingly, deficits were present only in female rats. The performance of PCB-exposed males was not affected. The pattern of the deficit observed in female rats was similar to the one we had observed previously in perinatally PCB-exposed monkeys. That is, performance of the PCB-exposed females rats was equally impaired at all delays including the shortest delay. As discussed earlier, this type of deficit is similar in pattern to the deficits seen following lesions of the prefrontal cortex[46] or destruction of the dopamine input to the prefrontal cortex.[7,8] We are currently conducting additional studies to determine whether or not the congeners used in this study alter dopamine receptor binding or dopamine levels in the prefrontal cortex and/or other dopamine rich brain regions.

Despite the differences in chlorine substitution patterns and chemical properties of the three congeners, there were no clear differences in the pattern of effects for the three congeners. Thus, the results illustrate that exposure to individual *ortho*-substituted PCB congeners can result in long-lasting spatial learning deficits similar to those seen in monkeys exposed to commercial PCB mixtures, but they do not provide any clues about possible structure-activity relationships. The data also suggest that there may be sex specific effects of PCBs on spatial learning. The mechanism for such effects is unknown. However, PCBs are known to modulate various hormonal systems including the gonadal steroids[39] and the thyroid hormones.[51] Changes in gonadal steroid function or thyroid function that occur during development have been shown to affect learning and memory in male and female rats differentially.[79]

267

Several other laboratories are currently testing rats exposed to various *ortho*-substituted and coplanar PCB congeners on similar tests of spatial learning and memory. It will be important to see if they replicate our finding of a sex specific spatial learning deficit. In our laboratory, we are currently testing rats exposed to coplanar PCBs (PCB 77 or PCB 126) or TCDD on the same two learning tests we used in the study of *ortho*-substituted congeners. In contrast to previous studies evaluating coplanar PCBs, we have used low doses which are not associated with any obvious signs of developmental toxicity. It will be interesting to see if our rats show facilitated spatial learning similar to that we previously observed in TCDD-exposed monkeys.[71]

5   Effects on CNS function: human studies

5.1   THE YUCHENG CHILDREN

Children born to YuCheng mothers during or after the poisoning episode have been assessed longitudinally for physical and neuropsychological effects (Table 9.5). One hundred and thirty-two children born between 1978 and 1985 were identified in 1985, and 118 of the children have been thoroughly evaluated for physical, cognitive, and behavioral effects on a yearly basis since that time. The YuCheng children were found to be smaller than controls, averaging 93% of control weight and 97% of control height, adjusted for age and sex.[57] They also exhibited various dermatological signs typical of PCB poisoning including chloracne, hyperpigmented skin, and nail deformities,[21] and they were delayed relative to control children on a number of developmental milestones, including the age at which they performed tasks such as saying phrases and sentences, turning pages, carrying out requests, pointing to body parts, holding a pencil, imitating a drawn circle, or catching a ball.[57]

The YuCheng children also scored lower on standardized intelligence tests than control children matched for neighborhood, age, sex, maternal age, parental education, and parental occupation.[10] Although statistically significant, the difference between the YuCheng children and their matched controls was small. The exposed children scored only 5 IQ points lower than the controls. Blood samples were collected from only a small subset of the children for PCB analysis, but there was no apparent relationship between a child's IQ scores and his or her serum PCB concentration. Blood samples were collected from most of the mothers a short time after the exposure occurred, but serum PCB concentrations in the mothers were also unrelated to the childrens' IQ scores. The data were also analyzed to see if children born during or shortly after the episode were more severely affected than those born later, but there did not appear to be any relationship between timing of exposure and severity of effect.

The neurophysiological functioning of a subset of 27 YuCheng children was assessed using auditory event related potentials (P300), visual evoked potentials, and short-latency somatosensory evoked potentials.[11] The 27 children and their matched controls also underwent thorough neurological examinations which included tests for signs of impaired motor development. The mean P300 latencies of the YuCheng children were prolonged compared with those of their matched controls, and the longest P300 latencies were observed in exposed children with the lowest IQ scores. The authors point out that a similar relationship between longer P300 latencies and lower IQ scores has been observed in other studies. The mean P300 amplitudes of YuCheng children were also reduced relative to controls. The authors point out that low P300 amplitudes have been reported in children with attention deficit disorder or reading disabilities. They conclude, based on the P300 findings, that children exposed to PCBs *in utero* may be at increased risk of cognitive dysfunction. No abnormalities were found in the general neurological exams and the YuCheng children did not differ significantly from matched controls on the motor tests. Visual and somatosensory evoked potentials were also normal, implying that these sensory pathways were intact.

TABLE 9.5

Summary of neurobehavioral effects of PCB exposure in children

Source of exposure	Age	Test	Behavioral effect	Reference
Maternal consumption of contaminated rice oil (Taiwan)	4-7 years	Stanford-Binet or WISC-R	lower IQ scores	Chen et al., 1992
	7-12 years	auditory evoked potentials	longer P300 latencies and lower P300 amplitudes	Chen, Hsu, 1994
	3-12 years	Rutter Child Behavior Scale; Werry-Weiss-Peters Activity Scale	more behavior problems; higher activity levels	Chen et al., 1994
Maternal consumption of contaminated Lake Michigan fish (Michigan)	birth	Brazelton Scale	underresponsive, hyporeflexive	Jacobson et al., 1984
	7 months	Fagan Visual Recognition Memory Test	impaired recognition memory (correlated with prenatal exposure)	Jacobson et al., 1985
	4 years	McCarthy Scales	lower verbal and numerical memory scores (correlated with prenatal exposure)	Jacobson et al., 1990a
	4 years	visual discrimination; short-term memory scanning; sustained attention	slower reaction time on visual discrimination; more errors on short-term memory scanning (correlated with prenatal exposure)	Jacobson et al., 1992
	4 years	Activity Rating Scale	lower activity levels (correlated with postnatal exposure)	Jacobson et al., 1990b
Maternal background environmental exposure (North Carolina)	birth	Brazelton Scale	underresponsive, hyporeflexive	Rogan et al., 1986b
	6, 12, 18 and 24 months	Bayley Scales	lower scores on motor scale (correlated with prenatal exposure)	Gladen et al., 1988; Rogan, Gladen, 1991
	3-5 years	McCarthy Scales	no effects	Gladen, Rogan, 1991

The full cohort of YuCheng children was assessed on several behavior rating scales, the Rutter Child Behavior Scale which is used to identify children with emotional or behavioral disorders and the Werry-Weiss-Peters Activity Scale which evaluates the child's activity level.[12] A higher score on the Rutter indicates a higher frequency of behavioral problems, whereas a higher score on the Werry-Weiss-Peters indicates a more active child. The YuCheng children consistently scored higher than their matched controls on both the Rutter and activity scales. The differences between the YuCheng children and their controls did not lessen as the children grew older and there was no evidence of decreased differences between exposed and control children as the interval between maternal exposure and year of birth increased. Interestingly, there was no correlation between physical signs of PCB intoxication and behavioral scores. Children with the severest physical symptoms were not necessarily those with the highest behavioral scores. Maternal serum PCB levels were also unrelated to the childrens' Rutter or activity scores. As discussed above, a similar lack of relationship between exposure indices and IQ test scores was observed.[10]

The lack of a relationship between indices of exposure (e.g., child's serum PCB level, maternal serum PCB level, severity of physical symptomatology) and neuropsychological outcomes is bewildering. It is possible that the measures of dose did not accurately reflect the child's actual *in utero* exposure. However, it is also possible that parents of exposed children perceived their children as "damaged" and thus reared them differently, or that the stress of growing up in a family that had endured the poisoning episode may have produced the behavioral differences.

As a final note, it is important to point out that the PCBs to which the YuCheng people were exposed were thermally degraded and thus contained unusually high concentrations of PCDFs and PCDDs.[44] The PCDFs in the contaminated oil were predominantly the extremely toxic 2,3,7,8-tetrachlorinated and 2,3,4,7,8-pentachlorinated congeners which animal studies suggest may be 100-500 times more toxic than the 48% chlorine mixture of PCBs that was in the oil.[44] The extreme toxicity of these compounds relative to PCBs has led to the assumption that PCDFs may actually be the causative agents in both Yusho and YuCheng. In support of this are studies of persons occupationally exposed to PCBs. The serum PCB concentrations following occupational exposure are typically as high or higher than those of Yusho and YuCheng patients, but the observed health effects are mild by comparison to Yusho and YuCheng.[24,50,52] Therefore, effects observed in the Yusho and YuCheng populations may not be representative of the effects that can be expected in other PCB-exposed populations.

5.2 THE MICHIGAN AND NORTH CAROLINA COHORT STUDIES

Jacobson, Jacobson and colleagues have conducted the only published, longitudinal prospective study relating maternal PCB exposure from the food chain (Lake Michigan fish) to neurodevelopmental outcomes in children.[18,30,31,35-37] They began the study by screening over 8000 women who delivered babies in four western Michigan hospitals in 1980–81. On the day following delivery, each women was interviewed about how much and what kinds of Lake Michigan fish she ate. All women who had consumed 26 or more pounds of Lake Michigan fish during the preceding six years were asked to participate in the study. Women who did not eat Lake Michigan fish were randomly selected and invited to serve as controls.

PCB exposure measures used in the study included the mother's estimated total lifetime Lake Michigan fish consumption, the infant's umbilical cord serum PCB level, the maternal serum PCB level at birth, and the breast milk PCB level at birth and five months for women who breast fed. The fish consumption measure was a weighted sum of annual Lake Michigan fish consumption. A value was assigned to each fish species based on the contaminant level for that species.[38] The amount of each fish species consumed was multiplied by the assigned value and the numbers were then summed across species to get the annual weighted fish consumption. Average fish consumption for the women who reported eating Lake Michigan fish was 6.7 kg/yr, equivalent to about two salmon or Lake trout meals/month. The women reported eating Lake Michigan fish for an average of 16 years.[18] Maternal serum PCB concentrations averaged 5.5 ng/ml and umbilical cord serum PCB concentrations averaged 2.5 ng/ml. The concentration in milk fat was 812 ng/ml in the first week after birth and 769 ng/ml at five months. Contaminated fish consumption was moderately correlated ($r = 0.30-0.38$) with PCB levels in maternal serum and milk.[69]

The children were evaluated at birth, 5 months, 7 months, and 4 years of age (Table 9.5). Outcome measures at birth included birth weight, body length, head circumference, and gestational age. Neonatal behavioral function was assessed using the Brazelton Neonatal Behavioral Assessment Scale (NBAS). PCB exposure, measured both by maternal contaminated fish consumption and by umbilical cord serum PCB levels was associated with lower birth weight, smaller head circumference, and shorter gestational age.[18] Exposed infants weighed 160–190 g less, their heads were 0.6–0.7 cm smaller, and they were born 4.9–8.8 days earlier depending on whether fish consumption or umbilical cord PCB level was used as the exposure measure. Head circumference was significantly smaller even after controlling for birth weight and gestational age. As the authors point out, the size deficits they observed are comparable to those associated with smoking during pregnancy.[33] However, offspring of women who smoke tend to catch up in the first few postpartum months, whereas higher umbilical cord serum PCB levels continued to be correlated with smaller size at 5 months of age.

Maternal contaminated fish consumption was also associated with several adverse behavioral outcomes on the NBAS.[30] Infants of women who ate the most fish exhibited motoric immaturity, poorer lability of states, a greater amount of startle, and more abnormally weak (hypoactive) reflexes. However, the other, more direct measure of exposure, umbilical cord serum PCB level, was not related to any adverse behavioral outcomes on the NBAS. This suggests that the adverse behavioral outcomes observed on the NBAS could be related to other contaminants present in the fish or to some other factor(s) that differed between the fish-eaters and non-fish-eaters.

Infant cognitive functioning was assessed at 5 and 7 months of age.[31,34] The Bayley Scales of Infant Development, a frequently used standardized test of infant cognitive development, was administered at 5 months of age.[34] Fagan's test of visual recognition memory was administered at 7 months of age.[31] In the Fagan test, the infant is shown a visual stimulus. Later, the familiar stimulus is paired with a novel one. The normative response for young infants is to spend more time looking at a novel stimulus. Therefore, infants who recognize the original stimulus will spend more time looking at the new one, and preference for the novel stimulus, as measured by the percentage of time the infant fixates the novel rather than the familiar stimulus, indicates the capacity to recall the original stimulus and discriminate it from the novel one. The Fagan test has become a popular assessment tool because it requires relatively complex information processing including stimulus discrimination, memory storage, and retrieval. Also, as discussed by Jacobson et al.[31] it can be performed on young infants, it is sensitive to a range of at-risk conditions and performance on the Fagan is a reasonably good predictor of later verbal IQ.

Neither maternal fish consumption, nor umbilical cord serum PCB level was related to scores on the Bayley Scales.[34] In contrast, both contaminated fish consumption and umbilical cord serum PCB level were associated with less preference for the novel stimulus.[31] In this case, the more direct measure of exposure (umbilical cord serum PCB level) was the stronger predictor. Preference for novelty decreased in a dose-dependent fashion as prenatal exposure to PCBs increased. The observed effects did not seem to be mediated by shorter gestation, reduced birth size, or poorer performance on the NBAS. In contrast to prenatal exposure, postnatal PCB exposure via breast milk had no effect on visual recognition memory.[31]

Children from the cohort were reassessed when they reached 4 years of age.[35-37] They were tested on the McCarthy Scales of Children's Abilities, which is a standardized test of cognitive function, and on several specific tests developed by researchers in the field of human information processing. These tests were designed to focus on specific aspects of cognitive processing including reaction time, short-term memory processing efficiency, visual discrimination, and sustained attention. Higher levels of prenatal PCB exposure, as measured by umbilical cord serum PCB levels, were associated with poorer scores on two subtests of the McCarthy Scales that measure verbal and numerical memory.[35] Higher maternal milk PCB concentrations were also associated with poorer performance on both of the subtests. In contrast, the quantity of breast milk consumed was not related to any of the outcomes. The authors conclude that the correlation with breast milk PCB levels probably derives from the fact that PCB concentrations in breast milk are representative of maternal body burden, and children of mothers with the highest PCB body burdens would have received the greatest transplacental exposure to PCBs. The fact that two exposure indices correlate with the same set of outcome variables lends strength to these findings.

The findings from the tests of cognitive processing efficiency were not as clear. The authors report that prenatal exposure to PCBs was associated with less efficient visual discrimination processing and more errors in short-term memory scanning.[37] However, in each case only one of the two exposure measures correlated with the outcome. Reaction time on the visual discrimination task was correlated with maternal milk PCB level, but not with umbilical cord serum PCB level, whereas the number of errors on the short-term memory scanning task was correlated with umbilical cord PCB level, but not with maternal milk PCB level. These inconsistencies suggest that these findings should be interpreted with caution. No effects on sustained attention were observed.

The physical growth and activity level of the children in the PCB fish exposure cohort were also assessed at 4 years.[36] Prenatal PCB exposure, as measured by umbilical cord serum PCB levels continued to be associated with lower body weight at 4 years, but there was no relationship between prenatal PCB exposure and height or head circumference at 4 years. The decrease in body weight was more pronounced in girls than in boys. A relationship between PCB exposure and activity level was also observed. However, unlike the effects on cognitive function and physical growth, the changes in activity level were related to the child's current PCB body burden, the principle determinant of which was postnatal exposure to PCBs via breast milk.[32] Children with the highest PCB body burdens at 4 years were less active than children with lower body burdens. These were all children who were breast fed for at least 1 year and had mothers with above average breast milk PCB levels. A similar relationship between increased PCB body burdens and decreased activity levels has been observed in school-aged children from farms with PCB-contaminated silos.[65] In contrast, the YuCheng children were found to be more active than control children.[12] However, YuCheng mothers were encouraged not to breast feed and very few of them did. Thus, exposure of the YuCheng children to PCBs was primarily *in utero*. As reviewed in Section 4.2, laboratory rodents exposed to PCBs postnatally exhibit hypoactivity,[16,54] whereas laboratory rodents exposed to PCBs prenatally tend to be hyperactive.[1,47] A group of monkeys that were exposed both pre- and postnatally were hyperactive initially, but became hypoactive as juveniles.[4]

Thus, both age of exposure and age of assessment could be important in determining the effects of PCBs on activity level.

A small group of children who were *in utero* or breast feeding during the PBB incident in Michigan have also been assessed on the McCarthy Scales.[68,70] At 2–3 years of age, children with higher PBB body burdens scored lower than children who were less heavily exposed.[70] A similar inverse relationship between PBB body burdens and McCarthy scores was observed when the children were 4–6 years of age.[68] However, these data must be interpreted with caution because the sample size was extremely small (19 children), no controls were included, and only some of the McCarthy subtests were administered.

Rogan, Gladen, and colleagues have also conducted a longitudinal, prospective study looking at the relationship between *in utero* and lactational PCB exposure and neurodevelopmental outcomes.[19,20,56,58,59] The women in their cohort were selected from the general population in North Carolina and had not been exposed to any known dietary source of PCBs other than the background levels that contaminate the general food supply. The original cohort consisted of 880 pregnant women recruited at term from three health centers in the Raleigh-Durham area.[58] The women were not a random sample and cannot be considered as a representative cross-section of the population in that area. Nearly all of them were Caucasian, over half had college educations, and two-thirds were either professionals or white-collar workers. An unusually high percentage (88%) breast fed their infants. A few women dropped out of the study almost immediately, but 807 women (856 children) continued to participate beyond the initial neonatal contact,[58] and over 700 of the children were still available for follow-up at 3-5 years of age.[19] Most of those that were lost to follow-up had relocated out of the area.

Maternal blood, umbilical cord blood, placenta, and milk/colostrum were collected at the time of birth for PCB analysis. Additional milk samples were collected from women who breast fed at 6 weeks, and at 3, 6, and 12 months if the woman was still breast feeding. The median maternal serum PCB concentration at the time of birth was 9.06 ng/ml.[58] The median PCB concentration in maternal milk fat at birth was 1.77 mg/ml. After 12 months of breast feeding, that was reduced to 1.17 mg/ml. These exposure levels are actually higher than those reported by Jacobson, Jacobson, and colleagues for their Great Lakes fish exposure cohort.[69] However, differences in analytical technique make it difficult to compare results from the two studies. As Jensen[40] discusses, the method used by Rogan and colleagues probably over-estimated the actual PCB concentration by about a factor of two. Thus, the exposure may actually be somewhat higher in the fish exposure cohort even though the reported values are lower.

The children in the North Carolina cohort were assessed shortly after birth and were followed up at 6 month intervals until 2 years of age and then at yearly intervals until 5 years of age (Table 9.5). Outcome measures at birth included birth weight, head circumference, and scores on the NBAS. The PCB levels in umbilical cord blood were nearly all below the detection limit. Therefore the investigators, assuming that mothers with higher PCB body burdens would transfer more PCB to their fetus's, used the PCB content of maternal milk fat at birth as an indicator of the child's prenatal exposure. Unlike the findings of Jacobson, Jacobson, and colleagues,[18] no association between birth weight or head circumference and PCB exposure was observed in the North Carolina cohort.[59] However, PCB exposure was associated with several adverse outcomes on the NBAS.[59] Infants whose mothers had the highest PCB concentrations in their milk fat (> 3.5 mg/ml) had less muscle tone, lower activity levels, and were hyporeflexive. As reviewed above, the infants in the Michigan cohort whose mothers ate the largest quantities of Great Lakes fish were also hyporeflexive.[30] However, in that cohort there was no relationship between actual PCB exposure, as measured by umbilical cord serum PCB levels, and NBAS scores which makes the results difficult to interpret.

Infant cognitive and motor development was assessed in the North Carolina study by administering the Bayley Scales of Infant Development at 6, 12, 18, and 24 months of age.[20,56] Higher transplacental exposure to PCBs was associated with lower psychomotor scores at 6, 12, and 24 months of age.[19,20] A similar trend was seen at 18 months, but the difference was not statistically significant at that age. There was no relationship between transplacental PCB exposure and scores on the Mental Development Scale, and postnatal exposure through breast feeding was unrelated to performance on either scale. The investigators concluded that there was a small, but persistent delay in motor maturation attributable to transplacental PCB exposure. Jacobson and Jacobson also tested their cohort on the Bayley Scales and did not observe any relationship between psychomotor scores and PCB exposure. However, they tested the children at 5 months of age and did not reassess them at later ages. Because of the infants still limited behavioral repertoire, 5 months may be too early to accurately assess small delays in psychomotor development. Many motor behaviors first begin to emerge around 6 months of age.

The North Carolina children were later assessed on the McCarthy Scales of Children's Abilities at 3, 4, and 5 years of age, and neither transplacental or breast-feeding exposure affected scores on any of the McCarthy scales.[19] Despite the early deficits in psychomotor performance on the Bayley Scales, there was no indication of a relationship between PCB exposure and scores on the McCarthy Motor Scale. This scale is not an exact analogue of the Bayley Psychomotor Scale, but it is similar in that it uses common, age-appropriate tasks to assess motor function.[19] These findings suggest that the initial delay in psychomotor development associated with transplacental PCB exposure does not persist beyond 2 years of age. There was also no indication of a relationship between PCB exposure and scores on the McCarthy Memory Scale. Thus, these authors were not able to confirm the relationship between prenatal PCB exposure and short-term memory deficits reported by Jacobson, Jacobson, and colleagues.[31,35,37]

In the North Carolina study, the difference between the lowest and highest PCB exposure groups on the Bayley Psychomotor Scale was 4–9 points depending on the age of assessment. These are small changes (less than half a standard deviation), and they were present only in children above the 95th percentile for PCB exposure. The effects observed by Jacobson, Jacobson, and colleagues on the McCarthy Memory Scale were of similar magnitude, and were also present only in the most highly exposed children.[35] As Rogan and Gladen[56] point out, changes of this small magnitude are of doubtful clinical significance for individual children. However, the public health implications of an effect of this magnitude could potentially be very significant. This point has been emphasized by Davis[14] with regard to low level lead exposure. At the population level, a shift in mean performance on the Bayley of just four points would result in a 50% increase in the number of children with subnormal scores (e.g., scores > 1 SD below the mean).

5.3   OTHER COHORT STUDIES

A number of other prospective longitudinal cohort studies are now under way in the United States, Canada, and Europe. These include a number of studies of Great Lakes fish-eaters. The neuropsychological functioning of both children and adults is being evaluated in several of these studies, and many of the same tests used by Jacobson and colleagues are being employed. This will allow for comparison of effects between various cohorts consuming fish from a number of sites on the Great Lakes.

Neuropsychological function of children is also being assessed in a Dutch study.[22] A cohort of approximately 400 Dutch children, half from an urban, industrialized area (Rotterdam) and half from a rural area (Groningen) are being studied.[22,43] Half of the children from each location were breast fed and the other half were bottle fed. A distinctive feature of this study is the extensive exposure assessments that are being done. Umbilical cord blood, maternal milk, and maternal blood samples have been obtained for detailed congener specific PCB analyses.[43] Polychlorinated dibenzofurans and dibenzodioxins are also being quantified. In addition, vitamin A status and thyroid function are being assessed. The neuropsychological measures being used include the Bayley Scales and an updated version of the Fagan test of visual recognition, tests which were also used in the earlier cohort studies conducted in the United States. Findings from assessments conducted up to 18 months of age will be available soon. Comprehensive follow-up assessments will be conducted when the children reach 3.5 years of age. The battery will include the Kaufman Assessment Battery which assesses attention, information processing and achievement, the Reynell Developmental Language Scales which assess language development, and the Groninger Behavioral Observation Scale which is used for detecting attention deficit disorder. Tests of visual discrimination, sustained attention and memory used by the Jacobsons in their Michigan fish exposure cohort[37] will also be repeated in this cohort of children. Recently a German cohort of 200–300 children was formed, and will be assessed in conjunction with the Dutch study.

6   Future directions

6.1   IDENTIFYING AT RISK POPULATIONS

The two major prospective studies summarized in this chapter, as well as the more recently initiated studies described above, have focused on populations whose body burdens of PCBs are similar to background or only slightly elevated relative to background. Although it is important to understand the potential impact background levels of PCBs and related compounds may be having on neuropsychological function, it may be difficult to reliably detect effects at such low levels of exposure. Therefore, an important goal of current and future research should be to identify and study populations that have unusually high exposure to these compounds. The Inuit people who reside in the Arctic regions of Quebec are one such example. For cultural and economic regions the Inuit rely on fish (mainly Arctic char and cod) and sea mammals (mainly seal, beluga whale, and walrus) as the primary sources of protein and lipids in their diet. As a result, they have unusually high body burdens of PCBs and other organochlorine compounds. In a recent study, breast milk samples collected from Inuit women had PCB concentrations seven times greater than PCB concentrations in breast milk samples collected from a control group of Caucasian women in southern Quebec.[15] Concentrations of DDE, HCB, dieldrin, and other organochlorines were also elevated in breast milk from the Inuit women. The surprisingly high PCB body burdens present in the Inuits identify them as an important at risk population. Epidemiologic studies are now underway to examine the putative effects of PCB exposure on neuropsychological development in Inuit infants.[15]

Another population that has unusually high dietary exposure to PCBs is Faroe Islanders, who rely on whale blubber as a staple of their diet. Because whales are at the apex of the marine food chain and because organochlorine

275

compounds such as PCBs accumulate in fatty tissue, the residue levels in blubber can be quite elevated, leading to considerable exposure for those who include it in their diets. Studies evaluating neuropsychological functioning in a cohort of approximately 1000 PCB-exposed Faroe Island children are currently underway.[22] Marine food sources also have elevated levels of mercury and, thus, concurrent mercury exposure will be a major issue in interpretation of the results from this study.

Until recently studies designed to assess the neurotoxic risk from dietary exposure to PCBs and related compounds have focused almost exclusively on infants and young children, the assumption being that *in utero* and/or lactational exposure to these chemicals would have the greatest impact. However, there are reasons to suspect that older individuals with elevated PCB body burdens may also be at increased risk. Animal studies have demonstrated rather convincingly that exposure to PCBs during adulthood lowers brain dopamine levels.[71] The changes are not dramatic, but decreases on the order of 20–30% were observed in some dopamine rich areas of the primate brain. Furthermore, the changes appear to be very long-term. Six months after the monkeys were removed from the PCB diets, their brain dopamine levels remained depressed to the same extent as they were during exposure. These findings may be important with respect to the elderly, because brain dopamine levels decrease dramatically with aging, potentially putting the elderly at greater risk of neurological dysfunction from PCB exposure.

With this in mind, we recently initiated a study evaluating neuropsychological functioning in a cohort of aging Lake Michigan fish-eaters and a group of age- and sex-matched non-fish-eating controls. The study participants were originally recruited by the Michigan Department of Public Health (MDPH) in 1980 and 1982 and have been followed longitudinally since that time. Thus, historical data regarding their fish consumption practices and body burdens of PCBs and other contaminants are available. PCB body burdens were clearly elevated in the fish-eaters at the time the study was initiated[28] and a follow-up assessment of a subset of the cohort in 1989-1991 indicated that body burdens had remained stable even though contaminant levels in the fish dropped significantly during that time.[26] These fish-eaters have PCB body burdens substantially higher than other fish-eating cohorts recruited more recently.[75] There are several unique features of the cohort that explain these differences and, at the same time, make them an ideal group for our study. First, they are relatively heavy consumers of sport-caught Great Lakes fish, eating on the average 56 fish meals/year. Second, they were recruited 14 years ago, in 1980. Thus, most had been fishing and consuming fish during the 1970s when the most prized sport fish, lake trout and salmon, were both more plentiful and more heavily contaminated with PCBs. And finally, as a group they are now quite advanced in age. Most have been consuming Lake Michigan fish for many, many years. We are currently conducting detailed neuropsychological assessments of the fish-eaters and their age- and sex-matched controls. Results should be available by late 1995.

6.2 UNDERSTANDING MECHANISMS FOR FUNCTIONAL EFFECTS

Another important goal of future research will be to learn more about the mechanisms responsible for PCB-induced changes in neurobehavioral function. Given the structural diversity and complexity of these compounds, it is unlikely that any one mechanism will emerge to explain all of the neurobehavioral effects. We are more likely to find that a number of factors interact in complex ways to produce the observed effects. As has been discussed in earlier sections of this chapter, PCBs are known to modulate or disrupt several hormonal systems that play important roles in brain development. In particular, some PCB congeners have estrogenic or antiestrogenic activity,[39] and some reduce circulating thyroid hormone concentrations.[51] There is also *in vitro* evidence to suggest that some PCBs have direct toxic effects on nerve cells, lowering cellular dopamine content[72] and/or altering the Ca^{2+} homeostasis of the cell.[42] The structure-activity relationships for these various actions appear to be different, but overlapping. For example, some congeners are antiestrogenic and reduce thyroid hormone levels, while others are estrogenic and reduce thyroid hormone levels. Still others are estrogenic and reduce dopamine levels, but do not affect thyroid hormones. The actions of many congeners are unknown, or only partially understood. Furthermore, when exposure involves PCB mixtures, the situation is likely to be even more complex. All of these mechanisms, and possibly others, may come into play, and the relative importance of each is likely to vary depending on the congener make-up of the mixture. Clearly, determining the mechanisms through which PCBs exert their neurobehavioral effects will be one of the challenges of the future.

7   Summary

PCBs and related compounds are ubiquitous in the environment and are present in foods consumed by humans, particularly fish and fish products from polluted waters such as the Great Lakes and the Baltic Sea. Studies in rodents, monkeys, and humans have demonstrated that *in utero* and lactational exposure to PCBs are associated with a number of neurobehavioral effects including delayed reflex development, changes in locomotor activity, and deficits in learning ability.

The various animal studies that have been conducted to date have used a wide array of PCB congeners and mixtures, exposure protocols, and behavioral testing strategies. The complexity of the resulting data set makes generalizations difficult. However, a few general conclusions can be draw. First, it seems clear that perinatal exposure to PCBs affects both locomotor activity and learning (Tables 9.3 and 9.4). Changes in locomotor activity seem to occur following either pre- or postnatal exposure, and the timing of the exposure seems to be important in determining the nature of the effect. Animals exposed either prenatally or both pre- and postnatally exhibit increased levels of activity. In contrast, animals exposed only during the postnatal period exhibit decreased levels of activity.

Learning deficits have also been observed in a number of studies using an array of different testing procedures. Several authors have reported deficits on tests requiring memory function. The prevailing opinion is that these learning deficits are related to prenatal exposure. However, only two animal studies have specifically addressed the issue of timing of exposure and they have generated conflicting results. Further studies looking at the importance of timing of exposure are clearly needed before we assume that postnatal exposure via breast milk is not of concern.

At this point, there are no real clues as to structure-activity relationships for neurobehavioral effects. Neurochemical studies using *in vitro* methods have shown that lightly chlorinated *ortho*-substituted congeners are the most potent in reducing cellular dopamine content and in altering Ca^{2+} homeostasis in nerve cells. However, complex PCB mixtures, individual *ortho*-substituted PCB congeners, and individual coplanar PCB congeners have all been shown to produce neurobehavioral effects, and there do not appear to be any clear differences in the pattern of effects for the different congeners and mixtures. This could be because each congener or mixture has been assessed in only one or two studies, and/or because different exposure protocols and behavioral testing methods have been used in the various studies, making comparisons difficult. Some of the most recent studies are beginning to address these issues by comparing the neurobehavioral effects of various individual congeners using the same exposure protocols and behavioral testing procedures. These studies should soon shed some light on this important issue.

Changes in activity level and in learning ability have also been observed in several human populations exposed to PCBs *in utero* and via lactation (Table 9.5). In humans, as in animals, increased activity has been associated with exposure that was primarily prenatal and reduced activity has been associated with postnatal exposure via breast feeding. Deficits in performance on learning tasks requiring memory function have also been documented in human infants and children. In the human studies, there is evidence that these effects are related to prenatal exposure rather than to postnatal lactational exposure. Despite the similarities between the animal and human findings, the results of the human studies that have been published to date have been subject to skepticism.[53] Data from the newer cohort studies that are currently underway should help to resolve the current uncertainties regarding the risk of neurobehavioral dysfunction in humans from low level PCB exposure. Other important directions for future research will include assessing "at risk" human populations that have unusually high PCB body burdens and/or are at increased risk because of other factors such as advanced age, as well as determining the structure activity relationships and mechanisms of action underlying the neurobehavioral effects of PCBs.

References

1. Agrawal, A., Tilson, H., Bondy, S. (1981) 3,4,3',4'- Tetrachlorobiphenyl given to mice prenatally produces long-term decreases in striatal dopamine and receptor binding sites in the caudate nucleus, *Toxicol. Lett.* **7**: p 417.

2. Ahlborg, U., Brouwer, A., Fingerhut, M., Jacobson, J., Jacobson, S., Kennedy, S., Kettrup, A., Koeman, J., Poiger, H., Rappe, C., Safe, S., Seegal, R., Tuomisto, J., van den Berg, M. (1992) Impact of polychlorinated dibenzo-*p*-dioxins, dibenzofurans, and biphenyls on human and environmental health, with special emphasis on application of the toxic equivalency factor concept. *Eur. J. Pharmacol.* **228**: p 179.

3. Bernhoft, A., Nafstad, I., Engen, P., Skaares, J. (1994) Effects of pre- and postnatal exposure to 3,3',4,4',5- pentachlorobiphenyl on physical development, neurobehavior and xenobiotic metabolizing enzymes in rats. *Environ. Toxicol. Chem.* **13**: p 1589.

4. Bowman, R. (1982) Behavioral sequelae of toxicant exposure during neurobehavioral development, *Banbury Rep. 11: Environmental Factors in Human Growth and Development* **11**: p 283.

5. Bowman, R.E. and Heironimus, M.P., Hypoactivity in adolescent monkeys perinatally exposed to PCBs and hyperactive as juveniles, *Neurobehav. Toxicol. Teratol.* **3**: p 15.

6. Bowman, R., Heironimus, M., Allen, J. (1978) Correlation of PCB body burden with behavioral toxicology in monkeys, *Pharmacol. Biochem. Behav.* **9**: p 49.

7. Brozoski, T., Brown, R., Rosvold, H., Goldman, P. (1979) Cognitive deficit caused by regional depletion of dopamine in prefrontal cortex of rhesus monkey, *Science* **205**: p 929.

8. Bubser, M. and Schmidt, W. (1990) 6-Hydroxydopamine lesion of the rat prefrontal cortex increases locomotor activity, impairs acquisition of delayed alternation tasks, but does not affect uninterrupted tasks in the radial maze. *Behav. Brain Res.* **37**: p 157.

9. Carter, L.J. (1976) Michigan's PBB incident: chemical mix-up leads to disaster, *Science* **192**: p 240.

10. Chen, Y.-C., Guo, Y.-L., Hsu, C.-C., Rogan, W. (1992) Cognitive development of Yu-Cheng ("oil disease") children prenatally exposed to heat-degraded PCBs. *J. Am. Med. Assoc.* **268**: p 3213.

11. Chen, Y.-J. and Hsu, C.-C. (1994) Effects of prenatal exposure to PCBs on the neurological function of children: a neuropsychological and neurophysiological study. *Dev. Med. Child Neurol.* **36**: p 312.

12. Chen, Y.-C., Yu, M.-L., Rogan, W., Gladen, B., Hsu, C.-C. (1994) A 6-year follow-up of behavior and activity disorders in the Taiwan Yu-Cheng children. *Am. J. Public Health* **84**: p 415.

13. Chou, S.M., Miike, T., Payne, W.M., Davis, G. (1979) Neuropathology of "spinning syndrome" induced by prenatal intoxication with PCB in mice. *Ann. N.Y. Acad. Sci.* **320**: p 373.

14. Davis, J.M. (1990) Risk assessment of the developmental neurotoxicity of lead, *Neurotoxicology* **11**: p 285.

15. Dewailly, E., Ayotte, P., Bruneau, S., Laliberte, C., Muir, D., Norstrom, R. (1993) Inuit exposure to organochlorines through the aquatic food chain in Arctic Quebec. *Environ. Health Perspect.* **101**: p 618.

16. Eriksson, P., Lundkvist, U., Fredriksson, A. (1991) Neonatal exposure to 3,4,3',4'-tetrachlorobiphenyl: changes in spontaneous behaviour and cholinergic muscarinic receptors in the adult mouse, *Toxicology* **69**: p 27.

17. Falandysz, J., Kannan, K., Tanabe, S., Tatsukawa, R. (1994) Organochlorine pesticides and polychlorinated biphenyls in cod- liver oils: North Atlantic, Norwegian Sea, North Sea and Baltic Sea, *Ambio* **23**: p 288.

18. Fein, G., Jacobson, J., Jacobson, S., Schwartz, P., Dowler, J. (1984) Prenatal exposure to polychlorinated biphenyls: effects on birth size and gestational age. *J. Pediatr.* **105**: p 315.

19. Gladen, B. and Rogan, W. (1991) Effects of perinatal polychlorinated biphenyls and dichlorodiphenyl dichloroethane on later development. *J. Pediatr.* **119**: p 58.

20. Gladen, B.C., Rogan, W.J., Hardy, P., Thullen, J., Tingelstad, J., Tully, M. (1988) Development after exposure to polychlorinated biphenyls and dichlorodiphenyl dichloroethene transplacentally and through human milk, *J. Pediatr.* **113**: p 991.

21. Gladen, B., Taylor, J., Wu, Y.-C., Ragan, N., Rogan, W., Hsu, C.-C. (1990) Dermatological findings in children exposed transplacentally to heat-degraded polychlorinated biphenyls in Taiwan. *Br. J. Dermatol.* **122**: p 799.

22. Golub, M.S. and Jacobson, S.W. (1995) Workshop on perinatal effects of dioxin-like compounds IV Neurobehavioral effects. *Environm. Health Perspect.*, **163** (suppl. 2): p 151.

23. Hansen, L. (1987) Food chain modification of the composition and toxicity of polychlorinated biphenyl (PCB) residues. In: *Reviews in Environmental Toxicology 3*. Elsevier Science Publishers BV, Amsterdam, p 149.

24. Hara, I. (1985) Health status and PCBs in blood of workers exposed to PCBs and of their children. *Environ. Health Perspect.* **59**: p 85.

25. Harada, M. (1976) Intrauterine poisoning. *Bull. Inst. Constit. Med.* **25**: p 38 .

26. Hovinga, M., Sowers, M., Humphrey, H. (1992) Historical changes in serum PCB and DDT levels in an environmentally-exposed cohort. *Arch. Environ. Contam. Toxicol.* **22**: p 362.

27. Hsu, S.-T., Ma, C.-I., Hsu, S.-H., Wu, S.-S., Hsu, N.-M., Yeh, C.-C., Wu, S.-B. (1985) Discovery and epidemiology of PCB poisoning in Taiwan: a four-year followup. *Environmental Health Perspect.* **59**: p 5.

28. Humphrey, H. (1983) Population studies of PCBs in Michigan residents, in *PCBs: Human and Environmental Hazards*. Butterworth, Boston, p 299.

29. Humphrey, H. (1988) Chemical contaminants in the Great Lakes: the human health aspect. In: *Toxic Contaminants and Ecosystem Health: A Great Lakes Focus*. John Wiley & Sons, New York, p 153.

30. Jacobson, J., Fein, G., Jacobson, S., Schwartz, P., Dowler, J. (1984) Prenatal exposure to an environmental toxin: a test of the multiple effects model. *Dev. Psychol.* **20**: p 523.

31. Jacobson, S., Fein, G., Jacobson, J., Schwartz, P., Dowler, J. (1985) The effect of intrauterine PCB exposure on visual recognition memory. *Child Dev.* **56**: p 853.

32. Jacobson, J., Humphrey, H., Jacobson, S., Schantz, S., Mullin, M., Welsh, R. (1989) Determinants of polychlorinated biphenyls (PCB's), and dichlorodiphenyl trichloroethane (DDT) levels in the sera of young children, *Am. J. Public Health* **79**: p 1401.

33. Jacobson, J. and Jacobson, S. (1988) New methodologies for assessing the effects of prenatal toxic exposure on cognitive functioning in humans. In: *Toxic Contaminants and Ecosystem Health; A Great Lakes Focus*, John Wiley & Sons, New York, p 373.

34. Jacobson, S., Jacobson, J., Fein, G. (1986) Environmental toxins and infant development. In: *Theory and Research in Behavioral Pediatrics*, Vol. 3. Plenum Press, New York, p. 96.

35. Jacobson, J., Jacobson, S., Humphrey, H. (1990a) Effects of *in utero* exposure to polychlorinated biphenyls and related contaminants on cognitive functioning in young children. *J. Pediatr.* **116**: p 38 .

36. Jacobson, J., Jacobson, S., Humphrey, H. (1990b) Effects of exposure to PCBs and related compounds on growth and activity in children, *Neurotoxicol. Teratol.* **12**: p 319.

37. Jacobson, J., Jacobson, S., Padgett, R., Brumitt, G., Billings, R. (1992) Effects of prenatal PCB exposure on cognitive processing efficiency and sustained attention, *Dev. Psychol.* **28**: p 297.

38. Jacobson, S., Jacobson, J., Schwartz, P., Fein, G. (1983) Intrauterine exposure of human newborns to PCBs: measures of exposure. In: *PCBs: Human and Environmental Hazards*. Butterworth, Boston, p. 311.

39. Jansen, H., Cooke, P., Porcelli, J., Liu, T., Hansen, L. (1993) Estrogenic and antiestrogenic actions of PCBs in the female rat: *In vitro* and *in vivo* studies. *Reprod. Toxicol.* **7**: p 237.

40. Jensen, A. (1987) Polychlorobiphenyls, polychlorodibenzo-*p*-dioxins and polychlorodibenzofurans in human milk, blood and adipose tissue. *Sci. Total Environ.* **64**: p 259.

41. Juarez de Ku, L., Sharma-Stokkermans, M., Meserve, L. (1994) Thyroxine normalizes polychlorinated biphenyl (PCB) dose-related depression of choline acetyltransferase (ChAT) activity in hippocampus and basal forebrain of 15 day old rats. *Toxicology* **94**, p 19.

42. Kodavanti, P., Shin, D., Tilson, H., Harry, J. (1993) Comparative effects of two polychlorinated biphenyl congeners on calcium homeostasis in rat cerebellar granule cells. *Toxicol. Appl. Pharmacol.* **123**: p 97.

43. Koopman-Esseboom, C., Huisman, M., Weisglas-Kuperus, N., Van der Paauw, C., Tuinstra, L., Boersma, E., Sauer, P. (1994) PCB and dioxin levels in plasma and human milk of 418 Dutch women and their infants. Predictive value of PCB congener levels in maternal plasma for fetal and infant's exposure to PCBs and dioxins. *Chemosphere* **28**: p 1721.

44. Kunita, N., Kashimoto, T., Miyata, H., Fukushima, S., Hori, S., Obana, H. (1984) Causal agents of Yusho. *Am. J. Ind. Med.* **5**: p 45.

45. Landrigan, P., Wilcox, K., Silva, J., Humphrey, H., Kauffman, C., Heath, C. (1979) Cohort study of Michigan residents exposed to polybrominated biphenyls: epidemiologic and immunologic findings. *Ann. N.Y. Acad. Sci.* **320**: p 284.

46. Levin, E., Schantz, S., Bowman, R. (1992) The lesion model of neurotoxic effects on cognitive function in monkeys. *Neurotoxicol. Teratol.* **14**: p 131.

47. Lilienthal, H., Neuf, M., Munoz, C., Winneke, G. (1990) Behavior effects of pre- and postnatal exposure to a mixture of low chlorinated PCBs in rats. *Fundam. Appl. Toxicol.* **15**: p 457.

48. Lilienthal, H. and Winneke, G. (1991) Sensitive periods for behavioral toxicity of polychlorinated biphenyls: determination of cross-fostering in rats. *Fundam. Appl. Toxicol.* **17**: p 368.

49. Luthman, J., Fredriksson, A., Lewander, T., Jonsson, G., Archer, T. (1989) Effects of d-amphetamine and methylphenidate on hyperactivity produced by neonatal 6-hydroxydopamine treatment. *Psychopharmacology* **99**: p 550.

50. Maroni, M., Colombi, A., Cantoni, S., Ferioli, E., Foa, V. (1981) Occupational exposure to polychlorinated biphenyls in electrical workers. I. Environmental and blood polychlorinated biphenyls concentrations. *Br. J. Ind. Med.* **38**: p 49.

51. Ness, D., Schantz, S., Moshtaghian, J., Hansen, L. (1993) Effects of perinatal exposure to specific PCB congeners on thyroid hormone concentrations and thyroid histology in the rat. *Toxicol. Lett.* **68**: p 311.

52. Ouw, H., Simpson, G., Siyali, D. (1976) Use and health effects of Aroclor 1242, a polychlorinated biphenyl, in an electrical industry. *Arch. Environ. Health* **31**: p 189.

53. Paneth, N. (1991) Human reproduction after eating PCB-contaminated fish. *Health Environ. Digest* **5**: p 2.

54. Pantaleoni, G., Fanini, D., Sponta, A.M., Palumbo, G., Giorgi, R., Adams, P.M. (1988) Effects of maternal exposure to polychlorobiphenyls (PCBs) on F1 generation behavior in the rat. *Fundam. Appl. Toxicol.* **11**: p 440.

55. Rogan, W.J. (1982) PCBs and cola-colored babies: Japan, 1968, and Taiwan, 1979. *Teratology* **26**: p 259.

56. Rogan, W.J., Gladen, B.C. (1991) PCBs, DDE, and child development at 18 and 24 months. *Ann. Epidemiol.* **1**: p 407.

57. Rogan, W., Gladen, B., Hung, K., Koong, S., Shih, L., Taylor, J., Wu, Y., Yang, D., Ragan, N., Hsu, C. (1988) Congenital poisoning by polychlorinated biphenyls and their contaminants in Taiwan. *Science* **241**: p 334.

58. Rogan, W.J., Gladen, B.C., McKinney, J.D., Carreras, N., Hardy, P., Thullen, J., Tingelstad, J., Tully, M. (1986a) Polychlorinated biphenyls (PCBs) and dichlorodiphenyl dichloroethene (DDE) in human milk: effects of maternal factors and previous lactation. *Am. J. Public Health* **76**: p 172.

59. Rogan, W.J., Gladen, B.C., McKinney, J.D., Carreras, N., Hardy, P., Thullen, J., Tingelstad, J., Tully, M. (1986b) Neonatal effects of transplacental exposure to PCBs and DDE. *J. Pediatr.* **109**: p 335.

60. Safe, S. (1983) Polychlorinated biphenyls (PCBs) and polybrominated biphenyls (PBBs): biochemistry, toxicology and mechanism of action. *CRC Crit. Rev. Toxicol.* **13**: p 319.

61. Safe, S. (1990) Polychlorinated biphenyls (PCBs), dibenzo-*p*-dioxins (PCDDs), dibenzo-furans (PCDFs), and related compounds: Environmental and mechanistic considerations which support the development of toxic equivalency factors (TEFs). *CRC Crit. Rev. Toxicol.* **21**: p 51.

62. Safe, S., Safe, L., Mullin, M. (1985) Polychlorinated biphenyls: congener-specific analysis of a commercial mixture and a human milk extract. *J. Agric. Food Chem.* **33**: p 24.

63. Schantz, S. and Bowman, R. (1983) Persistent locomotor hyperactivity in offspring of rhesus monkeys exposed to polychlorinated or polybrominated biphenyls, *Soc. for Neurosci. Abstr.* **9**: p 423.

64. Schantz, S., Jacobson, J., Humphrey, H., Jacobson, S., Welch, R., Gasior, D. (1994) Determinants of polychlorinated biphenyls (PCBs) in the sera of mothers and children from Michigan farms with PCB-contaminated silos. *Arch. Environ. Health* **49**: p 452.

65. Schantz, S., Jacobson, J., Jacobson, S., Humphrey, H. (1990) Behavioral correlates of polychlorinated biphenyl (PCB) body burden in school-aged children. *Toxicologist* **10**: p 303.

66. Schantz, S., Levin, E., Bowman, R. (1991) Long-term neurobehavioral effects of perinatal PCB exposure in monkeys. *J. Environ. Toxicol. Chem.* **10**: p 747.

67. Schantz, S., Moshtaghian, J., Ness, D.K. (1995) Spatial learning deficits in adult rats exposed to ortho-substituted PCB congeners during gestation and lactation, *Fundam. Appl. Toxicol.*, **26**: p 117.

68. Schwartz, E. and Rae, W. (1983) Effect of polybrominated biphenyls (PBB) on developmental abilities in young children. *Am. J. Public Health* **73**: p 277.

69. Schwartz, P., Jacobson, S., Fein, G., Jacobson, J., Price, H. (1983) Lake Michigan fish consumption as a source of polychlorinated biphenyls in human cord serum, maternal serum, and milk. *Public Health Briefs* **73**: p 293.

70. Seagull, E. (1983) Developmental abilities of children exposed to polybrominated biphenyls (PBB). *Am. J. Public Health* **73**: p 281.

71. Seegal, R. and Schantz, S. (1994) Neurochemical and behavioral sequelae of exposure to dioxins and PCBs. In: Schecter, A. (Ed.) *Dioxins and Health.* Plenum Press, New York, p 409.

72. Seegal, R. and Shain, W. (1992) Neurotoxicity of polychlorinated biphenyls: the role of ortho-substituted congeners in altering neurochemical function. In: *The Vulnerable Brain and Environmental Risks. Vol. 2: Toxins in Food.* Plenum Press, New York, p 169.

73. Shiota, K. (1976) Postnatal behavioral effects of prenatal treatment with PCBs (polychlorinated biphenyls) in rats, *Okajimas Fol. Anat. Jap.* **53**: p 105.

74. Skaare, J., Tuveng, J., Sande, H. (1988) Organochlorine pesticides and polychlorinated biphenyls in maternal adipose tissue, blood, milk, and cord blood from mothers and their infants living in Norway. *Arch. Environ. Contam. Toxicol.* **17**: p 55.

75. Sonzogni, W., Maack, L., Gibson, T., Degenhardt, D., Anderson, H., Fiore, B. (1991) Polychlorinated biphenyl congeners in blood of Wisconsin sport fish consumers. *Arch. Environ. Contam. Toxicol.* **20**: p 56.

76. Storm, J., Hart, J., Smith, R. (1981) Behavior of mice after pre- and postnatal exposure to Aroclor 1254. *Neurobehav. Toxicol. Teratol.* **3**: p 5.

77. Tilson, H., Davis, G., McLachlan, J., Lucier, G. (1979) The effects of polychlorinated biphenyls given prenatally on the neurobehavioral development of mice. *Environ. Res.* **18**: p 466.

78. Urabe, H., Koda, H., Asahi, M. (1979) Present state of Yusho patients. *Ann. N.Y. Acad. Sci.* **320**: p 273.

79. van Haaren, F., van Hest, A., Heinsbroek, R.P.W. (1990) Behavioral differences between male and female rats: effects of gonadal hormones on learning and memory. *Neurosci. Biobehav. Rev.* **14**: p 23.

80. Yamashita, F. and Hayashi, M. (1985) Fetal PCB syndrome: clinical features, intrauterine growth retardation and possible alteration in calcium metabolism, *Environ. Health Perspect.* **59**: p 41.

Contents Chapter 10

Neurotoxic food additives

Chapter 10

Neurotoxic food additives

Paul V. Kaplita

1   Introduction

Food ingredients (including food additives) have the potential to affect brain chemistry and function[68] either by contributing molecules into existing neurochemical pathways, or by altering the levels of hormones that normally regulate brain chemistry. In order for a food constituent to affect the central nervous system it must necessarily be able to be absorbed from the gastrointestinal tract, enter the circulation and then elevate its plasma concentration above a 'normal' range. Then, it must have the ability to either cross the blood-brain barrier (due to its lipophilicity), or to serve as a ligand in one of the various blood-to-brain transport mechanisms.

The acute effects of food ingredients on brain function (e.g., behavior) are usually quite subtle; chronic effects are often more indicative of neurotoxicity[26]. Nevertheless, transient elevations in blood and brain levels of an offending food constituent may be all that is required to damage neurons, particularly developing ones[41]. For example, neurotoxic effects in children may not be apparent at the time of exposure, but may result in subtle disturbances in neuronal function and behavior later in life. Thus, studies concerned with dietary constituents and behavior often pay particular attention to children.

Assessment of the potential neurotoxic effects of food additives is a complex undertaking[68]. A number of factors must be considered before implicating a food additive as neurotoxic:

1. Does dietary consumption of the additive increase blood levels above a normal range (or zero)?
2. Does this elevation in blood levels elicit a corresponding elevation in brain levels?
3. Does dietary consumption of the suspected ingredient alter neurotransmitter synthesis, release, levels, postsynaptic receptor interactions or metabolism/reuptake?
4. Does dietary consumption of the additive elicit autonomic, behavioral or subjective effects?
5. Can these functional effects be correlated with the neurochemical effects?

This chapter will discuss the cases of several food additives, each of which has been purported to be neurotoxic. We will consider the basis of each claim and review the clinical and experimental evidence for and against each additive.

2   Food additives and childhood behavior disorders

2.1   THE FEINGOLD HYPOTHESIS

For over a century, synthetic chemicals have been added to foods solely to enhance their physical appearance. An apparent lack of toxicity from artificial food colors was presumed to be due to a combination of low exposure levels, low gastrointestinal absorption, extensive hepatic degradation, and thorough biliary and renal tubular excretion. Only recently have rational risk-benefit assessments been considered for food colors.

In the mid-1970s, however, the late Dr. Benjamin Feingold, a respected clinical allergist in the San Francisco area of California, contended[14] that almost 50% of children diagnosed with childhood hyperkinesis and learning disorder (minimal brain dysfunction, attention deficit disorder) respond with normal behaviors when placed on a diet that eliminated all artificial colors and flavors, some artificial preservatives, and all foods containing "natural salicylate radicals". He further declared that any infraction in this diet would result in a recurrence of the hyperactive symptoms within hours, often lasting for several days. The diet was so effective, he claimed, that most children on the diet could forego their conventional medications for the disorder.

The original elimination diet was based on clinical case descriptions of adult patients who experienced allergic-like sensitivity symptoms in response to aspirin. When the subjects' conditions failed to improve after eliminating all exposure to aspirin-like compounds, they were placed on a salicylate-free diet that prohibited the ingestion of all salicylates, including those Feingold thought occur naturally in foods. The artificial color Food, Drug and Cosmetic (FD&C) Yellow No. 5 (tartrazine) was also eliminated from the diet because it often produced the same allergic-like symptoms in aspirin-sensitive patients as salicylates. Not all of his patients responded favorably to this diet, so Feingold expanded the diet to exclude all artificial food additives, even if they were chemically unrelated to aspirin. Feingold noted in anecdotal reports that this diet not only alleviated allergic-like sensitivity to aspirin in susceptible patients, but also improved the behavioral symptoms of some psychologically disturbed patients. He then began using the diet to treat children with various behavioral disturbances (including hyperactivity), eventually justifying his diet as a rational therapy by declaring that some causal relationship must exist between the increasing amounts of artificial additives used in foods over the last several decades and the increasing number of world-wide clinical reports of childhood hyperkinesis and learning disorder over the same period of time.

The original hypothesis and diet have undergone subsequent revisions. Most of the vegetables and fruits that Feingold claimed contained "natural salicylates" (e.g., apples, apricots, cherries, grapes and raisins, oranges, peaches, plums, cucumbers and tomatoes) actually have salicylic acid levels that vary between only 0.1 and 0.8 parts per million[11]. Most of the thousands of synthetic chemicals contained in artificial food flavors are, in fact, identical to those that occur naturally in foods. A modified diet subsequently excluded only artificial colors and flavors, although most Feingold diets that have been studied in clinical tests omit the prohibition of artificial flavors, owing to difficulties with disguising the flavors in placebo diets and challenges, isolating the suspected offending compounds from the many thousands of flavor constituents in use, and due to the fact that most "artificial flavors" aren't really artificial. Feingold has since altered his original claim, such that many cases of childhood hyperkinesis and learning disorder can be explained as idiosyncratic, not allergic, reactions to food colors, which manifest themselves as behavioral responses in genetically susceptible children. Later written versions of the

Feingold diet excluded the artificial preservatives butylated hydroxyanisole (BHA) and butylated hydroxytoluene (BHT), despite the fact that these preservatives were not cited in any of the earlier statements or writings of Feingold , nor do they bear any apparent relationship to aspirin sensitivity. Apparently, this restriction was suggested by others. These sorts of ad hoc embellishments are noteworthy if only to emphasize the vagueness of Feingold's claims.

Uncontrolled tests of the Feingold diet have shown it to be an apparently effective therapy. Indeed, 40–60% of children diagnosed as hyperkinetic showed improved behavior when placed on the diet. Feingold Associations, with some 20,000 member families, have appeared across the U.S. Feingold recommended that legislation be adopted that would require a complete ingredient statement on all food labels, and a symbol that would signify that no synthetic colors or flavors are present in a product. He also recommended that federally subsidized school lunch programs exclude additive-containing foods. All these test reports, however, were based on anecdotal observations, with no controls and no quantitative assessment of behaviors. Despite some gaps in logic and scientific rigor, the idea seemed to work and was deemed by the food industry, pediatric psychologists, and toxicologists to be worth testing. Experimental and clinical tests of the validity of the Feingold hypothesis have received numerous reviews[11, 63].

Childhood hyperkinesis, the behavioral response hypothesized by Feingold to be the result of an idiosyncratic hypersensitivity to food additives, has been studied for over 50 years. The condition can be best described as a cluster of symptoms. In the U.S., the syndrome has been reported in 3–10% of the child population, with 5–10 times greater incidence in males than in females. These rates are much higher than in most other countries; this could be a reflection of differences in medical and social perceptions, as well as occurrence. Children diagnosed as hyperkinetic usually show the symptoms listed in Table 10.1.

TABLE 10.1

Clinical signs of childhood hyperkinesis

Clinical signs of childhood hyperkinesis
• elevated levels of motor activity • poor control of impulsive behavior • restlessness • short attention span • low tolerance for frustration • irritability • excitability • aggressiveness

Not every child, however, exhibits each and every symptom. Thus, it is difficult to determine whether childhood hyperkinesis is a discrete illness, or a catch-all syndrome with several biochemical and physiological explanations. The etiology of hyperkinesis may involve genetically determined variations in behavioral development, psychogenic disorders related to childrearing, brain injury due to trauma, disease, allergy, or environmental toxicity. Some have even suggested that childhood hyperkinesis results from ingesting refined sugar, wearing tight underwear, or prolonged exposure to fluorescent lighting.

Treatment of childhood hyperkinesis can involve stimulant medications (e.g., methylphenidate, *d*-amphetamine, etc., which produce a "paradoxical" calming effect), behavioral modification therapy, and counseling. Some 75–80% of children who have been properly diagnosed with the disorder respond favorably to medication. The response rate is even higher if medication is used in combination with behavioral modification. These treatments are not without limitations, however. Stimulant medications are not a cure; the symptoms recur once the drugs wear off. There exists much controversy (some might say misunderstanding) over whether or not stimulant drugs actually improve learning, behavior, and visual-motor coordination in children. Likewise, behavior modification and counseling involve a long-term commitment and are difficult and expensive.

Food allergy and food sensitivity are well-documented to occur in susceptible individuals. Immediate food allergy produces an atopic response (including skin rash, asthma, and rhinitis), which is mediated by an antibody-antigen interaction, and can be detected with a positive skin test. Delayed food allergy results in diarrhea, vomiting, abdominal pains, dermatitis, or wheezing. It is mediated primarily via lymphocytes, and often produces equivocal results in a skin test. Food sensitivity, on the other hand, manifests itself through vague and variable symptoms including aches and pains, fatigue, and behavioral disturbances. These symptoms are thought to be the consequences of atypical or idiosyncratic biochemical reactions with chemical constituents of food consumed by a genetically susceptible population.

Feingold maintained that similar reactions can be produced by ingestion of food additives. Thus, the hypothesis and diet became an attractive option for families of behaviorally disturbed children, since:

1. The diet offered an alternative to drugs
2. The hypothesis relieved parental guilt by transferring the blame from the parents to the processed food industry
3. The hypothesis removed any blame from schools and teachers
4. The hypothesis and diet were consistent with the growing societal awareness regarding "natural" foods, ecology, etc.
5. The hypothesis empowered parents to control the child's diet, enabling the parents to become the principle therapeutic agents. Initial, anecdotal evidence suggested that the diet could be a successful therapeutic alternative to medication. Controlled clinical tests of the diet's efficacy, however, have indicated otherwise.

TABLE 10.2

Guidelines for clinical evaluation of food colors on behavior in children

---

Guidelines for clinical evaluation of food colors on behavior in children

– test population includes both normal and behaviorally disturbed children
– each child and its family should be interviewed by a child psychologist
– each child should be examined by a clinical psychiatrist for:
  • neurological function
  • behavior
  • achievement
  • intelligence
  • learning and development
– behavioral evaluations:
  • global behavioral ratings by parents and teachers
  • specific cognitive and learning tests by trained observers
  • performed on the children:
    for a baseline period before the study
    during the course of the diets
    during the course of the challenges

2.2   CLINICAL STUDIES OF THE FEINGOLD DIET

Two general types of clinical studies have been employed to test the Feingold hypothesis: a food color elimination or dietary intervention design, and a food color challenge or pharmacological design. An ideal, controlled, objective evaluation of the effects of food colors on behavior in children required that a homogeneous sample of children be placed, double-blind (i.e., unknown to both observers and subjects), on either a Feingold diet or a control diet and later be challenged, double-blind, with either food colors or a placebo. Table 10.2 describes critical aspects of the study protocols.

To maximize subject compliance, tasty nutritious diets should be administered either in a hospital inpatient setting or at home, with the entire family participating. The most appropriate food color challenge was thought to be a blend of the nine colors approved for use in the U.S. and Canada, with the proportion of each individual color in the blend based on the amount of each color actually used on a per capita basis in food production (see Table 10.3)[11].

TABLE 10.3
Blend of nine food colors used as a challenge in clinical trials

Color	% of blend
FD&C Blue No. 1	3.12
FD&C Blue No. 2	1.70
FD&C Green No. 3	0.13
FD&C Red No. 3	6.08
FD&C Red No. 4	0.50
FD&C Red No. 2 or No. 40	38.28
FD&C Yellow No. 5	26.91
FD&C Yellow No. 6	22.74
Orange B	0.54

The total amount of FD&C colors in this blend (27.3 mg) represented the amounts of colors consumed daily by the general population (calculated by dividing the total weight of food colors certified annually in the U.S. by the total U.S. population). Children (normal and hyperkinetic), however, typically consume higher levels of food colors than adults, simply due to the preponderance of colors in soft drinks and snack foods. Therefore the daily child consumption estimate was revised upwards, to 36 mg/day.

Several problems with this general study design, which Feingold advocates allege have obscured the results, have been noted. This design was extremely expensive, and many of the clinical studies were supported by the Nutrition Foundation, which is in turn supported by the U.S. food industry. This led to charges of subjective interpretation of the data from those particular studies. Often the validity of particular psychological and behavioral test methods were questioned. There was little attention paid to dose-response relationships in some of the studies. Many of the problems with dosages were due to the fact that it was often difficult for food technologists to incorporate 36 mg food color blend into a cookie or soft drink without revealing the presence of colors, thus negating the disguise of the placebo challenge. Finally, the use of chocolate to mask the presence of food colors in the challenge in many studies was criticized, since chocolate is thought by some to elicit behavioral effects of its own.

The earliest tests of Feingold's hypothesis involved simply placing children diagnosed as hyperkinetic on a Feingold diet and observing any changes in behavior. These trials tended to support Feingold's claim, i.e., typically 40–60% of hyperkinetic children placed on the Feingold diet responded favorably. One such study reported that 10 of 15 hyperkinetic children showed improved behavior while on the diet, according to questionnaires filled out by their parents[12]. Another laboratory noted that parents, teachers, and pediatricians observed improvements in behavior in 11 of 32 hyperkinetic children placed on the diet[4]. Other investigators attempted to identify idiosyncratic responses of 31 hyperkinetic children to four food colors by monitoring systolic blood pressure, pulse, incidence of headache, rash, pallor, perspiration, and activity. Eighteen children reacted to the colors; 15 of these children were placed on a Feingold diet for 4 weeks[54]. According to parental rating scores, the children's behavior improved by 93% while on the diet. Likewise, in smaller studies, clinicians noted that the behavior of two hyperkinetic girls deteriorated significantly following the ingestion of a mix of artificial food colors[52], and that a hyperkinetic boy responded positively to a Feingold diet, as observed by parents and teachers[6]. These clinical results also appeared to corroborate Feingold's anecdotal reports.

Conners[11], however, performed several controlled studies which refute Feingold's hypothesis. Fifteen children evaluated as hyperkinetic were observed by parents and teachers over a 4-week baseline period. The children then received a Feingold diet and a control diet, each for 4 weeks, in a double-blind, random, counterbalanced crossover design, with half the children receiving first one diet and then the other, and the other half receiving the diets in the opposite order. With this protocol design, each child served as its own control, and any placebo effect or order effect produced by the diet due to expectations or suggestibility of the child, parents, or other observers can be detected. Although the control diet contained levels of food colors found in the typical U.S. diet (see above), it was not a true control because it was explained to the subjects as an alternate elimination diet involving other food constituents. Both test diets were nutritionally equivalent to a normal diet, although the Feingold diet contained slightly less vitamin C and carbohydrates than normal. Using behavioral rating scales, teachers detected an improvement in behavior in children on the Feingold diet relative to the baseline and control diet periods. The differences between the effects produced during the Feingold and control diets, however, were observed only when the control diet preceded the Feingold diet. Parents detected improvements in behavior only between the baseline and Feingold diet periods, and again only when the control diet preceded the Feingold diet. These results are typical of an "order effect" produced when the effects of a treatment combine with a factor of suggestibility, which amplifies the effects of treatment when it follows a placebo.

In a second study, 16 children who showed a significant response to the Feingold diet were then observed by parents and teachers over a 1-week baseline and a 3-week diet period[11]. During 8 additional weeks on the Feingold diet, the children received twice daily challenges of food color or placebo, alternating every 2 weeks in a double-blind random, counterbalanced crossover design. The active challenge consisted of chocolate cookies that contained 13 mg (a relatively modest dose; see above) of the food color blend. The placebo cookies contained no color blend. Neither parents nor teachers detected any behavioral effects from the challenges, as measured by the ratings scales. The absence of effect may have been due to the low challenge dose, which was only 1/3 of the estimated daily food color intake in a typical diet.

In a study employing a tracking task that required the child to keep a spot of light on a moving target while having to respond to intermittent flashing lights on either side of the tracking task, however, the children performed less well within the first hour of consuming the active challenge than after placebo or in control test[11]. To determine whether or not the results of the tracking task indeed represented a short-term pharmacological effect of an acute dose of food colors, 13 children who showed a significant response to the Feingold diet were observed by parents and teachers for 3 hours following challenge with either food colors or placebo. After observing the children's behavior for a 2-week baseline period and for three weeks on the diet, the parents evaluated their children's behavior immediately following twice daily challenges with a cookie containing either 13 mg color blend or placebo. After 1 week the challenge was reversed. According to rating scales, four of the 13 children showed a significant worsening of behavior immediately after the color challenge. This study was then repeated, but with 30 children who, after initially responding to the Feingold diet, received either color challenge or placebo twice daily, alternating each week for 4 weeks in a double-blind, random, counterbalanced crossover design. The parental ratings scores revealed no differences in any child's behavior for the 3 hours following either active challenge or placebo. The challenge study was repeated once more, with behavior and restlessness evaluated by trained observers, activity levels measured by a counter, and paired-associated learning tasks performed before and during the 4 hours after either the color challenge or placebo. Nine children who responded favorably to the Feingold diet later received twice daily challenges or placebos in a double-blind, counterbalanced crossover protocol. None of these children showed any differences in any of the behavioral evaluations before or after the color challenge or placebo. The negative results of this test, with its higher challenge dosage and rigorous behavioral monitoring, are much more compelling.

In similar studies, another laboratory also reported subtle, inconsistent results with a Feingold diet. After being observed for a 2-week baseline period, 46 hyperkinetic children and their families received 4 weeks of the Feingold diet and 4 weeks of a control diet in a double-blind crossover protocol[22]. This particular control diet was designed in such a way as to make it indistinguishable from, yet nutritionally equivalent to, the Feingold diet. Although teachers and trained observers detected no differences in children's behavior during either the diet or baseline periods, the parents did notice an improvement in behavior during the Feingold diet, but again only when the control diet preceded the Feingold diet. Nine children who showed the most significant changes while on the Feingold diet were observed for an additional two weeks of baseline and 2 weeks of diet[23]. Then, over a 9-week period they received twice daily challenges of food colors or placebo, alternating every 2 weeks in a double-blind crossover design. The active challenge again consisted of cookies that contained 13 mg of the color blend. No significant differences were seen in the behavioral ratings of parents, teachers or observers between periods of active challenge or placebo, although one child did show some slight response to the colors.

Clinicians also examined the effects of larger (and perhaps more relevant) doses of food colors on children's behavior[38]. A hyperkinetic boy was placed on the Feingold diet and subsequently challenged twice weekly over a 10-week period with either 39 mg of the color blend or a placebo in a double-blind design. Although teacher and child could detect no behavioral changes following challenge or placebo, the child's mother often noted a deterioration of behavior following the color challenge. This experiment was extended to examine the effects of thrice daily challenges with either 78 mg of the color blend or placebo, administered to 13 hyperkinetic children in a double-blind crossover design. No differences were observed between reactions following color challenge or placebo in parental or teacher behavioral evaluations, or psychiatric tests. In another study 22 children, placed on a

Feingold diet for at least 3 months, received random, double-blind challenges over a 7 week period in the form of soft drinks containing either 35.6 mg of a blend of food colors or placebo. Twenty of the children showed no sensitivity to the challenge, but two (one of which was the smallest of the group) reacted after each color challenge with disturbed behavior, as measured by parents and teachers using a ratings scale.

A select sample of 40 children with behavioral symptoms suggesting childhood hyperkinesis were also examined[64]. Twenty of the children had previously responded favorably to stimulant medication and scored poorly on behavioral ratings scales, thus were considered "confirmed" hyperactives. The remaining 20 showed adverse reactions to stimulant therapy and scored high on behavioral ratings scales; their diagnosis was "unconfirmed". The basis of these diagnoses, however, has been questioned. After 5 days on the Feingold diet as hospital inpatients, 20 of the children (ten from each diagnosis group) received a capsule containing either 100 mg of the food color blend or placebo. The other 20 children received a capsule containing either 150 mg color blend or placebo. This bolus dose of the color blend falls within the 90th percentile for daily consumption of artificial food colors by 5–12 year-old-children. This color blend reflects estimated relative yearly consumption figures for the U.S., as used in the studies cited above. After the administration of the color blend, those children with "confirmed" hyperkinesis showed a significant impairment in paired-associated learning performance relative to their performance before dosing. Those children with an "unconfirmed" diagnosis showed no significant differences in learning performance before or after dosing. The reaction took over 30 min to become apparent and reached a maximum intensity after an hour and a half. The interpretation that these results demonstrate a short-term pharmacological effect on cognitive function has been questioned, however, since the learning responses of the "confirmed" group after placebo differed from the learning responses of the "unconfirmed" group after placebo. A placebo effect, after all, would not be expected to vary between groups. Interestingly, there was no difference between the behavioral ratings of either group after color challenge or placebo.

One laboratory studied the relative effects of a Feingold diet versus a control diet that contained food additives on 39 children diagnosed with moderate-to-severe learning disabilities[21]. Eighteen of these children were regarded as hyperkinetic, and 17 were being treated with various medications during the course of the study. In a double-blind design, the children were observed by teachers during one week on a Feingold diet, followed by one week on the control diet. No significant behavioral differences were detected by the teachers.

The specific effect of FD&C Yellow No. 5 (tartrazine) on children with hyperkinetic behavioral problems was also examined[31, 32]. After a period of baseline observations, thirty children were placed on a Feingold diet for 4 weeks. The children were then challenged daily over the next 2 weeks with either 4-5 mg tartrazine or placebo in a double-blind cross-over design. The children then continued on the Feingold diet for another 4 weeks. Teacher evaluations and cognitive and learning tests detected no effect on the children's behavior by the diet, the color challenge, or the placebo. Parents, however, noticed a significant improvement in behavior, but only during the first 4 weeks on the Feingold diet. Thirteen children who showed the most favorable responses to the diet underwent a second challenge period. Behavioral reactions to the tartrazine challenge were detected by parents, but not by the teachers nor by the cognitive and learning tests. The eight children that did react to tartrazine received a third challenge trial. This time, the parental evaluations detected no differences in behavior after either tartrazine or placebo.

Another group examined the effects of a Feingold diet, and subsequent challenge with tartrazine and the red food color, carmoisine, on children referred for suspected hyperactivity[53]. Of 220 children originally referred, 55 were enrolled in a 6-week open trial of the Feingold diet. Eight of these children responded favorably to the diet and were then enrolled in a double-blind crossover protocol, employing a single-subject repeated measures design. The children were challenged daily with either 50 mg tartrazine or carmoisine, or placebo for 18 weeks. Two children reacted to the color challenge with increased irritability, and restlessness, and disturbed sleep patterns.

The relative therapeutic effects of stimulant medication (methylphenidate) were compared with a Feingold diet on 26 hyperkinetic children[67]. After one week of baseline observations and five weeks on the Feingold diet, the children received either stimulant medication or placebo, and four challenges per week in the form of cookies containing 26 mg color blend or placebo, in a double-blind, random, counterbalanced crossover design. Behavioral responses were measured using parental and teacher evaluations. Stimulant medication produced significant, consistent improvement in the children's behavior. In this case, stimulant medication served as a standard of efficacy. Neither the active challenge nor the placebo had any effect while the children were on the stimulant medication. The effects of the diet were inconsistent, however, with parents noting no changes in behavior during the time the children were on the diet or undergoing challenges, whereas teachers observed that the children

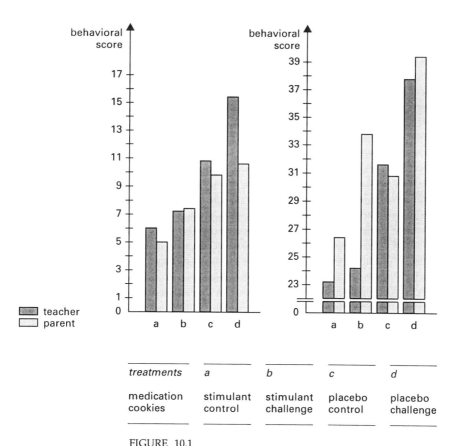

treatments	a	b	c	d
medication cookies	stimulant control	stimulant challenge	placebo control	placebo challenge

FIGURE 10.1

Effects of combining drug treatment with food-additives challenge

Left panel 17-item checklist ratings, right panel 39-item checklist ratings. (Adapted from Williams, J.I. et al. (1978) *Pediatrics* **61**: pp 811–817.)

TABLE 10.4
Summary of the controlled clinical trials on the Feingold hypothesis

Description	Subjects	Results	Reference
– double-blind, challenge-placebo or placebo-challenge cross-over design with 4-week periods; teacher and parent rating scales	15 hyperkinetic children	significant 'order' effects found by both parents and teachers in behavior	(11)
– 1-week baseline, 3-week diet period, 8-week diet period with challenges of color blend or placebo alternating over 2-weeks periods; tracking tasks, tracking task repeated	16 children with positive prior response on Feingold diet	no effects on any measures	(11)
– 2-week baseline, 3-week diet period with challenges of color blend or placebo over alternating 2-week periods; tracking task	13 children with positive prior response on Feingold diet	4 of 13 children exhibited worsening of behavior after challenge	(11)
– double-blind diet cross-over design, 4 weeks on diet, 4 weeks control; teacher, parent and clinician rating scales	46 hyperkinetic children	no effects observed by teachers or clinicians; 'order' effect observed by parents	(22)
– 2-week baseline, 2-week diet period with challenges of color blend or placebo alternating over 2-week periods	9 children with positive prior response on Feingold diet	no effects observed	(23)
– 1 week on diet, then challenge with a high-dose color blend; learning behavior	40 children with behavioral disorders	impairment in learning after challenge	(64)
– double-blind diet cross-over design, 1-week diet period, 1-week control	39 children with learning disabilities	no behavioral differences	(21)
– double-blind, challenge-placebo or placebo-challenge cross-over design, 1-week baseline, then 5-week diet period, challenge with color blend or placebo alternating over 2 weeks, then diet continues for 4 weeks; teacher and parental evaluations, cognitive and learning tests	30 hyperkinetic children	teachers observed no effects, parents noted behavioral deterioration after challenge	(31, 32)
– double-blind, challenge-placebo or placebo-challenge cross-over design, 1-week baseline, then 5-week diet or medication period, challenge with color blend or placebo alternating over 2 weeks, then diet continues for 4 weeks; teacher and parental evaluations	26 hyperkinetic children	stimulant medication produced more consistent behavioral improvements than diet	(67)

showed the worst behavior following an active challenge while on placebo medication. Since all challenges were administered during school hours, it may be possible that the behavioral deterioration observed by the teachers could have been a short-term reaction that disappeared by the time the children left school at the end of the day. Regardless, behavioral improvements following stimulant medication were much more consistently observed by parents and teachers, emphasizing yet again the tenuousness of Feingold's claim.

Thus, Feingold diets generally do not appear to work in controlled, blind studies. At best, less than 10% of hyperkinetic children who were enrolled in the various controlled trials responded to either the diet or the color challenges. Indeed, a U.S. National Institutes of Health Consensus Development Conference concluded that "the Feingold diet may be helpful for a small number of children with hyperkinesis" (25–250 per 100,000) " ... however, these decreases in hyperactivity were not observed consistently". The isolated favorable responses that have been reported suggest that clinical studies ought to be extended. The best responders to the diet and challenges were often the smallest, youngest children. Although this observation may reflect a developmental influence, the magnitude of responses to a consistent amount of color challenge in different children could be a function of body weight, with smaller children actually receiving a larger mg/kg dose than older, larger children. Dose-response relationships should be examined for color blends and individual FD&C dyes, perhaps using capsules instead of cookies as the challenge vehicle to minimize jeopardizing the placebo disguise at higher doses. The pharmacokinetics of the dyes, including whether or not they cross the blood-brain barrier, should also be studied. It might be of interest to alter the frequency of dosing with the challenges, in order to more accurately represent real-life daily consumption of food colors.

2.3   ANIMAL EXPERIMENTS WITH FOOD COLORS

Initial studies with rodents suggested that erythrosine (Erythrosin B, FD&C Red No. 3, E127, Figure 10.2) may be the pharmacologically active agent in food color blends. Rat pups (5 days old) treated with 6-hydroxydopamine (6-OHDA) typically exhibit a childhood hyperkinesis-like syndrome, including persistent learning deficits and an adolescent hyperactivity that abates with maturity[58]. Rat pups pretreated with 6-OHDA, then orally dosed daily for 25 days with 0.5-2.0 mg/kg color blend showed increases in motor activity relative to controls that received 6-OHDA alone, although no consistent dose-response relationship was observed[59, 60]. Moreover, the colors had no effect on the pups' learning performance in the T-maze and shuttle-box. But pups pretreated with water (as a control for 6-OHDA) and then administered 2 mg/kg color blend also showed increases in activity, and exhibited impaired learning performance when compared with controls. Unfortunately, the interaction between 6-OHDA and erythrosin B treatments was unclear or perhaps obscured by a "ceiling effect" present in the experimental design. Similar effects on activity and learning performance were seen in naive and 6-OHDA-pretreated rat pups that received continuous gastric infusion of 1 mg/kg/day color blend[18]. Daily oral administration of 1 mg/kg erythrosin B, however, inexplicably failed to elicit any significant effect on the motor activity or cognitive performance of developing rat pups[19]. Likewise, erythrosin B fed at up to 1% of the diet (which resulted in dosages substantially higher than those cited above) produced no consistent, dose-dependent behavioral effects or developmental toxicity in pregnant rats, rat dams, and rat pups[65].

In another study, intraperitoneal injection of erythrosin B at 50 mg/kg had no effect on locomotor activity in rat pups pretreated with 6-OHDA or a vehicle[36]. Rat pups treated with 50-300 mg/kg erythrosin B, however, showed significantly increased punished-response behavior relative to controls. This effect was dose-related and was similar to the change in punished-response behavior seen following barbiturate or benzodiazepine administration. Interestingly, barbiturates and benzodiazepines also exacerbate the symptoms of childhood hyperkinesis in humans, and this effect can be blocked by *d*-amphetamine.

While most studies have concentrated on the effects of erythrosin B, one laboratory examined the behavioral toxicity of FD&C Red No. 40 (Figure 10.2) in rats[66]. The animals were exposed to the dye at up to 10% of their rat chow diet. Red No. 40 decreased the onset of various developmental landmarks, decreased motor activity, and increased open-field rearing behaviors. Another laboratory[20] examined the behavioral toxicity of sulfanilic acid, a metabolite of the azo dyes (Figure 10.2) FD&C Yellow No. 5 (tartrazine) and FD&C Yellow No. 6 (sunset yellow) in rat pups. As with animals treated with the food color blend, rat pups injected intraperitoneally with 1 mg/kg sulfanilic acid exhibited elevated levels of motor activity. Sulfanilic acid also decreased T-maze performance, but had no effect on shuttle-box performance. The observed behavioral effects, while significant, are not necessarily predictive for attention deficit disorder in children. Although idiosyncratic responses to tartrazine have been shown in clinical studies to correlate with sensitivity to aspirin (which is what originally inspired Feingold's work; see above), and preliminary data suggest that sulfanilic acid can enter the brain, it is by no means clear that these azo food dyes (or their metabolites) or aspirin are behavioral or developmental toxins.

An added complication in any interpretation of these animal data would be the questionable equivalence between rodents and children. The obvious differences in species, as well as in dosages, rates of consumption and absorption, and in behavioral tests needs to be considered. Nevertheless, as with the clinical studies, these results were at best inconsistent.

Erythrosin B is a commonly used synthetic dye approved by the U.S. Food and Drug Administration for use in foods, drugs, and cosmetics. General toxicology studies indicated that erythrosin B is relatively nontoxic. Erythrosin B administered orally to Beagles at 0.5–5.0% of dietary levels for 2 years produced no apparent toxic effects, but in rats caused caecal distension, reduced spleen weight, and depressed growth at the highest dose that was administered. An acute LD_{50} for oral administration in rats was calculated between 2–7 g/kg. At toxic doses, the animals' internal organs, muscle, fat, ears,

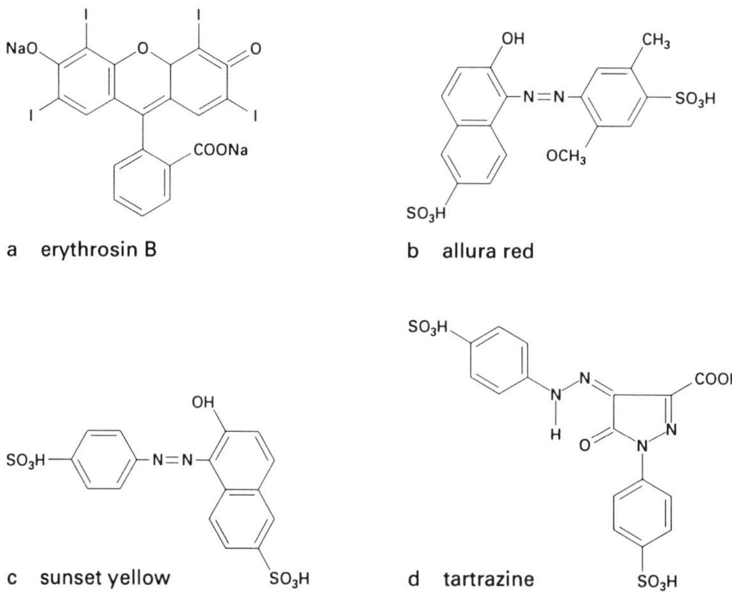

FIGURE 10.2

Chemical structures of (a) FD&C Red No. 3 (erythrosin B); (b) FD&C Red No. 40 (Allura Red); (c) FD&C Yellow No. 6 (Sunset Yellow); (d) FD&C Yellow No. 5 (tartrazine)

tails, paws, and noses took on a pink coloring. Pigs treated for 90 days with erythrosin B, but not rats treated for 85 weeks, showed alterations in thyroid weight and serum thyroxine levels. High doses of erythrosin B showed very slight but statistically significant mutagenic effects. The acceptable daily intake of erythrosin B was established as 2.5 mg/kg. An average daily dietary intake of erythrosin B has been estimated at 34.1 mg, which for a 70 kg human is equivalent to a daily exposure of 0.5 mg/kg. In rats, oral doses of erythrosin B are largely excreted in the feces, although a small amount is also secreted in bile. The distribution and metabolism of oral doses of erythrosin B in humans are not known, although it is thought that humans also excrete the dye largely in the feces. Any interpolation between behavioral studies and biochemical and physiological experiments must be made cautiously, because in rats erythrosin B does not penetrate the blood-brain barrier and enter the central nervous system following peripheral administration (except perhaps at heroic doses; see above). Likewise, there are no neuropathological data available for erythrosin B.

Erythrosin B and analogs exerted toxic effects on several species of insects, including house flies, boll weevils, and mosquitoes. Of these chemically-related dyes, only erythrosin B is approved for use in the U.S. as a food additive. Dietary levels of 0.5–5.0 mM erythrosin B produced nearly 100% lethality in sample insect populations. This corresponded to whole body concentrations of 100–200 nM. The relative order of insecticidal efficacy of the dyes was rose bengal $\geq$ erythrosin B > rhodamine B > phloxine B > eosin Y > fluorescein. The insecticidal effect was dependent on both dose and duration of exposure[15]. At least two mechanisms of action seem to be involved: a rapid, dye-sensitized photo-oxidation that uses energy from trapped photons of light to transform molecular oxygen into a toxic singlet oxygen state, and a slower, light-independent reaction. The dark reaction occurred at higher doses of the dye than those producing the light-dependent toxicity. The relative potency of the analogs to produce light-independent toxicity appeared to be related to the number of halogen atoms in each molecule. Thus, rose bengal (LC_{50} = 1.8 mM) with four iodine and four chlorine atoms was more toxic than erythrosin B (LC_{50} = 3.0 mM) with only four iodine atoms. The biological mechanism of toxicity is not known; there is no data available on the neurotoxicity or neuropathology of these dyes in insects.

In vitro, erythrosine has a variety of actions due in large measure to its ability to affect biological membranes and lipohilic sites of enzymes. Table 10.5 summarizes the main biochemical and pharmacological *in vitro* effects of erythrosine.

From this table it is quite clear that erythrosin B and related dyes, alone among the food colors, can alter a variety of membrane-mediated events, including several involved in neurotransmission. Whether or not any of these *in vitro* events can be directly linked to any behavioral syndromes in humans is, however, not clear. Indeed, Mailman and Lewis[37] suggest that the ability of erythrosin B to affect so many seemingly unrelated biochemical processes would argue against its neurotoxicity, since potent neurotoxins typically have very specific mechanisms of action.

2.4   CONCLUSION

When administered under controlled conditions, Feingold elimination diets generally produced subjective and equivocal changes in the behavior of disturbed children. Few instances of consistent, dramatic deterioration in behavior were seen after double-blind challenge with artificial colors during the duration of diet enrollments. The major remedial effect of the diet appears to be due to a placebo response arising from changes in the attitudes of parents and

TABLE 10.5

In vitro effects of erythrosin B (FD C3; E127) interfering with biological membranes and lipophilic sites of enzymes

Description of effect	Animal species/organ system	Supposed mechanism	Reference
– inhibition of egg fertilization	sea urchin (*Strongylocentrototus pur puratus*) eggs	↑ K+ membrane permeability	(8)
– hyperpolarization of egg plasma membrane and increase of plasma membrane conductance	sea urchin eggs	↑ K+ membrane permeability	(30)
– eliciting reversible increases in neuronal membrane potential and conductance in isolated ganglia	mollusc (*Navanax inermis*) buccal ganglia	↑ K+ membrane permeability	(29)
– producing an irreversible, concentration-dependent increase in the frequency of miniature end-plate potentials (mepps)	frog neuromuscular synapse	depletion of neurotransmitter stores and ↑ membrane permeability to ions	(1)
– inhibiting high-affinity [3H]ouabain binding to Na+/K+ ATPase without influence on cardiac glycoside-induced contractility	guinea-pig heart	interaction with ouabain binding site on ATPase	(16)
– inhibiting [3H]ouabain binding to 'digitalis receptor'	rat brain and guinea-pig heart	interaction with ouabain binding site on ATPase	(24)
– increasing membrane permeability to Ca2+ and other ions	planar phospholipid bilayers	formation of transmembrane pathways highly permeable to ions	(10)
– inhibiting accumulation of [3H] dopamine, choline, gamma-aminobutyric acid, glutamic acid, glycine, norepinephrine, serotonin and taurine	rat brain synaptosomes	uncompetitive inhibition of membrane transport processes	(33)
– inhibiting dopamine transport	rat caudate nucleus synaptosomes	nonspecific disruption of membrane physiology due to intercalation into lipid bilayer	(28)
– affecting neurotransmitter release	frog neuromuscular synapse	depletion of neurotransmitter stores and ↑ membrane permeability to ions	(1)
– inhibiting [86Rb] uptake and increasing contractile force	isolated guinea-pig heart	interaction with ouabain binding site on ATPase	(2)
– affecting electrical and mechanical activity	guinea-pig taenia coli	↑ membrane permeability to ions	(3)
– inhibiting Ca2+ transport and ATPase activity	muscle sarcoplasmic reticulum	interaction with membrane ATPase	(39)
– altering spontaneous release of neurotransmitter	frog neuromuscular junction	↑ membrane permeability to ions	(1)

children about the condition, positive expectations regarding the diet, and an underlying belief that certain components of processed foods are toxic. Over the course of the diet period, the child receives much more attention from family members, and any blame or guilt concerning the child's condition is deflected to the family's diet. Thus the diet acts almost as a form of behavior modification.

The biochemical studies, however, offer evidence that FD&C Red No. 3 may be neurotoxic. The behavioral effects observed in the animal studies and the limited number of responders in the clinical studies suggest that a very small percentage of hyperkinetic and learning disabled children may be adversely affected by one or more food colors. Feingold diets are safe as well as nutritionally adequate, so the U.S. National Institutes of Health (NIH) Consensus Committee has advised that affected children may try the diet as long as conventional therapeutic options, like medication, are not discarded. Although Feingold's earlier legislative recommendations have not been widely implemented, he served to focus public attention on food additives and food package labeling.

3   Aspartame

The artificial sweetener aspartame (L-aspartyl-L-phenylalanine methyl ester, Figure 10.3) has been an increasingly popular food additive in the United States since its introduction in the early 1980s[34]. Shortly after its approval for use in foods by the U.S. Food and Drug Administration (FDA), however, anecdotal reports of a variety of adverse effects began to surface. These included such central nervous system-related complaints as headache, behavioral disturbances, and enhanced seizure susceptibility.

At first glance, these reports seemed to corroborate the warning of Maher and Wurtman[35] that ingestion of aspartame, especially as part of a high carbohydrate diet, could elevate plasma and brain levels of phenylalanine above normal levels, and result in subtle neurotoxicity, or even a phenylketonuria-like syndrome.

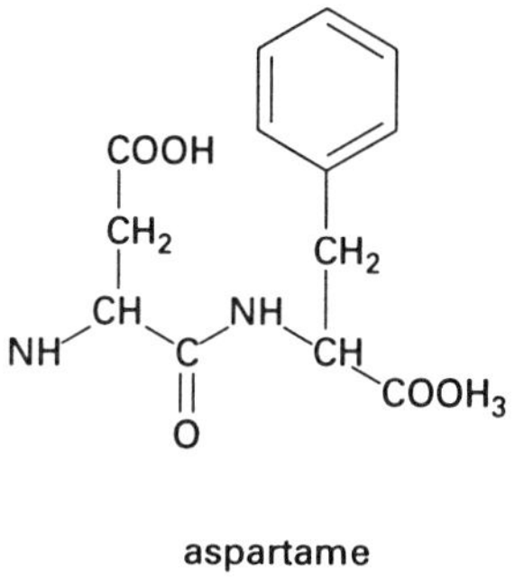

aspartame

FIGURE 10.3
Chemical structure of aspartame

Aspartame acts on the tongue to stimulate sensory taste receptors to elicit the perception of sweetness[41]. The dipeptide is subsequently metabolized in the gastrointestinal tract to aspartate, phenylalanine, and methanol. At U.S. FDA acceptable daily intake levels of 40–50 mg/kg, the amounts of methanol produced for ingestion of aspartame are sub-toxic. Likewise aspartate, although purported to be an excitotoxin (see below), does not attain significantly elevated plasma levels following ingestion of aspartame to be of concern, and is quickly metabolized to the nontoxic amino acid alanine. In contrast, there are claims that consumption of aspartame can result in elevated blood and brain levels of phenylalanine[34, 35]. At worst, the result could be a toxic syndrome similar to that seen with the autosomal recessive inherited disease phenylketonuria. Individuals who suffer from this condition have elevated levels of phenylalanine due to a decrease in the activity of the enzyme phenylalanine hydroxylase, which normally converts phenylalanine to tyrosine. The resulting adverse effects are abnormal electroencephalograph (EEG) activity, mental retardation, insomnia, seizures, behavioral, and neuroendocrine changes. Alternatively, elevated brain levels of phenylalanine can inhibit the enzyme tyrosine hydroxylase, thus depress the synthesis of the monoamine neurotransmitters norepinephrine, dopamine and serotonin, possibly resulting in an enhanced susceptibility to seizures.

Most clinicians regard aspartame as unlikely to be a health hazard. The reported adverse reactions are mild, common, and too varied to be particularly meaningful. In addition, these reactions are usually not temporally related to aspartame ingestion, nor can the reactions always be replicated by cessation of exposure to aspartame, followed by rechallenge. Most clinical studies report nonsignificant effects on behavior or cognition (e.g., motor activity or attention span) following aspartame ingestion by children or adults[27], despite earlier, anecdotal reports suggesting that aspartame evokes increases in activity and aggression. Similarly, double-blind challenges with 40–50 mg/kg aspartame demonstrated little or no effects on seizure threshold or EEG spike-wave discharge rates in children with histories of seizures[7,61,62].

In animal studies, large quantities of aspartame were required to be consumed before neurotoxic effects were noted, and at best the results were inconclusive. Olney[41] reported that oral administration of 3 g/kg aspartame induces pathologic lesions in rodent hypothalamus. Others, however, were unable to observe hypothalamic lesions in infant macaques that were dosed orally with 2 g/kg daily[50]. Likewise, oral aspartame at 1–2 g/kg elicited no behavioral effects in rodents or primates. Aspartame at 500 mg/kg did not promote pentylenetetrazol (PTZ)-induced seizures in mice[40]. At 1–2 g/kg, however, oral administration of aspartame increased the frequency of seizures in mice treated with PTZ, fluorothyl and electroshock[46]. This response could be mimicked by administration of equimolar doses of phenylalanine, and blocked by co-administration with aspartame of the neutral amino acid valine. At 1 g/kg aspartame slightly elevated blood and brain levels of phenylalanine and tyrosine, yet had no effect on brain monoamine levels in rats[17]. In contrast, 1–2 g/kg aspartame elicited no effect on the EEG recordings from primates[51]. Large doses of aspartame elicited dose-dependent effects on the development of rat pups. Female rats (and ultimately, their pups) consumed aspartame in their diet or drinking water prior to conception, and then for 40–90 days post partum[5,25]. Only at daily doses exceeding 5 g/kg were signs of developmental toxicity (e.g., increased infant mortality, reduced birth weights, delayed development of reflexes, hypoactivity) evident. These effects were also observed following equivalent doses of phenylalanine.

Thus, the neurotoxic potential for aspartame in the general population appears to be overstated. Any experimentally observed signs of toxicity appear only after doses well above typical consumption levels. The data suggest, though, that aspartame's toxicity may be due to its phenylalanine component. Therefore it does seem prudent for individuals with phenylketonuria to strictly limit their ingestion of this particular food additive.

4   Monosodium glutamate: excitatory neurotoxin

Monosodium glutamate (MSG) is perhaps the most commonly used food additive in the world, both in commercially processed foods, and as a condiment for use at home. Glutamate, a naturally occurring amino acid, is not only found in many foods, but is intentionally added as a flavor enhancer. Usually used as the monosodium salt, glutamate is thought to act by stimulating sensory taste receptors on the tongue.

Glutamate has long been on the GRAS (generally regarded as safe) listing of the U.S. FDA. For many years glutamate was included in processed baby foods, primarily to stimulate parental palates. In 1969, however, manufacturers began to voluntarily eliminate MSG from baby food formulas when the potential for glutamate to induce brain damage in rat pups was first demonstrated. At that point, MSG was replaced by hydrolyzed vegetable proteins, which are rich in glutamate and a related acidic amino acid, aspartate. Eventually hydrolyzed vegetable proteins were removed from baby foods as well. The U.S. FDA subsequently adopted the position that it is unsafe to add glutamate to foods that are intended for consumption by infants and children[41].

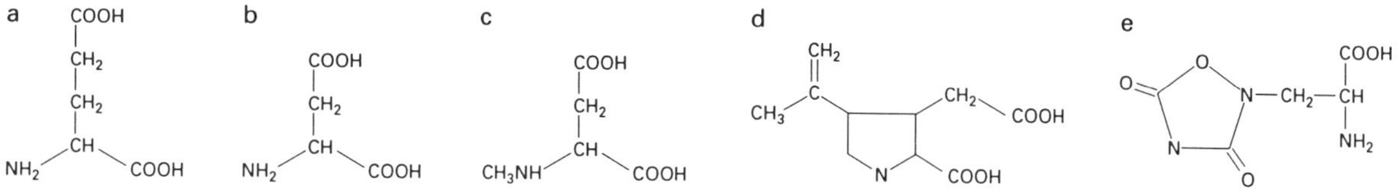

FIGURE 10.4
Chemical structures of excitatory amino acids (a) glutamate; (b) aspartate; (c) N-methyl-D-aspartate; (d) kainate; (e) quisqualate

Glutamate and aspartate occur naturally and abundantly in central nervous system tissue. Indeed, these amino acids are the chemical neurotransmitters for most excitatory synapses in the mammalian brain[9]. Glutamate and aspartate normally act to relay electrical signals through neuronal pathways by stimulating, or exciting, postsynaptic neurons. Glutamate and aspartate interact with 3 classes of postsynaptic dendrosomal membrane receptors, each displaying a preference for a particular pharmacological agonist: N-methyl-D-aspartate, quisqualate, or kainate (Figure 10.4; see also Chapter 4.). These amino acids, as well as various analogs, can also be neurotoxic when they excessively stimulate the same excitatory receptors, a phenomenon known as excitotoxicity. Receptor antagonists have anti-toxic activities that correlate in potency and specificity with their anti-excitation activities, confirming the causal relationship between excitotoxicity and excessive receptor stimulation. Although its precise molecular basis is not known, excitotoxicity is thought to occur through two processes:

1. A fast, extracellular Na^+ and Cl^--dependent neuronal swelling
2. A slow, extracellular Ca^{2+}-dependent neuronal degeneration

The neurotoxicity of endogenous glutamate and aspartate may well be a pathological cause (or contributor) in acute injury to the nervous system such as prolonged seizures, hypoxia-ischemia, glucose deprivation, and direct mechanical trauma, and in chronic neurodegenerative diseases such as Huntington's disease and Alzheimer's dementia[9].

The neurodegenerative syndrome lathyrism has been causally linked to the dietary consumption of a plant-derived excitatory amino acid analog of glutamate (see also Chapter 8)[44, 55]. These findings have been corroborated by animal feeding studies. Rodents and primates administered glutamate or aspartate orally (by feeding tube) or subcutaneously, starting at 500 mg/kg, exhibited neuronal destruction. Amino acid analogs of glutamate also induced neuronal necrosis at dose levels that correlated well with agonist (i.e., excitatory) potencies[43]. The damaged neurons were located in or near the circumventricular organs, brain regions that lack blood-brain barriers. The animals, all treated in infancy, exhibited a syndrome that included obesity, skeletal stunting, reproductive failure, hypoplasia of the adenohypophysis and gonads, low levels of luteinizing hormone, growth hormone, and prolactin.

These findings have been challenged. Many laboratories, while conceding the excitotoxicity of endogenous glutamate, claim that orally administered glutamate is nontoxic, since less glutamate is absorbed into the circulation during its consumption along with other foods (especially sugars and carbohydrates that yields glucose or pyruvate in the gastrointestinal tract) than if glutamate were consumed "neat". Attempts to rebut this argument include demonstrating that newly weaned mouse pups, which were withheld water overnight (but otherwise had free access to chow) voluntarily drank solutions of 10% glutamate or 5% glutamate/5% aspartate in water the following morning. These levels of consumption, equivalent to 1–3.5 g/kg, were sufficient to induce damage to neurons in the circumventricular organs[45]. Olney[41, 42] raises the possibility that infant animals (and humans?) may be more susceptible than adults, perhaps due to age-related differences in gut absorption of amino acids, in metabolic clearance of amino acids from the circulation, or in neuronal protective mechanisms (e.g., glutamate reuptake).

The link between consumption of MSG and neurotoxicity is not quite as apparent. Typical dietary consumption of MSG varies. For example, various commercial soup preparations contain around 1 gram MSG per 6 oz serving. Similarly, one 4 and 1/2 oz jar of baby food at one time contained about 0.8 g of MSG[41]. Several cases of children with shudders, which were apparently evoked by consumption of MSG, have been noted[49]. The symptoms did not respond to treatment with the anti-epileptic drug, diphenylhydantoin, but could be completely reversed by placing the children on a diet that excluded MSG. A curiously different effect was noted in a small clinical study, which demonstrated that 300 mg MSG administered to learning-disabled adults improved oral reading speed relative to control adults.

Behavioral effects of MSG in rats have been observed, albeit at very high dosages[13,47]. At 5 g/kg, MSG orally administered daily to neonatal rats resulted in suppression of growth, reduced levels of spontaneous motor activity and impaired discrimination learning in a maze. Likewise, subcutaneous injection of 4 g/kg MSG in female rat pups evoked dose- and age-related alterations in the monoamine levels of regions of the hypothalamus.

There are few other clinical reports of behavioral or neurological disturbances following ingestion of MSG, with the possible exception of the so-called "Chinese restaurant syndrome". This is a discrete set of symptoms, primarily pain and pressure sensations about the face, neck and torso, which appear seconds to minutes following the consumption of a Chinese restaurant meal[48]. The symptoms have also been reported following meals in Japanese, Italian, and French restaurants. The symptoms can be mimicked by oral consumption of MSG, and typically appear at a threshold of approximately 50 mg/kg. The syndrome is thought to be a benign interaction in sensitive individuals between MSG and peripheral sensory nerve endings[56,57].

The experimental evidence for the excitotoxicity of glutamate is now widely accepted. Hypotheses for the pathogenesis of several acute and chronic brain disorders, based on glutamate neurotoxicity, are being actively pursued. Clinical signs of acute or chronic toxicity of MSG in humans, however, are much less apparent. Nevertheless, the potential for subtle brain damage from consumption of MSG, particularly by children, cannot be ruled out. Clearly, more extensive data regarding the neuropathology of MSG in children is needed. Considering the risks involved, the benefits of adding large amounts of MSG to foods solely to enhance flavor are becoming less and less compelling.

5 Summary

Food additives carry with them the liability of adversely affecting brain chemistry and function.

Consumption of food colors has been linked to childhood hyperactivity, yet controlled trials of elimination diets typically resulted in equivocal changes in behavior. Biochemical studies indicated that certain food colors can alter a wide variety of membrane-mediated events, possibly due to the lipophilic nature of their molecular structures. The equally equivocal behavioral effects observed in numerous animal studies, taken with the limited number of responders in the clinical studies, suggest that a very small portion of the general population may be sensitive to one or more food colors.

The artificial sweetener, aspartame, has been reported to evoke subtle symptoms of neurotoxicity. Typical reactions, however, were mild, common and quite varied. Experimentally observed signs of toxicity, both in animals and humans, appeared only after artificially high doses of aspartame had been consumed.

Monosodium glutamate (MSG) has been implicated in several neurodegenerative syndromes. Although excessive levels of endogenous glutamate can result in neuronal death, dietary consumption of MSG appears to result in nothing more serious than the "Chinese restaurant syndrome".

Although the risks of neurotoxicity from any of these particular food additives is low, the controversies have proven rather useful in that they have focused public awareness on the need for accurate food package labeling, and for rational risk-benefit assessments on food additive use.

References

1. Augustine, G.J. and Levitan, H. (1980) Neurotransmitter release from a vertebrate neuromuscular synapse affected by a food dye. *Science* **207**: pp 1489–1490.

2. Bihler, I., LaBella, F.S., Sawh, P.C., Hnatowich, M. (1981) Erythrosin B inhibits 86Rb uptake and increases contractile force in the isolated guinea-pig heart. *Pharmacologist* **23**: p 552.

3. Braithwaite, I.J., Matthews, E.K., Mesler, D.E. (1983) Photodynamic effects of erythrosin B on the electrical and mechanical activity of guinea-pig taenia coli. *Br. J. Pharmacol.* **79**: p 206P.

4. Brenner, A. (1977) A study of the efficacy of the Feingold diet on hyperkinetic children. *Clin. Pediatrics* **16**: pp 652–656.

5. Brunner, R.L., Vorhees, C.V., Kinney, L., Butcher. R.E. (1979) Aspartame: assessment of developmental psychotoxicity of a new artificial sweetener. *Neurobehav. Toxicol.* **1**: pp 79–86.

6. Burlton-Bennet, J.A. and Robinson, V.M.J., A single subject evaluation of the K-P diet for hyperkinesis. *J. Learning Disabilities* 20: pp 331–346.

7. Camfield, P.R., Camfield, C.S., Dooley, J.M., Gordon, K., Jollymore, S., Weaver, D.F. (1992) Aspartame exacerbates EEG spike-wave discharge in children with generalized absence epilepsy: a double-blind controlled study. *Neurology* 42: pp 1000–1003.

8. Carroll, E.J. and Levitan, H. (1978) Fertilization in the sea urchin *Strongylocentrotus purpuratus* is blocked by fluorescein dyes. *Dev. Biol.* 63: pp 432–440.

9. Choi, D.W. (1988) Glutamate neurotoxicity and diseases of the nervous system. *Neuron* 1: pp 623–634.

10. Colombini, M. and Wu, C.Y. (1981) A food dye erythrosine B increases membrane permeability to calcium and other ions. *Biochim. et Biophys. Acta* 648: pp 49–54.

11. Conners, K.C. (1980) *Food Additives and Hyperactive Children.* Plenum Press, New York.

12. Cook, P.S. and Woodhill, J.M. (1976) The Feingold dietary treatment of the hyperkinetic syndrome. *Med. J. Aust.* 2: pp 85–90.

13. Dawson, R., Simpkins, J.W., Wallace, D.R. (1989) Age- and dose-dependent effects of neonatal mono-sodium glutamate (MSG) administration to female rats *Neurotoxicol. Teratol.* 11: pp 331–337.

14. Feingold, B.F. (1975) *Why Your Child Is Hyperactive.* Random House, New York.

15. Fondren, J.E., Norment, B.R., Heitz, J.R. (1978), Dye-sensitized photooxidation in the house fly *Musca domestica. Environmental Entomology* 7: pp 205–208.

16. Fricke, U. (1985) Erythrosin B inhibits high affinity ouabain binding in guinea-pig heart Na^+-K^+-ATPase without influence on cardiac glycoside induced contractility. *Br. J. Pharmacol.* 85: pp 327–334.

17. Garattini, S., Perego, C., Caccia, S., Vezzani, A., Salmona, M. (1987) Aspartame brain amino acids and neurochemical mediators. In: Williams, G.M. (Ed.) *Sweeteners: Health Effects.* Princeton Scientific Publishing, Princeton, NJ, pp 137–148.

18. Goldenring, J.R., Wool, R.S., Shaywitz, B.A., Batter, D.K., Cohen, D.J., Young, J.G., Teicher, M.H. (1980) Effects of continuous gastric infusion of food dyes on developing rat pups. *Life Sciences* 27: pp 1897–1904.

19. Goldenring, J.R., Batter, D.K., Shaywitz, B.A. (1981) Effect of chronic erythrosin B administration on developing rats. *Neurobehav. Toxicol. Teratol.* 3: pp 57–58.

20. Goldenring, J.R., Batter, D.K., Shaywitz, B.A. (1982) Sulfanilic acid: behavioral changes related to azo food dyes in developing rats. *Neurobehav. Toxicol. Teratol.* 4: pp 43–49.

21. Gross, M.D., Tofanelli, R.A., Butzirus, S.M. (1987) The effects of diets rich in and free from additives on the behavior of children with hyperkinetic and learning disorders. *J. Am. Acad. Child Adolescent Psychiatry* 26: pp 53–55.

22. Harley, J.P., Ray, R.S., Tomasi, L., Eichman, P.L., Matthews, C.G., Chun, R., Cleeland, C.S., Traisman, E. (1978) Hyperkinesis and food additives: testing the Feingold hypothesis. *Pediatrics* 61: pp 818–828.

23. Harley, J.P., Matthews, C.G., Eichman, P.L. (1978) Synthetic food colors and hyperactivity in children: a double-blind challenge experiment. *Pediatrics* 62: pp 975–983.

24. Hnatowich, M. and LaBella, F.S. (1982) Light-enhanced inhibition of ouabain binding to digitalis receptor in rat brain and guinea-pig heart by the food dye erythrosine. *Mol. Pharmacol.* 22: pp 687–692.

25. Holder, M.D. (1989) Effects of perinatal exposure to aspartame on rat pups. *Neurotoxicol. Teratol.* 11: pp 1–6.

26. Kruesi, M.J.P. and Rappoport, J.L. (1986) Diet and human behavior: how much do they affect each other? *Annu. Rev. Nutr.* 6: 113–130.

27. Kruesi, M.J.P. and Rappoport, J.L. (1987) Aspartame and children's behavior. In: Williams, G.M. (Ed.) *Sweeteners: Health Effects.* Princeton Scientific Publishing, Princeton, NJ, pp 173–178.

28. Lafferman, J.A. and Silbergeld, E.K. (1979) Erythrosin B inhibits dopamine transport in rat caudate synaptosomes. *Science* 205: pp 410–412.

29. Levitan, H. (1977) Food drug and cosmetic dyes: biological effects related to lipid solubility. *PNAS* **74**: pp 2914–2918.

30. Levitan, H. and Carroll, E.J. (1977) Sea urchin egg membrane properties are altered by fluorescein dyes. *J. Cell. Biol.* **75**: p 231a.

31. Levy, F., Dumbrell, S., Hobbs, G., Ryan, M., Wilton, N., Woodhill, J.M. (1978) Hyperkinesis and diet: a double-blind crossover trial with a tartrazine challenge. *Med. J. Aust.* **1**: pp 61–64.

32. Levy, F. and Hobbes, G. (1978) Hyperkinesis and diet: a replication study. *Am. J. Psychiatry* **135**: pp 12–16.

33. Logan, W.J. and Swanson, J.M. (1979) Erythrosin B inhibition of neurotransmitter accumulation by rat brain homogenate. *Science* **206**: pp 363–364.

34. Maher, T.J. (1986) Neurotoxicology of food additives. *Neurotoxicology* **7**: pp 183–196.

35. Maher, T.J. and Wurtman, R.J. (1987) Possible neurologic effects of aspartame a widely used food additive. *Environ. Health Perspect.* **75**: pp 53–57.

36. Mailman, R.B., Ferris, R.M., Tang, F.L.M., Vogel, R.A., Kilts, C.D., Lipton, M.A., Smith, D.A., Mueller, R.A., Breese, G.R. (1980) Erythrosin (Red No. 3) and its nonspecific biochemical actions: what relation to behavioral changes? *Science* **207**: pp 535–537.

37. Mailman, R.B. and Lewis, M.H. (1981) Food additives and developmental disorders: the case of erythrosin (FD&C Red #3) or guilty until proven innocent? *Appl. Res. Ment. Retard.* **2**: pp 297–305.

38. Mattes, J., and Gittelman-Klein, R. (1978) A crossover study of artificial food colorings in a hyperkinetic child. *Am. J. Psychiatry* **135**: pp 987–988.

39. Morris, S.J., Silbergeld, E.K., Brown, R.R., Haynes, D.H. (1982) Erythrosin B (USFD&C Red 3) inhibits calcium transport and ATPase activity of muscle sarcoplasmic reticulum. *Biochem. Biophys. Res. Comm.* **104**: pp 1306–1311.

40. Nevins, M.E., Arnolde, S.M., Haigler, H.J. (1986) Aspartame: lack of effect on convulsant thresholds in mice. *Fed. Proc.* **45**: p 1096.

41. Olney, J.W. (1980) Excitatory neurotoxins as food additives: an evaluation of risk *Neurotoxicology* **2**: pp 163–192.

42. Olney, J.W. (1984) Excitotoxic food additives — relevance of animal studies to human safety. *Neurobehav. Toxicol. Teratol.* **6**: pp 455–462.

43. Olney, J.W., Ho, O.L., Rhee, V. (1971) Cytotoxic effects of acidic and sulphur-containing amino acids on the infant mouse central nervous system. *Exp. Brain Res.* **14**: pp 61–76.

44. Olney, J.W., Sharpe, L.G., Feign, R.D. (1972) Glutamate-induced brain damage in infant primates. *J. Neuropathol. Exp. Neurol.* **31**: pp 464–488.

45. Olney, J.W., Labruyere, J., DeGubareff, T. (1980) Brain damage in mice from voluntary ingestion of glutamate and aspartate. *Neurobehav. Toxicol.* **2**: pp 125–129.

46. Pinto, J.M.B. and Maher, T.J. (1988) Administration of aspartame potentiates pentylenetetrazole- and fluorothyl-induced seizures in mice. *Neuropharmacology* **27**: pp 51–55.

47. Pradhad, S.N. and Lynch, J.F. (1972) Behavioral changes in adult rats treated with monosodium glutamate in the neonatal stage. *Arch. Int. Pharmacodyn.* **197**: pp 301–304.

48. Reif-Lehrer, L . (1976) Possible significance of adverse reactions to glutamate in humans. *Fed. Proc.* **35**: pp 2205–2211.

49. Reif-Lehrer, L. and Stemmermann, M.G. (1975) Monosodium glutamate intolerance in children. *N. Engl. J. Med.* **293**: pp 1204–1205.

50. Reynolds, W.A., Parsons, L., Stegink, L.D. (1984) Neuropathology studies following aspartame ingestion by infant nonhuman primates In: Stegink, L.D. and Filer, L.J. (Eds.) *Aspartame Physiology and Biochemistry*. Marcel Dekker, New York, pp 363–378.

51. Reynolds, W.A., Bauman, A.F., Stegink, L.D., Filer, L.J., Naidu, S. (1984) Developmental assessment of infant macaques receiving dietary aspartame or phenylalanine. In: Stegink, L.D. and Filer, L.J. (Eds.) *Aspartame Physiology and Biochemistry*. Marcel Dekker, New York, pp 405–423.

52. Rose, T.L. (1978) The functional relationship between artificial food colors and hyperactivity. *J. Appl. Behav. Anal.* **11**: pp 439–446.

53. Rowe, K.S. (1988) Synthetic food colourings and "hyperactivity": a double-blind crossover study. *Aust. Pediat. J.* **24**: pp 143–147.

54. Salzman, L.K. (1976) Allergy testing psychological assessment and dietary treatment of the hyperactive child syndrome. *Med. J. Aust.* **2**: pp 248–251.

55. Schainker, B. and Olney, J.W. (1974) Glutamate-type hypothalamic-pituitary syndrome in mice treated with aspartame or cysteate in infancy. *J. Neural. Trans.* **35**: pp 207–215.

56. Schaumberg, H.H. and Byck, R. (1968) Sin cib-syn: accent on glutamate *N. Engl. J. Med.* **279**: p 105.

57. Schaumberg, H.H., Byck, R., Gerstl, R., Mashman, J.H. (1969) Monosodium-L-glutamate: its pharmacology and role in the Chinese restaurant syndrome. *Science* **163**: pp 826–828.

58. Shaywitz, B.A,, Yager, R.D., Klopper, J.H. (1976) Selective brain dopamine depletion in developing rats: an experimental model of minimal brain dysfunction. *Science* **191**: pp 305–308.

59. Shaywitz, B.A., Goldenring, J.R., Wool, R.S. (1978) The effects of chronic administration of food colorings on activity levels and cognitive performance in normal and hyperactive developing rat pups. *Ann. Neurol.* **2**: p 196.

60. Shaywitz, B.A., Goldenring, J.R., Wool, R.S. (1979) Effects of chronic administration of food colorings on activity levels and cognitive performance in developing rat pups treated with 6-hydroxydopamine. *Neurobehav. Toxicol.* **1**: pp 41–47.

61. Shaywitz, B.A., Novotny E.J., Ebersole, J.S., Anderson, G.M., Sullivan, C.M., Gillespie, S.M. (1992) Aspartame does not provoke seizures in children with epilepsy. *Pediatr. Res.* **31**: p 354A.

62. Shaywitz, B.A. and Novotny, E.J. (1993) Aspartame and seizures. *Neurology* **43**: pp 630–631.

63. Silbergeld, E.K. and Anderson, S.M. (1982) Artificial food colors and childhood behavior disorders. *Bull. N.Y. Acad. Med.* **58**: pp 275–295.

64. Swanson, J.M. and Kinsbourne, M. (1980) Food dyes impair performance of hyperactive children on a laboratory learning test. *Science* **207**: pp 1485–1487.

65. Vorhees, C.V., Butcher, R.E., Brunner, R.L., Wootten, V., Sobotka, T.J. (1983) A developmental toxicity and psychotoxicity evaluation of FD&C Red Dye #3 (erythrosine) in rats. *Arch. Toxicol.* **53**: pp 253–264.

66. Vorhees, C.V., Butcher, R.E., Brunner, R.L., Wootten, V., Sobotka, T.J. (1983) Developmental toxicity and psychotoxicity of FD&C Red Dye No. 40 (Allura Red AC) in rats. *Toxicology* **28**: pp 207–217.

67. Williams, J.I., Cram, D.M., Tausig, F.T., Webster, E. (1978) Relative effects of drugs and diet on hyperactive behaviors: an experimental study. *Pediatrics* **61**: pp 811–817.

68. Wurtman, R.J. and Maher,T.J. (1984) Strategies for assessing the effects of food additives on the brain and behavior. *Fundam. Appl. Toxicol.* **4**: S318–S322.

Suggestions for further reading

Drake, J.J.P. (1975) Food colours — Harmless aesthetics or epicurean luxuries? *Toxicology* **5**: pp 3–42.

Kaplita, P.V. and Triggle, D.J. (1982) Food dyes: behavioural and neurochemical actions. *Trends Pharmacol. Sci.* **3**: pp 70–71.

Lipton, M.A., Nemeroff, C.B., Mailman, R.B. (1979) Hyperkinesis and food additives. In: Wurtman, R.J. and Wurtman, J.J. (Eds.), *Nutrition and the Brain*, Vol. 4. Raven Press, New York, pp 1–27.

Maher, T.J. (1987) Natural food constituents and food additives: the pharmacologic connection. *J. Allerg. Clin. Immunol.* **79**: pp 413–422.

Weiss, B. (1986) Food additives as a source of behavioral disturbances in children. *Neurotoxicology* **7**: pp 197–208.

Weiss, G. and Hechtman, L. (1979) The hyperactive child syndrome. *Science* **205**: pp 1348–1354.

Part III

Environmental contaminants

Contents Chapter 11

Behavioral toxicology of environmental contaminants – an overview

Chapter 11

Behavioral toxicology of environmental contaminants — an overview

Deborah C. Rice

1 Introduction

There are presently thousands of chemicals in our environment — with more being released every year. How many of them are neurotoxic? Tragic episodes of nervous system poisoning in humans as a result of environmental exposure have focused attention on the potential for synthetic and even natural chemicals to produce devastating neurological disease or functional impairment in both adults and infants[84] (Table 11.1).

For most of the chemicals released into the environment, the potential for neurotoxic damage is unknown or incompletely characterized. The workplace also represents a source of exposure to neurotoxic agents. Over 750 chemicals with direct or indirect effects on the nervous system have been identified, with millions of workers exposed to a great number of them. Allowable exposure levels for almost one-third of the chemicals in the workplace in the United States have been determined solely or in part on the basis of neurotoxicity. Many lessons have been learned from the characterization of the effects of known neurotoxic agents; the challenge for the future is to utilize this knowledge to identify agents that produce neurotoxicity and to characterize their effects before they are disseminated into global or local environments.

2 Neurotoxicants in the environment

Accidental acute exposure to a large amount of toxic material usually causes a well-defined syndrome that often involves other organs as well as the nervous system. Therefore, recognition of the deleterious neurobiological effects of most chemicals has evolved from anecdotal observation to studies of illness in persons exposed to high doses. On the other hand, if chronic exposure to potentially toxic materials is followed by illness, it may be difficult to establish exposure to a toxicant as the cause. Currently, the more subtle effects of exposures to environmental neurotoxicants are being documented: reduction in intelligence, impairment in reasoning ability, shortening of attention span, and sensory or motor impairment. Substances to which millions of persons are exposed occupationally and in the general environment that can result in such deficits include lead, organophosphorus pesticides, certain chlorinated hydrocarbons, carbon disulfide, solvents, and mercury. For example, amyotrophic lateral sclerosis (ALS) has developed in persons occupationally exposed to lead, mercury, and selenium, but a cause-effect relationship has not been established. The problem of discovering a cause-effect relationship becomes even greater for people who are exposed to hazardous materials that are released as pollutants into the environment away from the place of manufacture or primary use. For

311

TABLE 11.1

Selected major neurotoxicity events

(From U.S. National Research Council, *Environmental Neurotoxicology*, National Academic Press, Washington, D.C., 1992.)

Year(s)	Location	Substance	Comments
370 B.C.	Greece	Lead	Hippocrates recognizes lead toxicity in mining industry
1st Century A.D.	Rome	Lead	Pliny warns against inhalation of vapors from lead furnaces
1837	Scotland	Manganese	First description of five cases of chronic manganese poisoning in factory workers handling powdered manganese dioxide
1924	United States	Tetraethyllead	In incidents at two plants processing the gasoline additive, over 300 workers suffer neurologic symptoms and five die; nonetheless, its use in gasoline continues for over 50 years
1930	United States	TOPC	Compound often added to lubricating oils intentionally added to Ginger Jake, an alcoholic beverage substitute; more than 5,000 paralyzed, 20,000-100,000 affected
1930s	Europe	Apiol (with TOPC)	Abortion-inducing drug containing TOPC causes 60 cases of neuropathy (with TOPC)
1932	United States	Thallium	Barley laced with thallium sulfate, used as a rodenticide, is stolen and used to make tortillas; 13 family members hospitalized with neurologic symptoms; six die
1937	South Africa	TOPC	60 South Africans develop paralysis after using contaminated cooking oil
1946	England	Tetraethyllead	People suffer neurologic effects of varied degrees of severity after cleaning gasoline tanks
1950s	Japan (Minamata)	Methylmercury	Hundreds ingest fish and shellfish contaminated with mercury from chemical plant; 121 poisoned; 46 die; many infants with serious nervous system damage
1950s	France	Organotin	Medication (Stalinon) containing diethyltin diodide results in more than 100 deaths
1950s	Morocco	Manganese	150 ore miners suffer chronic manganese intoxication involving severe neurobehavioral problems
1956	Turkey	Hexachlorobenzene	Hexachlorobenzene, a seed-grain fungicide, leads to poisoning of 3,000-4,000; 10% mortality rate
1956-1977	Japan	Clioquinol	Drugs used to treat travellers' diarrhea found to cause neuropathy; as many as 10,000 affected over 2 decades
1959	Morocco	TOPC	Cooking oil deliberately and criminally contaminated with lubricating oil affects some 10,000 people
1960	Iraq	Methylmercury	Mercury used as fungicide to treat seed grain is used in bread; more than 1,000 people affected
1964	Japan	Methylmercury	Methylmercury affects 646 people
1968	Japan	PCBs	Polychlorinated biphenyls are leaked into rice oil; 1,665 people are affected
1969	Japan	*n*-Hexane	93 cases of neuropathy follow exposure to *n*-hexane, used to make vinyl sandals
1969	United States	Methylmercury	Ingestion of pork contaminated with fungicide-treated grain results in severe cases of human alkyl mercury poisoning – first instance of such poisoning in United States
1971	United States	Hexachlorophene	After years of bathing of infants in 3% hexachlorophene, the disinfectant is found to be toxic to nervous system and other systems

TABLE 11.1 (continued)

Selected major neurotoxicity events

(From U.S. National Research Council, *Environmental Neurotoxicology*, National Academic Press, Washington, D.C., 1992.)

Year(s)	Location	Substance	Comments
1971	Iraq	Methylmercury	Methylmercury used as fungicide to treat seed grain is used in bread; more than 5,000 severe poisonings and 450 hospital deaths occur; effects on many infants exposed prenatally not documented
1972	France	Hexachlorophene	204 children become ill and 36 die in an epidemic of percutaneous poisoning due to 6.3% hexachlorophene in talc baby powder
1973	United States	Methyl *n*-butylketone (MnBK)	Fabric-production plant employees exposed to solvent; more than 80 workers suffer polyneuropathy, and 180 have less severe effects
1974-1975	United States	Chlorodecone (Kepone)	Chemical-plant employees exposed to insecticide; more than 20 suffer severe neurologic problems, and more than 40 have less-severe problems
1976	United States	Leptophos (Phosvel)	At least nine employees suffer serious neurologic problems after exposure to insecticide during manufacturing process
1977	United States	Dichloropropene (Telone II)	24 people are hospitalized after exposure to pesticide Telone II due to traffic accident
1979-1980	United States	2-*t*-Butylazo-2-hydroxy-5-methylhexane (BHMH, Lucel-7)	Seven employees of plastic-bathtub manufacturing plant experience serious neurologic problems after exposure to BHMH
1980s	United States	1-Methyl-4-pentyl-1,2,3,6-tetrahydropyridine (MPTP)	Impurity in synthesis of illicit drug is found to cause symptoms identical with those of Parkinson's disease
1981	Spain	Toxic oil	20,000 persons poisoned by toxic substance in oil; more than 500 die; many suffer severe neuropathy
1985	United States and Canada	Aldicarb	More than 1,000 people in the U.S. and Canada experience neuromuscular and cardiac problems after ingestion of melons contaminated with the pesticide
1987	Canada	Domoic acid	Ingestion of mussels contaminated with domoic acid causes 107 illnessess and three deaths; form of marine vegetation found in estuaries off Prince Edward Island proves to be apparent source of contaminant
1988	India	TOCP	Ingestion of adulterated rapeseed oil causes about 600 cases of polyneuritis
1989	United States	L-tryptophan-containing products	Ingestion of a chemical contaminant associated with the manufacture of products containing L-tryptophan by one company results in outbreak of eosinophilia-myalgia syndrome; by 1990, over 1,500 cases have been reported nationwide

example it is uncertain what harm, if any, has come to North American aboriginals who consume large amounts of fish contaminated with methylmercury. Even more equivocal are situations such as the Love Canal incident of chemical dumping near Niagara Falls, defoliant spraying in Vietnam, and the aerial distribution of organophosphorus pesticides in the western United States.

313

It is clear that there is an increasing need for an objective and scientific approach, including dose-response data. Methods such as behavioral testing and other types of neurotoxicity are being added to the more traditional toxicological and biochemical techniques to discover possible consequences of chronic low-dose exposure to diverse materials such as lead, aluminium, solvents, and pesticides.

2.1   SOURCES AND ROUTES OF EXPOSURE

In our highly industrialized, urban society, exposure to neurotoxic agents may come from many sources, including industrial waste, agriculture, vehicular traffic, and agents added purposefully to food or water. Exposure may be through oral, inhalation, or dermal routes, and ingestion may be from food, water, air, or soil. For many agents, global cycling and/or bioaccumulation may be the major source of exposure for large segments of the population.

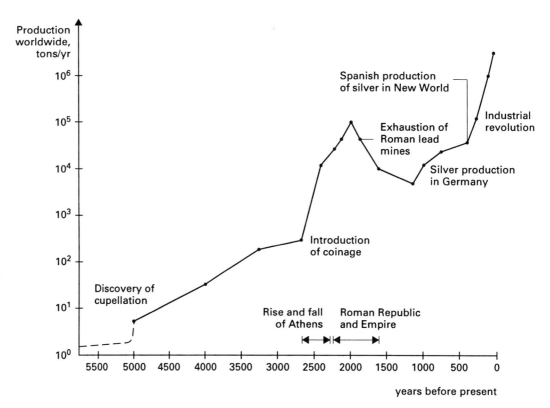

FIGURE 11.1

Amount of lead per year converted to bioavailable forms by human activity, from ancient times to the present
(From U.S. Environmental Protection Agency, Air Quality Criteria for Lead, EPA/8-83-028bF, 1986.)

For example, use of lead by humans is responsible for a dramatic increase in the amount of lead converted to bioaccessible forms (Figure 11.1), with the consequence that humans presently have 100–1000 times more lead in their bodies than in prehistoric times[72]. An example of the complicated pathway for human exposure to lead is provided in Figure 11.2 (which may serve to confuse rather than clarify)[83].

A major source of exposure to lead in the United States from the 1920s through the 1970s was lead added to gasoline as an octane booster. The

314

exhaustion of lead into the atmosphere resulted in widespread contamination of food, soil, and water. The average level of lead in blood in the United States closely parallelled the amount of lead used in gasoline, and decreased precipitously after removal of lead from gasoline in 1978 (Figure 11.3)[70].

A different pattern of environmental exposure is exemplified by methylmercury and PCBs. The main route of exposure to both of these agents is from fish (and marine mammals in populations that rely on subsistence hunting and fishing) as a result of bioaccumulation. Both agents are concentrated as they pass up the food chain, with the result that fish-eating humans, at the top of the food chain, may be exposed to high dietary levels. PCBs, a group of synthetic chemicals, are no longer released into the environment; nonetheless they persist in sediments and the food chain because they are not degraded biologically or chemically, are stored in body fat, and have very long biological half-lives[76]. Methylmercury is still being released into aquatic systems by human activity and natural processes, and is also persistent in the environment and in animals[78]. Both agents are circulated in the global atmosphere, with the result that substantial amounts are found thousands of miles from sources of contamination, including Arctic snows and animals such as polar bears and seals.

Agricultural chemicals such as pesticides and herbicides are used globally by the thousands of tons per year. For many of these agents, inhalation may be

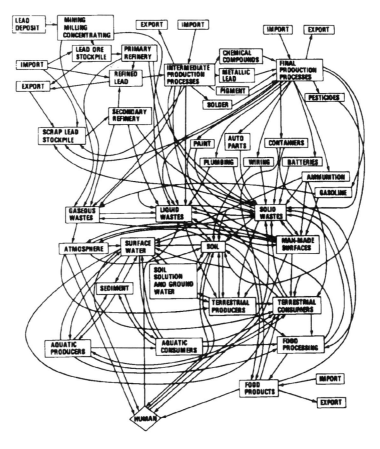

FIGURE 11.2

A schematic of the cycling of lead in the environment, including uses, products, transport media and pathways

(From U.S. National Academy of Sciences, *Lead in the Human Environment*, Washington, D.C., 1980.)

an important route of exposure for agricultural workers. Dermal absorption may also be a significant source of poisoning, depending on the degree of lipophilicity of the agent. For the more general population, pollution of water or residues on food represent the typically exposure routes. Some of these chemicals may also persist in the ecosystem. For example, despite the fact that the use of DDT was banned in North America more than a decade ago, the parent compound and its metabolites persist in wildlife and human tissue[15].

3 Testing for neurotoxicity

Behavior is the functional output of the nervous system, and as such represents the most easily interpretable measure of neurotoxicity. Behavioral changes following exposure to a neurotoxic chemical can be sensitive indicators of disturbed function of the nervous system, since they may be observed earlier and/or at doses lower than demonstrable clinical symptoms or structural lesions. On the other hand, there is the possibility that some structural loss associated with neurotoxicity may occur in the nervous system while the animal remains functionally normal owing to the reserve capacity of the nervous system[40].

3.1   TEST METHODOLOGY IN HUMANS

Many different behavioral functions may be affected by neurotoxic agents from environmental or occupational sources (Table 11.2).

Very often the family physician, or industrial hygienist, is faced with nonspecific complaints such as headache, nausea, lethargy, depression, decreased libido, or impaired reproductive function which many have many

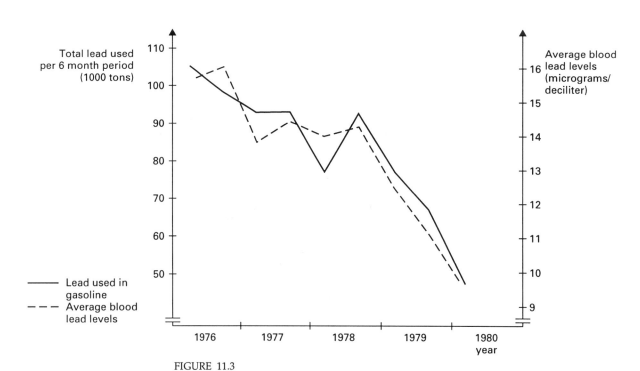

FIGURE 11.3

The amount of lead used in gasoline and the average blood lead levels in the American population from 1976 to 1980
(From U.S. Environmental Protection Agency, Regulation of fuel and fuel additives, *Fed. Reg.*, 47, 7812, 1982.)

316

TABLE 11.2

Neurotoxic chemical exposures and associated neurological signs and symptoms

Neurotoxicant	Major uses/Sources of exposure	Signs and symptoms	Suspected pathological target
METALS			
Arsenic	Pesticides Pigments Antifouling paint Electroplating industry Seafood Smelters Semiconductors	Acute: Encephalopathy Chronic: Peripheral neuropathy	Unknown (a) Axon (c)
Lead	Solder Lead shot Illicit whiskey Insecticides Auto body shop Storage battery manufacturing Foundries, smelters Lead-based paint Lead pipes	Acute: Encephalopathy Chronic: Encephalopathy and peripheral neuropathy	Blood vessels (a) Axon (c)
Manganese	Iron, steel industry Welding operations Metal-finishing operations Fertilizers Manufacturers of fireworks, matches Manufacturers of dry cell batteries	Acute: Encephalopathy Chronic: Parkinsonism	Unknown (a) Basal ganglia neurons (c)
Mercury	Scientific instruments Electrical equipment Amalgams Electroplating industry Photography Felt making	Acute: Headache, nausea, onset of tremor Chronic: Ataxia, peripheral neuropathy, encephalopathy	Unknown (a) Axon (c) Unknown (c)
Tin	Canning industry Solder Electronic components Polyvinyl plastics Fungicides	Acute: Memory defects, seizures, disorientation Chronic: Encephalo- myelopathy	Neurons of the limb system (a&c) Myelin (c)
SOLVENTS			
Carbon disulfide	Manufacturers of viscose rayon Preservatives Textiles Rubber cement Varnishes Electroplating industry	Acute: Encephalopathy Chronic: Peripheral neuropathy, parkinsonism	Unknown (a) Axon (c) Unknown (c)
n-hexane, methyl-butyl-ketone	Paints Lacquers Varnishes Metal-cleaning compounds Quick-drying inks Paint removers Glues, adhesives	Acute: Narcosis Chronic: Peripheral neuropathy	Unknown (a) Axon (c)
Perchloroethylene	Paint removers Degreasers Extraction agents Dry cleaning industry Textile industry	Acute: Narcosis Chronic: Peripheral neuropathy, encephalopathy	Unknown (a) Axon (c) Unknown (c)

317

TABLE 11.2 (continued)

Neurotoxic chemical exposures and associated neurological signs and symptoms

Neurotoxicant	Major uses/Sources of exposure	Signs and symptoms	Suspected pathological target
METALS			
Toluene	Rubber solvents Cleaning agents Glues Manufacturers of benzene Gasoline, aviation fuels Paints, paint thinners Lacquers	Acute: Narcosis Chronic: Ataxia, encephalopathy	Unknown (a) Cerebellum (c) Unknown (c)
Trichloroethylene	Degreasers Painting industry Varnishes Spot removers Process of decaffeination Dry cleaning industry Rubber solvents	Acute: Narcosis Chronic: encephalopathy, cranial neuropathy	Unknown (a) Unknown (c) Axon (c)
INSECTICIDES			
Organophosphates	Agricultural industry: Manufacturing and application	Acute: Cholinergic poisoning Chronic: Ataxia, paralysis, peripheral neuropathy	Acetylcholinesterase (a) Long tracts of spinal cord (c) Axon (c)
Carbamates	Agricultural industry: Manufacturing and application Flea powders	Acute: Cholinergic poisoning Chronic: Tremor, peripheral neuropathy	Acetylcholinesterase (a) Dopaminergic system (c)

Note: (a) acute, (c) chronic.

etiologies, including exposure to toxic agents. Similarly, neuroteratogens may produce decreased birth weight and failure to thrive, effects which themselves predispose toward neurological and other health problems. The task of the behavioral toxicologist is to identify which agents produce behavioral toxicity, and to characterize their effects.

Epidemiological research strategies involve identifying a human population at risk for exposure to a suspected neurobehavioral toxicant, and correlating exposure indices or levels in a tissue such as blood with preselected outcome variables. A similar strategy may be used in the workplace, where performance on a variety of measures of neurological or cognitive function may be assessed in workers with a known current or historical exposure to a particular agent, compared to an appropriately matched group of unexposed individuals.

A wide variety of testing methodologies and strategies are available to assess behavioral function (see Chapter 3). Standardized tests of intelligence, clinical tests of motor and sensory function, and various psychological instruments of affect, mood, and psychiatric disorders have a long history of use by physicians and psychologists. IQ tests have the advantage of being standardized for the population in which they are used; however, results are not valid for populations for which they were not designed. In addition, IQ tests provide a global index of impairment; they do not assess specific behavioral functions or processes that may be preferentially affected by exposure to a particular neurotoxicant. Assessment of such functions might prove more sensitive than IQ tests, as well as offering the advantage of identifying specific behavioral mechanisms responsible for deficits in cognitive functioning. Simple clinical tests of sensory or motor function usually are able to detect only very severe impairment; e.g., using a tuning fork or pin prick to assess somatosensory function, or observation of gait or grasping of the clinician's fingers to assess motor function. While more sophisticated techniques are available, they may not be used to test individual patients presenting such complaints to a physician. Detailed assessment is most typically performed in identified populations of exposed individuals, and often then only in functional domains identified from signs and symptoms in highly affected individuals. For example, assessment of visual fields is performed in patients thought to be suffering from methylmercury poisoning, because constriction of visual fields is known to be a common effect of exposure to methylmercury.[27] An example of ongoing assessment of one patient from Minamata, for whom an increasing constriction of visual fields was documented over a number of years, is presented in Figure 11.4. Such a test would not typically be part of a clinical work-up.

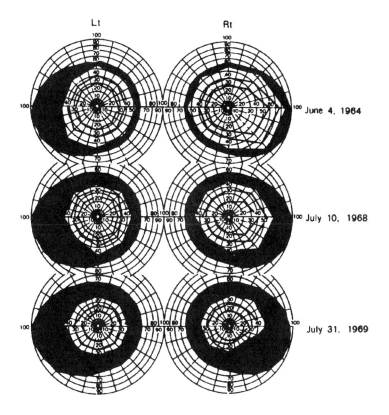

FIGURE 11.4

Documentation of progression of constriction of visual fields, as assessed in a Goldmann perimeter, of a patient with methylmercury poisoning

The filled area on each panel represents the difference in the field of vision between a normal individual (outside boundary) and an individual with methylmercury poisoning (inside boundary). Visual function deteriorated over the course of 5 years in this individual even through methylmercury exposure had ceased. Lt = left eye, Rt = right eye. (From Iwata, K. (1980) In: *Neurotoxicology of the Visual System*, Merigan, W.H. and Weiss, B. (Eds.), Raven Press, New York, p 165. With permission.)

TABLE 11.3

Neurobehavioral effects reported following chemical exposure in the workplace and functions assessed by standardized batteries

(Adapted from Anger, W.K. (1990) In: *Behavioral Measures of Neurotoxicity*, Russell, R.W., Flattau, P.E., Pope, A.M. (Eds.), National Academy Press, Washington, D.C., p 69. With permission.)

Neurobehavioral effects	Functions assessed by WHO, NES or FIOH batteries
Motor	
Activity changes	
Ataxia	
Convulsions	
Incoordination/unsteadiness/clumsiness	x
Paralysis	x
Pupil size changes	
Reflex abnormalities	
Tremor/twitching	x
Weakness	
Sensory	
Auditory disorders	
Equilibrium changes	
Olfaction disorders	
Pain disorders	
Pain, feelings of	
Tactile disorders	x
Vision disorders	x
Cognitive	
Confusion	
Memory problems	x
Speech impairment	
Affect/personality	
Apathy/languor/lassitude/lethargy/listlessness	x
Delirium	x
Depression	x
Excitability	
Hallucinations	
Irritability	x
Nervousness/tension	x
Restlessness	x
Sleep disturbances	
General	
Anorexia	
Autonomic dysfunction	
Cholinesterase inhibition	
Depression of the central nervous system	
Fatigue	x
Narcosis/stupor	
Peripheral neuropathy	x
Pathology	
Psychic disturbances	

In recognition of the limitations of clinical neurological examinations to detect effects of toxic exposure, a number of researchers and international organizations have devoted considerable effort to the design of test batteries to assess neurotoxic effects of chemicals in the workplace. Designers of these batteries have attempted to include assessment of all possible functional domains, from mood states to sensory function, in a comprehensive and detailed manner (Table 11.3). However, batteries presently developed do not assess several major categories of effects reported as a result of exposure to chemicals[3].

Assessment of developmental toxicity as a result of environmental exposure presents a particular set of challenges. In some instances, such as in prospective studies of the neurotoxic potential of environmental agents, subjects are assessed for a number of years beginning at birth. Very young infants may undergo standardized tests of neurological functioning such as the Brazelton Neonatal Assessment Battery; the ability of such early tests to predict later functional integrity is largely unknown. Cognitive tests are performed on babies as young as several months of age; results are expressed as measures of "IQ". Such tests have proven to have little value in predicting IQ assessed at later ages. (However, it must be stated that several prospective studies on the effects of lead on intelligence have consistently observed lead-induced impairment from infancy through at least the early school years. See Chapter 12). The preferential looking test (Fagan Visual Recognition Test) which relies on the propensity of the infant to gaze longer at novel rather than familiar scenes, has been shown to be more highly correlated with later IQ measures[20].

Investigators have also attempted to assess specific behavioral functions, chosen on the basis of clues provided by previous research. For example, early research identified short attention span, impulsive behavior, and restlessness in addition to poor school functioning as permanent sequelae of lead poisoning in children. Inclusion of teachers' rating scales, designed to assess such behaviors, identified similar effects in a number of studies in children exposed to much lower levels of lead[43]. In addition, measurement of performance on tasks designed specifically to measure attentional processes revealed robust deficits, sometimes in the absence of deficits in IQ[90-92]. These behavioral tasks also have the advantage, as pointed out by at least one investigator, of not being as influenced by the many variables that affect IQ.

The development of specific and appropriate testing strategies may result in at least two significant advantages. First, it enables identification and characterization of specific deficits as a result of exposure to a neurotoxic agent which may otherwise have gone undetected if only generalized tests were used. Second, functional deficits may be detected at lower exposure levels or body burdens as a consequence of the use of sophisticated and specific tests that assess performance on the most sensitive end points identified. It may be a truism that you can't find what you don't look for, but it is one well remembered when evaluating effects of chemicals, or determining the levels at which they produce toxicity.

Research in human populations is fraught with many problems that complicate interpretation of results. The exposure of members of the sample under study is usually not random. Exposure may be linked to race, socioeconomic status (SES), place of residence, age, or other factors that might themselves influence outcome variables. In most studies of the effects of lead on IQ in children, for example, lead levels in the blood of children were inversely correlated with both the SES of the family and the mother's IQ, two factors which are well-known to affect a child's performance on tests of intelligence. When these factors were controlled statistically, the apparent effect of lead decreased. In part in recognition of this problem, a group of researchers in Boston studied a group of high SES children, in which the more privileged families lived in a more urban and therefore more lead-polluted environment[7,8]. In that study, controlling for the typical confounders increased the apparent deleterious effect of lead on IQ. To cite another example from the lessons we have learned from studying the effects of lead on behavior in children: it was observed in a number of studies that increased lead levels in the mother resulted in premature delivery[17,35] or in babies that were smaller for gestational age[8,9]. Since these factors are known to put the infant at risk for later behavioral and other neurological problems, they were controlled for in assessments of the behavioral effects of lead. However, since decreased birth weight and premature delivery may be a direct effect of lead exposure, controlling for them statistically may be considered "over controlling", thereby underestimating the true magnitude of the effect of lead.

Another potential confound is the possibility of multiple exposures. It may be that an unidentified agent, with an exposure pattern in the population highly correlated to that of the agent under study, is actually responsible for the observed effects, or that the observed effects are the result of exposure to two or more agents. For example, attempts to assess the effects of developmental exposure to PCBs are hampered by the fact that the fish (and therefore the humans) with high PCB levels also may contain elevated levels of methylmercury, a known neurotoxicant, as well as other persistent industrial chemicals and pesticides. It may be extremely difficult if not impossible to separate the effects produced by the different agents. (It may also be unnecessary: a legitimate conclusion may be that contaminated fish pose a hazard when ingested above a certain amount.) The problem of multiple exposures also plagues research on the effects of industrial exposure to neurotoxic agents. It is rare that workers exposed to solvents are exposed to only one at a time, for example. In addition, an individual may have been exposed to other solvents or neurotoxicants previously, compromising interpretation of the results of a study based on current exposure. Another confound encountered in research on the effects of workplace exposure is that exposure duration is often highly correlated with age. Some of the effects of solvents, for example, may mimic effects of aging or diseases of aging, such as short-term memory and motor deficits. Because it can never be absolutely determined that variables not considered in the design of the study are responsible for effects observed in human research, observed correlations cannot unequivocally establish a "cause and effect" relationship. However, performing many studies in different populations, with the same result increases the confidence that the results are attributable to known rather than unknown variables.

3.2   BEHAVIORAL TESTING IN ANIMAL MODELS

Behavioral toxicology is receiving increasing recognition by various regulatory agencies around the world. Presently, the United Kingdom and Japan require neurotoxicity testing to evaluate the developmental neurotoxicity of new drugs[66], and the U.S. Environmental Protection Agency (EPA) has developed testing protocols to address specific aspects of behavioral toxicity testing[73,74,79,80]. Several important workshops have been convened to characterize available behavioral methodology in detail and to make recommendations on strategies for behavioral toxicity testing (Table 11.4). An important collaborative behavioral teratology study, organized by the U.S. National Center for Toxicology Research and performed in six independent laboratories in the United States, has demonstrated a high degree of interlaboratory reliability regarding behavioral methodology[10]. Such developments underscore the firm foundation upon which behavioral toxicological methodology is based and the increasing recognition of its importance in toxicity testing.

Research performed in animal models offers significant advantages in experimental design over epidemiological studies. Animals are randomly assigned to treatment groups and all subjects are treated in the same manner except for exposure to the toxicant under study. Environmental factors, such as temperature, light/dark cycles, and diet can be carefully controlled. Exposure level and duration, and the developmental period during which exposure

TABLE 11.4

U.S. and International Expert Panels or Committees Making Recommendations for Testing

(From Tilson, H.A. (1990) *Neurotoxicol. Teratol.* **12**: p 293. With permission.)

Study group	Year	Recommendation
NRC	1975	Proposed a tiered evaluation of chemicals for potential neurotoxicity; tests include motor activity, functional observational battery, complex learned functions; behavioral teratology
NRC	1977	Recommended conditioned (i.e., schedule-controlled behavior) and unconditioned behavior (i.e., motor activity) for assessment
OECD	1982	Adopted acute and subchronic assays for delayed neurotoxicity of organophosphorus esters
NRC	1984	Neurobehavioral toxicity testing should include studies on behavior (conditioned and unconditioned) and morphology (neuropathology)
NRC	1986	Identified research needs to measure exposure using biological markers and better laboratory techniques
WHO/IPCS	1986	Recommended two levels of neurobehavioral testing (primary level, including functional obervational battery and motor activity; secondary level, including schedule-controlled behavior, sensory function and cognition)
EPA/FIFRA	1987	SAP subpanel recommends motor activity, functional observational battery and neuropathology for pesticide testing
WHO/IPCS	1988	Steering committee develops protocol for international collaborative study on primary level testing
OTA	1989	Workshop on Neurotoxicology in the Federal Government; Interagency Committee on Neurotoxicology (ICON) formed
OECD	1990	Proposed testing guidelines for neurotoxicity testing, including functional observational battery and neuropathology; considered testing protocol for 14-28 or 90 day studies which includes neurobehavioral observations and neuropathology where appropriate

occurs, can also be tightly controlled. Groups will be similar with respect to age, genetic variables, etc., unless the influence of these variables is one of the issues under study. The mechanisms responsible for behavioral toxicity, at the organ or cellular level, may also be explored. The levels of exposure or body burden at which effects occur can be precisely determined in experimental models; in addition, the kinds of effects (motor, sensory, cognitive, etc.) can be exhaustively characterized, and the most sensitive indicator(s) of toxicity identified. While there may be differences between animal models and humans both in the precise pattern of neurotoxicity and the absolute levels at which toxicity occurs, identification of neurotoxic effects in animals provides reassurance that neurotoxicity observed in humans is not the result of unidentified covariates, and provides the opportunity to identify neurotoxic agents before they are disseminated for use.

Behavioral toxicity testing can easily be incorporated into traditional testing protocols, with methodologies appropriate to acute, subchronic, chronic, and reproductive phases of chemical (or drug) testing. Typically, observation of the animal is part of traditional toxicological protocols and can serve as an indicator of possible neurotoxic effects. Signs such as lethargy or hyperactivity, piloerection, convulsions, ataxia, abnormal gate, tremor, and abnormal maternal behavior may be indicative of neurotoxicity. If such signs are observed, more detailed analyses can be incorporated into further testing protocols, including chronic or reproductive studies. Animals or a subset of animals committed to other toxicity studies can often be used for behavioral testing. Thus, behavioral testing, like other specialized toxicity testing such as immune or endocrine, need not (indeed should not) be performed in isolation from the data gathered in traditional toxicity studies.

The choice of behavioral methods runs the gamut from gross, simple observation requiring no equipment to extremely sophisticated measurements of specific nervous system functions. Each level of testing has advantages and disadvantages and occupies an appropriate place in the hierarchy of toxicity testing. It has been suggested that a "tiered" approach be used in toxicity testing including behavioral testing[81, 82, 93]. Such a scheme can begin with simple, rapid, and inexpensive tests to determine if behavioral effects are present and progress to tests of increasing complexity, duration, and cost. For chemicals for which there are no behavioral data, testing should begin at relatively high doses and acute exposure, for example, as part of an acute toxicity study. A number of investigators have proposed observational batteries designed to assess major overt neurotoxic effects of chemicals[22, 25, 42]. These batteries consist of assessment of a number of endpoints such as tremor, convulsions, ataxia, autonomic signs, paralysis, surface righting, and posture. Such batteries are appropriate in initial toxicity testing because they use semiquantitative subjective measures and are rapid and easy to implement. In addition, food and water consumption and body weight should be considered in the context of behavioral assessment.

Following the initial observational battery, there are a number of somewhat more specific tests available that may still be considered "screening procedures" because they are relatively nonspecific. A variety of such screening test batteries has been proposed[1, 10, 41, 86] and may include measures such as negative geotaxis orientation to stimuli (visual, auditory, olfactory), auditory startle habituation, locomotor activity, and simple learning. The best strategy is to choose a number of tests that assess different functions such as motor, sensory, and "cognitive" functions. Although a well-developed battery is capable of yielding important information about "where to look next", it must be remembered that the results of screening tests are not useful for determining no-effect levels because they lack the necessary sensitivity.

If effects are observed on any of the screening tests, a more detailed behavioral analysis should be made. Testing will typically include examination of learning and memory, intermittent schedules of reinforcement, and perhaps more detailed evaluation of sensory and motor function[93]. At this level of testing, it is necessary to evaluate toxicity carefully in other organ systems. Performance on a behavioral task may be altered by the animal being ill (e.g., from liver damage) as well as by primary changes in nervous system function. Toxicity in other organ systems must be considered in the evaluation of behavioral data. Failure to recognize nonspecific effects may result in attribution of behavioral changes to specific deficits, such as memory or motor deficits, when the findings are, in fact, the result of toxicity in organ systems other than the nervous system.

The final level of testing is designed to characterize in detail the nature of the toxic effect and to determine the lowest level at which any effect is observable. The most sensitive methodology is used with the most appropriate species. Exposure should be relevant to the human situation in terms of level, duration, and route of chemical or drug administration. Such testing should be carried out for any agent to which large numbers of people are exposed, and for which there is good evidence of nervous system toxicity.

For many agents, some information will be available concerning behavioral effects. These data may be derived from animal studies or from human exposure. If the substance is a known neurotoxicant, there is no need to perform a screening battery. Other simple procedures may also be omitted. This will be especially true if the signs and symptoms of poisoning are known in humans. If a toxicant is known or strongly suspected to affect a certain neurotransmitter system, then behavioral tests can be chosen that are known to affect the pathways involved. This is particularly relevant for some pesticides, for example.

For screening procedures, the rat or mouse will typically be the species of choice. Some countries require that testing be done in two species. The number of animals should be relatively large because these techniques are relatively insensitive and often produce large between-subject and/or within-subject variability. During the initial stages of testing (such as during acute toxicity testing), doses will be high, often producing gross toxicity; for example, weight loss, changes in food and water consumption, changes in body temperature, and ataxia. Detailed behavioral testing at these doses is not warranted because behavioral effects will be due partly or totally to gross toxicity. For example, malnutrition is known to produce behavioral changes, including intellectual impairment, in rodents[45, 60], primates[44], and humans[12]. If behavioral changes are observed at high doses, more detailed behavioral testing must be done with dosing and exposure regimens relevant to human exposure levels. This will more than likely entail chronic exposure or chronic intermittent exposure, with the route of administration usually being oral ingestion or inhalation, or occasionally percutaneous. At the chronic level of testing, behavioral monitoring must be carried out in conjunction with monitoring of other measures of toxicity in order for the general health status of the animal to be known concurrently. Only in this way will it be possible to determine whether the behavioral effects are the result of a chemical's primary effect on the nervous system or the result of its toxic effect on other organ systems. In addition, such monitoring is necessary to determine whether or not the nervous system is the most sensitive (or one of the most sensitive) system to the effects of the toxicant. Sophisticated behavioral testing is contraindicated when doses produce gross toxicity or when it is obvious that other systems are more sensitive.

4 Risk assessment issues in neurotoxicology

Risk management is the process of weighing policy alternatives and selecting the most appropriate regulatory action by integrating the results of risk assessment with social, economic, and political concerns to reach a decision. The environmental regulatory tool for risk management is risk assessment, which attempts to provide quantitative estimates of any adverse effects associated with environmental exposure to an agent. Risk assessment is defined as the characterization of the potential adverse effects of human exposures to environmental hazards[85]. The process of risk assessment consists of four main steps: (1) hazard identification (2) dose-response assessment (3) exposure assessment, and (4) risk characterization. Hazard identification concerns determination as to whether or not a particular agent is causally linked to particular health effects; dose-response assessment is the process of elucidating the relationship between exposure, dose at the site of action, and biological effect; exposure assessment is the process of measuring or estimating the intensity, frequency, or duration of human exposure to agents currently in the environment or of estimating hypothetical exposures that might occur; risk characterization is the process of estimation the incidence of a health effect under the various conditions of human exposure. There are a number of issues that must be considered when evaluating data for risk assessment. Some of these issues have been discussed extensively[66, 77]; selected issues are discussed below.

4.1   ANIMAL TO HUMAN EXTRAPOLATION

The issues that must be considered in the extrapolation of data gathered using animal models to potential effects in humans are of critical importance in all areas of toxicology, but perhaps particularly in behavioral toxicology. As in all toxicological disciplines, species differences in toxicokinetics and metabolism must be considered in the evaluation of neurotoxic effects. For example, methanol ingestion produces severe visual pathology resulting in blindness in humans and other primates as a result of metabolic acidosis[63]. This is due to conversion of methanol to formate, which is poorly metabolized in primates, including humans. Formate does not accumulate in the blood of rodent species or other animals, so that primate species are the only useful model to study methanol-induced visual impairment. To cite another example, the brain to blood ratio of methylmercury in humans and other primates is between 2–5 to 1[19, 32, 48]. The rat, on the other hand, has a ratio of 0.06[46]. Therefore, the blood mercury level at which neurotoxicity was observed in rats would not be a good predictor of blood level at which effects would be observed in humans, since it would not be reflective of comparable target organ (i.e., brain) mercury levels. This does not mean, however, that the rat may not represent a useful model to study particular aspects of methylmercury-induced neurotoxicity; indeed, motor deficits produced by methylmercury have been reliably replicated in rats.

Another important issue with respect to extrapolation from animal models to humans is the degree of similarity between the nervous system of the model species and that of humans. Humans have a large convoluted brain with a large neocortex, whereas the brain of rodent species has a smooth surface with relatively more area devoted to subcortical structures. This difference in shape may be an important determinant of differential effects between species. To again use methylmercury as an example: in humans, methylmercury preferentially damages deep sulci[37]. One of the consequences of this preferential damage is constriction of visual fields, since peripheral vision is represented in the calcarine fissure at the back of the brain. Comparable visual deficits would therefore not be expected in rats or mice as a result of methylmercury exposure. On the other hand, visual deficits comparable to those observed in humans have been replicated in monkeys[38, 52].

The comparability of sensory and motor systems between species is also important in evaluation of neurotoxicity data, since sensory or motor disturbances are a frequent consequence of exposure to neurotoxic agents. The visual system is probably the sensory system that differs most across species. While the visual system of rodents may be affected by agents that also produce ocular toxicity in humans[39], the rodent may be a poor model for exploring the effects of known or suspected neurotoxic agents on specific visual functions. The rodent visual system represents about 20% of the primate visual system, being largely color-blind with low acuity (Figure 11.5).

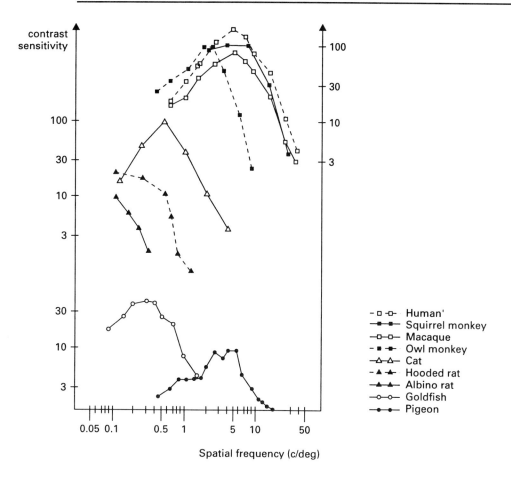

FIGURE 11.5

Spatial visual function of various species

Abscissa: spatial frequency. Ordinate: contrast sensitivity = reciprocal of contrast, which is a function of the difference between the light and dark bars of a series of parallel bars. The visual function of the albino rat is least like that of humans (note three scales of ordinate) while that of some monkey species is virtually identical to humans. (From Uhrich, D.J., Essock, E.A., Lehmkuhle, S. (1981) *Behav. Brain Res.* **2**: p 291. With permission.)

In fact, the rodent relies much more on olfaction and audition than on the visual system for information concerning its environment. In contrast, primates are highly dependent upon their ability for good spatial resolution and highly-developed color vision for much of the information about their world. Color vision is affected by some pesticides[46] and solvents[37], an effect that could not be detected using most common laboratory species including rodents. In contrast to the visual system, the basic anatomical and physiological substrates of audition vary little between mammalian species, even though many animals, including rodents, cats, dogs, and monkeys, can hear frequencies undetectable by humans. Similarly, there is a direct correspondence between species in the various modalities of the somatosensory system: light touch, pain, pressure, vibration, temperature, joint proprioception, and visceral sensation. These functional modalities are also subserved by similar or identical peripheral anatomical structures. The same may be said for the peripheral nerves subserving motor systems. Results from animal studies assessing peripheral neuropathy can probably in most instances be directly extrapolated to humans. Acrylamide produces a dying-back neuropathy in all animals, including humans[61], which is reflected in somatosensory and motor impairment. Lead produces peripheral motor and sensory neuronal impairment in both humans and animal models[61]. While central pathways also share many similarities between species, there are nonetheless some important species differences with respect to central damage that may underlie motor disease. Parkinson's disease is a disease characterized by tremors and lack of muscular control that stems from damage to dopaminergic systems. The neurotoxic agent MPTP (1-methyl-4-phenyl-1,2,3,6-tetrahydropyridine), a compound that can be created during production of synthetic heroin, produces a Parkinson-like syndrome in humans[85] and monkeys but not in rodents. Similarly, manganese produces a parkinsonian syndrome in workers exposed to high concentrations, such as miners. This syndrome is mimicked in monkeys, but not in rodents[75]. Damage to the substantia nigra, a small region deep in the center of the brain, is implicated in all of these diseases. As its name suggests, this structure is highly pigmented in normal humans and primates, but not in rodents. It has been hypothesized that the neuromelanin present in the substantia nigra of primates makes it peculiarly susceptible to damage[30], which may result in damage to brain structures to which the substantia nigra projects. This hypothesis is supported by evidence in dogs, which accumulate neuromelanin and suffer the same extensive cell loss in substantia nigra as primates after exposure to MPTP. However, rodents may nonetheless be acceptable models to study other neurochemical and neuropathological effects produced by manganese[75] or MPTP.

While the importance of the contribution of differences in anatomy and physiology to differences the manifestation of toxicity between species is a concern common to all disciplines in toxicology, an issue that is unique to neurotoxicology is the degree to which effects on tests of cognitive processes in animals are predictive of effects in humans. For example, are deficits observed on learning or memory tasks in animals indicative that similar effects would be observed in humans? Humans obviously have a much larger behavioral repertoire than does a rodent or even a monkey, and are capable of learning and performing much more complicated tasks. Despite this fact, it is nonetheless the case that all species must take in information about their environment, remember it, and synthesize its meaning in the context of previous experience in order to survive. In other words, all animals learn, remember, and seek to repeat positive experiences while avoiding negative ones. The behavioral processes that allow animals to accomplish these things show great similarities across species. It is very clear that developmental exposure to lead produce behavioral impairment in rodents, monkeys, and children[94]. It is also apparent that the underlying deficits responsible for the myriad of observed adverse effects are similar in humans and animals[50, 51] and include perseveration (persistence in a behavior after it is no longer adaptive), inability to inhibit inappropriate responding, and increased distractibility (attention deficits). Similarly, performance on intermittent schedules of reinforcement exhibits pronounced similarities across wide array of species, including humans under some conditions[28]. The effects of drugs and toxicants on such schedules is also similar across species[16, 26]. This lends considerable confidence that effects produced by exposure to a known or suspected neurotoxicant reflect impairment in behavioral processes common to many species including humans.

The degree to which animal species differ from each other or from humans in toxicokinetics, metabolism, physiology, anatomy, health, and nutritional status will be reflected in differences in the dose of a neurotoxic agent required to produce a neurotoxic effect. Often the dose required to produce even a very specific effect differs across species by orders of magnitude (Table 11.5).

TABLE 11.5

Dose (mg/kg) required to produce effects on fixed interval (FI) performance in various species
(Adapted from McMillan, D. (1990) *Neurotoxicol. Teratol.* **12:** p 523. With permission.)

	Phencyclidine ↓ FI	Chlorpromazine ↓ FI	D-amphetamine ↑ FI	Morphine ↓ FI
mouse	18		1	
pigeon	1.7	3-10	1-3	3
rat	3-5.6	1-1.5	1	3
squirrel monkey	0.3	0.1-0.3	0.3	0.3
rhesus monkey	0.025			0.3
chimpanzee		1	0.3-1	no effect

However, if levels in target organ (i.e., brain) are considered instead, levels at which similar effects are observed may be much more consistent across species (Table 11.6). Unfortunately, the relevant differences in toxicokinetics, metabolism, and tissue distribution are rarely known. Blood levels are usually not measured in animal studies, while blood is the most typically measured compartment in epidemiological studies. Brain levels are sometimes measured in animal studies, but are only available in postmortem tissue in human studies, and are rarely measured. For new chemicals, the toxicokinetics in humans will be unknown, rendering such comparisons impossible. These uncertainties have resulted in necessarily arbitrary strategies being adopted in regulatory efforts to utilize animal data to protect human health.

Confidence that behavioral deficits observed in animal models are predictive of effects in humans is also increased if underlying functional or anatomical substrates are identified[66]. If the neurochemical mechanism is known, the data may be directly evaluated in reference to species generality. Similarly, an understanding of the neuronatomical targets of a neurotoxicant provides important information regarding what effects should be expected in humans as a result of exposure to a particular agent. Differences in anatomy, physiology, and toxicokinetics must be included in the evaluation of such data.

4.2   ACUTE VS. CHRONIC EXPOSURE

The length and pattern of exposure may be an extremely important determinant of the kinds of effects produced by a toxic agent. Effects of acute exposure to a relatively high dose of an agent may produce effects different from those produced by lower-dose chronic exposure (Table 11.2). Acute high-dose exposure to lead in adults results in colic, anemia, and encephalopathy. Chronic exposure to lower levels, encountered in the workplace, produces impairment of motor and sensory function, cognitive and short-term memory deficits, and sexual dysfunction[18, 89]. Acute exposure to pesticides may produce ataxia, confusion, autonomic signs, unconsciousness, and even death. More chronic agricultural exposure may produce memory and personality disorders, and sensory system disturbances[5, 57]. The effects of exposure at environmental levels of many agricultural chemicals are not completely known, although some, such as DDT, have been banned on the basis of reproductive and behavioral effects observed in wildlife populations[15]. Exposure to solvents may also have differential effects depending on pattern of exposure. Acute high-dose exposure causes narcosis; intermittent lower dose exposure may produce effects including psychiatric symptoms, memory and intellectual impairment, and ataxia[4]. Acrylamide was known for many years to produce peripheral

TABLE 11.6

Comparison of neuropathological and neurobehavioral effects of developmental methylmercury exposure across species

(From Burbacher, T.M., Rodier, P.M., Weiss, B. (1990) *Neurotoxicol. Teratol.* **12**: p 191. With permission.)

Human	*Nonhuman primate*	*Small mammals*
	HIGH BRAIN DOSES **(12-20 ppm)**	
Neuropathology		
Decrease in size of brain, damage to cortex, basal ganglia and cerebellum, sparing of diencephalon, ventricular dilation, demyelinated fibers, ectopic cells, gliosis, disorganized layers, misoriented cells, loss of cells.	Decrease in size of brain, damage to cortex and basal ganglia, sparing of diencephalon, gliosis, loss of cells (sparing of cerebellum), ventricular dilation, ectopic cells, disorganized layers.	Decrease in size of brain, damage to cortex, basal ganglia, hippocampus, and cerebellum, sparing of diencephalon, ventricular dilation, loss of myelin, misoriented cells, loss of cells.
Neurobehavior		
Blindness, deafness, cerebral palsy, spasticity, mental deficiency, seizures	Blindness, cerebral palsy, spasticity, seizures.	Blindness, cerebral palsy, spasticity, seizures.
	MODERATE BRAIN DOSES **(3-11 ppm)**	
Neuropathology		
No data	No data	Decrease in size of brain, damage to cortex and cerebellum, loss myelin.
Neurobehavior		
Mental deficiency, abnormal reflexes and muscle tone, retarded motor development.	Retarded development of object performance, visual recognition memory, and social behavior, visual disturbances, reduced weight at puberty (males).	Abnormal on water maze, auditory startle, visual evoked potentials, escape and avoidance, operant tasks, activity, response to drug challenge.
	LOW BRAIN DOSES **(< 3 ppm)**	
Neuropathology		
No data	No data	Decrease in size of brain, and loss of cells.
Neurobehavior		
Delayed psychomotor development	No data	Response to drug challenge, active-avoidance, operant tasks.

neuropathies; more recently has come the recognition that extended exposure may result in central nervous system damage independent of peripheral damage[65].

The potential health ramifications of lifetime (including prenatal) exposure to the components of the chemical soup in which we all presently reside are a matter of increasing concern to toxicologists. To return to the example of lead, as previously stated, all of us are presently carrying levels of lead far above natural background levels. Over 90% of this is stored in bone[6], available for mobilization during pregnancy and lactation[64]. It is established that prenatal exposure produces behavioral deficits, as does exposure during infancy and early childhood (see next chapter). Bone density often decreases in old age, particularly after menopause in women; this resorption of bone may result in the mobilization of stored lead, thereby re-exposing the aging individual to increased lead levels. The effects of this "cradle to grave" exposure are unknown. To take another example, environmental contaminants such as dioxin, PCBs, and DDT mimic some of the effects of the natural hormone estrogen, potentially affecting reproductive function or producing other unpredictable effects[15]. The synthetic estrogen diethylstilbesterol (DES), given to pregnant women to prevent miscarriage, produced cancer in daughters and reproductive problems in sons when they reached adulthood.

The recognition that effects following high-dose acute exposure may be qualitatively different from those following longer exposure to lower doses must be considered in testing strategies designed to identify and characterize neurotoxicity using animal models. It has been suggested that in the first tier of testing, which is aimed at identifying acute hazards, a short exposure period and a wide range of doses be utilized[85]. Second and third tier testing would use repeated or continuous exposure. This dosing strategy would allow the characterization of delayed neurotoxicity, observation of the development of tolerance, and characterization of the reversibility of effects.

4.3   DELAYED NEUROTOXICITY

The term "delayed neurotoxicity" has been used to refer to several different processes. It is well known that some organophospherous (OP) pesticides produce delayed neurotoxicity days to weeks after a single exposure. This is the result of axonal degeneration in the peripheral nervous system and selected tracts of the central nervous system, resulting in weakness or paralysis[61]. The initiation of this effect is the result of inhibition of a particular esterase system, dubbed "neurotoxic esterase". OP-induced delayed neurotoxicity is replicable in animal models; the chicken is particularly sensitive.

An issue that is of considerable concern to neurotoxicologists is the potential for agents to produce delayed neurotoxicity years after cessation of exposure, or as a result of low-level exposure over a large portion of the lifespan. The possibility of an interaction between aging and exposure to neurotoxic agents is also of critical concern: this possibility was postulated two decades ago[88] and has been raised repeatedly since[84, 85, 87]. As the normal brain ages, there is a decrease in the number of cells in certain regions, as well as a decline in neurotransmitter levels and repair mechanisms. If this process were accelerated by chronic or historic exposure to a neurotoxicant, the effect as the individual ages would be a marked decrease in functional capacity (Figure 11.6).

Alternatively, damage that decreased the reserve capacity of the brain at any point in life might also hasten the appearance of functional deficits. For example, it has become apparent over last decade that up to 25% of people who contracted polio as children, apparently recovered and functioning normally, suffer a recurrence of clinical symptoms as they moved into middle age. There is evidence of degeneration in peripheral nerves of these individuals; this may reflect premature senescence of nerves compromised during the acute phase of the disease, or accelerated aging produced by increased metabolic demands on spared normal nerves. Polio may also offer evidence of a different sort of delayed neurotoxicity. Researchers in Great Britain compared the local incidence of polio in the 1930s with the incidence of motoneurone disease in the 1970s[33]. They observed a significant correlation, which they interpreted as a delayed consequence of polio resulting from loss of motor neurons that was not severe enough to cause motor symptoms at the time of the acute illness. Age-related diseases may also have as a contributing component past exposures to environmental toxicants. It has been hypothesized that various agents contribute to Alzheimer's disease (AD), Parkinson's disease (PD), and motoneurone disease (amyotrophic lateral sclerosis or ALS)[13]. As many as 80% of the neurons in the substantia nigra must be destroyed before the symptoms of Parkinson's disease become apparent. Acceleration of the aging process of the brain might result in the critical level of cell loss being reached in individuals who otherwise might not have exhibited impairment. The production of a Parkinson-like syndrome by the synthetic heroin contaminant MPTP suggests that exogenous substances may play a role in at least some cases of idiopathic Parkinson's disease. Severe cognitive dysfunction similar to AD has also resulted from occupational exposure to solvents, metal vapors[21,] or aluminum[95]. A toxin in the seeds of the cycad plant may be related to the syndrome of amyotrophic lateral sclerosis/parkinsonism dementia (ALS/PD) in various sites in the Western Pacific[61]. In some cases, victims were affected as long as 30

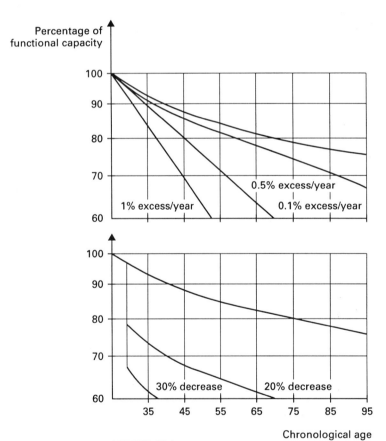

FIGURE 11.6

Delayed neurotoxicity

Top: The percent of functional capacity vs. chronological age associated with different degrees of acceleration in the decline of the functional capacity of the brain, beginning at 25 years of age. Bottom: Functional capacity vs chronological age as affected by a 20 or 30 percent loss of capacity at 30 years of age and normal aging thereafter. (Adapted from Weiss, B. and Simon, W. (1975) In: *Behavioral Toxicology*, Weiss, B. and Laties, V.C. (Eds.), Plenum Press, New York, p 429. With permission.)

years after ingestion of the seeds had presumably ended. Increased prevalence of ALS has also been linked to environmental lead exposure[58].

The neurotoxic agent for which evidence for delayed neurotoxicity is strongest is methylmercury. Two decades ago it was revealed that mice exposed developmentally to methylmercury, displaying no overt signs when young, began to display signs of neurotoxicity and other impairments as they aged[62]. Effects observed in individual animals included kyphosis, muscular atrophy, neuromuscular deficits, impaired immune function, and obesity. Delayed neurotoxicity has also been observed in monkeys exposed to methylmercury from birth to young adulthood[49]. When these monkeys reached middle age, six years after cessation of dosing, they began exhibiting clumsiness not present when they were younger. Follow-up examination revealed decreased response to light touch or pinprick in hands and feet, and a longer response time to retrieve raisins from a series of square compartments. Objective assessment of vibration thresholds in the finger tip revealed decreased sensitivity across a range of frequencies[53]. These effects may represent an interaction of methylmercury exposure and the aging process, even though monkeys were only middle aged when overt signs of toxicity became apparent. The most compelling evidence for accelerated aging as a result of neurotoxic exposure comes from a study of persons who had been diagnosed with Minamata disease, some as long as three decades previously[29]. 1144 persons over 40 years of age were compared with an equal number of matched controls. One aspect of the study included an inventory of activities of daily living, including the ability to eat, bathe, dress, wash the face, and use the toilet independently. There was an age-related decrement in the ability of people with Minamata disease to perform these functions (Figure 11.7). These results are extremely important, as they represent concrete evidence for accelerated aging produced by neurotoxic exposure much earlier in life.

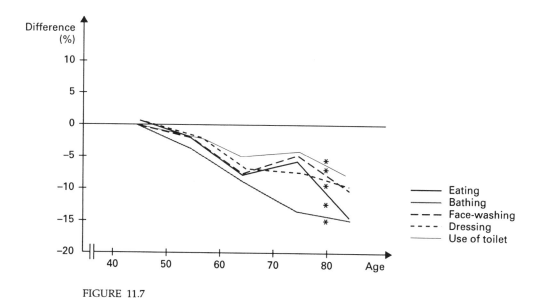

FIGURE 11.7

Difference between controls and persons diagnosed with Minamata disease, in activities of daily living (ADLs) as a function of age

Ordinate: the difference in ADL activity (subjects in no need of assistance) for persons with Minamata disease minus controls (%). (From Kinjo, Y., Higashi, H., Nakano, A., Sakamoto, M., Sakai, R. (1993) *Environ. Res.* **63**: p 241. With permission.)

4.4   SUSCEPTIBLE POPULATIONS

Particular subsets of the population may be differentially sensitive to adverse effects produced by neurotoxic agents. The developing organism and aging populations may be more sensitive than the young adult. Gender or other genetic factors may also influence susceptibility to neurotoxic damage, as may nutritional and health status.

The developing organism (fetus, infant, and child) may be particularly sensitive for a number of reasons. The developing brain is undergoing cell differentiation and migration, synaptogenesis, and subsequent pruning of the synaptic tree in response to ongoing neuronal activity. These processes may be selectively affected by exposure to neurotoxic agents[55]. For example, exposure to methylmercury interrupts normal cell migration[11]. Such effects, specific to the developing nervous system, may be at least partly responsible for the fact that the types of neurotoxicity observed as a result of developmental exposure may be quite different from those observed in adults. The developing organism may also be more vulnerable because of differences from the adult in absorption and other toxicokinetic factors. Both methylmercury and PCBs are readily excreted in breast milk[24, 76], with the result that the infant may receive large doses of these toxicants. In a study of the half-life of methylmercury in blood in women in Iraq, lactating women had half-lives that were 55% of those of non-lactating women, presumably because much of the mercury was being

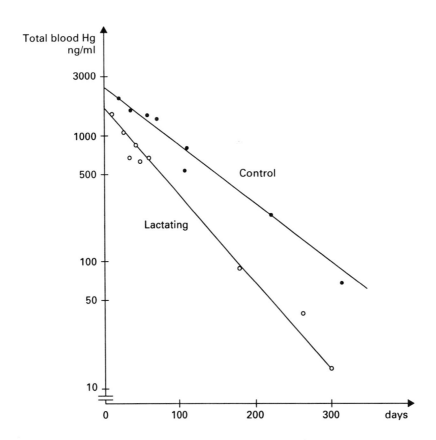

FIGURE 11.8

Half-lives of methylmercury in blood in lactating and non-lactating women
Lactating women excrete methylmercury more quickly, presumably in the milk. (From Greenwood, M.R., Clarkson, T.W., Doherty, R.A., Gates, A.H., Amin-Zaki, L., Elhassani, S., Majeed, M.A. (1978) *Environ. Res.* **16**: p 48. With permission.)

excreted into milk[24] (Figure 11.8). Similarly, a nursing woman may decrease her own PCB stores by 30–50%[5]. Young children are especially vulnerable to undue exposure from environmental lead because they absorb a much larger fraction by the gastrointestinal route than do adults (90% vs 10%)[69]. In addition, lead in dust and soil represents a significant exposure source because young children put their hands and other objects into their mouths. These factors contribute to the generally observed pattern of blood lead levels in humans peaking at around two years of age[69].

The aging population may also represent a group at special risk. Changes in brain function including decreased cell density and compromised repair mechanisms decrease the capacity of the brain to protect against or recover from toxic injury. Impairment of function in other organ systems such as kidney, liver, or GI tract may result in increased absorption, decreased metabolic capacity, and slower excretion. Many of these same issues are also relevant for members of the population with acute or chronic illness. In addition, drugs may change metabolism, tissue distribution, and absorption or excretion of other agents, including environmental toxicants.

Nutritional status may also influence toxicity. Selenium may protect against some aspects of methylmercury toxicity, although the mechanism is not established[14]. Absorption of lead is affected by various dietary constituents including particularly calcium, iron, vitamins C, D, and E, and fat[69]. Many children, especially those in the inner city who may live in environments with high lead concentrations, consume diets deficient in calcium, iron, and vitamins. In additions, iron deficiency itself produces neurotoxic effects, which may be exacerbated by elevated lead exposure.

Individual genetic variability may also be an important determinant of toxic effect, but one which is difficult to address, particularly in the human population. Perhaps the easiest difference to identify is variations in toxicokinetics. It has been understood for many years that both the therapeutic and toxic effects of drugs may be influenced by metabolic factors. Isoniazide, a drug used for many years to treat tuberculosis, may produce peripheral

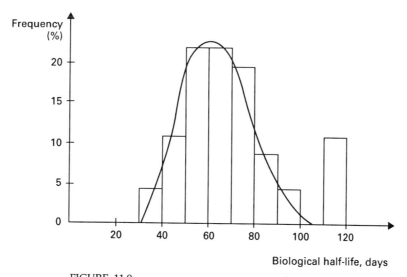

FIGURE 11.9

Distribution of biological half-lives of methylmercury in adult humans
(From Al-Shahristani, H. and Shihab, K.M. (1974) *Arch.Environ. Health* **28**: p 139. With permission from the Helen Dwight Reid Educational Foundation, 1319 Eighteenth St. N.W., Washington, D.C. 20036-1802. ©1974.

neuropathy and other side effects. An inability to metabolize the drug rapidly is associated with an increased risk for such toxicity. The ability to rapidly add an acetyl group to the drug is dependent upon a single gene, which has a differential distribution according to race: almost 100% of Inuit are rapid acetylators, as are 90% of Japanese, but only 40–45% of African and European individuals[61]. For reasons as yet undefined, the biological half-live of methylmercury in adults has a bimodal distribution: 95% of people have half-lives normally distributed with a mean of about 70 days, while a small subset of the population has half-lives of 110–120 days[2] (Figure 11.9). The contribution to neurotoxicity of other genetic variables, such as differences in neurochemistry or neuroanatomy, have typically not been specifically explored, even in animal models.

Gender-related differences in susceptibility have also been demonstrated. The brain is a sexually dimorphic structure, at least in some specific brain areas, so it is not surprising that males and females may respond differentially to toxic insult. Male mice showed more mitotic arrest and cerebellar loss in response to methylmercury than did females even though they did not have higher brain mercury concentrations[56]. Development effects of methylmercury on motor tests were more pronounced in boys than girls in one study[34]. Differential effects of gender on various behavioral scores as a result of childhood lead exposure have also been observed[31].

4.5   STATISTICAL ISSUES

Two inter-related issues that must be considered in interpretation of a study or a body of literature are the reliance on statistical significance as an indication of a "real" effect, and the correspondence between statistical significance and biological relevance. In studies performed on a very large number of individuals, an effect may be highly statistically significant but of no biological relevance. Usually, however, the problem is the opposite: the sample may not be large enough to detect a small or even moderate effect even if it exists. The power of a study to detect an effect of a particular size may be calculated, and has received attention in epidemiological research[23] (with respect to neurotoxicity produced by developmental lead exposure). This issue is often not taken into account in studies in which negative results (often defined as $\alpha >$ 0.05) are reported, however. The issue of insufficient statistical power is almost universally ignored with respect to animal research, despite the fact that experimental studies are performed with relatively small numbers of subjects. The "solution" to this problem is usually to choose doses sufficiently high to produce an effect in the face of little statistical power, a strategy which has implications for interspecies extrapolation. One way of addressing this issue is the use of the meta-analysis, which is a technique for combining results from a number of independent studies to achieve reasonable statistical power to detect an effect if one is present.

Another way to address this same issue is to calculate the degree of difference that must be produced by a treatment in order to be detected in a particular experiment. This coefficient of detection may be calculated after the experiment is completed, and is dependent upon the number of subjects, the variability of the data, and the significance level chosen (usually 0.05). In a collaborative behavioral teratology study in which six independent laboratories participated, coefficients of detection for most measures ranged between 10 and 50%, suggesting that the studies as designed were adequate to detect a moderate treatment effect if one existed[10]. On the other hand, performance on an intermittent schedule of reinforcement by a group of lead-treated monkeys and controls[54] required at least a 400% difference in performance between the groups in order to be detected[47], given the number of animals (four per group) and variation between animals. This is a huge difference, such that if no effect had been detected, these results would require cautious interpretation.

An issue that has created much controversy in the field of neurotoxicology is the apparent failure to recognize that a small effect may be an important effect. Many etiological factors may contribute to a particular adverse health effect. This may be particularly true in developmental studies, where measures such as early neurological development and later measures of intelligence are correlated with such variables as birth order, parental IQ, socioeconomic status, maternal nutrition, and drug use during pregnancy, to name but a few. It is not surprising that the addition of a toxicant, particularly at low levels, may result in a consistent but small effect. Again, lead is the best example available for illustrative purposes because the consequences of developmental exposure to lead have been extensively studied in the human population. Blood lead levels in children have been found to be inversely correlated with IQ in a number of studies. A recent meta-analysis revealed a decrement of about three points in IQ for every 10 μg/dl increase in blood lead level[59]. This has been referred to as a small effect, particularly since lead may predict only a few percent of the variance in IQ between children. (Other factors mentioned above predict most of variance from known sources.) However, what must be recognized is that a small shift in IQ of the *population* represents an effect of enormous magnitude (Figure 11.10).

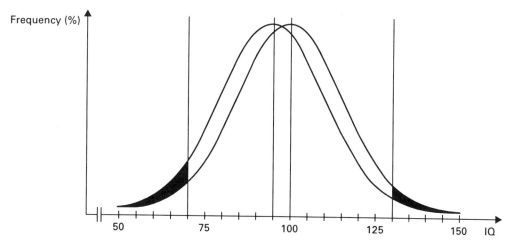

FIGURE 11.10

Normal distributions of IQ with a mean of 100 (right curve) and shifted to the left by 5 points (left curve)

This shift has an enormous effect on the tails of the distribution, greatly decreasing the number of people with high IQs and increasing those with low IQs. The filled area on the right tail of the distribution represents the decrease in the fraction of people with IQs above 130, while the filled area on the left tail represents the increase in the fraction of people with IQs below 70.

IQ tests are designed to yield a normal distribution, with a mean of 100 and a standard deviation of 15 points. In a population of 100 million children, a shift of five points for the population would result in a decrease of children with IQs above 130 from 2.3 million to 990 thousand, with a concomitant increase in the number of children with IQs below 70. The cost to society of the loss of exceptionally bright people will become increasingly important as the basis of the economy shifts from manufacturing to service and high-technology. The cost to society of the intellectual impairment as a result of the ubiquitous

337

exposure to lead has been estimated in the hundreds of millions of dollars per year[71]. The example of lead serves as a warning against complacency regarding the exposure of large segments of the population to chemicals for which the neurotoxic potential may be unknown.

5   Conclusion

It is clear that exposure to certain chemicals present in the environment has resulted in neurotoxicity; the potential for many other agents present in the environment to produce neurotoxicity is unknown. It is also apparent that the developing organism is often more sensitive to the effects of neurotoxic agents than is the young adult. The degree to which the aging individual represents a particularly susceptible population is less clear. Of critical concern to toxicologists and regulatory agencies is the potential for delayed neurotoxicity or accelerated aging as a result of exposure to toxic agents earlier in life or over a significant portion of the lifespan. Answers to many of these questions must rely on identification and characterization of neurotoxicity using animal models. The confidence with which data from animal research can be extrapolated to humans depends on the comparability in toxicokinetics, metabolism, physiology and anatomy between the animal model and humans, as well as how well the chosen behavioral paradigm(s) assesses the behavioral domain(s) which may be affected in humans. Since many of these variables will be incompletely understood, regulatory agencies have adopted various safety factors in the risk assessment process designed to protect human health. We still have much to learn about the contribution of environmental agents to human disease, and the best methods to protect the public from neurotoxic damage.

References

1.   Adams, J. and Buelke-Sam. J. (1991) Behavioral assessment of the postnatal animal: testing and methods development. In: Kimmel, C.A., Buelke-Sam, J. (Eds.) *Developmental Toxicology*. Raven Press, New York, p 233.

2.   Al-Shahristani, H. and Shihab, K.M. (1974) Variation in biological half-life of methylmercury in man. *Arch Environ Health* **28**: p 139.

3.   Anger, W.K. (1990) Human neurobehavioral toxicity testing. In: Russell, R.W., Flattau, P.E., Pope, A.M. (1990) *Behavioral Measures of Neurotoxicity*. National Academy Press, Washington, D.C., p 69.

4.   Anger, W.K. (1990) Worksite behavioral research. *Neurotoxicology* **11**: p 629.

5.   Anger, W.K., Moody, L., Burg, J., Brightwell, W.S., Taylor, B.J., Russo, J.M., Dickerson, N., Setzer, J.V., Johnson, B.L., Hicks, K. (1986) Neurobehavioral evaluation of soil and structural fumigators using methyl bromide and sulfyl fluoride. *Neurotoxicology* **7**: p 137.

6.   Barry, P.S.I. (1975) A comparison of concentrations of lead in human tissue. *Br. J. Indust. Med.* **32**: p 119.

7.   Bellinger, D., Leviton, A., Waternaux, C., Needleman, H., Robinowitz, M. (1987) Longitudinal analyses of prenatal and postnatal lead exposure and early cognitive development. *N. Engl. J. Med.* **316**: p 1037.

8.   Bellinger, D., Needleman, H., Leviton, A., Waternaux, C., Robinowitz, M., Nichols, M. (1984) Early sensory-motor development and prenatal exposure to lead. *Neurobehav Toxicol Teratol* **6**: p 387.

9.   Bornschein, R., Succup, P., Dietrich, K. (1987) Prenatal lead exposure and pregnancy outcomes in the Cincinnati lead study. In: Lindberg, S. and Hutchinson, T. (Eds.), *Proceedings of the 6th International Conference on Heavy Metals in the Environment*, Vol 1. CEP Consultants, Ltd., Edinburgh, p 156.

10. Buelke-Sam, J., Kimmel, C.A., Adams, J. (Eds.) (1985) Design considerations in screening for behavioral teratogens: results of the collaborative behavioral teratology study. *Neurobehav. Toxicol. Teratol.* **7(6)**.

11. Burbacher, T.M., Rodier, P.M., Weiss, B. (1990) Methylmercury developmental neurotoxicity: a comparison of effects in humans and animals. *Neurotoxicol. Teratol.* **12**: p 191.

12. Cabak, V. and Najdanvic, R. (1965) Effects of undernutrition in early life on physical and mental development. *Arch. Dis. Child* **40**: p 532.

13. Calne, D.B., Eisen, A., McGeer, E., Spencer, P.S. (1986) Alzheimer's disease, Parkinson's disease and motoneurone disease: a biotrophic interaction between aging and the environment? *Lancet* **II**: p 1067.

14. Chang, L.W., Dudley, A.W., Dudley, M.A., Ganther, H.E., Sunde, M.L. (1977) Modification of neurotoxic effects of methylmercury by selenium. In: Raisin, L., Shiraki, H., Groevic, N. (Eds.), *Neurotoxicology*, Vol. 1. Raven Press, New York, p 275.

15. Colborn, T. E. and Expert Group (1990) *Great Lake, Great Legacy*, The Conservation Foundation, Washington, D.C. and the Institute for Research on Public Policy, Ottawa, Ontario.

16. Cory-Slechta, D.A. (1995) Neurotoxicant-induced changes in schedule-controlled behavior. In: Chang, L.W. (Ed.) *Principles of Neurotoxicity*. Marcel Dekker, New York, p 313.

17. Dietrich, K., Krafft, K., Bier, M.. Succop, P., Berger, O., Bornschein, R. (1986) Early effects of fetal lead exposure: neurobehavioral findings at 6 months. *Int. J. Biosocial Res.* **8**: p 151.

18. Ehle, A.L. and McKee, D.C. (1990) Neuropsychological effect of lead in occupationally exposed workers: a critical review. *Crit. Rev. Toxicol.* **20**: p 237.

19. Evans, H.L., Garman, R.H., Weiss, B. (1977) Methylmercury: exposure duration and regional distribution as determinants of neurotoxicity in nonhuman primates. *Toxicol. Appl. Pharmacol.* **41**: p 15.

20. Fagan, J.F. and McGrath, S.K. (1981) Infant recognition memory and later intelligence. *Intelligence* **5**: p 121.

21. Freed, D.M. and Kandel, E. (1988) Long-term occupational exposure and the diagnosis of dementia. *Neurotoxicology* **9**: p 391.

22. Gad, S.C. (1982) A neuromuscular screen for use in industrial toxicology. *J. Toxicol. Environ. Health* **9**: p 691.

23. Gatsonis, C.A. and Needleman, H.L. (1992) Recent epidemiological studies of low-level lead exposure and the IQ of children: a meta-analysis review. In: Needleman, H.L. (Ed.) *Human Lead Exposure*. CRC Press, Boca Raton, FL, p 243.

24. Greenwood, M.R., Clarkson, T.W., Doherty, R.A., Gates, A.H., Amin-Zaki, L., Elhassani, S., Majeed, M.A. (1978) Blood clearance half-times in lactating and nonlactating members of a population exposed to methylmercury. *Environ. Res.* **16**: p 48.

25. Irwin, S. (1968) Comprehensive observational assessment in a systematic quantitative procedure for assessing the behavioral and physiological state of the mouse. *Psychopharmacology* **13**: p 222.

26. Iverson, S.D. and Iverson, L.L. (1981) *Behavioral Pharmacology*, 2nd edition. Oxford University Press, New York.

27. Iwata, K. (1980) Neuroophthalmologic indices of Minamata disease in Niigata. In: Merigan, W.H. and Weiss, B. (Eds.), *Neurotoxicology of the Visual System*. Raven Press, New York, p 165.

28. Kelleher, R.T. and Morse, W.H. (1969) Determinants of the behavioral effects of drugs. In: Tedeschi, D.J. and Tedeschi, R.E. (Eds.), *Importance of Fundamental Principles in Drug Evaluation*. Raven Press, New York, p 383.

29. Kinjo, Y., Higashi, H., Nakano, A., Sakamoto, M., Sakai, R. (1993) Profile of subjective complaints and activities of daily living among current patients with Minamata disease after 3 decades. *Environ. Res.* **63**: p 241.

30. Kopin, I.J. and Markey, S.P. (1988) MPTP toxicity: Implications for research in Parkinson's disease. *Annu. Rev. Neurosci.* **11**: p 81.

31. Leviton, A., Bellinger, D., Allred, E.N., Robinowitz, M., Needleman, H., Schoenbaum, S. (1993) Pre- and postnatal low-level lead exposure and children's dysfunction in school. *Environ. Res.* **60**: p 30.

32. Magos, L. (1987) The absorption, distribution, and excretion of methylmercury. In: Eccles, C.U. and Annau, Z. (Eds.), *The Toxicity of Methylmercury*. Johns Hopkins, Baltimore,MD, p 22.

33. Martyn, C.N., Barker, D.J.P., Osmond, C. (1988) Motor neurone disease and past poliomyelitis in England and Wales. *Lancet* **1**: p 1319.

34. McKeown-Eyssen, G.E., Ruedy, J., Neims, A. (1983) Methylmercury exposure in northern Quebec. II. Neurological findings in children. *Am. J. Epidemiol.* **118**: p 470.

35. McMichael, A., Vimpani, G., Robertson, E., Baghurst, P., Clark, P. (1986) The Port Pirie cohort study: maternal blood lead levels and pregnancy outcome. *J. Epidemiol. Community Health* **40**: p 18.

36. McMillan, D. (1990) The pigeon as a model for comparative behavioral pharmacology and toxicology. *Neurotoxicol. Teratol.* **12**: p 523.

37. Mergler, D. and Blain, L. (1987) Assessing color vision among solvent-exposed workers. *Am. J. Indust. Med.* **12**: p 195.

38. Merigan, W.H. (1980) Visual fields and flicker thresholds in methylmercury-poisoned monkeys. In: Merigan, W.A. and Weiss, B. (Eds), *Neurotoxicology of the Visual System*. Raven Press, New York, p 149.

39. Merigan, W.H. and Weiss, B. (Eds.) *Neurotoxicology of the Visual System*. Raven Press, New York.

40. Mitchell, C.L. and Tilson, H.A. (1982) Behavioral toxicology in risk assessment: problems and research needs, *Crit. Rev. Toxicol.* **10**: p 265.

41. Mitchell, C.L., Tilson, H.A., Cabe, P.A. (1982) Screening for neurobehavioral toxicity: factors to consider. In: Mitchell, C.L. (Ed.), *Nervous System Toxicology*. Raven Press, New York, p 237.

42. Moser, V. (1989) Screening approaches to neurotoxicity. A functional observation battery. *J. Am. Coll. Toxicol.* **8**: p 85.

43. Needleman, H.L., Gunnoe, C., Leviton, A., Reed, R., Peresie, H., Maher, C., Barret, P. (1979) Deficits in psychologic and classroom performance of children with elevated dentine lead levels. *N. Engl. J. Med.* **300**: p 689.

44. O'Connell, M., Yeaton, S.P., Stroebel, D.A. (1978) Visual discrimination in the protein malnourished rhesus. *Physiol. Biochem. Behav.* **20**: p 251.

45. Peters, D.P. (1979) Effects of prenatal nutrition on learning and motivation in rats. *Physiol. Behav.* **22**: p 1067.

46. Plèstina, R. and Piukovíc-Plèstina, M. (1978) Effects of anticholinesterese pesticides on the eye and on vision. *Crit. Rev. Toxicol.* **1**.

47. Rice, D.C. (1980) Neurotoxicity and behavior: behavioral aberrations. *Toxicol. Forum* **183**.

48. Rice, D.C. (1989) Brain and tissue levels of mercury after chronic methylmercury exposure in the monkey. *J. Toxicol. Environ. Health* **27**: p 189.

49. Rice, D.C. (1989) Delayed neurotoxicity in monkeys exposed developmentally to methylmercury. *Neurotoxicology* **10**:p 645.

50. Rice, D.C. (1993) Lead-induced changes in learning: evidence for behavioral mechanisms from experimental animal studies. *Neurotoxicology* **14**: p 167.

51. Rice, D.C. (1995) Behavioral effects of lead: commonalities between experimental and epidemiological data. *Environ. Health Perspect.* **103** (Suppl. 9).

52. Rice, D.C. and Gilbert, S.G. (1990) Effects of developmental exposure to methylmercury on spatial and temporal visual function in monkeys. *Toxicol. Appl. Pharmacol.* **102**: p 151.

53. Rice, D.C. and Gilbert, S.G. (1995) Effects of developmental methylmercury exposure or lifetime lead exposure on vibration sensitivity function in monkeys. *Toxicol. Appl. Pharmacol.* **134**. p 161.

54. Rice, D.C., Gilbert, S.G., Willes, RF. (1979) Neonatal low-level lead exposure in monkeys: locomotor activity, schedule-controlled behavior, and the effects of amphetamine. *Toxicol. Appl. Pharmacol.* **51**: p 503.

55. Rodier, P.M. (1986) Time of exposure and time of testing in developmental neurotoxicology. *Neurotoxicology* **7**: p 69.

56. Sager, P.R., Aschner, M., Rodier, P.M. (1984) Persistent, differential alterations in developing cerebellar cortex of male and female mice after methylmercury exposure. *Dev Brain Res* **12**: 1.

57. Savage, E.P., Keefe, T.J., Mounce, L.M., Heaton, R.K., Lewis, J.A., Burcar, P.J. (1988) Clinical neurological sequelae of acute organophosphate pesticide poisoning. *Arch. Environ. Health* **43**: p 38.

58. Scarpa, M., Columbo, A., Panzetti, P., Sorgato, P. (1988) Epidemiology of amyotopic lateral sclerosis in the province of Modina, Italy. Influence of environment lead exposure. *Acta Neurol. Scand.* **77**: p 456.

59. Schwartz, J. (1994) Low-level lead exposure and children's IQ: a meta-analysis and search for a threshold. *Environ. Res.* **65**: p 42.

60. Sobotka, T.J., Cook, M.P., Brodie, R.E. (1976) Effects of neonatal malnutrition and perinatal exposure to various pesticides. *Mater. Med. Pol.* (English ed.) **8**: p 152.

61. Spencer, P.S. and Schaumburg, HH. (Eds.) (1980) *Experimental and Clinical Neurotoxicology.* Williams & Wilkins, Baltimore, MD.

62. Spyker, J.M. (1975) Assessing the impact of low level chemicals in development: behavioral and latent effects. *Fed. Proc.* **34**: p 1835.

63. Tephly, T.R. (1991) The toxicity of methanol: minireview. *Life Sci.* **48**: 1031.

64. Thompson, G.N., Robertson, E.F., Fitzgerald, S. (1985) Lead mobilization during pregnancy. *Med. J. Aust.* **143**: p 131.

65. Tilson, H.A. (1981) The neurotoxicity of acrylamide: an overview. *Neurobehav. Toxicol. Teratol.* **3**: p 445.

66. Tilson, H.A., MacPhail, R.C., Crofton, K.M. (1995) Defining neurotoxicity in a decision-making context. *Neurotoxicology* **16**: p 363.

67. Tilson, H.A. (1990) Neurotoxicology in the 1990s. *Neurotoxicol. Teratol.* **12**. p 293.

68. Uhrich, D.J., Essock, E.A., Lehmkuhle, S. (1981) Cross-species correspondence of spatial contrast sensitivity functions. *Behav. Brain Res.* **2**: p 291.

69. U.S. Agency for Toxic Substances and Disease Registry (1988) *The Nature and Extent of Lead Poisoning in Children in the United States: A Report to Congress.* U.S. Department of Health and Human Services, Washington. D.C.

70. U.S. Environmental Protection Agency (1982) Regulation of fuel and fuel additives, *Fed. Reg.* **47**: 7812.

71. U.S. Environmental Protection Agency (1985) *Costs and Benefits of Reducing Lead in Gasoline: Final Regulatory Impact Analysis.* EPA-230-05-85-006. Washington, D.C.

72. U.S. Environmental Protection Agency (1986) *Air Quality Criteria for Lead*, EPA/8-83-028bF, U.S. EPA, Washington, D.C.

73. U.S. Environmental Protection Agency (1988) *Fed. Reg.* **53**: p 5932.

74. U.S. Environmental Protection Agency (1989) *Fed. Reg.* **54**: p 13472.

75. U.S. Environmental Protection Agency (1991) Manganese and methylcyclopentadienyl manganese tricarbonyl (MMT) conference. Summary.

76. U.S. Environmental Protection Agency (1993) *Workshop Report on Developmental Neurotoxic Effects Associated with Exposure to PCBs*, EPA/630/R-92/004, Washington, D.C.

77. U.S. Environmental Protection Agency (1994) Principles of neurotoxicity risk assessment. *Fed. Reg.* **59**: p 42360.

78. U.S. Environmental Protection Agency, Office of Water (1995) *National Forum on Mercury in Fish*, Washington, D.C.

79. U.S. Environmental Protection Agency, Office of Toxic Substances (1985) Toxic substances control act test guidelines: final rule. 40 CFR part 798: Health effect test guidelines. Subpart G: Neurotoxicity. *Fed. Reg.* **50**: p 39458.

80. U.S. Environmental Protection Agency, Office of Toxic Substances (1986) Triethylene glycol monomethyl, monoethyl, and monobutyl ethers: Proposed test rule. Developmental neurotoxicity screen. *Fed. Reg.* **51**: p 17890.

81. U.S. Food and Drug Administration (1986) *Predicting Neurotoxic and Behavioral Dysfunction from Preclinical Toxicologic Data.* Life Sciences Research Office, Bethesda, MD.

82. U.S. National Academy of Sciences (1975) *Principles for Evaluating Chemicals in the Environment.* National Academy of Sciences, Washington, D.C.

83. U.S. National Academy of Sciences (1980) *Lead in the Human Environment*, Washington, D.C.

84. U.S. National Research Council (1992) *Environment Neurotoxicology*. National Academy Press, Washington, D.C.

85. U.S. Office of Technology Assessment (1990) *Neurotoxicity: Identifying and Controlling Poisons in the Nervous System*. U.S. Government Printing Office, Washington. D.C.

86. Vorhees, C.V. (1986) Methods for assessing the adverse effects of foods and other chemicals on animal behavior. In: Olson, R.E. (Ed.), *Nutrition Reviews Diet and Behavior: A Multidisciplinary Evaluation*. Vol. 44. Nutrition Foundation, Washington, D.C., p 185.

87. Weiss, B. (1990) Risk assessment: The insidious nature of neurotoxicity and the aging brain. *Neurotoxicology* **11**: p 305.

88. Weiss, B. and Simon, W. (1975) Quantitative perspectives on the long-term toxicity of methylmercury and similar poisons. In: Weiss, B. and Laties, V.C. (Eds.), *Behavioral Toxicology*. Plenum Press, New York, p 429.

89. Williamson, A.M. and Teo, R.K.C. (1986) Neurobehavioral effects of occupational exposure to lead. *Br. J. Indust. Med.* **43**: p 374.

90. Winneke, G., Brockhaus, A., Ewers, U. et al. (1990) Results from the European multicenter study on lead neurotoxicity in children: implications for risk assessment. *Neurotoxicol. Teratol.* **12**: p 553.

91. Winneke, G., Brockhaus, A., Ewers, U., Kujanek, H., Lechner, H., Janke, W. (1983) Neuropsychologic studies in children with elevated tooth-lead concentration. II. Extended study. *Int. Arch. Occup. Environ. Health* **51**: p 231.

92. Winneke, G. and Kraemer, V. (1984) Neuropsychological effects of lead in children: interaction with social background variables. *Neuropsychobiology* **11**: 195.

93. World Health Organization (1986) *Environmental Health Criteria 60: Principles and Methods for the Assessment of Neurotoxicity Associated with Exposure to Chemicals*. World Health Organization, Geneva.

94. World Health Organization (1995) *Environmental Health Criteria on Inorganic Lead* (in press).

95. Yokel, R.A., Provan, S.D., Meyer, J.J., Campbell, S.R. (1988) Aluminum intoxication and the victims of Alzheimer's disease: similarities and differences. *Neurotoxicology* **9**: p 429.

Contents Chapter 12

Metals

Chapter 12

Metals

Gerhard Winneke and Hellmuth Lilienthal

This chapter describes both human epidemiological and experimental animal studies on the neurobehavioral toxicology of selected metals. Instead of covering the full literature in this field, which would be an impossible task, an effort will be made to use a selected number of typical examples to illustrate basic strategies, findings, and problems encountered in studying the neurobehavioral toxicity of lead, mercury, manganese, and aluminum, as well as the neurochemical and neuropathological background of such toxicity. Lead and mercury were selected because of the wealth of information available on behavioral effects in the developing and the adult nervous system, and manganese was chosen because behavioral alterations associated with exposure resemble those observed in Parkinson's disease. Aluminum is mentioned, because it is an abundant metal with documented neurotoxicity under special conditions.

I   Human studies

Epidemiological design issues will briefly be discussed before entering into a more detailed description and discussion of typical studies. Both cross-sectional as well as prospective studies have been and are being used in studying neurobehavioral effects of neurotoxic metals, particularly lead, in humans. Both approaches have their advantages and disadvantages. In cross-sectional studies neurobehavioral functions in exposed subjects are measured at one point in time and typically related to concurrent exposure either with respect to control (unexposed) subjects or, if the exposure gradient is sufficiently broad to cover low and high levels of exposure, with respect to one exposure axis using regression models. The advantage of this approach is that data is collected within a rather brief period of time, and that loss of subjects upon follow-up is avoided, so that the full sample is usually available for analysis. The distinct disadvantage of this approach is that the degree of comparability of control and exposed subjects in terms of extraneous variables is extremely crucial in order to be able to discuss results in cause-effect terms. In prospective or longitudinal study protocols associations are established between changes in neurobehavioral functions and the natural history of exposure. This approach offers the opportunity to identify critical periods of exposure and to clarify, if the observed deficit is persistent or reversible. In addition, problems related to the discussion of causality, particularly reverse causality, are more easily handled by means of prospective approaches. It remains true, however, that prospective study protocols require long-term and costly study-efforts, that repeated functional assessments are typically Subject to changes due to learning and memory-interactions, and that loss of subjects upon follow-up, which may even be selectively due to exposure, is likely to reduce the sample-size for final analysis.

1   Lead

Lead has been used since antiquity, and its detrimental health effects have been well known for centuries. It is probably the best studied toxic element, and comprehensive reviews covering chemical, environmental, and biomedical aspects (e.g., the recent Environmental Health Criteria (EHC) document of the World Health Organization[148]) should be consulted for more detailed information. From a chemical point of view this metal occurs both in organic and inorganic form, namely in the form of lead salts of widely different water solubilities. Although the organometallic lead species, due to their lipid solubility, have been found to be highly neurotoxic in acute occupational exposure, chronic low level exposure to inorganic lead constitutes the more serious public health issue. Inorganic lead enters the body by way of inhalation or ingestion; quantitatively uptake through the gastrointestinal tract is the more important route of exposure. Absorption is better in infants than in adults. Lead concentration in whole blood (PbB), normally expressed as μg 100 $\mu m l^{-1}$ or $\mu mol\ l^{-1}$, is a representative marker of current lead exposure, whereas tooth lead concentrations have also been used as markers of past exposure[108,153]. Both placental transfer and transfer across the blood-brain barrier occur easily in the developing brain, so that prenatal exposure as well as CNS involvement are possible. Lead serves no known physiological function. Its toxicity may largely be explained by interference with various enzyme systems; lead inactivates these systems by binding to sulfhydril (SH) groups or by competitive interaction with other essential metal ions. Therefore, almost all organs or organ systems are potential targets for lead; depending on the duration and degree of exposure, diverse biological effects have been described, the more critical of which are those on heme biosynthesis, erythropoiesis, and the nervous system.

Both the peripheral (PNS) and the central nervous system (CNS) are targets of lead, although PNS effects (wrist drop, slowing of nerve conduction velocities) have, so far, preferentially been described in occupational settings in adults. Lead encephalopathy has been reported to occur in cases of acute symptomatic lead poisoning in adults and in children and, in the latter case, appears to be associated with PbB exceeding 80–100 μg 100 ml^{-1} resulting from swallowing of lead-based paint. The clinical picture of lead encephalopathy is characterized by some or all of the following signs and symptoms[30]: coma, seizures, ataxia, apathy, incoordination, vomiting, clouded consciousness, and loss of previously acquired skills. Since neuropsychological sequelae are reported to occur in individuals surviving acute lead encephalopathy the hypothesis has been proposed that even asymptomatic lead exposure might be associated with subclinical neurobehavioural deficit which, due to its sublety, may often go undetected. The following text describes typical epidemiological studies in occupational and environmental exposure settings dealing with this basic hypothesis.

1.1   NEUROBEHAVIORAL OBSERVATIONS IN ADULTS

Cognitive as well as sensorimotor functions have typically been studied. In most studies cognitive functions are usually measured by standard intelligence tests, such as the "Wechsler Adult Intelligence Scale (WAIS)" or parts of such clinical tests. Despite their established psychometric qualities in terms of reliability, validity, and normative data bases such tests are of limited usefulness for studying cognitive functions in detail, because performance involves several subfunctions and it is, therefore, difficult to identify the basis for poor performance from a particular test score. For this reason, in some studies more specific tests for measuring specific cognitive functions have been used. Tests for the assessment of sensorimotor functions include different reaction time paradigms, tasks of sensorimotor speed and coordination, as well as tasks of sustained attention. Details of such tests will be given in the subsequent study descriptions.

1.1.1   Cross-sectional studies

Several studies have been conducted generally not showing neurobehavioral deficits to occur at blood lead levels below 40–50 μg 100 ml^{-1} [141,142,149]. From this set of studies those from Stollery et al.[141,142] will be described in more detail, because they: (1) studied more workers than most of the other studies (2) selected tasks on theoretical considerations based on models of cognitive information processing models rather than on purely clinical grounds, and (3) looked at three rather than only one marker of lead exposure.

Stollery et al.[141] examined 91 workers of mean age 41.5 years from battery and printing industries with low (< 20 μg 100 ml^{-1}), medium (21-40 μg 100 ml^{-1}), and high (41-80 μg 100 ml^{-1}) blood lead concentrations (PbB). Workers were given computer-based tests assessing sensorimotor serial reaction time, visual memory, attention, verbal reasoning, and spatial processing. In addition to PbB, zinc protoporphyrin (ZPP) as well as aminolevulinic acid in urine (ALA-U), were measured as biochemical responses to exposure. In the serial reaction time task subjects (Ss) extinguished one of five peripheral lights by touching the adjacent response disc (decision time) and started the next trial by touching a center disc (movement time). In the visual memory task up to six circles (the memory set) were shown at randomly selected positions within a large square on a video display. One second after erasure of the memory set, a question mark probe was presented, and the subject had to decide whether or not the memory probe was at a memory set position. In the verbal or syntactic reasoning task subjects were asked to verify statements claiming to describe the sequence of a pair of letters, for example "A is followed by B - AB" by true-false responses. In another verbal task, the category search task, subjects decided whether or not a series of 80 nouns, presented individually, belonged to the same semantic category (true-false). Later on subjects were given 2 minutes of written free recall to remember as many of the previously presented nouns as possible. In addition to these tests a mood checklist was given as well.

For the statistical analysis a number of non-lead (extraneous) variables, as, e.g., years of lead exposure, years of schooling, alcohol consumption, work demands, and stress at work, were controlled for by analysis of covariance, if they exhibited associations with outcome. Exposure response-associations were evaluated by means of multiple regression analysis. Significant exposure-related neurobehavioral deficit was found for decision and movement times, as well as for decision gap rates of the serial reaction time task and, although less compelling, for classification speed and accuracy in the category search task. In both cases only subjects with PbBs in excess of 40 μg 100 ml^{-1} exhibited functional deficit. It was concluded that the sensorimotor, rather than the cognitive requirements of the neurobehavioral tasks provided the most sensitive indicator of the effects of chronic occupational lead exposure, and that the effect on gap rates could be interpreted in terms of lead-related attentional difficulties. From the similarity of lead-related behavioral disruptions to those observed under the influence of chlorpromazine and amphetamine it is proposed that the dopaminergic system might be involved in producing these effects.

Stollery et al.[142] describe results from a follow-up study of the same cohort as above using the same battery of computer-based cogNitive tests. During the follow-up period of 8 months lead exposure had not changed so that the same groups of low (< 20 µg 100 ml⁻¹), medium (21-40 µg 100 ml⁻¹), and high (41–80 µg 100 ml⁻¹) lead exposure, respectively, remained, although the sample size was reduced to only 70 Ss. The results of the previous study were supported upon follow-up. Impairment tended to be restricted to subjects with PbBs above 40 µg 100 ml⁻¹. Again, the main deficit was a slowing of sensorimotor reaction time, which was most evident if the cognitive task-load was low. In the simple five choice serial reaction time-task, decision- and motor response-times as well as lapses of attention exhibited associations with blood lead levels. In the other cognitive task subjects of the high lead groups also tended to be slower, but the primary focus on cognitive rather than sensorimotor demands weakened exposure-effect associations. Subjects of the high lead groups also performed poorly in recalling those nouns from the previous semantic classification task, which were unrelated to the search category (unrelated distracters); this deficit increased over repeated sessions. From these findings the conclusion is drawn that the spectrum of lead-induced neurobehavioral impairment is characterized by sensorimotor slowing coupled with deficits in recalling incidental information.

1.1.2   Prospective studies

Apparently, only one prospective study has been conducted in occupationally exposed adults[93]. Twenty-four workers from a storage battery plant were studied over a period of 2 years with a battery of psychodiagnostic tests before and after entering lead-works, but only 16 of them remained in the study for final follow-up; 33 subjects from cable and power plants served as controls. Blood lead levels in the exposed group increased from 15.3 µg 100 ml⁻¹ to 30.5 µg 100 ml⁻¹ on the average over the 2-year-period, whereas those of the control subjects remained stable between 10.5 and 10.3 µg 100 ml⁻¹. The psychodiagnostic test battery was composed of tests measuring visual intelligence, namely Block design and Digit span from the Wechsler Adult Intelligence Scale (WAIS) (a visual reproduction test for the assessment of memory functions) the Bourdon-Wiersma test for the assessment of sustained attention, and the Santa-Ana sensorimotor coordination test. Although there was no initial performance deficit between the exposed and the control Ss, the latter group tended to improve upon follow-up, whereas the neurobehavioral performance of the exposed subjects continued to deteriorate more often than it improved from initial to final testing. Thus, group differences favoring control subjects were significant in five out of nine different performance measures. Despite the small number of Ss, the complex demand characteristics of the clinical measures, and the intrinsic interaction between learning-induced performance improvement due to repeated testing and exposure-induced functional impairment, the results of this study suggest that increasing average blood lead levels from 15 to 31 µg 100 ml⁻¹ is sufficient for neurobehavioral deficit to become apparent.

1.2   NEUROBEHAVIORAL OBSERVATIONS IN CHILDREN

Children rather than adults are considered a particularly sensitive risk group for lead because, on a body weight basis, their dietary intake, absorption, and total retention is markedly higher. In addition, the blood-brain barrier is not yet fully developed in young children, and specific behavioral characteristics, such as hand-to-mouth and playground activities, put them at higher risk. It should, furthermore, be mentioned that the developing rather than the mature brain is assumed to be particularly vulnerable to the effects of chemical exposures. For these reasons both cross-sectional and prospective epidemiological studies have been conducted to test the assumptions that children may be particularly at risk for neurobehavioral effects even of environmental lead exposure.

Blood lead concentrations (PbB) and/or lead concentrations in shed deciduous teeth (PbT) have been used as markers of lead exposure. Children with different degrees of environmental lead exposure, namely general population exposure due to lead-based paint, airborne lead from traffic, or lead in drinking water, or specific exposure due to lead-emitting industrial sources such as lead-zinc smelters, have already been studied in the early and mid-seventies using neuropsychological tests for the assessment of neurobehavioral deficit in the context of cross-sectional or clinical studies.

This review will, however, be restricted to more recent work with more sophisticated control for confounding variables. Confounders are variables which, due to their association with both the effect and the exposure measures, may give rise to spuriously high or low exposure/effect associations. Socioeconomic status, parental education, parental intelligence, and quality of the home environment are examples of potent confounders. Multiple linear regression analysis is typically being used as the statistical tool to control for confounding.

1.2.1   *Cross-sectional studies*

Following the pioneering work of Needleman et al.[108] several cross-sectional studies have primarily relied on PbT as a marker of long-term retrospective cumulative lead exposure[153, 113], whereas others have primarily, although not always exclusively, used blood lead concentrations as the measure to characterize internal exposure[54, 154, 67].

Both cognitive as well as sensorimotor outcome measures have been used to characterize lead-induced neurobehavioral deficit in these studies. Cognitive measures in most instances were partial or full-size clinical tests of intelligence such as the Wechsler Intelligence Scale for Children (WISC) or the Wechsler Preschool and Primary Scale of Intelligence (WPPSI), the British Ability Scales (BASC), and the Stanford-Binet Intelligence Scale. Measures of sensorimotor performance include different types of simple and choice reaction time-paradigms, tapping speed, and memory for design-tasks, such as the Benton test or the Bender Gestalt test. Two studies will be described in order to illustrate typical approaches and findings.

Out of 1210 eligible children, aged 6–9 years, 501 children from 18 primary schools in central Edinburgh were tested for correlations between blood lead, intelligence and school attainment[54]. The IQ measure was taken from the British Ability Scales, and the geometric mean PbB was 11.5 µg 100 ml^{-1} (range: 3.3-34.0). After adjustment for 31 potential confounders a significant negative association between PbB and BASC-IQ was found. In order to illustrate dose-response contingencies between PbB and IQ 10 groups of about 50 children each were formed on the basis of PbB; a dose-dependent relationship (Figure 12.1) was found with 5.8 IQ points difference between the lowest (5.6 µg 100 ml^{-1}) and the highest PbB-group (22.1 µg 100 ml^{-1}). No threshold was identified within the dose-effect curve. The effect-size to be explained by lead was only 0.9% of total variance relative to 45.5% of the full regression-model. Potent confounders within the model exhibiting pronounced relationship with outcome and strongest associations with blood lead levels were length, parental qualifications and social class, as well as mother's and father's performance in a vocabulary test.

Results from the European Multicenter Study were published by Winneke et al.[154] and combined eight individual studies from eight European countries (Bulgaria, Denmark, Greece, Germany, Hungary, Italy, Romania, Yugoslavia), tied together by a common study protocol with inherent quality assurance elements. Six to 11 year old children ($n = 1879$) were studied altogether, although for some of the tests the number of children was reduced to only 971. The range of PbB values was about 5–60 µg 100 ml^{-1}; a smaller number of PbT values was available in some of the studies, but was not evaluated in a systematic manner. Tests used in all or part of the individual studies included the full or partial WISC (four subtests), a serial choice reaction-time paradigm, the Bender-Gestalt test, the trail making test, as well as behavior rating-scales for teachers and parents. The overall statistical analysis was done using a uniform predetermined regression model with age, gender, occupational status of the father, and maternal education as confounders or covariates. The most salient findings of that study are given in Table 12.1. An inverse association between PbB and IQ was of only borderline significance ($P < 0.1$), and an IQ decrement of 3 points was estimated for a PbB increase from 5 to 20 µg 100 ml^{-1}, and of 5.3 points from 5 to 50 µg 100 ml^{-1}. The degree of inverse association between the error-scores of the Bender-Gestalt test and the serial choice reaction-paradigm were statistically significant and more consistent across individual studies, although the outcome-variance explained by lead never exceeded 0.8% of the total variance. No obvious threshold could be located on the dose-effect curves.

In summary, clinical measures of intelligence, due to their satisfactory psychometric quality, have been the preferred outcome measure in all of the cross-sectional studies although, as mentioned above, this global measure does not carry information about the specific cognitive functions primarily influenced by lead. Since, however, the degree of comparability of findings between studies is higher for the IQ measure relative to the different sensorimotor paradigms, efforts have been made to estimate the overall effect-size across studies by means of meta-analysis, concentrating on more recent blood lead-studies and on those studies which have given regression coefficients as

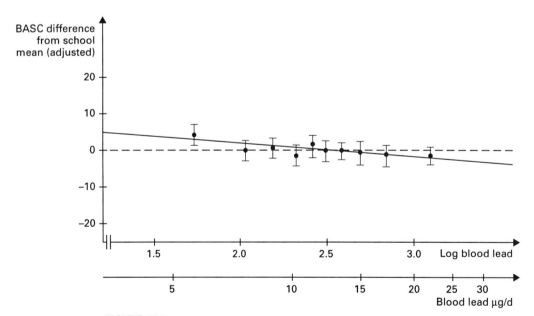

FIGURE 12.1

Results from the Edinburgh-study

The blood lead-variable (abscissa was subdivided into 10 categories of equal size. The mean (± CI95) confounder-adjusted BASC-score (British Ability Scales) is plotted for each PbB-category. The difference between the highest and the lowest category is 5.8 points . Values on the abscissa are natural logarithms. (Adapted from Fulton, M. et al. (1987) *Lancet* I: 1221–1226. With permission.)

TABLE 12.1

Summary of main results from the European Multicenter Study based on overall multiple regression analyses of the full set of data across studies

For the different variables the slopes of the regression lines (b), the *t*-values of testing for statistical significance, as well as the proportion of total variance explained by lead is given. Different sample sizes are due to the fact, that in some instances the full set of data was not available. (Modified from Winneke, G. et al., (1990) *Neurotoxicol. Teratol.*, 12: 553–559. With permission.)

Neurobehavioral endpoint		N of children	Slope of regression	Significance t-value	% Variance explained by PbB
WISC		1698	−0.53	−1.62*	0.19%
Bender Gestalt-Test (errors)[1]	easy	1698	1.64	2.17**	0.31%
	difficult	1584	2.03	2.54***	0.46%
Reaction performance (errors)	easy	971	0.15	1.68**	0.37%
	difficult	971	0.20	2.37***	0.66%
Delayed Reaction Time[2]		986	0.01	0.41	0.02%
Behavior ratings by teachers (Needleman-Scales)[3]		1012	−0.00	−0.03	0.00%

* p < 0.1; ** p < 0.05; *** p < 0.01 (one-tailed p)
[1] x' = log(x + 0.5); [2] log msec; [3] sum of 6 items

estimates of effect-size[148, 112]. Although, for the majority of studies, non-significant outcome was found, the overall evaluation revealed a highly significant average IQ decrement of 1–3 points for a change of PbB from 10 to 20 g 100 ml^{-1}.

The cross-sectional character of these studies, as well as the fact that only a single measure of current lead exposure was used in most cases, limit their value in answering questions related to the natural history of the association between lead exposure and neurobehavioral outcome. This includes the inability of identifying sensitive periods of exposure and of answering questions related to the reversibility or persistence of lead-induced neurobehavioral deficit. The more promising approach in this regard is the prospective study protocol.

1.2.2   Prospective studies

Recent prospective studies have attempted to clarify the role of environmental lead exposure as a contributing factor in developmental neurotoxicity. Such studies have been performed in Boston[16], Cincinnati[43, 44], Cleveland[50], Glasgow[102], Port Pirie[12], Sydney[33], and the Kosovo area in Yugoslavia[145]. The strength of this set of studies, apart from their prospective design, is given by the fact that initial agreement was reached on the main features of a core protocol covering independent, dependent, and main confounding variables. Essential neurobehavioral tests in these studies included the Bayley Scales of Infant Development until age 2 years, yielding a "Mental Development Index (MDI)" and a "Psychomotor Development Index (PDI)"; the McCarthy Scales or the Kaufman Assessment Battery beyond age 3 years; and the WPPSI or the WISC-R at or beyond 5–6 years of age. It was also agreed to monitor blood lead levels at regular intervals in order to document the natural exposure history from pre-/perinatal stages until or beyond school age. Important confounders within the core protocol included maternal IQ and the HOME scale for a semi-quantitative assessment of the quality of the home environment.

Despite these efforts towards harmonization of study protocols the final results were not fully consistent across studies. Only three of these studies reported inverse associations between indices of pre-/perinatal lead exposure (prenatal or cordblood Pb) and reduced MDI and/or PDI-scores of the Bayley up to 6 or 12 months[50, 43], or up to 2 years of age[16], but these associations were no longer detectable after 2 years of age. The other studies did not report significant negative MDI/PDI-associations with prenatal lead exposure after confounder control. As for preschool age assessments, both the Boston and the Port Pirie studies, using the McCarthy Scales (General Cognitive Index (GCI)) at 57 or 48 months, respectively, did report significant inverse associations between GCI-scores and blood lead levels measured at different postnatal ages. Apart from the Kosova study[145] none of the other studies did so after confounder adjustment. At school age, using the WISC-R, significant inverse associations between postnatal blood leads and IQ were observed in the Boston cohort[140], the Cincinnati sample[44], and in the Port Pirie cohort, as well[12]. Apparently in none of the other studies the cohorts were followed into school age.

In summary, despite some inconsistencies described above, the overall outcome of prospective studies confirms the conclusions already drawn from the cross-sectional studies (see above), namely an inverse association between environmental lead exposure during development with cognitive ability in children. Quantitatively, as confirmed by meta-analysis[148, 112] and expressed in probabilistic terms, an increase of developmental lead exposure from 10 to 20 µg 100 ml^{-1} is likely to be associated with a cognitive deficit of between 1 and 3 IQ points. It should also be mentioned, however, that the particular strength of the developmental approach, namely the potential of identifying sensitive periods of exposure and excluding the possibility of reverse causality, could not be demonstrated here because peak effects occurred at different time-points in different studies, and because postnatal rather than prenatal exposure was most consistently related to neurodevelopmental outcome. It should be noted, however, that quantitatively both the cross-sectional and the prospective studies are in remarkably good agreement.

1.3   SENSORY EFFECTS

The effects of lead intoxication on visual functions are known and well described[62]; little is known, however, about visual dysfunction at low levels of exposure. Effects on the visual flicker fusion test have been studied in lead-exposed workers. The critical flicker fusion frequency (CFF) characterizes the temporal resolution of the visual system for flickering light. CFF is that frequency of intermittent light at which intermittency fuses into steadiness. It also reflects CNS function. Lowered CFF was found in workers with PbB above 60 µg 100 ml^{-1} relative to those below 20 µg 100 ml^{-1} [155]. Similar changes were seen in 59 lead workers at average PbBs of 48 µg 100 ml^{-1} [149].

Ototoxicity in lead-exposed workers has been reported in several studies summarized by Repko and Corum[115], although many of these reports are lacking methodological precision for final conclusions to be drawn. Based on audiometric data from a subsample of 4519 children and adolescents aged 4–19 years from the "2nd National Health and Nutrition Survey (NHANES II)" Schwartz and Otto[128] found a highly significant linear increase of pure tone hearing thresholds between 0.5 and 4 kHz . These observations were confirmed in 3262 subjects between ages 6–19 years from the Hispanic Health and Nutrition Survey[129]. An increase of PbB from 7 to 18 µg 100 ml^{-1} was associated with about 2 dB loss of pure tone hearing at frequencies between 0.5 and 4 kHz. Regression models with careful confounder-adjustment were used to establish these associations.

1.4   SUMMARY AND CONCLU.S.IONS FOR LEAD

Cognitive, visualmotor, and sensory deficits have been shown to be associated with environmental lead exposure in children at blood lead concentrations below 20 μg 100 ml^{-1}. In comparing lead-related neurobehavioural findings in adults and children striking sensitivity-differences are obvious. Whereas in lead-exposed workers neuropsychological impairment has not been demonstrated at blood lead concentrations below 40 μg 100 ml^{-1}, subtle cognitive deficit has been shown in children that are associated with PbBs as low as 10–15 μg 100 ml^{-1}. The reasons for this difference in vulnerability are not entirely clear at present, although the developing blood-brain barrier in young children as well as the particular vulnerability of the developing brain are hypothesized as contributing to the particular susceptibility to lead of the immature versus the mature central nervous system.

The subtlety of lead-associated cognitive impairment in children should also be emphasized. For both cross-sectional and prospective studies average IQ decrement between 1 to 3 points has been shown to be associated with increasing PbB from 10 to 20 μg 100 ml^{-1}. This corresponds to about 1/5 of a standard deviation of the typical IQ distribution, which is 15 around a mean of 100. This, by any standard, must be qualified as a small population-effect, which applies to the individual child only in a probabilistic sense. It has however been argued that even small average IQ differences between populations of children with low and elevated lead exposure, respectively, may have substantial effects at the upper and lower ends of the IQ distribution. This 'population-shift hypothesis' still awaits more general validation, however.

2   Mercury

Like lead, mercury belongs to those metals known to, and used by, humans since ancient times; for centuries it has primarily been used for therapeutic purposes. For a more detailed treatment of chemical, environmental, and biomedical aspects, more comprehensive reviews should be consulted (e.g., 17).

There are different physical and chemical forms of mercury. It has three oxidative states, namely metallic (Hg^0) , mercurous (Hg_2^{2+}), and mercuric mercury (Hg^{2+}), each of which has unique characteristics of target organ toxicity. Exposure to the vapors of metallic mercury is largely limited to occupational settings. Of more general public health significance is the ability of mercury to form stable organic compounds with alkyl groups, such as CH_3Hg^+ and CH_3HgCH_3 (monomethyl- and dimethyl-mercury), which have come to be known under the common term "methyl mercury (MMC".

The classic symptoms from undue exposure to elemental mercury vapor ($Hg°$) and methyl mercury (MMC) are due to central nervous system (CNS) involvement, while the kidney is the target organ for the mono- and divalent salts of mercury (Hg^+ and Hg^{++}). It is accepted today that the qualitative and quantitative differences in toxicity among inorganic mercury compounds is largely determined by their physical properties and redox potentials, whereas the ability of MMC to cross the blood-brain barrier is responsible for its accumulation in the brain and for the clinical picture which is dominated by neurological and neurobehavioral dysfunction. The present section will be concerned with neurobehavioral effects of elemental mercury vapors in occupational exposure settings on the one hand, and with those associated with environmental MMC exposure on the other. Mercury exposure due to amalgam fillings, which has received much public and scientific attention recently, will not be considered here because, although clearly elevated Hg concentrations in blood and urine have been found in humans with such fillings, these are far below those levels associated with early signs of impaired health[14].

2.1   INORGANIC MERCURY

Several investigators have investigated the effects of mercury vapors on peripheral and central nervous system functions of occupationally exposed workers[80, 88, 123, 138]. Hg concentrations in urine (U-Hg), based on volume or creatinine, and/or in blood (B-Hg) were usually measured to indicate degree of exposure, and tremor tests as well as sensorimotor or cognitive outcome measures were taken.

Older investigations studied workers with relatively high U-Hg concentrations well in excess of 50 µg l^{-1}. At average U-Hg of 96 µg g^{-1} creatinine and B-Hg of 29 µg l^{-1}, Roels et al.[123] found deficits of arm-hand steadiness and of eye-hand coordination in 43 chloralkali workers relative to 47 matched controls. These effects were independent of signs of renal dysfunction but did not exhibit a clear-cut dose-response relationship. A threshold limit value of 50 µg Hg l^{-1} creatinine was proposed to safeguard against preclinical CNS-effects of Hg. On a group basis, although not necessarily for the individual, creatinine-adjusted U-Hg values are comparable to volume-based concentrations.

In a more recent study, Soleo et al. (138) used the WHO core test battery to compare 28 Hg-exposed workers with 22 non–exposed workers; average U-Hg in the exposed group had been 30-40 µg l^{-1} over the years. Short term memory as assessed by means of the digit span test from the WAIS was impaired in the exposed group, whereas other sensorimotor functions (simple reaction time, visual recognition memory, eye hand coordination) were unaltered. Eighty-nine Hg-exposed chloralkali workers (mean U-Hg = 25.4 µg g^{-1} creatinine) and 75 matched controls were studied by Langworth et al.[80] using computerized psychometric tests, a tremor test, as well as a symptom and a personality questionnaire. Whereas no group differences in terms of performance changes were observed, a higher degree of neuroticism and higher symptom reporting (tiredness, confusion) were found in the chloralkali-exposed group. You-Xin Liang et al.[88] compared 88 Hg-exposed workers (mean U-Hg: 25 µg l^{-1}) with 97 carefully matched non-exposed controls using a computerized performance test-system and a subjective mood questionnaire. Significant performance-deficit was seen in 8 out of 15 tests (e.g., mental arithmetic, perceptual speed, attentive performance, choice reaction time, finger tapping). In addition, as with Langworth et al. (see above), increased fatigue and confusion was reported in the Hg-exposed group.

In summary, it is concluded that, although the evidence is by no means fully consistent, both neurobehavioral performance changes as well as altered mood states have been reported to occur at low levels of occupational exposure to inorganic mercury. Since all of these studies are of a cross-sectional nature, observed inconsistencies may well be due to insufficient matching. Apparently no systematic studies have been done to examine the reversibility of functional impairment following cessation of exposure.

2.2   ORGANIC MERCURY

The most important species in this category is methylmercury (MMC). It is through these stable organometallic compounds that increased environmental exposure may occur in segments of the general population, because inorganic Hg entering the environment from a variety of sources is methylated by microorganisms in fresh and ocean waters, thus entering the food chain. Highest MMC concentrations have been measured in large predatory fish, such as shark and tuna.

Knowledge about the neurological and neurobehavioral effects of high MMC exposure was primarily gained in two episodes of mass poisoning, namely the Minamata Bay tragedy in Japan in 1950[66], and another tragedy in Iraq 20 years later[13]. Exposure of humans in the Minamata incident was through contaminated fish from Minamata Bay, which had been polluted for years by metallic mercury from industrial sources. MMC entered the food chain after inorganic mercury was methylated by microorganisms in the environment. Thousands of people were exposed, and several hundred cases of MMC poisoning have been documented. The Iraq poisoning episode occurred by ingestion of seed grain treated with an MMC-containing fungicide; the grain had been ground into flour to make bread. About 7000 poisoned victims were hospitalized, and over 400 died.

The clinical picture of MMC poisoning is characterized by sensory, motor, and cognitive deficits. Prominent signs of neurotoxicity are constricted vision ("tunnel vision"), deafness, dysarthria, and ataxia. Mental disturbances and impairment of the chemical senses (taste, smell) may occur as well. An important feature of MMC neurotoxicity is the particular vulnerability of the developing CNS, which has been observed in human cases and in animal models as well (see below). In both the Minamata and the Iraq incidents, pregnant women gave birth to children with severe CNS damage. At high MMC levels in maternal blood, microcephaly, hyperreflexia, and severe motor and mental impairment were prominent. For lower levels of exposure, subtle deficits were difficult to diagnose shortly after birth but became increasingly striking later on. The mildest cases in the Iraq incident presented with signs of the minimal brain dysfunction syndrome, characterized by hyperactivity and attention deficit[6]. The likelihood of mental retardation increased with increasing maternal MMC hair concentrations. In the Minamata tragedy, follow-up studies revealed strong associations between cord blood MMC levels and mental retardation in 20-year-old victims of prenatal exposure.

Although metal concentrations in hair, because of the possibilities of external contamination, are generally not considered to be valid markers of body burden, MMC in hair is a remarkable exception to this rule. MMC accumulates and concentrates in hair as soon as the hair is being formed in the hair follicle. MMC concentrations in newly formed hair are closely related to concurrent MMC blood levels at a ratio of about 250. Following uptake MMC concentrations remain constant during subsequent hair-growth.

Clinical observations of the Iraq victims have also been used for more quantitative risk assessment[38] with Hg concentrations in maternal hair at parturition as biomarkers of *in utero* MMC exposure. By means of different dose-response models for prenatal MMC exposure, "estimated lowest effect levels" for motor retardation (retarded walking) and neurological CNS signs (e.g., increased limb tone, deep tendon reflexes with persisting extensor plantar responses, ataxia, hypotonia) were located at about 10 ppm Hg in hair.

So far only a few studies have looked at neurological and neurobehavioral effects of elevated environmental methyl mercury exposure. Two such studies in New Zealand looked at subtle neurobehavioral deficits in children born to mothers with MMC hair levels in excess of 6 ppm[72, 73]. MMC exposure in New Zealand is primarily due to consumption of shark meat. The first report[72] describes results from a pair-matching approach to study mental retardation associated with prenatal MMC exposure. From a basic cohort of 11,000 mother-child pairs, 31 with elevated maternal MMC hair levels between 6 and 20 ppm were compared with pair-matched referents. Neurodevelopmental status of 4-year-old children was assessed by means of the Denver Developmental Screening Test. Matching criteria was based ont the mother's ethnic group (Pacific Islanders, Maori, Europeans), age, as well as place and date of birth. About 50% of the high MCC children had abnormal or borderline results, whereas this was true for only 17% of the reference children. This difference was highly significant. No influence of confounding by socio-economic factors, maternal health, or smoking habits was found. A significant dose-response association was found between mean hair mercury levels during pregnancy and developmental status on the Denver test.

The second report[73] describes the results from a follow-up study of these plus additional children at 6 years of age. This time 61 children with high Hg levels in maternal hair (6–86 ppm) were compared with a low level group (3–5.99 ppm), and two control groups with high and low fish consumption, but low Hg hair levels (0–3 ppm). These groups were fully matched for ethnic group of the mother, sex of child, maternal age, smoking habits, place of residence, and duration of residence in New Zealand before the child's birth. Testing was done by means of a battery of scholastic and neuropsychological tests, including the Test of Language Development (TOLD), the Wechsler Intelligence Scale for Children (WISC-R), and the McCarthy Scales of Children's Abilities (MSCA). Borderline or significant inverse associations were found between hair Hg levels exceeding 6 ppm and performance on all of the above mentioned tests after adjustment for confounding. The relatively largest deficit was seen for children with maternal Hg hair levels exceeding 10 ppm.

2.3   SUMMARY AND CONCLU.S.IONS FOR MERCURY

Both neurobehavioral performance decrement and altered mood states have been reported in adults at relatively low levels of occupational exposure to airborne inorganic mercury, and threshold limit values have been proposed based on such evidence. Neurological findings from mass poisoning episodes as well as neurobehavioral observations in children at elevated environmental MMC exposure from excessive fish-consumption converge on "critical" MMC hair levels between 10 and 20 ppm. The particular vulnerability of the developing brain and the irreversible nature of the deficit, as observed in the neurodevelopmental clinical studies, is confirmed by animal models to be described later.

3   Manganese

Manganese (Mn) is an essential trace element that, in cases of excessive exposure, induces signs and symptoms of CNS involvement, which bears similarities to Parkinson's disease[20, 59].

3.1   OCCUPATIONAL EXPOSURE

Most of what is known today about the neurotoxicity of manganese comes from neurological and neurobehavioral studies in occupationally exposed workers. Contrary to lead and mercury, no biochemical markers have as yet been found to relate the target dose to neurotoxicity following long term exposure. Instead Mn concentrations in respirable and/or total dust are typically used to characterize occupational exposure. Three recent studies on neurobehavioral effects of long term occupational Mn exposure will be used to illustrate the degree of convergence of findings and the sensitivity of neurobehavioral tests to detect subclinical CNS dysfunction resembling parkinsonism.

Thirty workers from two steel smelting works with moderate Mn exposure (190-1390 μg m^{-3}; total dust) were compared with 60 age-matched unexposed controls[147]. Neurophysiological measures, such as the electroencephalogram (EEG), auditory evoked potentials (AEP), brainstem auditory evoked potentials (BAEP), diadochokinesis, as well as sensorimotor (reaction time, tapping, eye-hand coordination, vigilance performance, perceptual speed) and cognitive (mental arithmetic, short term memory, verbal comprehension), as well as mood scales and psychiatric ratings were used to differentiate Mn-exposed and unexposed workers. There were no group differences for most neurophysiological, psychiatric and cognitive outcome-measures, but the Mn-exposed workers, relative to those of the control group, exhibited slower diadochokinesis, longer reaction times, and impaired tapping- as well as digit-span performance (short term memory). These findings are discussed as being partly indicative of impaired extrapyramidal functions and functions related to efferent signal processing, alterations also seen in patients suffering from Parkinson's disease.

Ninety-two workers from a dry alkaline battery factory exposed to MnO$_2$-dust (948 μg Mn m^{-3} total dust) were compared with 101 matched controls[124]. Since exposure conditions had not changed for 15 years, total integrated lifetime exposure (TILE) to Mn was calculated for each worker by multiplying degree of exposure with duration of exposure. Apart from neurobehavioral measures, namely reaction time, eye-hand-coordination, and hand steadiness, lung ventilatory parameters and biochemical parameters (calcium, iron, luteinizing hormone, prolactin concentrations in serum, blood counts, and Mn levels in blood and urine) were measured. Educational level, age, coffee, and alcohol consumption as well as smoking habits did not differ between both groups. Measures of internal Mn exposure differed markedly between groups but did not correlate with TILE; there was a correlation between current Mn exposure and Mn in urine but not in blood. Whereas lung function parameters were not related to Mn exposure, highly significant group-differences were observed for sensorimotor test performance: Reaction times were markedly prolonged, eye-hand coordination impaired, and hand-steadiness (tremor) exhibited borderline deficit. Pronounced dose-response associations with TILE were shown for these neurobehavioral parameters. Based on these results a drastic downward revision of the current time weighted average exposure to Mn dust and regular surveillance of exposed workers with simple validated sensorimotor tests was proposed.

Mergler et al.[98] administered a battery of neurobehavioral tests to 115 workers in Mn alloy production with long-term exposure to average Mn-dust concentrations of 890 μg m^{-3}. A non-exposed group of workers pair-matched with regard to age, educational level, smoking status, and number of children served as controls. There were pronounced group differences for symptom-reporting, such as tiredness, forgetfulness, somnolence, agitation, nightmares, aggressive feelings, impotentia coeundi, lower back pain, joint pain, tinnitus, or numbness. Sensorimotor functions, such as tapping speed, hand steadiness, diadochokinesis, but not simple and choice reaction time, and cognitive flexibility (Stroop test), were found to be impaired in Mn-exposed workers. Intellectual functions largely remained unaltered. It is concluded that the manifestations of early manganism can be detected by means of sensitive testing methods.

3.2   GENERAL POPULATION EXPOSURE

Significant population exposure is rare but may occur through excessive Mn levels in drinking water[78]. In this study random samples of between 49 and 77 aged subjects (over 50 years old) from three regions in Greece with low (3.6-14.6 µg l⁻¹), moderate (81.6-252.6 µg l⁻¹), and high (1800-2300 µg l⁻¹) Mn concentrations in drinking water, respectively, were clinically examined by a trained neurologist, who was unaware of the Mn concentrations of the drinking water in these areas. A list of 35 neurological signs and symptoms, such as irritability, insomnia, loss of libido, depression, gait disturbances, monotonous speech, rigidity, tremor, akinesia/dyskinesia served as the clinical guideline; each item was weighted according to intensity of appearance (0 = absent to 3 = strong) and to its value in the diagnosis of idiopathic Parkinsonism using scores from 1 (e.g., insomnia, irritability) through 3 (e.g., static tremor, rigidity). The sum total of these weighted items was used as the individual neurological score.

Highly significant area differences of neurological scores, following the progression of Mn concentrations in drinking water, were observed both in female and male Ss, and increasing Mn concentrations in hair, but not in blood, were highly significant, too. Prevalence of the cardinal signs of idiopathic Parkinsonism, such as tremor, rigidity, or bradykinesia, are not given. It is concluded that elevated Mn concentrations are associated with higher area averages of neurological signs of chronic manganese poisoning. These findings do not support the largely negative results of studies in workers occupationally exposed to Mn in air, in which subtle sensorimotor impairment in the absence of neurological symptoms was reported (see below). The authors conclude that the higher age of the subjects of their study (over 50 years) relative to the younger subjects in occupational exposure studies may have exacerbated the Parkinson-like neurological symptomatology in the areas with elevated Mn levels in drinking water.

3.3   SUMMARY AND CONCLU.S.IONS FOR MANGANESE

In a general population study with elderly subjects exposed to elevated Mn concentrations in drinking water, increases of Parkinson-like neurological symptomatology with dose-effect characteristics were reported. On the other hand, no parkinsonian symptomatology was found in younger workers with long-term occupational exposure to Mn in air. Taken together, the results from these workplace studies are sufficiently consistent. In demonstrating Mn-related sensorimotor deficits, compatible with deficient extrapyramidal functions however, existing discrepancies of findings might well be due to the chemical form of manganese in the different occupational settings[98]. Although a discussion of sensorimotor impairment in the framework of Parkinsonism is tempting, additional support regarding the involvement of nigro-striatal dopaminergic structures with high levels of neuromelanin would be needed. Such information could be provided by neurobehavioral animal studies with neurochemical and neuropathological support, to be presented below.

4   Aluminum

Aluminum (Al) is the most abundant metal in the earth's crust. Its toxicity for humans has traditionally been rated low in the past because it was considered almost nonabsorbable from the gastrointestinal tract. There is now evidence, however, that both ingested and inhaled aluminum is absorbable to some extent. Al exists in inorganic and organic form; the inorganic aluminum salts are differing in terms of water solubility. Chemical, environmental, toxicological, and biomedical aspects of Al are reviewed in ref. 48.

The neurotoxicity of aluminum was first studied in animals after Al-injection (see below). First information about Al-induced neurotoxicity in humans was obtained from clinical observations in dialysis patients developing a progressive dementing illness, which often proved fatal if untreated by chelation. Clinical signs of this type of brain damage include speech and motor disturbances, memory deficit, personality changes, dementia, and seizure disorders. Although, initially, there was debate about the etiological contribution of aluminum to this disease, which has been termed dialysis dementia, it is now accepted that the use of Al-containing phosphate-binding gels or of water with high Al-content was the cause of this disorder. Impaired renal function may also result in a significant accumulation of aluminum in the body, associated with 'dialysis dementia' in patients who had never received dialysis treatment, but who had taken large doses of aluminum hydroxide for other purposes.

A possible role of aluminum in the pathogenesis of Alzheimer's disease is being discussed (e.g., ref. 39). This hypothesis rests on partial similarity of dialysis dementia to presenile and senile features of Alzheimer's disease, both in clinical and pathological respects. It has been shown, for example, that in Alzheimer patients aluminum selectively accumulates in the nucleus of the brain cells that form the neurofibrillary tangles that are typical of Alzheimer's disease. Neurofibrillary tangles have also been abserved subsequent to injection of aluminum salts in cats and rabbits in addition to behavioral alterations (see below). In some studies, elevated aluminum levels in the gray matter of Alzheimer patients with normal kidney function were found as well. There are, however, several inconsistent findings, so that the evidence supporting an association between environmental Al exposure and Alzheimer's disease is still a matter of controversy[48]. This is why aluminum, although certainly neuroxic if it reaches the nervous tissue, has not been given extensive consideration in the present review.

5 Summary and conclusions from human studies

The administration of neurobehavioral tests covering cognitive and sensorimotor functions in cross-sectional and prospective studies in children and adults has helped to document the risk for neurotoxicity associated with occupational or environmental exposure to low levels of inorganic lead, organic or inorganic mercury, and manganese. In many instances such observations have been used to establish or revise exposure limits, particularly in occupational settings. It should, however, be pointed out that often such observations have not been sufficiently consistent to be fully acceptable within the scientific community and/or in regulatory bodies. This is particularly true if, for example in the case of inorganic lead, the critical neurobehavioral effects were small and embedded in a complex background of confounding variables. In such cases questions as to the true cause-effect contingencies have been, and are still being raised. In these critical issues neurobehavioral studies in animal models have helped substantially to corroborate epidemiological observations, contribute to elaborating the underlying neural mechanisms, and to identifying critical periods of vulnerability. This important approach will be discussed in the following sections.

II Animal studies

1 Lead

1.1 EFFECTS AT DIFFERENT DEVELOPMENTAL STAGES

Adult organisms tolerate comparatively high levels of lead. Therefore, in establishing a rodent model of lead exposure, the study by Pentschew and

Garro[110] describing lead encephalopathy in the offspring from dams with dietary lead exposure during lactation was a significant contribution. The higher sensitivity to lead in early development stages has inspired many following studies.

In comparing developmental exposure in different species, the equivalency of the developmental periods must be considered. While the time course of neuronal development is largely similar in mammals, the time when birth takes place during development differs across species (Figure 12.2).

When the rat is born its brain is in the same developmental stage as the human brain at 5–6 months of gestation. As a consequence, the stage of the human brain at birth is achieved by the rat not before postnatal day (PD) 10. In the rat cerebral neurogenesis is nearly complete at birth and gliogenesis starts shortly after birth; thus, the postnatal increase in cerebral cells by 50% in the rat

Rat

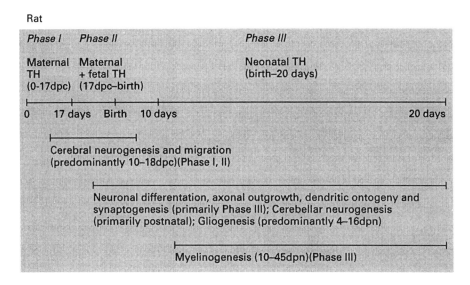

Human

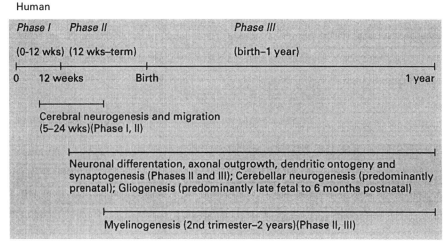

FIGURE 12.2

Comparison of developmental phases in the nervous system of the rat and the human

TH, thyroid hormones; dpc, days postconception; dpn, days postnatal. (Adapted from Porterfield, S.P. and Hendrich, C.E. (1993) *Endocr. Rev.* **14**: 94–106.)

is due to glial cell proliferation. The period in which neuronal differentiation, ontogeny of axons and dendrites, and synaptogenesis as well as neurogenesis in the cerebellum occur, begins at the end of gestation in the rat. Development of nerve terminals is complete on PD 30, while on about PD 45 glial cells and myelination are fully developed. In the human cerebral cortex neurogenesis is complete by the 7th month of pregnancy and gliogenesis, neuronal differentiation, synaptogenesis, and cerebellar development start prior to birth extending into the first years after birth. For more detailed information the reader is referred to textbooks of neuronal development (e.g., see ref. 19).

Following the pioneering work by Pentschew and Garro[110] in studies of developmental lead exposure most authors preferred treatment of the offspring via the dam after parturition until weaning with or without subsequent exposure, although, in order to approach environmental exposure conditions, some studies also used exposure paradigms including pre-conception, in utero and pre- postweaning exposure in rats[152] and in monkeys[89].

In contrast, direct comparisons of exposure in different developmental periods and their behavioral sequelae are less frequent. In one of these studies exposure periods of 21 days started at conception, at parturition, or after weaning in rats[127]. Internal exposure in the brain after 1090 mg Pb l^{-1} water varied with the start of treatment: the earlier the start, the higher the brain lead level. Neonatal lead caused increases in densities of adrenoreceptors in the cortex, of striatal dopamine (DA) D2-receptors, and of hippocampampal serotonin (5HT) receptors. In contrast, striatal muscarinic receptors were more decreased by prenatal than by postnatal lead. Post-weaning exposure failed to exert any effects on receptor densities, although some functional changes were observed by examination of receptor binding. Prenatal lead (545 mg l^{-1}) during gestational days (GD) 14–21 in rats also induced de-masculinized behavior in male offspring as well as morphological changes in the sexual dimorphic areas of the hypothalamus and irregular patterns of hormone release in both sexes[96].

At high doses of lead rat pups treated during the first week after birth developed neuropathologic disorders in the entire brain. When exposure started during the second week postnatally pathologic effects were confined to the cerebellum and only minimal effects were observed following postweaning exposure. Thus, these findings correlate well with the neurodevelopmental events during the periods examined in this study. In guinea pigs, which are born at a rather advanced level of maturation in comparison to rats and mice, exposure from GD 22 to GD 52 or 62 resulted in diminished activities of marker enzymes for astroglia and oligodendroglia in dams and fetuses. Blood lead levels were below 30 or 40 g 100 ml^{-1} for termination of exposure on GD 52 and 62, respectively.

In a cross-fostering study on rats decreased synaptic densities in the parietal cortex were detected in groups with prenatal lead exposure (the prenatally exposed group and the pre- and postnatally exposed group), while no changes were found after only postnatal treatment[95]. Since formation of synapses in the rat predominantly occurs after birth, this outcome might be indicative of a disturbance of factors initiating synaptogenesis.

Other developmental effects reported are delays in the postnatal increase of cerebral cytochrome concentration as well as effects on NCAM (neural cell adhesion molecule) which acts as a morphoregulator during neural growth and differentiation[114]. Recently, it has been shown by both in vitro and in vivo studies that lead inhibits the function of the NMDA receptor channel and that this effect can be found only in rats with developmental exposure, but not in rats treated when adults.

Behavioral studies also reveal the particular sensitivity of early developmental periods for lead-induced effects. Using a cross-fostering design

361

delays in exploratory and locomotor activity were reported for the prenatal and the pre- and postnatal rats, but not for the postnatal group[40]. Similarly, no differences in spatial learning deficits were observed between maternally exposed rats and rats with lifetime exposure on PD 165, while on PD 500 effects were more pronounced in the group with permanent exposure indicating an additional influence by lead in aging rats[105, 106]. Additionally, in drug discrimination learning there was no difference between rats with exposure during lactation and rats with permanent exposure. In other reports on drug discrimination differential effects of preweaning and postweaning exposure involving different DA receptors were observed (described below).

In addition, effects of postweaning or adult lead treatment were found on other types of operant behavior using fixed interval schedules (e.g. see ref. 35), conditioned suppression[107], Sidman avoidance[132], or a multiple FI-FR (fixed interval-fixed ratio) schedule[7] in rats. In the latter study a differential influence of the exposure period was described on interresponse time. Preweaning exposure decreased while postweaning exposure increased the interresponse time. In a cross-fostering experiment in rats the pre- and postnatally exposed group received fewer reinforcements than groups with either prenatal or postnatal treatment or the controls. In addition, spatial behavior and probability learning in a runway[42] were found to be affected by postweaning or adult lead treatment, respectively.

Thus, these studies indicate effects by lead also after exposure in ontogenetic stages after weaning. However, after treatment in early developmental stages lead effects persist for long periods after the exposure, while persistence of lead influences after administration in later development is less clear.

In comparison to rats and mice monkeys are born in a more mature stage, although differentiation in the brain is known to extend into the first 3 years of life[1, 56]. Most of the work in primates used postnatal lead exposure which was sometimes terminated after about 1 year. Few studies tried to evaluate sensitive periods for lead in monkeys. In one of these attempts rhesus monkeys with either pre- or postnatal exposure were compared on the Hamilton search task which is a test for spatial memory and, therefore, should be sensitive to hippocampal dysfunction. The test was conducted more than 3 years after termination of the treatment. A lead-related deficit was seen only after postnatal exposure, while prenatal exposure was not effective[84]. However, these groups were not directly compared, but tested in two different experimental parts. After prenatal exposure impairments of a form discrimination learning and response inhibition were found in cynomolgus monkeys at 6–18 months and 19–26 months of age, respectively[68]. The differences in the outcome of these studies may be related to the different tasks used. Prenatal lead exposure during the last 17 or 12 weeks of gestation in squirrel monkeys at levels which were comparable to the blood lead levels in the macaques examined did not cause maternal toxicity. However, a high incidence of stillborns and neonatal deaths together with severe encephalopathy and other overt signs of toxicity was reported in this species[91]. From this, it appears that the developing squirrel monkey is extremely sensitive to lead, but the reason for this is unknown.

In another series of studies in postnatally exposed cynomolgus monkeys early treatment from PD 0 to PD 400 was compared to later exposure from PD 300 on. In addition, the late exposure period used in these studies started at a time when development of the monkey brain is not complete. No differences between different exposure periods were seen on delayed spatial alternation or on FI operant performance by testing at an age of 6–7 and 7–8 years, respectively. In contrast, nonspatial discrimination reversal groups with late and permanent (early and late) lead exhibited deficits when tested at an age of

7-8 years, while early treatment was without effects[120]. Using the same protocol impairment was detected on spatial discrimination reversal after permanent treatment when the monkeys were trained in the absence of irrelevant cues and in all groups when irrelevant cues were present. The permanent group also exhibited more pronounced effects on concurrent visual discrimination at the age of 8-9 years[119].

Taken together, these studies found the strongest behavioral effects by permanent postnatal exposure to lead, but early or later postnatal treatment also caused impairments. This suggests that current low level lead exposure in monkeys aggravates deficits on learning tasks after treatment during developmental stages. As in rodents, there is little information about the persistence of these effects following exposure in subadult or adult phases. However, after early postnatal exposure during the first year of life, long-lasting lead-related deficits were detected on discrimination reversal[27], the Hamilton search task[84], delayed spatial alternation, and operant conditioning[97, 85] up to 9 years of age.

1.2   MECHANISMS OF NEUROBEHAVIORAL TOXICITY

1.2.1   *Interactions with dopamine*

The role of dopamine (DA) in cognitive and motor functions has provoked a large number of studies in recent years dealing with various aspects of dopaminergic processes from molecular biology of different receptors to behavioral endpoints in experimental approaches and clinical symptoms in patients. The outcome of these studies is summarized in several excellent reviews devoted to many of these aspects[94, 122, 127]. For background information the reader is referred to these. The early literature on the neurochemical actions of lead is also covered by reviews[11, 131, 151] and the effects on neurotransmitter release in general were summarized by Bressler and Goldstein (22). The focus of this chapter is on the relation between lead-induced behavioral effects and alterations in the dopaminergic system.

This relation has attracted attention since an interaction of lead with (+)-amphetamine (+AP) was reported in mice after exposure during lactation. Although not selective in its action, effects by +AP are in part mediated by an influence on the dopaminergic system. +AP was found to ameliorate lead-induced hyperactivity[134, 135]. These studies thus seemed to confirm the therapeutic use of psychostimulants in hyperkinetic children[146]. In accordance with the assumption of a dopaminergic mediation of lead effects, lead-induced activity changes were reported together with decreases in DA concentrations in the cortex, the midbrain, and the hypothalamus as well as altered levels of other neurotransmitters.

However, hyperactivity in lead-exposed animals and +AP effects on this behavior remained controversial (review in ref 21). It appears that in the early studies important confounders like undernutrition were neglected. In only part of these studies internal exposure to lead was measured, but the dosages employed and the lead values, when determined, point at high levels in different matrices (e.g., according to Bornschein et al.[21] about 150 µg 100 ml^{-1} blood for the protocol used by Silbergeld and Goldberg[134]). Such exposure levels result in body weight reductions by undernutrition which in turn alters activity levels. In a better controlled study a pair-fed group was included to separate effects by undernutrition from lead-induced effects[150]. The dosage used (4% in the diet given to dams during lactation and 40 mg kg^{-1} diet to the offspring thereafter) caused hyperactivity in lead-exposed rats, but locomotor behavior was not changed by the DA agonist apomorphine or +AP.

An additional factor which influences the outcome of activity measurements and which can explain some of the contradictory results is the type of device and the procedure used for the determination. It has been pointed out that a placement of an animal in a device like an open field and a short observation period provide a measure of reactivity to the handling and unfamiliar environment rather than a measure of the activity level as such, since in the former case locomotion is strongly influenced by an emotional response[21]. In contrast, longer observation periods provide a measure of basal motor activity, particularly in devices to which the animal is accustomed. Both types of evaluations can be combined by longer observations in formerly unknown devices or by repeated measures for short periods as used by Winneke et al. (1982). This allows to assess the course of habituation and such a procedure can detect a differential influence of an agent on either phase.

Despite the problems in the early activity studies, subsequent research has continued to pay attention to a dopaminergic mediation of lead-induced behavioral effects, and although controversial in detail, there is an increasing evidence for an involvement of DA in lead exposure.

When the DA agonist apomorphine (APO) which causes aggression in normal untreated male rats, a lead-induced reduction of fighting behavior was detected[45]. After postnatal dietary exposure to lead at levels which did not result in overt toxicity like undernutrition, weight-matched pairs of male rats were tested on PD 90. During the observation period the elapsed time to the first attack, the total number of attacks, and the total fighting time were recorded. After the injection of the vehicle both the lead-treated and the control pairs engaged in social behaviors. Aggressive bouts were rare and short and there were no differences between groups. After APO injection, however, control rats exhibited more and longer attacks than lead-treated animals and the latency to the first attack was decreased in controls, while it was increased in exposed subjects.

Further studies suggested correlations between neurochemical changes, in particular, in concentrations of DA and its metabolites, and stereotypic behavior, pole-climb active avoidance as well as operant responding on an FI schedule.

While in the latter studies associations between several types of behavior and DA levels were described, the use of drug discrimination learning (DD) allows a more direct examination of the relationship between behavior and neurotransmitter action. In this paradigm animals learn to discriminate the stimulus properties of drugs for which this procedure is most sensitive when conducted using operant techniques[109]. The first DD study in lead exposed animals assessed the sensitivity to +AP in rats after 0, 109, or 2725 mg Pb/l water during lactation with or without postweaning exposure yielding four treated groups and the controls. In comparison to controls all exposed groups were less sensitive to +AP. This effect was suggested to be mediated via the dopaminergic system[158]. A later experiment confirmed the subsensitivity to +AP in lead-exposed rats.

However, it is known that +AP is not a selective dopaminergic drug, but acts on other neurotransmitter systems as well. An indication that the reported subsensitivity to +AP was not due to its stimulus effects on dopaminergic processes is derived from other DD studies[36, 37]. In these experiments an attempt was made to differentiate between lead-induced effects on the D_1-type and the D_2-type of DA receptors. For this, the stimulus properties of the selective D_1-agonist SKF 38393 and the D_2/D_3-agonist quinpirole were compared in lead-exposed rats and supersensitivity due to lead was reported for both dopaminergic drugs after postweaning exposure[36] and for quinpirole after exposure during lactation[37]. The effects on DD were detected at low to moderate exposure levels (31-73 µg 100 ml^{-1} blood in the first study, 16–34 µg 100 ml^{-1} blood in the second study).

These studies are consistent with reports of an inhibited DA release by lead, a decreased DA turnover, and an altered regulation of DA synthesis[81]. A reduced DA release and availability may result in increased sensitivity to dopaminergic stimulation as revealed by DD, because of an upregulation of the DA receptors. In accordance with this, a lead-induced increase in the density of D_2-receptors was reported, while D_1-mediated receptor binding and adenyl cyclase activity as well as D_1-mediated grooming behavior were not affected after pre- and postnatal exposure[103].

The assumption of a reduced availability of DA is supported by results in rhesus monkeys with early postnatal lead treatment. The lead-induced impairment on a spatial delayed alternation task was found to be ameliorated by the DA precursor L-DOPA. This effect was observed 6–8 years after termination of the exposure[86]. In addition, lead effects on the electroretinogram in rhesus monkeys were related to effects on tyrosine hydroxylase, the key enzyme in DA synthesis[76, 90].

In conclusion, the present behavioral and neurochemical results indicate an involvement of dopaminergic processes in lead-induced behavioral alterations. The effects appear to be more clearly expressed on processes mediated by the D_2-type of DA receptors, since information indicating an involvement of the D_1-receptor is scarce. Other neurotransmitter systems may be also affected. These effects may be directly or indirectly linked to altered functions of DA as in the case of decreases in GABA activity in the striatum or decreased striatal enkephalin levels in developing rats after pre- and postnatal treatment. Cholinergic effects appear to occur rather unspecific at more elevated dose levels[151]. Recently, impairment of the activation of the NMDA glutamate receptor was detected which was dependent on the age and the brain region studied. Furthermore, a decreased NMDA receptor density in the cerebral cortex was found in rats at PD 14, while in adult rats following pre- and postnatal exposure increases in receptor densities were reported[63]. The implications of these findings for long-term potentiation (LTP) will be discussed below.

1.2.2   *Relations to lesions in the prefrontal cortex*

Most of our knowledge about lead effects on the prefrontal cortex derives from behavioral results in monkeys which resemble effects of lesions in this brain region (review in 87). In monkeys the prefrontal cortex is differentiated into two major subregions, the dorsolateral and the medial orbital areas[58]. These areas differ in their patterns of ascending and descending fiber connections and the behavioral impairments following lesions in either part (Figure 12.3). Whereas depletion in the medial orbital cortex results in deficits in object reversal learning and on object alternation tasks, lesions in the dorsolateral area impair delayed spatial alternation. Spatial reversal learning is affected by lesions in both areas. However, destruction in the dorsolateral part causes deficits on the first and second reversal, while subsequent reversals are affected by medial orbital lesions.

There are further developmental differences between both parts of the prefrontal cortex. The medial orbital area is mature by 1 year and lesions prior to this age do not cause permanent behavioral effects. In contrast, maturation of the dorsolateral cortex is not complete before 3 years of age and after lesions during infancy adverse effects on behavior are delayed[56,57].

A very similar pattern of behavioral disturbances is found after lead exposure. A deficit on the first reversal on spatial and object reversal learning was observed in rhesus monkeys exposed to lead and trained during the first year of life[26]. This deficit in spatial reversal persisted when these monkeys were tested 3 years later after internal exposure in the blood had declined to control levels[27]. In another cohort of monkeys no impairment of object reversal was found at the age of 16 months; however, a deficit was found on the first reversal in spatial reversal learning in groups exposed to transient high pulses of lead[82]. In general, these results are corroborated by studies in cynomolgus monkeys following exposure in different postnatal periods (see above). In addition to deficits on early reversals, late reversals were also affected, both on object[121] and spatial reversal tasks[55]. There were also impairments of spatial delayed alternation[85] and delayed matching-to-sample[118].

In conclusion, the general pattern of behavioral deficits after lead exposure affecting both object and spatial reversal and delayed spatial alternation is consistent with effects by lesions in the medial orbital cortex and in the dorsolateral cortex and/or damage in subcortical structures connected to the prefrontal cortex like the caudate nucleus and the thalamic medial dorsal nucleus[87]. It is also in accordance with action by lead on dopaminergic processes, since the described cognitive deficits also emerge after depletion of prefrontal DA[23, 136]. However, a contribution of limbic structures like the hippocampus and the amygdala cannot be excluded since late reversals and the Hamilton search task (see below) were also impaired in lead-exposed monkeys. Furthermore, deficits were also found on a concurrent discrimination task[119] and on learning set formation[89] for which an implication of the prefrontal cortex has not been reported.

1.2.3   Effects on the hippocampus

Another site in the brain which has received much attention, in particular in rodent studies of lead treatment, is the hippocampus. It is known from work in human patients that damage to this area causes severe impairments in memory. Recently, it has been shown that neurochemical inhibitors of LTP (long-term potentiation) which is suggested to form the cellular basis for learning and

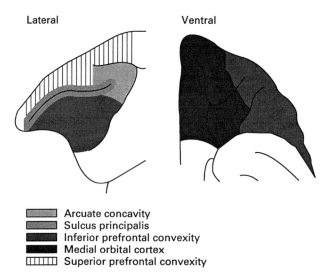

Lateral   Ventral

▨ Arcuate concavity
▨ Sulcus principalis
▨ Inferior prefrontal convexity
■ Medial orbital cortex
⨅ Superior prefrontal convexity

FIGURE 12.3

Orbital, dorsolateral, and prefrontal cortical areas in the rhesus monkey
Different areas are given in different shading patterns according to the explanation in the figure. (Adapted from Rosenkilde, C.E. (1979) *Behav. Neur. Biol.* **25**: pp 301-345.)

memory also impair spatial learning in a manner like hippocampal lesions. These studies emphasize the importance of NMDA receptor function in the CA1 region of the hippocampus for LTP and efficient spatial performance. Interrelations between these phenomena have inspired many experiments following the initial work by Morris et al.[104].

The first reports of lead effects on the hippocampus described morphological alterations. The main morphological structures of the hippocampus are shown in Figure 12.4. A reduced thickness of the pyramidal and the granular cell layers were observed together with decreases in the activity of cholinergic enzymes in rats following neonatal exposure until PD 10[92]. Internal exposure was not reported in this study, but a similar treatment with 7.5 mg Pb acetate kg bw^{-1} caused blood lead levels of about 33 and 29 µg 10 g^{-1} in rat pups on PD 11 and PD 21, respectively, and brain levels ranging in different areas from about 0.13 to 0.51 µg g^{-1} tissue on PD 11 and from 0.22 to 0.39 µg g^{-1} on PD 21 [75]. Interestingly, in the latter study brain levels in the cerebral cortex and in the cerebellum were higher on PD 11 than on PD 21, whereas in the brainstem and hippocampus the reverse was found.

Other authors reported a retardation of synaptogenesis in the dendritic layers of the infrapyramidal cells in the dentate gyrus as well as delayed glial development in the dentate hilus on PD 15 after lactational exposure in rats. In adult animals an increased number of glia cells in the pyramidal cell layer were detected together with an increased areal density of the profiles of mossy fiber boutons[28, 29]. In another study bimodal effects depending on the dose were found in the mossy fiber zone, the granular cell layer, the dentate molecular layer, and in the pyramidal cell layer[137]. Thus, lead appears to exert rather com-

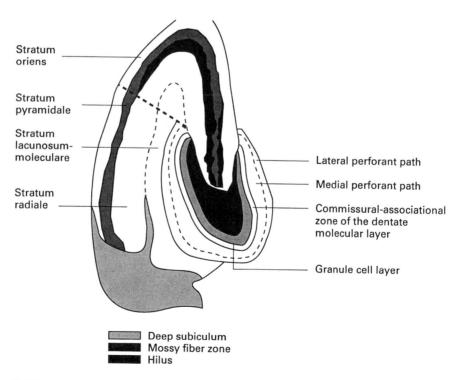

FIGURE 12.4

Hippocampal region of the rat
(Adapted from Slomianka, L. et al. (1989) *Neurotoxicology* **10**: 177-190. With permission.)

plex effects on hippocampal morphology depending on the exposure level and regimen used. In a series of other studies, morphological alterations in the hippocampus were described together with a diminished cholinergic plasticity[3]. These effects were observed at blood lead levels of about 1400 μg 100 ml^{-1}, a level at which severe signs of intoxication must have been present. In this series of experiments behavioral effects on spatial learning were also reported[2]. However, at lower levels of exposure the effects by lead are less pronounced. Certainly, an increased number of days to criterion as well as performance changes such as an increased number of days missing rewards and a reduced number of visited arms per minute were found in a radial arm maze in rats with maternal or lifetime low level lead exposure[105], but these behavioral effects were clearly different from the effects caused by chemical lesions with ibotenic acid of the dorsal hippocampus. The number of days to criterion were not altered in this positive control group in comparison to sham-lesioned controls, while the arms per minute were changed in opposite directions by lead and by lesions. The only similarity between the lesioned and the lead-treated groups in comparison to controls was in the retention phase of the experiment.

In contrast, much clearer effects were found on LTP in the hippocampus both after acute and chronic treatment[5]. This study reported an impairment of LTP at blood lead levels below 20 μg 100 ml^{-1}. Since a close relation between LTP and spatial learning, particularly in the Morris water tank, was detected after lesions or treatment with pharmacological agents acting on the NMDA receptor[104], this poses the questions as to whether or not low level lead effects on spatial learning are to be re-evaluated as lead is known to alter NMDA receptor function[4, 63] and to inhibit NMDA-activated currents in the hippocampus[144].

1.3   SUMMARY FOR LEAD EFFECTS IN ANIMALS

Although behavioral effects have been also detected after exposure in adult animals, the early developmental stages are particularly sensitive to lead. Exposure in early life leads to long-lasting changes which can be found even when lead tissue levels have declined to control values and the impairments are detected at lower dose levels. Neurochemical data and the behavioral pharmacology point at an involvement of the dopaminergic system in lead-induced effects. The results indicate a mediation by the D_2-type of the DA receptors rather than by the D_1 receptor. The pattern of the learning impairments in primates is largely similar to the pattern found after lesions in the prefrontal cortex and, moreover, after depletion of DA in that area. In addition, several lines of evidence indicate effects on the hippocampus. The morphology and the function of this part of the brain are affected by lead. Reduced LTP was detected in the hippocampus of rats after acute and chronic lead treatment which might be related to effects on the NMDA receptor.

2   Mercury

Mercury in its organic and inorganic ionized and metallic forms causes neurotoxic effects which have been repeatedly reviewed in comprehensive articles that the reader is referred to for more information[32, 71]. Whereas organic forms of mercury, like methylmercury (MMC), and also metallic mercury vapor because of its lipophilic nature enter the brain readily, passage of inorganic mercury to the CNS is hindered by the blood brain barrier. However, neurobehavioral alterations have been described for all forms. Among the effects, impairments of sensory functions are particularly striking.

2.1   BEHAVIOR

In squirrel monkeys no impairment of visual discrimination learning and learning set behavior was observed after MMC exposure in the adult stage until visual disturbances emerged[18, 69]. In cynomolgus monkeys treated pre- and postnatally with MMC there were no differences to controls in reversal learning, but monkeys in the high dose group (peak level 2 µg Hg g blood^{-1}, steady state level 1µg g^{-1}) required more trials to learn the original discrimination[117]. In the latter species there was also an impairment of the development of object permanence together with sensorimotor disturbances after prenatal exposure[24]. In addition, these monkeys exhibited a deficit in visual recognition memory[64], a test which was originally developed for human infants and which is regarded as a valid predictor of later intelligence in children[51]. It appears from these studies that cognitive functions are impaired by MMC only after prenatal exposure and often together with sensorimotor effects, while after postnatal treatment sensory deficits become more striking and learning behavior is affected as a consequence of this.

One of the first studies of behavioral effects by MMC which inspired many following scientists was conducted by Spyker et al.[139] in mice. These authors examined the development of activity in the open field and of swimming behavior on PD 30 following maternal exposure on GD 7 or GD 9. The doses used (0.16 mg MMC dicyandiamide/20 g bw) did not result in overt toxic signs. In the open field latency to movement onset, backward stepping, urination, and defecation were increased by MMC suggesting an influence on emotional reactivity in this test. In treated mice there were also signs of neuromuscular impairments like motionless floating, excessive movements of all legs and the tail, as well as a vertical position in the water. No effects were found on the activity of cholinergic enzymes in the brain.

Examination of sensitive periods for exposure to MMC revealed deficits in the acquisition of a spatial discrimination in a water T-maze in rats exposed prenatally or after weaning until testing on PD 30. On a retention test 3 weeks later only the groups with prenatal or maternal exposure were impaired. In contrast, postnatal exposure by cross-fostering pups from control dams to treated dams failed to exert an influence on either acquisition or retention[157]. This result is in general accordance with the primate studies as it demonstrates the importance of the prenatal exposure period for the development of learning deficits.

Prenatal and/or early postnatal exposure to MMC also affects tactile discrimination, operant performance on a DRH (differential reinforcement of high rates) schedule in rats, extinction of taste aversion learning in mice, and detour learning in chicks hatched from MMC treated eggs.

Further studies tried to evaluate the sensitivity in different prenatal phases by administration of MMC to rat dams on GD 8 or GD 15. On PD 65 the offspring exhibited deficits in the acquisition and relearning of active avoidance in the shuttle box which were more pronounced after treatment during late gestation (GD 15). In these animals there were also effects on preweaning locomotor activity, but the difference between both exposure phases was less clear. Performance on a operant task using a DRL (differential reinforcement of low rates) schedule was not different from controls. However, after a challenge with +AP controls exhibited shorter interresponse times and, thus, more disruption of DRL performance at the higher +AP dose than MMC treated rats suggesting an altered sensitivity to this drug and a shift in the dose-response curve[46, 47]. There are further effects by MMC on sleep-waking rhythms[9, 10], sleep disorders, and other behavioral paradigms some of which are summarized in reviews[25, 133].

Behavioral alterations were also reported following exposure to inorganic mercury in rats[74] and after inhalation of metallic mercury vapor. In a pole-climbing chamber the latencies of escape responses were increased and the percentage of avoidance responses was decreased with the duration of exposure to mercury vapor at a level of 17 mg/m^3 on 2 hours per day for a total of 30 days in rats. In addition, shock-elicited fighting was increased[15]. The behavioral effects were reported to recover after termination of exposure. Recovery of behavioral changes after exposure in adult rats was also observed in another study. However, after prenatal exposure to 1.8 mg Hg/m^3 on 1 or 3 hours per day levels of locomotor activity were found to be affected up to 14 months of age in rats and spatial learning in a radial arm maze was impaired at an age of 3 months[41].

2.2   AUDITION

The ototoxic effects by mercury differ from the effects by many other compounds like aminoglycoside antibiotics and cis-platinum by causing the most severe damage to the apical coils of the cochlea related to low-frequency hearing. Outer hair cells in the apical cochlea of guinea pigs were found to be affected by MMC[52, 53] and by inorganic mercury[8]. Cellular changes in the guinea pig cochlea include derangement of the stereocilia pattern of outer hair cells, cytoplasmic vesiculation, and damage to cell membranes following exposure to 2 mg kg bw^{-1} daily for periods from 1 to 11 weeks. However, the testing of hearing function using Preyer's reflex failed to reveal any abnormalities[52, 53]. In another experiment in guinea pigs using acute treatment with 10-25 mg kg bw^{-1}, subacute treatment with 7.5 mg kg^{-1}, or chronic exposure to 2.5-5.0 mg kg^{-1} daily for several weeks degeneration of hair cells was detected only in animals with chronic exposure. In addition, degenerations of afferent and efferent nerve endings were found as well as vacuolization of marginal cells in the stria vascularis. In this study effects on nerve terminals seemed to preceed damage in hair cells. After acute exposure Preyer's reflex was transiently lost, while in chronic animals it disappeared after prolonged exposure. The disturbances after chronic exposure were observed in the presence of substantial decreases in body weights. The initial cochlear lesions after MMC exposure are followed by retro-cochlear lesions in the auditory pathway[101].

2.3   VISION

Most studies examining the visual system have used MMC as the toxicant. Apart from studies in human patients with mercury intoxications, most of our knowledge about visual disturbances by mercury derives from studies in nonhuman primates. This model was chosen by most investigators because of the similarity between the human and the monkey visual system.

Visual field constriction, a symptom observed in poisoned patients was also detected in the macaques at Hg blood levels of 300 µg 100g blood^{-1} [99] and also reported in a more qualitative form in squirrel monkeys. However, this effect can be found only in severe cases of poisoning. In the early stages deficits in scotopic vision are more obvious. MMC exposure resulted in a reduced discrimination performance at scotopic conditions in macaques. In squirrel monkeys a progressive increase in the luminance levels at which a flickering light of 10 Hz could be discriminated from steady light was reported with increasing mercury levels in the blood. In accordance with this, in macaques a reduced temporal contrast sensitivity (TCS) was described[99]. TCS is a measure of the threshold at which a flickering light stimulus with a given depth of modulation can be distinguished from steady light of equal luminance. However, reductions of TCS may only be found at higher exposure levels, since at lower levels a better performance after MMC treatment was found[117].

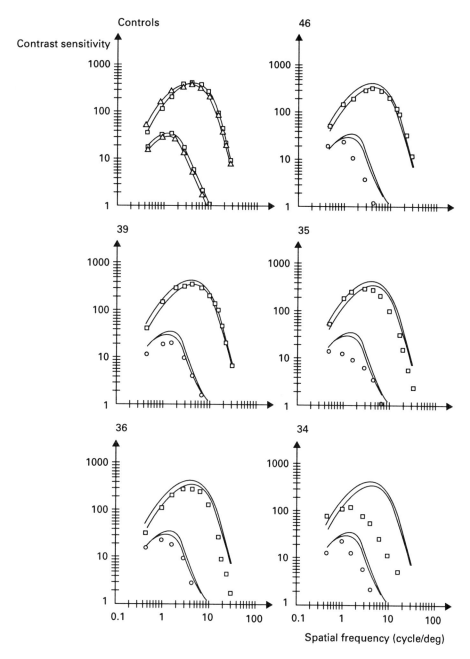

FIGURE 12.5

Spatial contrast sensitivity functions for two controls and five monkeys treated with MMC at high and low luminance conditions, respectively

Values for the individual controls are given by squares and triangles, respectively. For each exposed monkey, squares give the thresholds at high luminance, circles at low luminance, while the solid lines represent the envelopes of the control thresholds. (Redrawn from Rice, D.C. and Gilbert S.G. (1982) *Science* **216**: 759-761. Copyright 1982. American Association for the Advancement of Science.)

The most pronounced visual changes by MMC were detected on spatial contrast sensitivity (SCS) in cynomolgus macaques. SCS resembles TCS in the spatial domain and is determined using sinusoidal gratings of various spatial frequencies and modulation depths. In the absence of overt toxic symptoms at peak Hg blood levels of about 1.3 µg g^{-1} and steady state levels of 0.6-0.9 µg g blood^{-1} three of five monkeys were impaired at high frequencies at the photopic luminance level used, while all exposed monkeys exhibited deficits at the low luminance level[116] (Figure 12.5).

Taken together, these studies demonstrate effects of all mercury species on several types of behavior in addition to an impairment of sensory function. While behavioral effects in adults at levels that which do not cause overt toxicity appear to be transient and even at higher levels predominantly affect sensory and motor processes, the particular sensitivity in early developmental phases results in long-lasting and more generalized deficits.

2.4   MECHANISMS

In human patients suffering from MMC poisoning effects on the sensory cells in the dorsal root ganglion of the spinal cord were detected leading to peripheral neuropathy. In addition focal necrosis in the visual areas of the occipital cortex and in the granular layer of the cerebellum were observed[32]. These findings fit well to the reported visual and motor disturbances by mercury at higher exposure levels. The described primary events which may lead to damage in the central nervous system include the formation of reactive oxygen species[83] as well as effects on mitochondrial DNA synthesis in cerebral and cerebellar cell types prone to undergo neurodegeneration, in particular in the granular cell layer of the cerebellum[100].

Inorganic mercury was shown to affect NA, 5-HT, and DA levels in various parts of the brain as well as ACh-esterase activity in developing rats[79]. Moreover, it was found to increase the binding of the stimulant methylphenidate to the DA transporters at low concentrations *in vitro*, while an inhibitory effect was detected at more elevated levels[130].

2.5   SUMMARY FOR MERCURY

Mercury enters the organism in organic and inorganic forms. The most prevalent organic compound MMC and metallic mercury vapor are more readily transferred to the brain than inorganic mercury salts. Most experimental studies are devoted to MMC. Prenatal exposure to MMC causes behavioral changes including the development of locomotor activity and swimming behavior in mice, operant conditioning, spatial learning in the T-maze, and altered sleep-waking rhythms in rats. Prenatal exposure to mercury vapor impaired spatial learning in rats. Cognitive deficits on various tasks were detected after prenatal MMC treatment in primates. In contrast, postnatal exposure affects predominantly sensory systems. At higher levels of chronic exposure to MMC or inorganic mercury, degenerations of nerve terminals and hair cells in the cochlea were found in guinea pigs, in some cases together with impaired auditory function. In primates high levels of MMC result in visual field constriction, whereas lower levels affect particularly scotopic vision. The most sensitive visual function is SCS at scotopic conditions. Mercury exerts influences on several neurotransmitter systems in the brain and causes both the formation of reactive oxygen sprecies and disruption of DNA synthesis in mitochondria in nerve cells, events which finally lead to neuronal degeneration.

3   Manganese

3.1   NEURONAL AND BEHAVIORAL EFFECTS

In comparison to lead and mercury information about manganese is limited. From human studies the induction of parkinson-like symptoms is known suggesting an interference with the nigrostriatal dopaminergic system. In general, this could be confirmed by experimental studies in animals. In comparison to divalent manganese salts like $MnCl_2$, $MnCO_3$, and manganese acetate, manganese(IV) dioxide (MnO_2) was found to be more potent in reducing DA concentrations in the striatum, the hypothalamus, and the midbrain, while it caused increases in the concentration of the DA metabolite HVA in the striatum, the midbrain, and in the cerebellum in mice. In contrast, DA levels were increased by MnO_2 in the cerebral cortex, the cerebellum, and in the medulla oblongata. In addition, concentration changes in norepinephrine and epinephrine were observed in mice, the direction of changes depending on the brain region studied. These results were detected together with a progressive body weight loss during the duration of exposure. The most prominent elevations of manganese levels in the brain were found in the hypothalamus and in the cerebral cortex. Motor activity was depressed by MnO_2 and $MnCO_3$[77].

In juvenile monkeys injected with a total dose of 8 g MnO_2 during 5 months manganese concentrated most in the following brain regions: globus pallidus > putamen > caudate nucleus > substantia nigra > cerebellum > cerebral cortex. Neurochemical alterations were also most pronounced in the globus pallidus and the putamen. DA depletion was most severe in these regions followed by the caudate nucleus. Reductions were also observed in concentrations of the DA metabolites DOPAC and HVA as well as in levels of 5HT and 5HIAA. Moreover, decreases in DOPA decarboxylase activity indicative of a reduced number of DA terminals was reported. Behavioral observation revealed initial hyperactivity which turned to hypoactivity during the exposure period. In addition, unsteady gait, tremor, weakness of upper and lower limbs, and clumsiness were described[49]. The reported changes which suggest a degeneration of DA terminals predominantly in the globus pallidus and the putamen are similar to disorders in Parkinson's disease and Parkinsonism induced by MPTP. In accordance with this, a depletion of neuromelanin in the substantia nigra was detected in another primate study[65].

In weanling rats exposed daily to 50 g $MnCl_2$ per subject for 60 days a marked increase in monoamine oxidase activity was found which suggests disturbance of catecholamine metabolism and may explain decreases in DA levels. During the treatment period encephalopathy with neuronal degeneration in the cerebral and cerebellar cortex followed by neuroglial proliferation in the cerebral cortex and the caudate nucleus was found. In comparison to adult animals, neurotoxic signs developed earlier and at lower dose levels in growing rats[31]. Since the dopaminergic system develops predominantly before weaning in rats (see above), a stronger influence of exposure during early development is reasonable and preweaning exposure to manganese should be even more effective.

3.2   SUMMARY FOR MANGANESE

Manganese interferes with the nigrostriatal system, thereby causing disturbances in the dopaminergic system. Manganese concentrates in the basal ganglia and neurochemical effects on the levels of DA and serotonin and their metabolites were detected in these areas. Behavioral studies revealed locomotor activity changes as well as disturbances in motor coordination. Higher exposure levels lead to neurodegeneration. The developmental stages seem to be more sensitive to manganese treatment than later ages. The general pattern of neuronal dysfunctions is similar to the one seen in Parkinson's disease.

4   Aluminum

4.1   NEURONAL AND BEHAVIORAL EFFECTS

Aluminum (Al) is known to cause the dialysis encephalopathy syndrome in patients with chronic renal deficiency who received dialysate with high levels of Al. In addition, Al exposure may be a risk factor in neurodegenerative diseases like Morbus Alzheimer. Thus, it was used as a model substance in experimental studies to investigate the mechanistic processes underlying neurodegeneration.

Behavioral studies in rabbits revealed deficits in active avoidance after intraventricular infusion of 5 µM Al[111] and an impairment of the conditioned nictitating membrane extension after 20 SC injections of 200 or 400 µmol Al within 4 weeks[156]. In adult rats Al exposure resulted in impaired passive avoidance behavior, while locomotor activity, active avoidance, and spatial learning in a radial-arm maze were not changed[70]. The exposure (0.3% in drinking water for 4 weeks) caused a tenfold increase in serum Al levels in comparison to controls.

However, prenatal exposure to Al may result in a different profile of behavioral effects, since exposed mice (200 mg/kg IP from GD 10 to GD 13) exhibited a decreased number of correct responses and an increased number of trials to enter all eight arms in a radial maze[126], pointing to an implication of the hippocampus in AL effects on the brain.

In a cross-fostering study in mice forelimb grip strength was affected by prenatal exposure, whereas negative geotaxis was influenced by postnatal exposure, and hindlimb grasp strength as well as temperature deficiency were altered by both gestational and lactational exposure to a 1000 mg Al/kg diet[60]. Feeding the same diet to growing mice for 90 days starting at an age of about 5 weeks resulted in decreased motor activity, reduced grip strength, and a decreased auditory startle response[61]. Brain Al levels were elevated threefold by the exposure.

The mechanisms underlying these behavioral effects are largely unknown. There may be a relation to an altered activity of choline acetyltransferase (ChAT) which depending on the brain area studied was found to be reduced or elevated after prenatal exposure to Al[34]. Damage to the cholinergic system has been implicated in many neurodegenerative diseases including Morbus Alzheimer.

Other effects by Al are increased levels of nerve growth factor (NGF) in the hippocampus of adult mice[126], which suggests a stimulating influence on brain repair mechanisms, and enhanced neurite outgrowth in treated neuroblastoma cells[143], that was not reversable after a change to control medium and which is similar to neurite sprouting described in the brains of Alzheimer patients.

Other animal studies reported Al effects on second messenger systems and increases in the phosphorylation of the cytoskeletal proteins MAP-2 (microtubule-associated protein 2) and NF-H (neurofilament protein). Abnormal phosphorylation may lead to impaired axonal transport resulting in altered concentrations of these structural proteins. Severe decreases in the concentration of MAP-2 were detected in the brainstem, cortex, and hippocampus as well as a reduced level of spectrin in the hippocampus of exposed rats. Cytoskeletal proteins are affected in many neurodegenerative diseases.

4.2   SUMMARY FOR ALUMINUM

There is a pronounced similarity between the neurodegenerative effects by Al and in certain neurological disorders. After Al exposure defics in classical conditioning, passive avoidance, grip strength, and auditory startle were detected at the behavioral level and at certain conditions impairment of spatial learning and altered motor activity. Similarities to certain neurodegenerative disorders were described such as effects on the cholinergic system, neurite outgrowth, and the cytoskeletal proteins. This suggests a possible implication of Al in diseases like Morbus Alzheimer, amyotrophic lateral sclerosis, and also Morbus Parkinson. Alternatively, there may be no causative relation to the development of disorders of this kind, so that the neurogenerative effects by Al exposure may simply resemble the symptoms of these diseases. In this way , Al exposure may serve as a model substance and also, can help to gain insight in the mechanism underlying the toxic effects in patients with renal failure who were intoxicated by dialysis fluids.

5   General summary and conclusions for animal studies

Animal studies help to clarify questions which are not possible or difficult to answer in epidemiological studies of the neurobehavioral effects by environmental exposure. Among these are the question of the cause-effect relationship, the dose-effect relationship, the mechanism by which effects occur, the determination of critical exposure periods, and the question of reversibility of effects. Control for confounding variables is more easily achieved under experimental conditions. Mechanisms can be isolated, the exposure regimen is exactly controlled for level and time, and comparative approaches offer the opportunity for a separation of species-specific and general findings. All these issue are important for the understanding of the development and persistence of environmental effects in humans, and help to identify possible risks and the levels at which effects might occur.

References

1. Alexander, G.E. and Goldmann, P.S. (1978) Functional development of the dorsolateral prefrontal cortex: an analysis utilizing reversible cryogenic depression. *Brain Res.* **143**: pp 233–249.

2. Alfano, D.P. and Petit, T.L. (1981) Behavioral effects of postnatal lead exposure: possible relationship to hippocampal dysfunction. *Behav. Neur. Biol.* **32**: pp 319–333.

3. Alfano, D.P. et al. (1983) Development and plasticity of the hippocampal cholinergic system in normal and early lead exposed rats. *Dev. Brain Res.* **10**: pp 117–124.

4. Alkondon, M. et al. (1990) Selective blockade of NMDA-activated channel currents may be implicated in learning deficits caused by lead. *FEBS Lett.* **261**: pp 124–130.

5. Altmann, L. et al. (1993) Impairment of long-term potentiation and learning following chronic lead exposure, Toxicol. Lett., **66**: pp 105–112.

6. Amin-Zaki, L. et al. (1974) Intra-uterine methylmercury poisoning in Iraq. *Pediatrics* **54**: pp 587–595.

7. Angell, N.F., Weiss, B. (1982) Operant behavior of rats exposed to lead before and after weaning. *Toxicol. Appl. Pharmacol.* **63**: pp 62–71.

8. Anniko, M. and Sarkady, L. (1978) Cochlear pathology following exposure to mercury. *Acta Otolaryngol.* **85**: pp 213–224.

9. Arito, H. et al. (1982) Changes in circadian sleep-waking rhythms of rats following administration of methylmercury chloride. *Indust. Health* **20**: pp 55–65.

10. Arito, H. et al. (1983) Effect of methylmercury chloride on sleep-waking rhythms in rats. *Toxicology* **28**: pp 335–345.

375

11. Audesirk, G. (1985) Effects of lead exposure on the physiology of neurons. *Prog. Neurobiol.* **24**: pp 199–231.

12. Baghurst, P.A., McMichael, A,J., Wigg, N.R. et al. (1992) Life-long exposure to environmental lead and childrens's intelligence at age seven: the Port Pirie cohort study. *N. Engl. J. Med.* **327**: pp 1269–1284.

13. Bakir, F.. et al. (1973) Methylmercury poisoning in Iraq. *Science* **181**: pp 230–241

14. Begerow, J. (1992) Quecksilberbelastung durch Amalgamfuellungen. In: Ges. z. Foerd. Lufthygiene- und Silikoseforschung (Ed.) *Umwelthygiene* **24**: pp 71–91.

15. Beliles, R.P. et al. (1968) The effects of exposure to mercury vapor on behavior of rats. *Toxicol. Appl. Pharmacol.* **12**: pp 15–21.

16. Bellinger, D., Leviton, A., Waternaux, C. et al. (1987) Longitudinal analyses of prenatal and postnatal lead exposure and early cognitive development. *N. Engl. J. Med.* **316**, pp 1037–1043.

17. Berlin, M. (1986) Mercury. In: Friberg, L., Nordberg, G.F., Vouk, V.B. (Eds.) *Handbook on the Toxicology of Metals.* Elsevier, Amsterdam, pp 387–435.

18. Berlin, M. et al. (1975) Neurotoxicoty of methylmercury in squirrel monkeys. *Arch. Environ. Health* **30**: pp 340–348.

19. Berry, M. (1974) Development of the cerebral neocortex of the rat. In: Michael, R.P. (Ed.) *Studies on the Development of Behavior and the Nervous System: Aspects of Neurogenesis,* Vol. 2. Academic Press, New York, pp 7–67.

20. Bleeker, M.L. (1988) Parkinsonism: a clinical marker of expsoure to neurotoxins. *Neurotoxicol. Teratol.* **10**: pp 475–478.

21. Bornschein, R., Pearson, D., Reiter, L. (1980) Behavioral effects of moderate lead exposure in children and animal models, *Crit. Rev. Toxicol.* **7**: pp 43–152.

22. Bressler, J.P. and Goldstein, G.W. (1991) Mechanisms of lead neurotoxicity. *Biochem. Pharmacol.* **41**: pp 479–484.

23. Brozovski, T.J. et al. (1979) Cognitive deficit caused by regional depletion of dopamine in the prefrontal cortex of rhesus monkeys. *Science* **205**: pp 929–932.

24. Burbacher, T.M. et al., Retarded object permanence development in methylmercury exposed Macaca fascicularis infants. *Dev. Psychol.* **22**: pp 771–776.

25. Burbacher, T.M. et al. (1990) Methylmercury developmental neurotoxicity: a comparison of effects in humans and animals. *Neurotoxicol. Teratol.* **12**: pp 191–202.

26. Bushnell, P.J. and Bowman, R.E. (1979) Reversal learning deficits in young monkeys exposed to lead. *Pharmacol. Biochem. Behav.* **10**: pp 733–742.

27. Bushnell, P.J. and Bowman, R.E. (1979) Persistence of impaired reversal learning in young monkeys exposed to low levels of dietary lead. *J. Toxicol. Environ. Health* **5**: pp 1015–1023.

28. Campbell, J. et al. (1982) Morphometric effects of postnatal lead exposure on hippocampal development of the 15-day-old rat. *Dev. Brain Res.* **3**: pp 595–612.

29. Campbell, J. et al. (1984) Morphometric effects of preweaning lead exposure on the hippocampal formation of adult rats. *Neurotoxicology* **5**: pp 125–148.

30. Centers for Disease Control (CDC) (1991) *Preventing Lead Poisoning in Young Children.* U.S. DHHS, Public Health Service.

31. Chandra, S.V. and Shukla, G.S. (1978) Manganese encephalopathy in growing rats. *Environ. Res.* **15**: pp 28–37.

32. Clarkson, T.W. (1983)Methylmercury toxicity. In: Sarkar, D. (Ed.) *Biological Aspects of Metals and Metal-Related Diseases.* Raven Press, New York, pp 183–197.

33. Cooney, G.H., Bell, A., McBride, W., Carter, C. (1989) Neuro-behavioral consequences of prenatal low level exposures to lead. *Neurotoxicol.Teratol.* **11**: pp 195–104.

34. Clayton, R.M. et al., Long-term effects of aluminum on the fetal mouse brain. *Life Sci.* **51**: pp 1921–1928.

35. Cory-Slechta, D.A. and Thompson, T. (1979) Behavioral toxicity of chronic postweaning lead exposure in the rat. *Toxicol. Appl. Pharmacol.* **47**: pp 151–159.

36. Cory-Slechta, D.A. and Widzowski, D.V. (1991) Low level lead exposure increases sensitivity to the stimulus properties of dopamine D1 and D2 agonists. *Brain Res.* **553**: pp 65–74.

37. Cory-Slechta, D.A. et al. (1992) Postnatal lead exposure induces supersensitivity to the stimulus properties of a D2-D3 agonist. *Brain Res.* **598**: pp 162–172.

38. Cox, C., Clarkson, T.W., Marsh, D.O. et al. (1989) Dose-response analysis of infants prenatally exposed to methyl mercury: an application of a single compartment model to single- strand hair analysis. *Environ. Res.* **49**: pp 318–332.

39. Crapper, D.R. and De Boni, U. (1980) Aluminum. In: Spencer, P.S. and Schaumburg, H.H. (Eds.), *Experimental and Clinical Neurotoxicology*. Williams & Wilkins, Baltimore, MD, pp 326–335.

40. Crofton, K.M. et al. (1980) Developmental delays in exploration and locomotor activity in male rats exposed to low level lead. *Life Sci.* **26**: pp 823–831.

41. Danielson, B.R.G. et al. (1993) Behavioral effects of prenatal metallic mercury inhalation exposure in rats. *Neurotoxicol. Teratol.* **15**: pp 391–396.

42. Davis, S.F. et al. (1993) The effects of chronic lead exposure on reactivity to frustrative nonreward in rats. *Toxicol. Lett.* **66**: pp 237–246.

43. Dietrich, K.N., Krafft, K.M., Bornschein, R.L. et al. (1987) Low-level fetal lead exposure effect on neurobehavioral development in early infancy. *Pediatrics* **80**: pp 721–730.

44. Dietrich, K.N., Berger ,O.G., Succop, P.A., Hammond, P.B., Bornschein, R.L. (1993) The developmental consequences of low to moderate lead exposure: intellectual attanment in the Cincinnati lead study cohort following school entry. *Neurotoxicol. Teratol.* **15**: pp 37–44.

45. Drew, W.G. et al. (1979) Effects of neonatal lead exposure on apomorphine-induced aggression and stereotypy in the rat. *Pharmacology* **18**: pp 257–262.

46. Eccles, C.U. and Annau, Z. (1982) Prenatal methyl mercury exposure: I. Alterations in neonatal activity. *Neurobehav. Toxicol. Teratol.* **4**: pp 371–376.

47. Eccles, C.U. and Annau, Z. (1982) Prenatal methyl mercury exposure: II. Alterations in learning and psychotropic drug sensitivity in adult offspring. *Neurobehav. Toxicol. Teratol.* **4**: pp 377–382.

48. Elinder, C.G. and Sjogren, B. (1986) Aluminum. In: Friberg, L., Nordberg, G.F., Vouk, V.B. (Eds.), *Handbook on the Toxicology of Metals*. Elsevier, Amsterdam.

49. Eriksson, H. et al. (1987) Effects of manganese oxide on monkeys as revealed by a combined neurochemical, histological and neurophysiological evaluation. *Arch. Toxicol.* **61**: pp 46–52.

50. Ernhart, C.B., Morrow-Tlucak, M., Marler, M.R., Wolf, A.W. (1987) Low-level lead exposure in the prenatal and early preschool periods: early preschool development. *Neurotoxicol.Teratol.* **9**: pp 259–270.

51. Fagan, J.F. and McGrath, S.K. (1981) Infant recognition memory and later intelligence. *Intelligence* **5**: pp 121–130.

52. Falk, S.A. et al. (1973) The ototoxicity of methyl mercury. *Toxicol. Appl. Pharmacol.* **25**: pp 465–466.

53. Falk, S.A. et al. (1974) Acute methyl mercury intoxication and ototoxicity in guinea pigs. *Arch. Pathol.* **97**: pp 297–305.

54. Fulton, M., Raab, G., Thompson, G. et al. (1987) Influence of blood lead on the ability and the attainment of children in Edinburgh. *Lancet* **I**: pp 1221–1226.

55. Gilbert, S.G. and Rice, D.C. (1987) Low level lifetime lead exposure produces behavioral toxicity (spatial discrimination reversal) in adult monkeys. *Toxicol. Appl. Pharmacol.* **91**: pp 484–490.

56. Goldman, P.S. (1971) Functional development of the prefrontal cortex in early life and the problem of neuronal plasticity. *Exp. Neurol.* **32**: pp 366–387.

57. Goldman, P.S. (1974) Recovery of function after CNS lesions in infant monkeys. *Neurosci. Res. Progr. Bull.* **12**: pp 211–222.

58. Goldman-Rakic, P.S. (1987) Circuitry of primate prefrontal cortex and regulation of behavior by representational memory. In: Plum, F. and Mountcastle, V. (Eds.), *Handbook of Physiology: The Nervous System*. American Physiological Society, Bethesda, MD, pp 373–417.

59. Goldsmith, J.R., Herishanu, Y., Abaranel, J.M., Weinbaum, Z. (1990) Clustering of Parkinson's disease points to environmental origin. *Arch. Environ. Health* **45**: pp 88–94.

60. Golub, M.S. et al. (1992) Effects of dietary aluminum excess and manganese deficiency on neurobehavioral endpoints in adult mice. *Toxicol. Appl. Pharmacol.* **112**: pp 154–160.

61. Golub, M.S. (1992) Neurodevelopmental effect of aluminum in mice: fostering studies. *Neurotoxicol. Teratol.* **14**: pp 177–182.

62. Grant, W.M. (1974) *Toxicology of the Eye*, 2nd edition. Charles C Thomas, Springfield, IL.

63. Guilarte, T.R. et al. (1993) Chronic prenatal and postnatal Pb^{2+} exposure increases [^{3}H]MK801 binding sites in adult rat forebrain. *Eur. J. Pharmacol. [Environ. Toxicol. Pharmacol Sect.]* **248**: pp 273–275.

64. Gunderson, V.M. et al. (1986) The effect of low level prenatal methylmercury exposure on visual recognition memory in infant crab-eating macaques. *Child Dev.* **57**: pp 1076–1083.

65. Gupta, S.K. et al. (1980) Neuromelanin in manganese-exposed primates. *Toxicol. Lett.* **6**: pp 17–20.

66. Harada, Y. (1966) Study group on Minamata disease. In: Katsuma, M. (Ed.), *Minamata Disease*. Kumamoto University, Kumamoto.

67. Hatzakis, A., Kokkevi, A., Maravelias, C. et al. (1989) Psychometric intelligence deficits in lead-exposed children. In: Smith, M., Grant, L., Sors, A. (Eds.), *Lead Exposure and Child Development: An International Assessment*. Academic Publications, Dordrecht, pp 211–223.

68. Hopper, D.L., et al. (1986) The behavioral effects of prenatal and early postnatal lead exposure in the primate *Macaca fascicularis*. *Toxicol. Ind. Health* **2**: pp 1–16.

69. Joiner, F.E., Hupp (1978) Behavioral observations in squirrel monkeys (*Saimiri sciureus*) following methylmercury exposure. *Environ. Res.* **16**: pp 18–28.

70. Jope, R.S., Johnson, G.V.W. (1992) Neurotoxic effects of dietary aluminum. *Ciba Foundation Symposium. Aluminum in Biology and Medicine* **169**: pp 254–267.

71. Junghans, R.P. (1983) A review of the toxicity of methylmercury compounds with application to occupational exposures associated with laboratory uses. *Environ. Res.* **31**: pp 1–31.

72. Kjellström, T., Kennedy, P., Wallis, S., Mantell, C. (1986) *Physical and Mental Development of Children with Prenatal Exposure to Mercury from Fish. Stage 1: Preliminary Tests at Age 4.* National Swedish Environmental Protection Board, Report 3080.

73. Kjellström, T., Kennedy, P., Wallis, S. et al. (1989) *Physical and Mental Development of Children with Prenatal Exposure to Mercury from Fish. Stage 2: Interviews and Psychological Tests at Age 6.* National Swedish Environmental Protection Board, Report 3642.

74. Klein, S.B. and Atkinson, E.J. (1973) Mercuric chloride influence on active-avoidance acquisition in rats. *Bull. Psychonom. Soc.* **1**: pp 437–438.

75. Klein, A.W. and Koch, T.R. (1981) Lead accumulations in brain, blood, and liver after low dosing of neonatal rats. *Arch. Toxicol.* **47**: pp 257–262.

76. Kohler, K. et al. (1993) Low level lead exposure results in a chronic decrease of tyrosin hydroxylase-like immunoreactivity in the rhesus monkey retina. *Invest. Ophthalmol. Vis. Sci.* **34** (Suppl.): p 754.

77. Komura, J. and Sakamoto, M. (1992) Effects of manganese forms on biogenic amines in the brain and behavioral alterations in the mouse: long-term oral administration of several manganese compounds. *Environ. Res.* **57**: pp 34–44.

78. Kondakis, X.G., Makris, N., Leotsinidis, M. et al. (1989) Possible health effects of high manganese concentration in drinking water. *Arch. Environ. Health* **44**: pp 175–178.

79. Lakshmana, M.D., Desiraju, T., Raju, T.R. (1993) Mercuric chloride-induced alterations of levels of noradrenaline, dopamine, serotonin and acetylcholine esterase activity in different regions of rat brain during postnatal development. *Arch. Toxicol.* **67**: pp 422–427.

80. Langworth, S., Almkvist, O., Söderman, E., Wikström, B.O. (1992) Effects of occupational exposure to mercury vapour on the central nervous system. *Br. J. Ind. Med.* **49**: pp 545–555.

81. Lasley, S.M. et al. (1988) Diminished regulation of mesolimbic dopaminergic activity in rat after chronic inorganic lead exposure. *Toxicol. Appl. Pharmacol.* **95**: pp 474–483.

82. Laughlin, N.K. et al. (1983) Neurobehavioral consequences of early exposure to lead in rhesus monkeys: effects on cognitive behaviors. In: Clarkson, T.W. et al. (Eds.), *Reproductive and Developmental Toxicity of Metals*. Plenum Press, New York, pp 497–515.

83. LeBel, C.P. et al. (1992) Deferoxamine inhibits methyl mercury-induced increases in reactive oxygen species formation in rat brain. *Toxicol. Appl. Pharmacol.* **112**: pp 161–165.

84. Levin, E.D. and Bowman, R.E. (1983) The effect of pre- and postnatal lead exposure on Hamilton search task in monkeys. *Neurobehav. Toxicol. Teratol.* **5**: pp 391–394.

85. Levin, E.D. and Bowman, R.E. (1986) Long-term lead effects on the Hamilton search task and delayed spatial alternation in monkeys. *Neurobehav. Toxicol. Teratol.* **8**: pp 219–224.

86.   Levin, E.D. et al. (1987) Psychopharmacological investigations of a lead-induced long-term cognitive deficit in monkeys. *Psychopharmacology* **91**: pp 334–341.

87.   Levin, E.D. et al. (1992) Use of the lesion model for examining toxicant effects on cognitive behavior. *Neurotoxicol. Teratol.* **14**: pp 131–141.

88.   Liang ,Y.X., Sun, R.K., Sun, Y. et al. (1993) Psychological effects of low exposure to mercury vapour: application of a computer-administered neurobehavioral evaluation system. *Environ. Res.* **60**: pp 320–327.

89.   Lilienthal, H. et al. (1986) Pre- and postnatal lead exposure in monkeys: effects on activity and learning set formation. *Neurobehav. Toxicol. Teratol.* **8**: pp 265–272.

90.   Lilienthal, H. et al. (1994) Persistent increases in scotopic b-wave amplitudes after lead exposure in monkeys. *Exp. Eye Res.* **59**: pp 203–209.

91.   Lögdberg, B. et al. (1988) Congenital lead encephalopathy in monkeys. *Acta Neuropathol.* **77**: pp 120–127.

92.   Louis-Ferdinand, R.T. et al. (1978) Morphometric and enzymatic effects of neonatal lead exposure in the rat brain. *Toxicol. Appl. Pharmacol.* **43**: pp 351–360.

93.   Mantere, P., Hänninen, H., Hernberg, S.. (1982) Subclinical neurotoxic lead effects: two-year follow-up study with psychological test methods. *Neurotoxicol. Teratol.* **4**: pp 725-727.

94.   Markham, C.H. (Ed.), Parkinson's disease. *Clin. Neurosci.* **1**: pp 1–64.

95.   McCauley, P.T. et al. (1982) The effect of prenatal and postnatal lead exposure on neonatal synaptogenesis in rat cerebral cortex. *J. Toxicol. Environ. Health* **10**: pp 639–651.

96.   McGivern, R.F. et al. (1991) Prenatal lead exposure in the rat during the third week of gestation: long-term behavioral, physiological, and anatomical effects associated with reproduction. *Toxicol. Appl. Pharmacol.* **110**: pp 206–215.

97.   Mele, P.C. et al. (1984) Prolonged behavioral effects of early postnatal lead exposure in rhesus monkeys: fixed-interval responding and interactions with scopolamine and pentobarbital. *Neurobehav. Toxicol. Teratol.* **6**: pp 129–135.

98.   Mergler, D., Huel, G., Bowler, R. et al. (1994) Nervous system dysfunction among workers with long-term exposure to manganese. *Environ. Res.* **64**: pp 151–180.

99.   Merigan, W.H.. (1980) Visual fields and flicker thresholds in methylmercury-poisoned monkeys. In: Merigan, W.H., Weiss, B. (Eds.) *Neurotoxicity of the Visual System.* Raven Press, New York, pp 149–165.

100.   Miller, C.T., Krewski, D., Tryphonas, L. (1985) Methylmercury-induced mitochondrial DNA synthesis in neural tissue of cats. *Fundam. Appl. Toxicol.* **5**: pp 251–264.

101.   Mizukoshi, K. et al. (1975) Neurological studies upon intoxication by organic mercury compounds. *Oto-Rhino-Laryngology* **37**: pp 74–94.

102.   Moore, M.R., Bushnell, I.W.R., Goldberg, A. (1989) A prospective study of the results of changes in environmental lead exposure in children in Glasgow. In: Smith, M.A., Grant, L.D., Sors, A.I. (Eds.), *Lead Exposure and Child Development. An International Assessment.* Kluwer, Dordrecht, pp 371–378.

103.   Moresco, R.M. et al. (1988) Lead neurotoxicity: a role for dopamine receptors. *Toxicology* **53**: pp 315–322.

104.   Morris, R.G.M. et al. (1986) Selective impairment of learning and blockade of long-term potentiation by an *N*-methyl-D-aspartate receptor antagonist, AP5. *Nature* **319**: pp 774–776.

105.   Munoz, C. et al. (1988) Significance of hippocampal dysfunction in low level lead exposure in rats. *Neurobehav. Toxicol. Teratol.* **10**: pp 245–254.

106.   Munoz, C. et al. (1989) Neuronal depletion of the amygdala resembles the learning deficits induced by low level lead exposure in rats. *Neurotoxicol. Teratol.* **11**: pp 257–264.

107.   Nation, J.R. et al. (1982) Conditioned suppression in the adult rat following chronic exposure to lead. *Toxicol. Lett.* **14**: pp 63–67.

108.   Needleman, H.L., Gunnoe, C., Leviton, A. et al. (1979) Deficits in psychologic and classroom performance of children with elevated dentine lead levels. *N. Engl. J. Med.* **300**: pp 689–695.

109.   Overton, D.A. (1991) Historical context of state dependent learning and discriminative drug effects. *Behav. Pharmacol.* **2**: pp 253–264.

110.   Pentschew, A. and Garro, F. (1966) Lead encephalo-myelopathy of the suckling rat and its implications on the porphyrinopathic nervous diseases. *Acta Neuropathol.* **6**: pp 266–278.

379

111. Petit, T.L. et al. (1980) Neurofibrillary degeneration, dendritic dying back, and learning-memory deficits after aluminum administration: implications for brain aging. *Exp. Neurol.* **67**: pp 152–162.

112. Pocock, S.J., Smith, M., Baghurst, P. (1994) Environmental lead and children's intelligence: a systematic review of the epidemiological evidence. *Br. Med. J.* **309**: pp 1189–1197.

112a. Porterfield, S.P. and Hendrich, C.E. (1993) The role of thyroid hormones in prenatal and neonatal neurological development - current perspectives. *Endocrinol. Rev.* **14**: pp 94–106.

113. Rabinowitz, M.B., Wang, J-D., Soong, W.T. (1991) Dentine lead and child intelligence in Taiwan. *Arch. Environ. Health* **46**: pp 351–360.

114. Regan, C. (1989) Lead-impaired neurodevelopment. Mechanisms and threshold values in the rodent. *Neurotoxicol. Teratol.* **11**: pp 533–537.

115. Repko, J., Corum, C. (1979) Critical review and evaluation of the neurological and behavioral sequelae of anorganic lead absorption. *Crit. Rev. Toxicol.* **6**: pp 135–187.

116. Rice, D.C. and Gilbert, S.G. (1982) Early chronic low level methylmercury poisoning in monkeys impairs spatial vision. *Science* **216**: pp 759–761.

117. Rice, D.C. (1983) Central nervous effects of perinatal exposure to lead or methylmercury in the monkey. In: Clarkson, T.W. and Nordberg, G.F. (Eds.), *Reproductive and Developmental Toxicity of Metals*. Plenum Press, New York, pp 517–539.

118. Rice, D.C. (1984) Behavioral deficit (delayed matching to sample) in monkeys exposed from birth to low levels of lead. *Toxicol. Appl. Pharmacol.* **75**: pp 337–345.

119. Rice, D.C. (1992) Effect of lead during different developmental periods in monkeys on concurrent discrimination performance. *Neurotoxicology* **13**: pp 583–592.

120. Rice, D.C. and Gilbert, S.G. (1990) Sensitive periods for lead-induced behavioral impairment (nonspatial discrimination reversal) in monkeys. *Toxicol. Appl. Pharmacol.* **102**: pp 101–109.

121. Rice, D.C. and Willes, R.F. (1979) Neonatal low level lead exposure in monkeys (*Macaca fascicularis*): effect on two-choice non-spatial form discrimination. *J. Environ. Pathol. Toxicol.* **2**: pp 1195–1203.

122. Robbins, T. (Ed.), Milestones in dopamine research. *Semin. Neurosci.* **14**(2): pp 93–190.

123. Roels, H., Gennart, J., Lauwerys, R. et al. (1985) Surveillance of workers exposed to mercury vapor. *Am. J. Ind. Med.* **7**: pp 45–71.

124. Roels, H.A., Ghyselen, P., Buchet, J.P. et al. (1992) Assessment of the permissible exposure level to manganese in workers exposed to manganese dioxide dust. *Br. J. Ind. Med.* **49**: 25-34.

124a. Rosenkilde, C.E. (1979) Functional heterogeneity of the prefrontal cortex in the monkey: a review. *Behav. Neur. Biol.* **25**: pp 301-345.

125. Roussow, J. et al. (1987) Apparent central neurotransmitter receptor changes induced by low level lead exposure during different developmental phases in the rat. *Toxicol. Appl. Pharmacol.* **91**: pp 132-139.

126. Santucci, D. et al. (1994) Early exposure to aluminum affects eight-arm maze performance and hippocampal nerve growth factor levels in adult mice. *Neurosci. Lett.* **166**: pp 89-92.

127. Schmidt, W. et al. (1992) Behavioural pharmacology of glutamate in the basal ganglia. *J. Neural Transm.* Suppl. **38**: pp 65–89.

128. Schwartz, J. and Otto, D. (1987) Blood lead, hearing thresholds, and neurobehavioral development in children and youth. *Arch. Environ. Health* **42**: pp 153–160.

129. Schwartz, J. and Otto, D. (1991) Lead and minor hearing impairment. *Arch. Environ. Health*, **46**: pp 300–305.

130. Schweri, M.M. (1994) Mercuric chloride and p-chloromercuriphenylsulfonate exert a biphasic effect on the binding of the stimulant [^{3}H]methylphenidate to the dopamine transporter. *Synapse* **16**: pp 188–194.

131. Shellenberger, M.K. (1984) Effects of early lead exposure on neurotransmitter systems in the brain. A review with commentary. *Neurotoxicology* **5**: pp 177–212.

132. Shigeta, S. et al. (1979) Effects of lead on Sidman avoidance behavior by lever pressing in rats. *Jpn. J. Hyg.* **34**: pp 677–682.

133. Shimai, S. and Satoh, H. (1985) Behavioral teratology of mercury. *J. Toxicol. Sci.* **10**: pp 199–216.

134. Silbergeld, E.K. andGoldberg, A.M. (1974) Lead-induced behavioral dysfunction: an animal model of hyperactivity. *Exp. Neurol.* **42**: pp 146–157.

135. Silbergeld, E.K. and Goldberg, A.M. (1975) Pharmacological and neurochemical investigations of lead-induced hyperactivity. *Neuropharmacology* **14**: pp 431–444.

136. Simon, H. et al. (1980) Dopaminergic A10 neurones are involved in cognitive functions. *Nature* **286**: pp 150–151.

137. Slomianka, L. et al. (1989) Dose-dependent bimodal effect of low level lead exposure on the developing hippocampal region in the rat: a volumetric study. *Neurotoxicology* **10**: pp 177–190.

138. Soleo, L., Urbano, M.L., Petrera, V., Ambrosi, L. (1990) Effects of low exposure to inorganic mercury on psychological performance. *Br. J. Indust. Med.* **47**: pp 105–109

139. Spyker, J.M. et al. (1972) Subtle consequences of methylmercury exposure: Behavioral deviations in offspring of treated mothers. *Science* **177**: pp 621–623.

140. Stiles, K.M. and Bellinger, D.C. (1993) Neuropsychological correlates of low level lead exposure in school-age children: a prospective study. *Neurotoxicol. Teratol.* **15**: pp 27–35

141. Stollery, B.T., Banks, H.A., Broadbent, D.E., Lee, W.R. (1989) Cognitive functioning in lead workers. *Br. J. Indust. Med.* **46**: pp 698–707.

142. Stollery, B.T., Broadbent, D.E., Banks, H.A., Lee, W.R. (1991) Short-term prospective study of cognitive functioning in lead workers. *Br. J. Indust. Med.* **48**: pp 739–749.

143. Uemura, E. et al. (1992) Enhanced neurite growth in cultured neuroblastoma cells exposed to aluminum. *Neurosci. Lett.* **142**: pp 171–174.

144. Ujihara, H. and Albuquerque, E.X. (1992) Developmental change of the inhibition by lead of NMDA-activated currents in cultured hippocampal neurons. *J. Pharmacol. Exp. Ther.* **263**: p 868.

145. Wasserman, G., Graziano, J.H., Factor-Litvak, P. et al. (1994) Consequences of lead exposure and iron supplementation on childhood development at age 4 years. *Neurotoxicol. Teratol.* **16**: pp 233–240.

146. Wender, P.H. (1971) *Minimal Brain Dysfunction in Children.* Wiley-Interscience, New York.

147. Wennberg, A., Iregren, A., Struwe, G. et al. (1991) Manganese exposure in steel smelters: A health hazard to the nervous system. *Scand. J. Work. Environ. Health* **17**: pp 255–262.

148. WHO (1996) *Environmental Health Criteria for Inorganic Lead.* World Health Organization, Geneva.

149. Williamson, A. and Teo, R.K.C. (1986) Neurobehavioural effects of occupational exposure to lead. *Br. J. Indust. Med.* **43**: pp 374–380.

150. Wince, L.C. et al. (1980) Alterations in the biochemical properties of central dopamine synapses following chronic postnatal $PbCO_3$ exposure. *J. Pharmacol. Exp. Ther.* **214**: pp 642–650.

151. Winder, C. and Kitchen, I. (1984) Lead neurotoxicity: a review of the biochemical, neurochemical and drug induced behavioural evidence. *Prog. Neurobiol.* **22**: pp 59–87.

152. Winneke, G. et al. (1977) Neurobehavioral and systemic effects of long-term blood lead elevation in rats. *Arch. Toxicol.* **37**: pp 247–263.

153. Winneke, G., Krämer, U., Brockhaus, A. et al. (1983) Neuropsychological studies in children with elevated tooth-lead concentrations. II. Extended study. *Int. Arch. Occup. Environ. Health* **51**: pp 231–252.

154. Winneke, G., Brockhaus, A., Ewers, U. et al. (1990) Results from the European multicenter study on lead neurotoxicity in children: Implications for risk assessment. *Neurotoxicol. Teratol.* **12**: pp 553–559.

155. Wooler, K, and Melamed, Y. (1978) Simple tests for monitoring excessive lead-absorption. *Med. J. Aust.* **1**: pp 163–169.

156. Yokel, R.A. (1983) Repeated systemic aluminum exposure effects on classical conditioning of the rabbit. *Neurobehav. Toxicol. Teratol.* **5**: pp 41–46.

157. Zenick, H. (1974) Behavioral and biochemical consequences in methylmercury chloride toxicity. *Pharmacol. Biochem. Behav.* **2**: pp 709–713.

158. Zenick, H. and Goldsmith, M. (1981) Drug discrimination learning in lead-exposed rats. *Science* **212**: pp 569–571.

Contents Chapter 13

Neurobehavioral toxicology of pesticides

382

Chapter 13

Neurobehavioral toxicology of pesticides

Robert C. MacPhail, William F. Sette,
and Kevin M. Crofton

1   Introduction

The term pesticides refers to a broad class of chemicals that are used to kill pests. By itself, the term provides little information other than the intended use of the chemical. Pesticides are often classified on the basis of the types of pests they are used to control, for example insecticides, fungicides, herbicides, acaricides, rodenticides, etc. Within each of these classes, however, there are generally numerous chemicals that have been used (See Table 13.1).

Pesticides are essential in the management of crops, food supplies and disease. In fact, pesticides are almost the only class of chemical that is intentionally introduced into the environment. While the benefits of pesticides are clear, considerably less is known about their adverse effects on health and the environment. Numerous episodes of accidental releases and/or exposures have underscored the fact that as a whole, pesticides are far from benign. For example, pesticide exposures have been implicated in a number of adverse health effects in humans[3], in wildlife kills[8], and in altered reproduction and development in many animals. In laboratory studies, pesticide exposures have been shown to produce a wide array of adverse effects on a variety of organ systems[3]. As a consequence, while the benefits of pesticides may be clear, they must be weighed against their potential adverse effects on non-target organisms and on environmental resources.

Pesticides are sometimes distinguished from pestistats. Pestistats are chemicals designed not to kill but to disrupt behavior patterns in target species. For example, 4-aminopyridine and methiocarb are pestistats that repel birds from their perches in, for example, city parks[20]. Semiochemicals such as insect sex attractant pheromones are also pestistats[13].

Pesticide exposures in humans and laboratory animals can produce damage to a number of organ systems. Pesticides can damage chromosomes, reproductive systems, the respiratory system, the immune system, the nervous system, and many can produce cancerous tumors. There is, in addition, currently considerable concern over the potential adverse effects of pesticides on the fetus and neonate[16]. The level of concern over potential adverse effects is particularly high when it comes to insecticides, as they are specifically designed to adversely affect the nervous system of a pest. Unfortunately, the nervous system has been highly conserved across many species. As a consequence, many insecticides have also been shown to adversely affect the nervous system of laboratory animals and humans. In fact, inadvertent exposure to insecticides has been one of the leading causes of neurotoxicity in the human population world-wide. For example, of the 34 episodes of human poisoning by neurotoxic substances compiled by the USA National Research Council, twelve involved

TABLE 13.1

Classification and examples of pesticides

Pesticide class	Chemical class	Example(s)
INSECTICIDES		
– anticholinesterase insecticides	Organophosphate insecticides	Chlorpyrifos
		Parathion
	Carbamate insecticides	Carbaryl
		Physostigmine
– organochlorine insecticides	Dichlorodiphenylethanes	DDT
		Dicofol
		Methoxychlor
	Cyclodienes	Aldrin
		Dieldrin
		Chlordane
		Endrin
		Endosulfan
		Heptachlor
		Camphene
	Hexachlorocyclohexanes	Lindane
		HCB
	Alicyclic hydrocarbons	Chlordecone (Kepone)
– pyrethroid insecticides	– Type I pyrethroids	Permethrin
	– Type II pyrethroids	Cypermethrin
– formamidine pesticides		Armitraz
		Chlordimeform
FUNGICIDES		
– contact fungicides		cupric calcium sulfate ('Bordeaux mixture')
	Dimethyldithiocarbamic acid	Ferbam (iron salt)
		Ziram (zinc salt)
	Captan	
– systemic fungicides	Benomyl	
	Benzimidazole fungicides	Thiabendazole
	Triazole fungicides	Triadimenol
		Triadimefon
	Carboxin ('Vitavax')	
HERBICIDES	Dinitro-O-cresol	
	Phenoxyacetic acids	
OTHER PESTICIDES		
Plant antiviral agents		Blasticidin S
Agricultural anthelmintics		1,2-dibromo-3-chloropropane (BDCP) ('Nemagon')
Molluscicides		'Bayluscid'
		N-Tritylmorpholine (Tridemorph)
Rodenticides	Indanediones	
	Hydroxycoumarins	Warfarin

inadvertent exposures to pesticides[15]. The largest episode of human food poisoning in the world was due to consumption of watermelons contaminated with the insecticide aldicarb. In addition, as many as 60,000 people in the USA were adversely affected by exposure to a medicament contaminated with tri-ortho-cresyl phosphate (TOCP), an organophosphate that has many similar actions with organophosphate insecticides. Finally, it is estimated that the insecticide parathion has been responsible for more deaths world-wide than any other manufactured chemical. These and many other episodes underscore the need for a firm understanding of the neurotoxic potential of pesticides

384

While concern over the effects of pesticides on the nervous system is justifiable, understanding the risks they may pose for human health requires knowledge of the exposure levels actually encountered. Pesticide exposures are, however, varied and can include acute, recurrent, perinatal, and chronic exposures. Nevertheless, a continuum of exposure may be generally described as follows. *Intentional* (as in suicide attempts) or *accidental acute exposures* typically occur to high levels of pesticides. High levels of exposure may also occur briefly during the manufacture, application, and disposal of pesticides. Farm workers may be exposed to lower levels (unless entry into a pesticide-treated field occurs shortly after application) but for a longer period of time. Finally, the general population may be expected to be exposed to much lower levels, for example, through raw or processed foods derived from pesticide-treated crops or through drinking water. These *low-level exposures* may occur for prolonged periods of time, and there is always the possibility that there may be critical "windows" of vulnerability such as during development or senescence. Accurate estimation of the risks of pesticides therefore depends on understanding their health hazards as well as the conditions of exposure.

Many pesticides seem to affect particular cell types, processes and/or regions within the nervous system. Many other pesticides appear to act on neurotransmitter systems, as will become evident. In the latter case, knowledge gained from studies on the role of neurotransmitters in the control of behavior, particularly from studies on drugs of abuse, may create a framework in which to study pesticides. Drugs of abuse and neuropharmacological probes may also be used as positive control compounds in conjunction with the pesticides to establish similarities in their behavioral actions. In this way basic research in neurobehavioral pharmacology can be used to expedite our understanding of the mechanisms by which pesticides can adversely affect neurobehavioral function.

The sections that follow describe our current understanding of the effects of several pesticide classes. For each class, prototype pesticides are identified as well as their uses. Most of the pesticides covered are insecticides, although there are exceptions. Effects on animal behavior are next described, as are effects in humans when known, as well as their possible mechanisms of action. When available, reviews of the different classes of pesticide are referenced, as well as some of the relevant literature. The chapter ends with pertinent source references for further information.

2   Carbamates and organophosphates

Carbamates and organophosphates are insecticides that are extensively used world-wide to treat infestations on a wide variety of crops.

FIGURE 13.1   physostigmine

Chemical structures of carbaryl (a), chlorpyrifos (b) and physostigmine (c)

Carbaryl can be considered a prototype carbamate (Figure 13.1) and chlorpyrifos a prototype organophosphate. Also shown in Figure 13.1 is the structure of physostigmine, which is an extensively studied carbamate that has been used both experimentally and clinically. Carbamates and organophosphates are known to inhibit acetylcholine esterase, the enzyme responsible for the rapid breakdown of acetylcholine neurotransmitter after release by nerve stimulation. Carbamates produce a reversible inhibition of cholinesterases while organophosphates produce an irreversible inhibition of the enzymes. Acute cholinesterase inhibition is associated with a variety of autonomic signs including salivation, lacrimation, urination and defecation, and neuromuscular signs including tremors and convulsions, as a consequence of cholinergic overstimulation. Death may occur after high acute exposures. The acute behavioral effects of cholinesterase inhibitors are generally reflected in response suppression.

Tolerance

Some evidence suggests that *tolerance* does not develop to all of the behavioral effects of repeated organophosphate exposure. For example, Overstreet et al.[17] determined the effect of repeated diisopropylfluorophosphate (DFP) exposure on food-reinforced alternation performance in rats. DFP initially decreased the frequency of responding on reinforced trials and increased the frequency of responding on non-reinforced trials. With continued exposure, tolerance developed to the effect of DFP on reinforced trials, but not to its effect on non-reinforced trials. McDonald et al.[12] determined the effect of repeated exposure of rats to DFP and to disulfoton, another organophosphate. Repeated exposure to the organophosphates reduced the intensity of the clinical signs obtained initially. A small but significant deficit in spontaneous alternation in a T-maze was obtained in the otherwise tolerant rats. More recently, Bushnell et al.[5] showed that repeated exposure of rats to DFP produced a gradual time-dependent deficit in the accuracy of responding under a delayed matching-to-position procedure that was not apparent when dosing began. Interestingly, the effect of repeated DFP on accuracy was similar to the effect of acute scopolamine administration. These results indicate that tolerance does not develop to all the behavioral effects of organophosphates and further suggest the existence of time-dependent changes in the pattern of behavioral effects produced by repeated exposures. These results also highlight the importance of defining tolerance in terms of the behavioral endpoint(s) of interest, rather than as a generalized state of the organism engendered by repeated exposure.

3   Organochlorines

Organochlorines are the oldest of the modern-day insecticides. DDT is perhaps the best known organochlorine insecticide that was discovered by Muller in 1939, for which he received the Nobel Prize in 1948. DDT has played an extremely important role in the control of many insect-borne diseases including malaria and typhus. Increased use, however, led to numerous instances of human poisonings and wildlife kills. While DDT is banned from use in many countries, including the USA, it is still used to control disease in many developing countries.

Four classes of organochlorines

Four classes of organochlorines have been described:

1. Dichlorodiphenylethanes such as DDT, dicofol and methoxychlor
2. Cyclodienes such as chlordane, heptachlor and endosulfan
3. Hexachlorocyclohexanes such as lindane and its metabolites
4. Alicyclic hydrocarbons such as chlordecone (see Figure 13.2)

As a whole, organochlorines are highly lipophilic compounds that are slowly metabolized and excreted. As a consequence, they are likely to accumulate in lipid-rich tissues (e.g., brain and liver) and produce a time-dependent cumulative toxicity with repeated exposure. Organochlorines have also been shown to persist in the environment and accumulate in the food chain, further amplifying their adverse effect on human health and the environment.

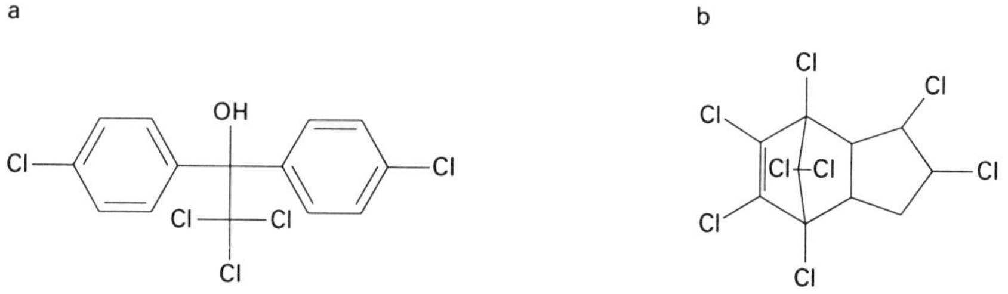

a   b

FIGURE 13.2
Chemical structure of dicofol (a) and chlordane (b)

Animal studies have routinely shown that organochlorine insecticides produce sensorimotor changes including increased reactivity to external stimuli, tremors, and convulsions. Tremors and convulsions are of variable onset and duration, sometimes occurring for days or weeks following exposure. Organochlorines have been the subject of intensive electrophysiological and neurochemical investigations into their mechanism of action. Current evidence indicates that organochlorines have prominent effects on sodium conductance in nerve axons and on the picrotoxin binding site of the GABA receptor complex in the central nervous system (as do many of the Type II pyrethroids described in the following section). Studies have also shown that neonate rats are generally less sensitive than are adults to the neurotoxic effects of organochlorines, presumably due to the lack of metabolic enzyme activity in neonates. Human exposures are also characterized by sensorimotor changes including increased reactivity to sensory stimuli, tremors, and convulsions. Death in humans is typically due to respiratory arrest. Management of poisoning in humans is symptomatic and typically focuses on the control over seizures. Recent concern over organochlorines in the environment has focused on their estrogenic effects and potential adverse effects on reproduction and development in both humans and in wildlife.

4   Pyrethroids

Pyrethroid insecticides are synthetic derivatives of pyrethrums. Pyrethrums are alkaloid extracts from chrysanthemum flowers. Both pyrethrums and pyrethroids are potent insecticides with increasing worldwide usage. Since their introduction a little over 2 decades ago, they have attained 30% of the world market for insecticides, to a large extent replacing chlorinated hydrocarbons. Increased pyrethroid usage is due to several factors including:

1.   A high selectivity factor (insect/mammalian toxicity ratio)
2.   Rapid metabolism and excretion in mammals
3.   Limited soil persistence

The major target site of pyrethroids is undoubtedly the nervous system (for recent reviews, see refs. 1, 19, and 22).

Pyrethroids are widely used in agriculture and the home to control a variety of insects.

Pyrethroid insecticides are classified as Type I or Type II based upon several factors:

1. *In vivo* toxic signs in rats, mice and insects
2. Duration of the prolongation of the sodium current in nerve axons
3. Putative interactions at the gamma-aminobutyric acid (GABA) receptor complex

Structurally, Type II pyrethroids usually contain an alpha-cyano-phenoxybenzyl moiety, whereas the Type I pyrethroids do not. Figure 13.3 shows the structure of two pyrethroids; permethrin, a Type I compound, and cypermethrin, a Type II compound. Based upon toxic signs in the rat, the pyrethroids have been divided into two types:

1. Compounds that produce a syndrome of aggressive sparring, increased sensitivity to external stimuli, and fine tremor progressing to whole body tremor and prostration (Type I)
2. Compounds that produce a syndrome of pawing and burrowing, profuse salivation, and coarse tremor progressing to choreoathethosis and clonic seizures (Type II)[21]

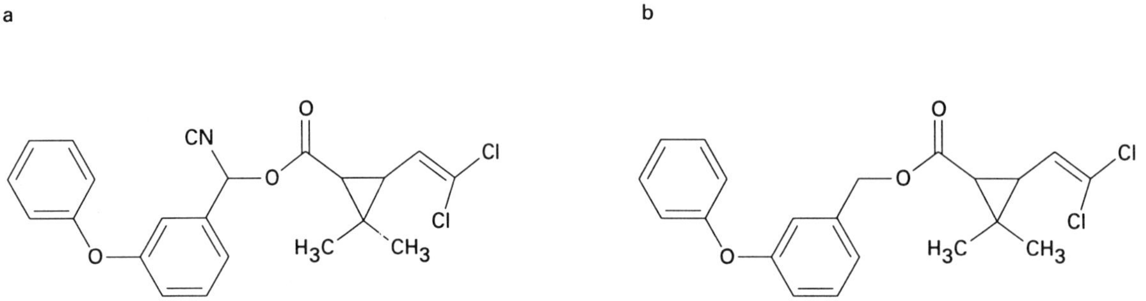

a   b

FIGURE 13.3
Chemical structures of the pyrethroids cypermethrin (a) and permethrin (b)

Most data support the conclusion that there are no chronic effects of pyrethroids, although there is some controversial data indicating that long-term exposure to high dosages may induce pathological damage to the peripheral nervous system[1].

Human exposure to pyrethroids is limited to agricultural workers, production staff, and treatment of pediculosis (body lice). Effects of dermal exposure in humans include contact dermatitis and transient paresthesias. Acute exposures to high concentrations, due to accidental ingestion and/or inhalation, have been reviewed by He *et al.*[9]. Symptoms include dizziness, headache, fatigue, nausea, muscle fasciculations, and convulsions. The use of permethrin to treat pediculosis has led to some anecdotal reports of toxic side effects, although no effects were reported in controlled clinical trials (see also ref. 10).

5   Formamidines

Formamidine pesticides are mainly used as insecticides and acaricides (that is, they kill mites and ticks). Two of the most widely used formamidines are amitraz (see Figure 13.4a) and chlordimeform, although registration of these chemicals for general use in the USA has been cancelled due to potential carcinogenicitiy. These two pesticides are used world wide to control eggs and mobile forms of ticks and mites on livestock and crops. Amitraz has also been formulated for use in dog shampoos for the controls of ticks[4].

The acute toxicity of formamidines is moderate with oral LD_{50}s in the range of 400–1500 mg/kg for amitraz and 200–600 for chlordimeform, for a number of mammalian species. Signs of toxicity following acute exposure to amitraz include decreased body temperature, decreased motor activity, aggressive responses to handling, piloerection, and exophthalmus. Signs of acute exposure to chlordimeform include hyperactivity, tremors, convulsions, and respiratory arrest. Repeated exposure to both amitraz and chlordimeform results in an increased sensitivity to seizures, decreased growth rates and hematocrit and/or hemoglobin levels, hyperphagia, as well as signs of CNS depression including depressed motor activity, and decreased fixed-interval rates of schedule-controlled responding. There have been a number of proposed mechanisms for the toxic effects of these chemicals including: monoamine oxidase inhibition, alpha-2-adrenergic agonist activity, anticholinergic activity, and inhibition of prostaglandin synthesis (for review see ref. 11). There iscurrently no data to suggest that either short-term or long-term exposures to formamidines will produce pathological effects in the PNS and/or CNS.

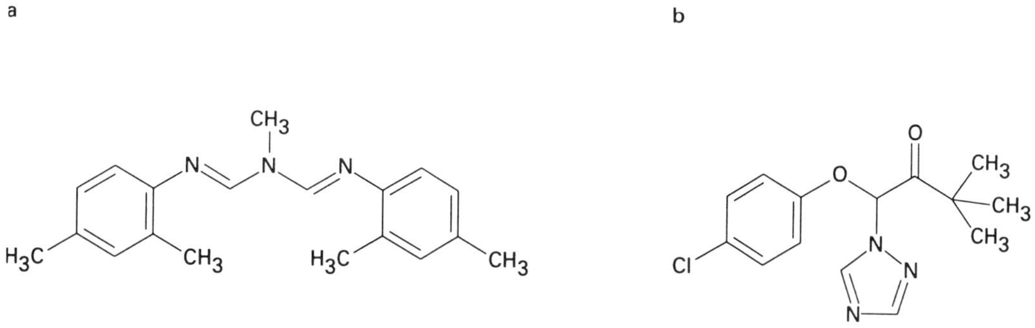

a   b

FIGURE 13.4

Chemical structures of the formamidine amitraz (a) and the triazole fungicide triadimefon (b)

Human exposure to formamidines has been limited to agricultural and production workers, as well as intentional ingestion in suicide attempts. Limited data from human oral exposures provides these signs of acute toxicity: lethargy, vomiting, muscle weakness, headaches, decreased MAO activity, and blurred vision. Effects of dermal exposure to amitraz may be due to the effects of the xylene solvent vehicle used in the formulation of the pesticide. Workers exposed in chlodimeform packaging plants have exhibited lack of appetite, fatigue, dizziness, swollen livers, dermatitis, dysuria, and abdominal pains. Urinary symptoms associated with chlordimeform exposure appear often and include: proteinuria, hematuria, and decreased bladder capacities (for review see ref. 11).

6   Triazole fungicides

Triazoles are widely used to prevent and treat a number of fungal infestations in plants. Figure 13.4b shows the structure of the triazole fungicide triadimefon. Triadimenol is a fungicide that resembles triadimefon except that the ketone (= 0) moiety is replaced with a hydroxy (-OH) group. Triadimenol is also a metabolite of triadimefon in both plants and animals. Both triadimefon and triadimenol are classified as systemic fungicides because they are taken up by plants in order to exert their fungicidal activity. Human exposures have not been documented, although the manufacturer's material safety data sheet for triadimefon states that acute exposure may produce "hyperactivity followed by sedation."

Evidence exists that triadimefon produces many behavioral effects in laboratory rats that resemble those of the psychomotor stimulants such as *d*-amphetamine, cocaine, and methylphenidate. For example, triadimefon increases motor activity under a variety of experimental conditions (e.g., ref. 6). Triadimefon also has effects on schedule controlled behavior that resemble those of the psychomotor stimulants, including rate-dependent effects on fixed-interval performance[2,14]. Discriminative stimulus control over operant behavior can be established in rats with triadimefon in a drug-discrimination paradigm, and there is a bidirectional cross-generalization with methylphenidate[18]. Finally, high doses of triadimefon induce stereotyped behaviors including repetitive movements and backward locomotion[14,23]. These and related findings suggest that triadimefon may be able to serve as a reinforcer in laboratory animals, and have abuse potential in humans, although there is currently no evidence for these effects.

The neurochemical effects of triadimefon are a topic of great interest. Crofton *et al.*[7], for example, showed that the hyperactivity produced by triadimefon could be blocked by pretreatment of rats with the amine-depleting agent reserpine but not with the tyrosine hydroxylase inhibitor alpha-methyl-para-tyrosine. Similar effects were obtained with methylphenidate but not with *d*-amphetamine. Walker *et al.*[23] showed effects of triadimefon on dopamine utilization in rats at low to intermediate doses, and effects on serotonin utilization at the highest dose. More recently, Walker *et al.*[24] have shown *in vitro* that triadimefon is a potent inhibitor of dopamine uptake in the central nervous system and is ineffective as a releaser of stored dopamine. These results further suggest a close similarity between the actions of triadimefon and the psychomotor stimulants.

Comparatively little is known about the effects of repeated triadimefon exposures on behavior. Based on its similarity with the psychomotor stimulants, however, one might predict sensitization to its activity-increasing effects, and tolerance to its effects on operant behavior, especially when acute exposure results in decreased reinforcement.

References

1. Aldridge, W.N. (1990) An assessment of the toxicological properties of pyrethroids and their neurotoxicity. *Crit. Rev. Toxicol.* **21**: pp 89–104.

2. Allen, A.R. and MacPhail, R.C. (1991) Effects of triadimefon on a multiple schedule of fixed-interval performance: comparison with methylphenidate, d-amphetamine and chlorproma-zine. *Pharmacol. Biochem. Behav.* **40**: pp 775–780.

3. Baker, S.R., Wilkinson, C.F. (Eds.) (1990) *The Effect of Pesticides on Human Health.* Princeton Scientific Publishing, New Jersey.

4. Briggs, S.A. (1992) *Basic Guide to Pesticides: Their Characteristics and Hazards.* Taylor and Francis, Washington, D.C., p 283.

5. Bushnell, P.J., Padilla, S.S., Ward, T., Pope, C.N., Olszyk, V.B. (1991) Behavioral and neurochemical changes in rats dosed repeatedly with diisopropylfluorophosphate. *J. Pharmacol. Exp. Ther.* **256**: pp 741–750.

6. Crofton, K.M., Howard, J.L., Moser, V.C., Gill, M.W., Reiter, L.W., Tilson, H.A., MacPhail, R.C. (1991) Interlaboratory comparison of motor activity experiments: implications for neuro-toxicological assessments. *Neurotoxicol. Teratol.* **13**: pp 599–609.

7. Crofton, K.M., Boncek, V.M., MacPhail, R.C. (1989) Evidence for aminergic involvement in triadimefon-induced hyperactivity. *Psychopharmacology* **97**: pp 326–330.

8. Flint, R.W., Vena, J. (eds) (1991) *Human Health Risks from Chemical Exposure: The Great Lakes Ecosystem*. Lewis Publishers, Chelsea, MI.

9. He, F., Wang, S., Liu, L., Chen, S., Zhang, Z., Sun, J. (1989) Clinical manifestations and diagnosis of acute pyrethroid poisoning. *Arch. Toxicol.* **63**: 54-58

10. Kaloyanova, F.P. and ElBatawi, M.A. (1991) *Human Toxicology of Pesticides*. CRC Press, Boca Raton, FL, pp 101–110.

11. Knowles, C.O. (1991) Miscellaneous pesticides. In: Hayes, W.J. and Laws, E.R. (Eds.) *Handbook of Pesticide Toxicology*. Vol. 3. *Classes of Pesticides*. Academic Press, New York, pp 1471–1526.

12. McDonald, B.E., Costa, L.G., Murphy, S.D. (1988) Spatial memory impairment and central muscarinic receptor loss following prolonged treatment with organophosphates. *Toxicol. Lett.* **40**: pp 47–56.

13. Mitchell, E.R. (Ed.) (1981) *Management of Insect Pests with Semiochemicals: Concepts and Practice*. Plenum Press, New York.

14. Moser, V.C. and MacPhail, R.C. (1989) Neurobehavioral effects of triadimefon, a triazole fungicide, in male and female rats. *Neurotoxicol. Teratol.* **11**: pp 285–293.

15. National Research Council (1992) *Environmental Neurotoxicology*. National Academy Press, Washington, D.C.

16. National Research Council (1993) *Pesticides in the Diets of Infants and Children*. National Academy Press, Washington D.C.

17. Overstreet, D.A., Russell, R.W., Vazquez, B.J., Dalglish, F.W. (1974) Involvement of muscarinic and nicotinic receptors in behavioral tolerance to DFP. *Pharmacol. Biochem. Behav.* **2**: pp 45–54.

18. Perkins, A.N., Eckerrnan, D.A., MacPhail, R.C. (1991) Discriminative stimulus properties of triadimefon: comparison with methylphenidate. *Pharmacol. Biochem. Behav.* **40**: pp 757–761.

19. Ray, D.E. (1991) Pesticides derived for plants and other organisms. In: Hayes, W.J., Laws, E.R. (Eds.) *Handbook of Pesticide Toxicology*. Vol. 3. *Classes of Pesticides*. Academic Press, New York, pp 585–636,

20. Schafer, Jr., E.W., Brunton, R.B., Cunningham, D.J. (1972) A summary of the acute toxicity of 4-aminopyridine to birds and mammals. *Toxicol. Appl. Pharmacol.* **26**: pp 532–538.

21. Verschoyle, R.D. and Aldridge, W.N. (1980) Structure-activity relationships of some pyrethroids in rats. *Arch. Toxicol.* **45**: pp 325–329.

22. Vijverberg, H.P.M. and van den Bercken, J. (1990) Neurotoxicological effects and the mode of action of pyrethroid insecticides. *Crit. Rev. Toxicol.* **21**: pp 105–126.

23. Walker, Q.D., Lewis, M.H., Crofton, K.M., Mailman, R.B. (1990) Triadimefon, a triazole fungicide, induces stereotyped behavior and alters monoamine metabolism in rats. *Toxicol. Appl. Pharmacol.* **102**: pp 474–485.

24. Walker, Q.D., Lewis, M.H., Mailman, R.B. (1991) Triadimefon selectively inhibits dopamine uptake *in vitro*. *Soc. Neurosci. Abstr.* **17**: p 1465.

Source references

Baker, S.R. and Wilkinson, C.F. (Eds.) (1990) *The Effect of Pesticides on Human Health*. Princeton Scientific Publishing, New Jersey.

Ballantyne, B. and Marrs, T.C. (Eds.) (1992) *Clinical and Experimental Toxicology of Organophosphates and Carbamates*. Butterworth and Heinemann, Oxford.

Briggs, S.A. (1992) *Basic Guide to Pesticides: Their Characteristics and Hazards*. Taylor and Francis, Washington D.C.

Ecobichon, D.J., Davies, J.E., Doull, J., Ehrich, M., Joy, R., McMillan, D., MacPhail, R., Reiter, LW., Slikker, Jr., W., Tilson, H. (1990) Neurotoxic effects of pesticides. In: Baker, S.R. and Wilkinson, C.F. (Eds.) *The Effect of Pesticides on Human Health*. Princeton Scientific Publishing, New Jersey.

Ecobichon, D.J. and Joy, R.M. (1992) *Pesticides and Neurological Disease*, 2nd edition. CRC Press, Boca Raton, FL.

Farm Chemicals Handbook '94 (1994) Meister Publishing Company, Willoughby OH.

391

Hayes, Jr., W.J. and Laws, E.R. (Eds.) (1991) *Handbook of Pesticide Toxicology*, Vol. 1–3. Academic Press, New York.

Saunders, D.S. and Harper, C. (1994) Pesticides. In: Hayes, A.W. (Ed.) *Principles and Methods of Toxicology*. Raven Press, New York.

Tordoir, W.F., Maroni, M., He, F. (Eds.) (1994) Health surveillance of pesticide workers. *Toxicology* **91**: pp 1–115 (special issue).

Contents Chapter 14

Neurobehavioral toxicology of organic solvents

Chapter 14

Neurobehavioral toxicology of organic solvents

Philip J. Bushnell and Kevin M. Crofton

1 Overview

Organic solvents comprise a class of chemicals loosely defined by their ability to dissolve other compounds which are hydrophobic, that is, which do not mix appreciably in water. Organic solvents typically contain carbon and hydrogen, to be electrochemically nonpolar, to be liquid at room temperature, and to have low molecular weight. Their miscibility with water also varies with their polarity, from the relatively polar alcohols, which mix well in water, to the nonpolar ethers, which do not. Relative to water, the more nonpolar solvents also have low hydrogen binding and low surface tension. These properties yield a high vapor pressure and volatility, which become important factors in considering their potential toxicity.

While the number of organic solvents is potentially infinite, a limited number find common use in industry and in the household. Commercial products that contain solvents include paints, inks, waxes, cleaning solutions, and pesticides. "Dry-cleaning" processes for clothing substitute organic solvents for water, and large volumes of solvents are used as degreasing agents in the electronic industry and for large-scale chemical synthesis and plastics manufacture. Thus the greatest exposure tends to occur in employees of those industries.

1.1 CLASSES OF COMMON SOLVENTS

Categorizing solvents by their chemical structure helps to understand their toxicity as well as their chemistry, because their chemical properties dictate their metabolic interactions with the body. One research goal is to be able to predict the toxicity of a chemical on the basis of its structure alone: few such "structure-activity relationships" have yet been determined, however.

The short hydrocarbons (alkanes (C_nH_{2n+2}) and alkenes (C_nH_{2n}) are not commonly used as solvents because they are not liquid at room temperature. The longer ones (e.g., octane) comprise a substantial fraction of liquid fuels. Halogenated solvents — usually containing chlorine — are less volatile and flammable, and find frequent use as degreasing agents. The various classes of oxygen-containing hydrocarbons — alcohols, ketones, glycols, and ethers — differ considerably in chemistry and toxicity. Among the alcohols, ethanol possesses relatively low toxicity and is of course a commonly ingested intoxicant. Due to its social acceptability, its recreational use and abuse far exceed those of any other solvent today. With the exception of methanol in primates (see section on the visual system below), the toxicity of the other straight-chain alcohols increases with the length of the carbon chain. Ketones (e.g., acetone and methyl isobutyl ketone) and glycols (e.g., ethylene and propylene glycols) comprise two other classes of oxygen-containing solvents.

Because of their low freezing point, high heat capacity and miscibility with water, the glycols find common use as automobile coolants. Aromatic solvents – those containing a benzene ring – comprise the most abundant solvents. Since discovery of its carcinogenicity, the use of benzene itself has declined and some of its methylated derivatives, notably toluene and xylene, have been adopted as substitutes. There exists also the unique solvent carbon disulfide, which is particularly important as a solvent for latex in the manufacture of rubber products, and for dissolving methyl cellulose in the production of viscose rayon. Prolonged exposure to carbon disulfide has been known for 150 years to cause changes in personality and sensory, motor, and cognitive functions. It appears to act primarily in the central nervous system (CNS)[6].

In addition, although some compounds possess chemical properties of solvents, they are primarily used as starting materials for synthesis of other chemicals. Nevertheless, they are often considered as "solvents" for the purpose of toxicological inquiry. A good example of such a compound is styrene, which is most commonly polymerized to make polystyrene plastic containers and "Styrofoam" insulation. In its monomeric (non-polymerized) form, it possesses chemical and toxicological properties very much like the aromatic solvents it resembles, while the polymer is relatively nontoxic and chemically inert.

Finally, it must be recognized that solvents are usually used in complex mixtures. Heating and motor fuels including gasoline and kerosene, and paint thinners including "white spirit" contain mixtures of chemicals with solvent properties. Characterizing the mechanisms of toxicity of mixtures presents a particularly difficult problem due to the complexity of the possible interactions with the body.

1.2   USE, ABUSE, AND ADDICTION

Ethnopharmacology, the study of the use of drugs by various cultures, has revealed a strong human tendency to explore and use the medicinal properties of naturally occurring chemical compounds. Given this tendency, it is not surprising that people in industrialized cultures should engage in exploration of the psychoactive effects of synthetic compounds as well. The invention of many sophisticated preparations of ethanol and the long history of their ingestion in Western and in Asian cultures testifies to the ingenuity applied toward the synthesis of novel and more pleasurable forms of this drug. Further more, the chemical similarity of ethanol to other organic solvents is mirrored by similarities of physiological and psychoactive effects[24]; exploration of the psychoactive properties of solvents thus follows readily from the use of ethanol.

Because inhalation of solvent vapors produces immediate pleasurable effects, their use is likely to be repeated. Failure to maintain control over this behavior may result in *dependence*. For some classes of chemicals, dependence takes the form of *addiction*: in these cases, strong psychological and physiological responses are engendered not only by the drug itself, but also by its removal (withdrawal). Lesser signs and symptoms typically follow cessation of use of other chemicals, including solvents. Differentiating dependence from addiction is both a matter of degree and of debate; compared to opiates, withdrawal from solvent use is relatively mild.

As with any other chemical, the toxicity of a solvent will increase with exposure level. This basic tenet of toxicology was stated in the 16th century by the physician-alchemist Paracelsus: "All substances are poisons; there is none that is not a poison. The right dose differentiates a poison from a remedy." In other words, any chemical is toxic at a high enough level in the body, and innocuous at a low enough level.

Thus, the desired effects of a drug, its *pharmacological actions*, are usually superseded by unwanted effects at some higher dose. These unwanted effects may be considered its toxicity. The toxicity of a chemical depends upon the amount that is necessary to cause a deleterious, or adverse effect. Further more, because no chemical exerts a single effect on the body, drugs may also produce unwanted "side effects", which are usually considered adverse. Finally, because chemicals in the workplace or in the environment are not intended to affect human physiology or health, any effect they cause is likely to be deemed adverse. Regardless of the source or intent of exposure, however, adversity remains a matter of opinion: fundamentally, *an adverse effect is an unwanted and/ or deleterious effect*.

Abuse of a chemical occurs when its use persists despite the occurrence of adverse effects, either in the individual consuming the chemical or in his/her social environment. It is the adverse societal effects that typically lead to legal controls and proscriptions. The rationale for these controls derives from the marked and permanent debilitation that can occur in users after prolonged exposure to solvents[62]. Occupational exposure to solvents may also induce more subtle, but still adverse, effects on sensory, motor, and cognitive functions. The evidence for these kinds of effects will be considered below.

Recreational use of solvent vapors dates to the early 1800s with the discovery of the psychoactive and anesthetic properties of nitrous oxide, ether, and chloroform. Intentional inhalation of these agents for the purpose of inducing short-term exhilaration, giddiness and euphoria was practiced among physicians and their students[66]. This practice faded from view in the latter part of the century, but began to resurface in the 1930s with the advent and availability of novel petrochemicals, including gasoline. This use of solvents emerged as a public health problem in the 1960s with reports of "glue sniffing" among teen-aged children. The primary solvent in glues at that time was toluene, which possesses a pleasant odor and causes little irritation; because of its high abuse potential, it was subsequently removed from most household adhesives.

Typical methods of solvent inhalation involve breathing the air in a bag or cloth containing the solvent vapor, or inhaling the vapor directly from a container of the liquid solvent. High concentrations of the vapor produce a rapid elevation of brain levels of the solvent, which generates a desirably rapid onset of effect. For this reason, a user may heat the liquid solvent to increase its vapor pressure. The flammability of many solvents makes this practice dangerous.

There is a clear progression of effect of a solvent vapor as its concentration increases in the brain; with increasing concentration, excitation develops and then is surpassed by lethargy, stupor, anesthesia, respiratory failure, and death at very high levels. This sequence of acute intoxication is thought to reflect a facet of the neurochemical organization of the central nervous system (see Section 3.2 below). In addition, deaths associated with solvent abuse have also been attributed to cardiac arrhythmia, probably due to a direct effect on the heart[40].

Some anecdotal evidence exists for tolerance to the immediate behavioral effects of solvents, but experimental studies in animals are needed to confirm this possibility. Similarly, evidence for physiological dependence upon solvents is weak, nor is it known whether combinations of solvents produce greater effects than an equivalent of any single one alone *(synergism)*. Indeed, studies in animals indicate that, in terms of ototoxicity (toxicity to the auditory system) at least, the effect of a combination of two solvents at a given total concentration is equivalent to that caused by either solvent alone at the same total concentration[58]. A non-synergistic relationship between two biologically active compounds is known as *additivity*.

397

See Section 3.1.3 for a more thorough discussion of the ototoxic effect of solvents.

In contrast, non-additive interactions with ingested ethanol have been recognized for decades. For example, workers engaged in degreasing operations with trichloroethylene will experience abnormally intense vasodilation of superficial veins upon drinking alcohol after work, a condition known as "degreaser's flush"[70]. This response appears to involve changes in the metabolism of ethanol induced in the liver by TCE[50]. This particular interaction has not been reported with other solvents; nevertheless, those engaged in intentional solvent inhalation may supplement the CNS effects of the solvent with alcoholic beverages. Finally, heavy alcohol consumption may increase the risk of toxicity from occupational or recreational solvent exposure[13].

The long-term consequences of single, acute, or short-term solvent exposure — as from induction of anesthesia for surgical purposes, or from a low-level industrial accident — are probably negligible. In contrast, it is clear that long-term, repeated exposure to the high solvent concentrations typically inhaled for recreational purposes will eventually cause severe neurological damage. For example, toluene can cause permanent loss of white matter (nerve fibers) in large regions of the brain[63] as well as cognitive changes resembling dementia and depression[25,72]. In addition, permanent damage to other organs including liver, kidney, heart, and lung have beer observed[40]. n-Hexane and some butylketones produce a peripheral neuropathy in which axonal swelling leads to degeneration of the long nerve fibers innervating the hands and feet[67]. Hypesthesia (lack of sensation), paresthesia (phantom sensations including tingling and pain), and weakness are symptoms of this neuropathy (see Section 3.2).

Perhaps the most problematic aspect of recreational solvent use arises from the fact that the population of users is most often young, disadvantaged boys, often in their early teens. Solvents from glues, paints, and gasoline are cheap and legally available, in contrast to alcohol and the illicit drugs more frequently used by adults. Better understanding of the social, economic, and cultural factors which promote solvent abuse among this group, and the long-term behavioral and physiological effects of this abuse are needed to reduce the negative personal and societal impact of this behavior.

The remainder of this chapter will explore the neurobehavioral consequences of acute and long-term exposure to organic solvents. As a rule, this work is directed toward understanding the hazards to human health associated with exposure to organic solvents. For reasons detailed in Section 2 below, however, most of this information has been obtained from experimental studies of the effects of solvents on the nervous system of laboratory animals.

2 Toxicological principles

Three principles apply with particular relevance to the study of organic solvents.

— Acute versus long-term exposure
Acute exposure refers to the immediate interaction between a subject and a solvent, either by inhalation, ingestion, or by dermal contact. Acute exposures are by definition brief. They may happen by accident or by intent; in the latter case, they may occur at concentrations approaching saturation in air (on the order of 10,000 to 100,000 ppm). Except in the case of repeated, high-level abuse, long-term exposure usually occurs at low concentrations over a protracted period of time, either intermittently or (rarely) continuously. Typical occupational exposures occur periodically (during the working hours of the work week) to low or moderate concentrations (on the order of 100

Concentrations of solvents at saturation vary widely. Some examples include toluene: 37,000 ppm; trichloroethylene: 75,1000 ppm; 1,1,1-trichloroethane, 132,000 ppm; methyl ethyl ketone 102,000 ppm; styrene: 8,550 ppm.[5]

ppm) of one or more solvents. Further more, this occupational exposure scenario rarely stays constant, but often varies with changes in jobs, production, and time of year (which may affect the degree of ventilation in the workplace). Finally, environmental chemicals are usually encountered either intermittently or at relatively stable, low concentrations (of the order of 1 ppm) of long duration.

— Acute versus long-term effects

In addition to the conditions of *exposure*, the nature of the *effects* of a solvent may also be transient or persistent. As a rule, a single, acute exposure to a sub-lethal concentration of a solvent will likely produce short-term effects which pass as the compound is metabolized and excreted. (Nevertheless, as noted above, these pharmacological effects may be quite pronounced: if the exposure is of sufficiently high concentration and duration, it may cause debilitation, narcosis, coma, and death.) While less frequently debilitating, the neurobehavioral effects of long-term solvent exposure are more likely to be irreversible, although strong evidence for such permanent effects is limited to a few compounds (e.g., carbon disulfide) and a few effects (e.g., hearing loss). Whether long-term occupational exposure to solvents causes subtle cognitive impairment remains an open question of current research.

— Exposure versus experiment

Human exposure to organic solvents occurs under a wide variety of conditions, which are never easy to define. Recreational use is usually covert, with neither the source (e.g., the fumes from spray paint or glue), the concentration, the frequency, nor the duration easily determinable. Use of other psychoactive drugs, particularly ethanol, may complicate interpretation of the effects of recreational solvent use as well. Industrial exposure typically involves more than one compound because the solvents and resins frequently encountered in the workplace rarely consist of a single chemical. Accidental exposures are invariably difficult to characterize in terms of concentration and duration; further more, they often involve more than one route (e.g., inhalation and dermal contact).

In contrast, toxicological experiments almost always involve well-controlled, precisely defined exposures to a single compound. Studies in human volunteers are ethically constrained to low-concentration, low-duration conditions, from which neither debilitating nor permanent effects will result. Less severe constraints apply to animal studies, so they can yield causal information regarding the effects of the compound on the animal, with corresponding limitations on its relevance to its effects in humans. For these reasons, most of the information regarding the neurobehavioral effects of solvents has been collected from experimental animals in studies designed to evaluate the risks of exposure and the mechanisms of the action of solvents on the CNS.

3   Functions of the nervous system affected by solvents

Functions of the nervous system may be divided conveniently into sensory, motor, and cognitive processes. The effects of a particular solvent on the nervous system may be expressed as a change in one or more of these functions. At a high enough concentration, all functions of the nervous system are likely to be suppressed by a solvent, reflecting the anesthetic properties exhibited by these compounds. At lower concentrations, different solvents may affect each process to a different degree or at different exposure levels. For example, those ketones which are metabolized to 2,5-hexanedione produce

399

a peripheral axonopathy expressed primarily as a sensori-motor dysfunction, whereas the toxicity of carbon disulfide involves primarily the central nervous system (e.g., cognitive and visual disturbances).

In the following sections, the actions of solvents on these processes will be discussed, not in encyclopedic fashion, but rather by example, to illustrate the kinds of effects typically observed during or after intoxication with exemplary compounds. In doing so, behavioral methods that are used to assess these processes will be introduced to provide insight into the study methods and their strengths and limitations for documenting toxicity at the behavioral level. Of course, non-behavioral methods are also useful in characterizing intoxication, and the real richness of neurotoxicology devolves from the interplay of observations at the behavioral, physiological, biochemical, and structural levels. Because this interplay lies beyond the scope of this text, such interactions can be only briefly noted when they clarify the issue at hand.

3.1   EFFECTS OF SOLVENTS ON SENSORY SYSTEMS

3.1.1   Olfactory and trigeminal systems

Most solvent vapors are detectable at very low concentrations by their stimulation of the olfactory epithelium of the nose. Many of the "aromatic" compounds comprise essential components of natural and synthetic flavors due to their potent and pleasant fragrances. Some 1,000 chemicals are available from "Flavour Houses" as synthetic flavoring agents; their safety in this regard falls under the category "Generally Regarded As Safe" by the U.S. Food and Drug Administration. For example, isopropyl and amyl alcohols, methyl isobutyl ketone, and ethyl acetate provide fruity flavors to some baked goods and candies; turpentine enhances the spiciness of some baked foods, and a styrene-butadiene copolymer provides the base material for chewing gum.

Airborne odors are detected by receptors of the nasal olfactory epithelium. The nerve cells in this region project via the olfactory nerve to the olfactory bulb of the brain, where characterization of the odors occurs. Detection of the presence of a solvent vapor can be accomplished at a concentration roughly three times lower than that necessary for its identification[4]. Thus one must distinguish between the *detection threshold* and *identification threshold* of a solvent vapor. The former refers to the lowest concentration that can be reliably identified as different from clean air, while the latter refers to the lowest concentration at which a subject can identify, or name, the odor. As with most threshold determinations, an accuracy level of 75% typically defines the reliability of the response. For purposes of industrial hygiene, detection thresholds are adequate. For food manufacturing purposes, concentrations sufficient to affect the overall flavor of the product are necessary.

Detection of a volatile compound may or may not be associated with *irritation* of the eyes, nose, and other mucous membranes. This irritation occurs upon stimulation of sensory endings of the afferent branches of the trigeminal nerve, which are diffusely spread throughout the mucous membranes of the eyes and oral and nasal cavities. For most solvent vapors, the detection threshold is lower than the level associated with irritation; however, both differ considerably across solvents, as is illustrated for a selected number of compounds in Table 14.1.

The irritant potency of a number of solvents has been quantified by an animal bioassay which utilizes the propensity of mice to slow their breathing when exposed to an airborne irritant[2]. Mice are placed in a plethysmograph, which permits measurement of their ventilation rate, and then exposed to an airstream containing a known concentration of a solvent vapor. The degree

TABLE 14.1

Olfactory and irritant properties and permissible exposure levels for selected solvents

All values are in ppm.

Solvent name	Olfactory threshold [a]	Mouse RD_{50} [b]	Irritation level [c]	TLV [d]	Criterion effect [e]
Benzene	12	None	–	10	C
Toluene di-isocyanate	0.17	0.39	0.012	0.005	P
Chlorine	0.31	9.34	0.28	0.5	I
Ammonia	5.2	303	9.1	25	I
Ethyl acetate	3.9	614	18.4	400	I
Styrene	0.32	666	20	50	C
Cyclohexanone	0.88	756	22.7	25	I
Ethylbenzene	2.3	1432	43	100	I
Xylene	1.1	1467	44	100	I, N
Methylisobutylketone	0.68	3195	96	50	I
n-Butyl alcohol	0.83	4784	144	50	N
Isopropyl alcohol	22	5000	150	400	I
Toluene	2.9	5300	159	100	N
Methylethylketone	5.4	9000	270	200	I
n-Propyl alcohol	2.6	12704	381	200	I
Ethyl alcohol	84	27314	819	1000	N
Methyl alcohol	100	41514	1245	200	N
Acetone	13	77516	2325	750	I

[a] Olfactory detection threshold (human)[4].

[b] RD_{50}: The concentration causing Respiratory Depression of 50%[3,18,52].

[c] Threshold for irritation in humans, defined as 0.03 times the mouse RD50 in col. 3.

[d] The TWA-TLV (Time-Weighted Average Threshold Limit Value) established by the American Council of Governmental Industrial Hygienists: the maximum concentration for an 8-hour work day at which no adverse effects are expected[1].

[e] The toxic effect upon which TWA-TLV is based: C = carcinogenesis; H = hematopoiesis; I = Irritation; N = neurotoxicity; P = pulmonary function; R = reproduction.

and duration of the respiratory depression (reduction in breathing rate) during a few minutes' exposure to a solvent is taken as a measure of the irritancy of the vapor. Irritancy is then expressed in terms of respiratory depression, i.e., the concentration of the vapor sufficient to depress respiration by 50% (RD_{50}). With a few exceptions, RD_{50} scores (Table 14.1, col. 3) correlate well with allowable industrial exposure levels (Table 14.1, col. 5), most of which are based on industrial hygiene reports of irritation in workers (Table 14.1, col. 6). These correlations indicate that, for these solvents, the assay reflects well the health concerns used to set standards for occupational exposure.

The irritant property of solvents thus provides a means by which exposure can be regulated. For some solvents, toxicity in other organ systems (e.g., the induction of cancer or neurotoxicity) occurs at exposure concentrations which are not irritating. For example benzene, which does not depress respiration at sub-anesthetic concentrations in mice, is carcinogenic. In these cases, regulation of exposure must be based on the most potent known effect of the compound. One must also be cautious about relying on irritation as an index of the neurotoxicity of a solvent. For example toluene, which has low irritant potency (RD_{50} = 5300 ppm), causes hearing loss in rats and in man, whereas p-xylene, which is more irritating (RD_{50} = 1467 ppm), does not.

Another way to assess the aversiveness of a solvent vapor involves induction of a conditioned flavor aversion. If a subject experiences a novel flavor just prior to induction of gastrointestinal discomfort (e.g., infection with a gastrointestinal parasite or exposure to X-irradiation), the subject will avoid consumption of that flavor for long periods thereafter. This phenomenon was termed the *"sauce béarnaise syndrome"* by Dr. Martin Seligman after he enjoyed a fancy meal featuring this exotic sauce immediately before succumbing to a stomach virus. Despite knowledge that the sauce was unrelated to his illness, its association with the nausea caused a strong aversion to its flavor, that persisted for months.

A related phenomenon may occur upon repeated exposure to solvent vapors. Some workers (e.g., spray painters) experience nausea after breathing a solvent vapor. Anecdotal reports suggest that this nausea may become conditioned to cues associated with the workplace. For example, a worker may initially experience nausea after inhaling the vapors. As repetition of this response occurs in this environment, the nausea may appear in anticipation of actual exposure to the vapor, for example upon, entering the building or while driving to work.

Conditioned flavor aversions have been well-documented for agents which produce toxicity when given by the oral route[61]. It is also possible to condition aversions in rats by pairing intake of a novel flavor (e.g., saccharin) with inhalation of a solvent vapor. Again using the comparison of toluene and *p*-xylene, it is difficult to induce an aversion to saccharin by pairing its flavor with exposure to toluene vapor[47], whereas inhalation of *p*-xylene induces a strong aversion to saccharin at concentrations above 200 ppm, and a weaker aversion at concentrations as low as 50 ppm (Figure 14.1)[12]. This difference in efficacy suggests that irritation may play a role in the induction of conditioned aversions to inhaled vapors.

It should also be noted that exposure to solvent vapors may affect the sense of smell itself. Because of its physical location, the olfactory epithelium may be exposed to toxicants both in the stream of inhaled air and circulating in the blood, and may therefore be especially vulnerable to the toxicity of solvent vapors. Occupational exposure to solvents involved in paint manufacture has been shown to impair identification of common odors presented in a standardized battery[65]. Experimental exposure of human volunteers to vapors of toluene and / or xylene elevated the detection thresholds for toluene but not for OPT-carbinol (phenylethyl methyl ethyl carbinol), a solvent not containing an aromatic benzene ring[43]. This data indicates a reduced capacity to *identify* and *detect* odors of vapors within a chemical class, but not of a different class. It is likely that saturation of sub-populations of receptors in the nasal epithelium is involved in the specificity of the detection threshold shifts observed. Because these studies involved concentrations of test agents lower than those needed to cause trigeminal irritation, it is unclear whether or not the subjects' sensitivity to irritation was also reduced.

3.1.2   *Visual system*

Intoxication with many solvents gives rise to vaguely defined symptoms of visual disturbance. "Blurred vision", for example, is a common complaint among intoxicated people that may have several sources, including lacrymation (tearing), oculomotor difficulties (inability to fixate, accommodate, or move the eyes normally), or reduced contrast sensitivity, acuity, or color vision. The characterization of visual impairment thus involves defining more precisely the nature of the visual deficit and specifying the mechanism(s) underlying these symptoms[44]. Visual dysfunction of all of these forms has long been known to occur after exposure to methanol and carbon disulfide, two organic solvents well known for their effects on vision. In addition, evidence is accumulating that exposure to other solvents may also affect vision; the slow progression of these changes may lead to subclinical impairment, i.e., reduced

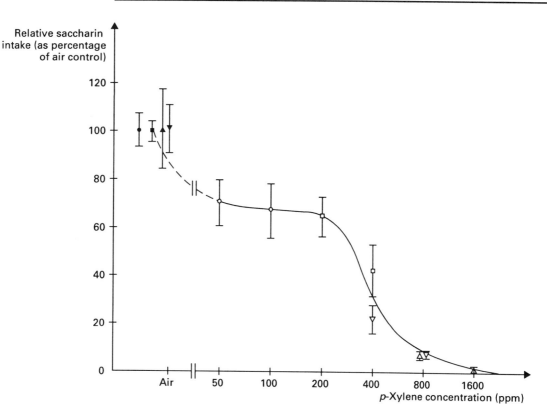

FIGURE 14.1

An aversion to the flavor of saccharin conditioned by inhalation of *p*-xylene

The aversion was indexed by the suppression of saccharin intake. Rats exposed only to air preferred saccharin-flavored water to tap water; this preference (relative saccharin intake) was reduced in rats whose first taste of saccharin was followed by exposure to *p*-xylene vapor. Rats exposed to *p*-xylene concentrations below 400 ppm showed a weaker preference for saccharin compared to controls (relative intake greater than 50%), whereas those exposed to higher concentrations preferred tap water (relative intake less than 50%). Symbol shapes reflect independent replications of the effect of *p*-xylene; filled symbols represent control performance; open symbols performance after *p*-xylene. (Adapted from Bushnell P.J. et al. (1988) *Neurobehav. Toxicol. Teratol.* **10**: 273–277.)

capacity not readily apparent to the subject, but detectable with sensitive test methods.

Ingestion of methanol ("wood alcohol") can cause severe visual disturbances, including complete blindness, in man, apes and monkeys, but not in rodents or guinea pigs due to differences in metabolism in these species. Dehydrogenation of methanol produces formaldehyde which is rapidly converted to formic acid. Because this acid is only slowly removed from the blood in primates, the resulting metabolic acidosis damages the retina. The visual symptoms of methanol intoxication in man include blurred vision, loss of acuity, photophobia, and loss of foveal vision (central scotoma). The photophobia probably results from mydriasis (dilation of the pupils) and loss of the pupil reflex. Symptoms of carbon disulfide poisoning are similar, but lack the photophobia. Motor disturbances resulting from either solvent are typically transient, occurring only during the acute phase of intoxication. Changes in retinal morphology and function may persist after the high level intoxication has passed, and lead to loss of contrast sensitivity and visual acuity, two characteristics of normal sight whose degradation can reduce clarity of vision. In addition, abnormal color vision may occur in parallel with these other changes, reflecting degeneration of neurons in the retina.

Visual contrast sensitivity and visual acuity have been quantified in humans and experimental animals both psychophysically and electrophysiologically. Psychophysical tests require instructing (or training, in the case of an experimental animal) the subject to discriminate between two images by making an appropriate motor response, while electrophysiological tests involve measurement of evoked electrical potentials from the surface of the scalp (in humans) or the visual cortex (in animals)[20]. The goal of these tests is to define the threshold as the lowest light intensity which produces reliable discrimination between the images.

Visual contrast sensitivity refers to the subject's ability to detect a visual image based on the amount of contrast (the difference in brightness) between it and its background. Visual patterns consisting of light and dark bars ("gratings") are often used as discriminative stimuli for this purpose. The ability to discriminate between a grating and a homogeneous field depends both upon the contrast between the bars of the grating (the difference in luminance between the light and dark bars divided by the sum of their luminances) and the width of the bars. In primates, sensitivity (defined as the reciprocal of the threshold) is optimal for patterns which repeat 2–8 times per degree of visual angle and falls for both wider and narrower gratings (Figure 14.2). The number of bars in each degree of visual angle is termed the *spatial frequency* of the grating. The higher the spatial frequency, the finer the grating; that is, the more bars per unit of width in the visual field. Spatial frequency is reported in units of *cycles per degree*, where each cycle represents one pair of bars (one light and one dark), and degree refers to one of the 360 degrees of visual angle in any circle whose center is located at the cornea of the eye. Loss of contrast sensitivity is revealed psychophysically as an increase in the threshold, i.e., the contrast necessary to elicit reliable discrimination between a grating and a blank field. Electrophysiologically, the threshold is defined as the lowest contrast level which evokes a definable electrical potential when a grating stimulus is presented to the subject.

A special case of contrast sensitivity, *visual acuity*, refers to the finest grating pattern (highest spatial frequency) which can be differentiated from homogeneity at high contrast. This measurement is used in routine testing of visual function, for example with the black-on-white letters of the Snellen eye chart, and is adequate for detecting acuity loss due to optical abnormalities of the eye.

Loss of visual acuity is thus the selective loss of contrast sensitivity for stimuli with high spatial frequency, and is experienced as an inability to resolve fine images, even under bright light, and preserved ability to detect images with coarse features. Loss of contrast sensitivity involves a reduced ability to discriminate between images of any size when their contrast is low (as in dim light). In other words, a subject with low contrast sensitivity requires a greater difference between the brightness of an object in relation to its background to resolve it visually. Loss of contrast sensitivity thus becomes important when vision is challenged under conditions of low illumination. Toxicant-induced changes in contrast sensitivity may be constant across spatial frequencies or vary with spatial frequency (Figure 14.2).

Color vision may be assessed in humans in a number of ways[35]. Color arrangement tests are often used to detect loss in both the red-green and blue-yellow dimensions. The most sophisticated of these tests is the Farnsworth-Munsell 100-Hue test, in which two subsets of hues, 85 saturated and 15 desaturated, are used. Smaller, more portable and less time-consuming variants, including the Lanthony desaturated D-15 series, have been applied successfully in the field to assess color vision in occupational settings. In these tests, the color stimuli are displayed on small moveable caps, which are first shown to the subject in spectral order and then presented in a random arrangement. The subject is

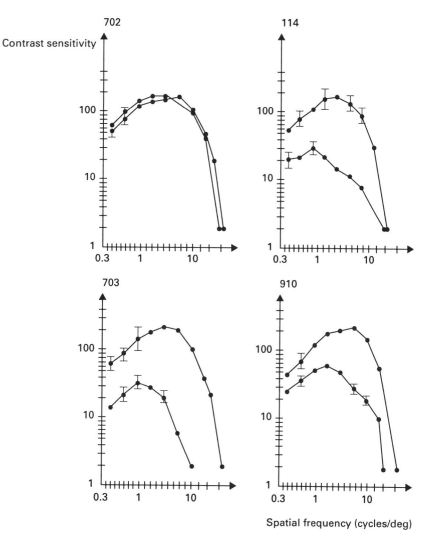

FIGURE 14.2

Contrast sensitivity functions of monkeys before and after long-term exposure to carbon disulfide, shown as contrast sensitivity plotted as a function of spatial frequency (width of the bars in a grating used as the test stimulus)

Each panel shows results from one subject. Maximum sensitivity prior to treatment (circles/upper curves) showed a peak at 2-8 cycles/degree and fell at lower and higher spatial frequencies. Monkey #702 was a control. Sensitivity functions after CS_2 (circles/lower curves) show a reduction in overall sensitivity and a shift in the peak toward wider stimuli (1-3 cycles/degree). Early in intoxication, monkey #114 showed reduced contrast sensitivity at all except the highest spatial frequency, indicating impaired sensitivity with wide stimuli but normal sensitivity at the highest spatial frequency (i.e., normal acuity). Later in intoxication, monkeys #703 and #910 showed reduced sensitivity at all frequencies, indicating loss of acuity as well as contrast sensitivity. (Adapted from Merigan, W.H. et al. (1988) *Invest. Ophthalmol. Vis. Sci.* **29**: 512-518. With permission.)

asked to rearrange the caps in sequence from one fixed extreme. Subjects with normal color vision order the colors correctly, corresponding to a color circle, while those with *dyschromatopsia* (deficient color vision) arrange the caps with specific errors characteristic of vision loss along blue-yellow and/or red-green color confusion axes. Congenital deficiencies in color vision typically involve confusion along red-green axes (Type I dyschromatopsia), whereas acquired deficits most often begin with confusion along the blue-yellow axis (Type III dyschromatopsia) and may progress into a mixture of Types III and I (see below).

Damage to the retina or other parts of the nervous system involved in vision may reduce color vision, contrast sensitivity, and acuity independently or in concert. For example, monkeys exposed to carbon disulfide by inhalation[45] showed primarily a loss of contrast sensitivity (Figure 14.2). Acuity was also reduced in all monkeys after prolonged exposure, but was preserved early in exposure, after contrast sensitivity had fallen. Some combination of loss of contrast sensitivity and/or acuity probably accounts for the "blurred vision" reported by humans poisoned with solvents. Humans exposed to carbon disulfide for prolonged periods also report loss of vision at the fixation point (foveal scotoma), which is consistent with the loss of retinal ganglion cells in the fovea of these monkeys after long-term exposure to this solvent.

Recently, Mergler[42] proposed that solvent-induced acquired dyschromatopsia proceeds in a orderly fashion from an early blue-yellow (Type III) defect to a more complex pattern involving a red-green defect (Type II) after prolonged occupational exposure to solvents. In addition to the behavioral evidence for this pattern, these types of functional impairment appear to correspond to different patterns of morphological change: the Type II defect appears to be associated with loss of cells in the retina, whereas the Type III defect may also involve changes in the optic nerve. Finally, high correlations between loss of color vision, near acuity, and contrast sensitivity have been documented in paint shop workers, suggesting that ophthalmotoxic damage from solvents may produce all three types of functional impairment, though whether or not these changes occur simultaneously and by the same mechanism remains to be determined.

3.1.3 *Auditory system*

Some of the first evidence for solvent-induced hearing loss (SIHL) was reported in epidemiological studies of workers exposed to high concentrations of trichloroethylene and toluene. Auditory deficits were also noted in clinical studies of patients with a history of solvent abuse[23,46]. The following paragraphs describe exemplary findings of SIHL in more detail to illustrate the major points regarding this serious and irreversible impairment of hearing.

Data collected from humans exposed to solvents either occupationally or intentionally provide an equivocal picture of the effects of solvents on auditory function. Two methods are commonly used to assess hearing in humans. Audiometric threshold testing is a standard clinical procedure used to detect hearing loss behaviorally[19]; it relies on psychophysical techniques analogous to those described above for olfaction and vision. Electrophysiological methods, usually measurements of brainstem auditory evoked potentials, are also used clinically to assess hearing. Research with these two methods has led to findings of sensorineural hearing loss (i.e., hearing loss resulting from damage to the inner ear) in humans exposed to trichloroethylene, carbon disulfide, styrene, toluene, and mixed exposures to paint products[55,64].

Another potentially important factor involved in occupational SIHL is noise-induced hearing loss, particularly in light of consistent findings that noise and solvent exposure may act together to elevate auditory thresholds[48,49]. This finding gains importance due to high noise levels typical of many occupational settings.

A number of methods have been used to detect and characterize auditory dysfunction in animals. Behaviorally, SIHL may be measured by the use of *conditioned avoidance responding* in which a warning tone is presented to a rat and followed after a few seconds by an aversive electric shock to the feet. The rat is given a means of escaping the shock, e.g., by climbing a pole suspended from the ceiling of the test chamber. After several pairings of tone and shock, the rat learns to climb the pole between onset of the tone and onset of the shock: on these trials, no shock is received, and the rat is said to have successfully avoided the shock.

Because successful avoidance behavior requires the ability to detect the tone, manipulation of the intensity and frequency of the tone can be used to assess the animal's hearing. For example, if a rat avoids shock when warned by a 4-kHz tone but fails to do so when warned with a 32-kHz tone, it may be said to exhibit high-frequency hearing loss. Further more, because hearing loss rarely involves a complete inability to detect sound, an elevation of the auditory threshold is more frequently observed than is complete inability to avoid shock. Finally, toxicant-induced threshold elevation is rarely uniform across the frequency spectrum, instead occuring to varying degrees at different frequencies. In the first animal model of SIHL, Pryor and colleagues[56] used this method to demonstrate auditory threshold elevation with warning tones of 8, 16, and 20 kHz in rats exposed to toluene for 5 weeks (Figure 14.3). Similar results have subsequently been obtained for many other solvents by this group using this procedure.

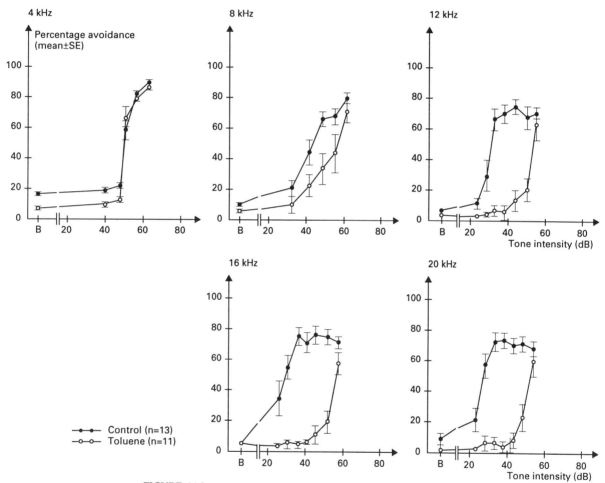

FIGURE 14.3

Toluene-induced hearing loss in rats as assessed by the conditioned avoidance procedure

Weanling rats inhaled toluene for 14 hr/day, 7 days/week, for 5 weeks, 2.5 months prior to testing. Percent avoidance (number of trials with successful avoidance/total trials) increased with tone intensity (abscissae) and tone frequency from 4 to 20 kHz. At intermediate intensities, the avoidance rate was lower in animals exposed to toluene than in controls, but reached levels close to control at high intensities. Normal performance under some conditions (i.e., with low-pitch or high-intensity stimuli) indicates that the treated rats could perform the task when the stimuli were audible. The rightward shift in the curves for the toluene rats shows the degree of threshold elevation, which was small at 4 and 8 kHz (low frequency), and large at 12, 16, and 20 kHz (higher frequency). The maximum shift (at 16 kHz) was about 40 dB. B = blank condition. (Adapted from Pryor, G.T. et al. (1993) *Neurobehav. Toxicol. Teratol.* **5**: 53-57.)

Another behavioral procedure used to quantify solvent-induced hearing loss is *reflex modification audiometry*. In rodents, reflex modification makes use of the whole body startle response, a characteristic sequence of unconditioned, reflexive muscle movements elicited by a sudden, intense sensory stimulus. Reflex modification refers to a change in the reflex response due to a perceptible, and antecedent, change in the sensory environment[14].

Findings of solvent-induced hearing loss with this procedure[16,17,30] are consistent with previous reports using avoidance tests. Moreover, use of this procedure has documented the magnitude of the threshold shift produced by exposure to solvents. For example, 5-day exposures to trichloroethylene, toluene, styrene, or mixed xylenes produced an approximately 25–30 dB threshold shift at 8 and 16 kHz. More importantly, this data clearly documents that the threshold elevations induced by trichloroethylene were restricted to the mid-frequency hearing range of the rat (Figure 14.4). Until this time, most reports of SIHL tested frequencies only in the low and middle range of the rat's hearing spectrum, which normally extends above 40 kHz (human hearing extends only to about 20 kHz).

Electrophysiological procedures have played major role in characterizing the effects of solvents on auditory function in animals. Early work in this area was conducted by Rebert and colleagues[59] using the *brainstem auditory evoked response* (BAER). The BAER is a field potential recorded from skull electrodes that are generated in the eighth cranial nerve and brainstem auditory structures[68]. An example of a BAER waveform is shown in Figure 14.5. Determinations of auditory thresholds are made in a manner analogous to that for visual contrast sensitivity (see above): the weakest auditory stimulus reliably eliciting recognizable components of the BAER waveform is taken as the auditory threshold for a given frequency. Recent BAER thresholds have confirmed that the auditory effects of toluene occur in the mid-frequencies for the rat.

Auditory deficits have been demonstrated from a variety of solvents (Table 14.2); however, an equal number of solvents have been shown not to be ototoxic. It is not yet possible to predict which solvents will be ototoxic: for example, *p*-xylene induces hearing loss, whereas the *o*- and *m*- isomers of xylene do not. As the lists of ototoxic and non-ototoxic solvents grow, *structure-activity relationships* may emerge to enable prediction of the mechanism by which these solvents exert their effects on hearing, thus which are likely to be ototoxic.

Indeed, the site and mechanism of action of solvents in causing hearing loss is a focus of current research which extends beyond the view of purely behavioral methods. Strong histopathological and electrophysiological evidence indicates a cochlear site of damage in SIHL; other evidence (e.g., effects on cortical potentials and speech perception in the absence of effects on auditory thresholds) suggests that solvents damage the CNS as well.

The first line of evidence regarding a cochlear site of action in SIHL involves structural changes in the cochlea associated with exposure to solvents. Damage can be visualized histopathologically with a surface preparation of the organ of Corti, the spiral structure within the cochlea that contains the inner and outer hair cells. (Hair cells are transducers in the cochlea which convert the sound-induced vibrations of the middle ear to nerve signals). Careful dissection of the temporal bones yields a preparation in which individual hair cells, both intact and damaged, can be counted[19]. Maps, called *cytocochleograms*, are constructed to provide an anatomical view of the damage. These maps are particularly instructive due to the tonotopic organization of the cochlea, in which the auditory frequencies (pitches) to which hair cells are sensitive decreases as the distance from base to the apex of the cochlea increases. Thus damage near the base of the cochlea is associated with high-frequency hearing loss, while damage near the apex reflects low-frequency loss.

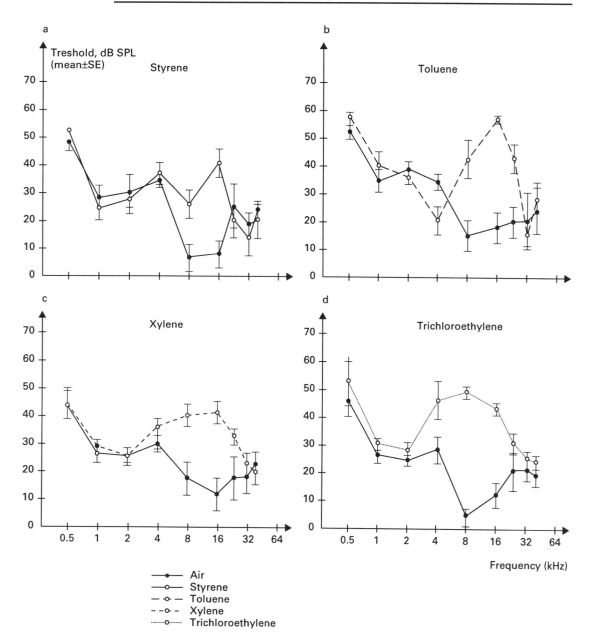

FIGURE 14.4

Hearing loss induced in rats by four solvents as assessed by prepulse modification of the acoustic startle response

Thresholds (determined as described in the text) for control rats show a typical 'tuning curve': lowest threshold (highest sensitivity) at frequencies at 8 and 16 kHz, and higher thresholds at lower and higher frequencies. Rats exposed to styrene, toluene, xylene, and trichloroethylene all showed threshold elevations in the mid-frequency range of 30 to 50 dB. Note that normal thresholds were observed with stimuli lower than 4 kHz and higher than 20 kHz, the highest frequency tested with the conditioned avoidance response in Figure 14.3. (Adapted from Crofton, K.M. et al. (1994) *Hearing Res.* **80**: 129–140.)

Although difficult and tedious to prepare, the cytocochleogram provides an excellent method of describing toxicant-induced damage to the organ of Corti. This procedure has been used to document the loss of hair cells following exposure to toluene and styrene. Consistent with the functional deficits reported to occur in the mid-frequency hearing range, these solvents appear to damage outer hair cells in the regions of the cochlea associated with mid-

409

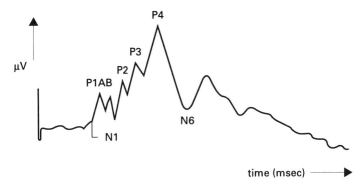

FIGURE 14.5

A representative BAER waveform from a normal adult Long Evans rat as recorded from electrodes located on the surface of the cortex

The peaks are numbered sequentially as positive (P) or negative (N) deflections in voltage. Each peak is thought to reflect the electrical activity of a brainstem generator in the pathway from the cochlea to the auditory cortex after activation by an auditory stimulus. (Adapted from Crofton, K.M. et al. (1994) *Hearing Res.* **80**: 129-140.)

frequency hearing. Figure 14.6 illustrates these results with data from animals exposed to toluene.

Other lines of evidence regarding a cochlear site of action of SIHL involve changes in the acoustics and electrophysiology of the inner ear. For example, *distortion product otoacoustic emissions* (DPOEs) are sounds generated by the cochlea in response to acoustic stimuli which can be recorded from the ear canal. The diagnostic value of DPOEs relies on their reflection of active processes in outer hair cells of the cochlea[28]. A recent study demonstrated reduced DPOE amplitudes in rats exposed to toluene, suggesting that its effects may involve these cells that is consistent with the structural changes noted above.

Animal models have also played an important role in determining the potential confounding effects of noise in SIHL. Cochlear damage can follow exposure to loud noise alone, and most experimental solvent exposures take place in flow-through chambers, which tend to be noisy. Johnson[31] has shown that the effects of toluene vapor are amplified when damaging levels of noise

TABLE 14.2

Solvents for which there is evidence of ototoxicity in animals or in humans
(Adapted from: Crofton et al. (1994), *Hearing Res.* **80**: 129–140.)

Animals	Humans
Carbon disulfide	Styrene
1-Chloro-4-fluorobenzene	toluene
Ethylbenzene	Trichloroethylene
4-Fluorotoluene	Mixed solvents
Methoxybenzene	Glue
1-Methoxy-1,4-cyclohexadiene	
α-Methylstyrene	
Monochlorobenzene	
n-Propylbenzene	
Styrene	
Toluene	
Trichloroethylene	
p-Xylene	

410

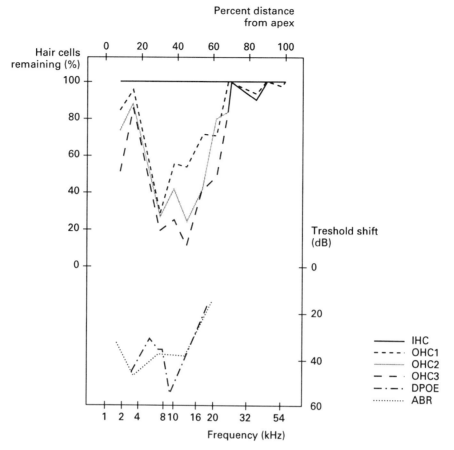

FIGURE 14.6

Cytocochleogram (upper panel) showing hair cell loss in a single rat 4 days after inhalation of toluene at 1400 ppm for 16 hr/day × 8 days

Virtually no loss of inner hair cells (IHC) was observed, while loss of outer hair cells in each row (OHC1 - OHC3) depended upon the distance of the cell from the apex of the cochlea (scale at top of figure), which in turn is related to the frequency of sound to which the cells are tuned (scale at bottom of figure). Lower curves show threshold shifts previously recorded in this animal using the distortion product otoacoustic emission (DPOE) or auditory brainstem response (ABR) technique (the ABR is equivalent to the BAER; see text for descriptions of these techniques). Both techniques demonstrated 20 to 40-dB shifts in post-exposure threshold at intermediate frequencies, relative to pre-exposure thresholds (set at 0 dB). (Adapted from Johnson, A.C. and Cablon, B. (1994) *Hearing Res.* **75**: 201–209.)

follow toluene exposure and that when toluene follows noise, their effects are additive. Evidence that noise is not essential for toluene-induced SIHL has also been reported: hearing loss has been demonstrated in rats after subcutaneous exposure to toluene, avoiding the noise present in inhalation chambers[57].

There are at this date a number of important research questions concerning SIHL that remain to be determined. Whereas it is clear that toluene and styrene damage outer hair cells, the mechanism is unknown. The atypical mid-frequency nature of the effect suggests the possibility of a unique mechanism of action compared to the high-frequency loss induced by many "classical" ototoxicants (e.g., aminoglycoside antibiotics). Any attempt to define a mechanism of action must account for the non-linear nature of these lesions. Anatomical evidence from solvent-exposed rats also suggests a greater vulnerability of outer hair cells over inner hair cells. This selectivity must also be considered in any mechanistic studies of solvent-induced ototoxicity.

411

3.1.4   Somatosensory system

Another common complaint associated with solvent intoxication is loss of sensation (*hypoesthesia*), or "phantom" sensations (*paresthesia*), usually reported as occurring first in the extremities (fingers and/or toes). Decreased sensitivity may go unnoticed until revealed by psychophysical testing. Testing humans for such deficits typically involves use of a vibrotactometer, a device with a small (~1 mm diameter) wand, mounted in a stand, through which vibrations are transmitted to the fingers or toes of the subject[41]. A *vibrotactile threshold* is determined as the magnitude of the vibration-inducing oscillation of the wand which the subject can detect reliably. Loss of sensitivity (elevation of threshold) has been documented in persons exposed to carbon disulfide, workers in paint manufacturing and application[7,] and in microelectronic workers[8]. In addition, workers exposed to styrene or toluene exhibited slower somatosensory evoked potentials from median and tibial nerves than controls, suggesting damage to the sensory afferents innervating the hands and feet.

Reduction in vibrotactile sensitivity is generally taken as a sign of damage to the peripheral sensory nerves, which suggests that workers exposed to solvents also suffer impairment of the peripheral nerves. Such a conclusion is not without precedent, as there is good evidence that *peripheral neuropathy* can be induced by organic solvents. The primary exemplar of solvent-induced neuropathy involves hexacarbon solvents *n*-hexane and methyl *n*-butylketone, which are both metabolized to 2,5-hexanedione (*2,5-HD*)[69]. 2,5-HD impairs axonal transport, causing intra-axonal accumulation of neurofilaments and characteristic swelling of nerves proximal to the nodes of Ranvier (gaps between the Schwann cells of myelinated axons). Because these effects occur primarily in long afferent nerve fibers subserving distal sensory receptors and in long efferent fibers innervating muscles, the signs and symptoms of 2,5-HD toxicity initially appear as sensorimotor impairment in fingers, toes, hands, and feet. This "stocking-glove" pattern of dysfunction is not uncommon among peripheral neuropathies with other etiologies.

Summary of effects on sensory systems
A listing of the solvents that have been found to affect sensory function in humans and in experimental animals is shown in Table 14.3. Since the olfactory and auditory information has previously been tabulated (Tables 14.1 and 14.2), these results are not shown again. Several points need to be made regarding the information in this table (and Tables 14.4 and 14.5). First, no differentiation is made between effects of acute and repeated exposure, nor between transient and persistent effects of the solvents. Second, no information regarding the exposure dose or route is presented. Third, the experimental animal species used is not specified. Fourth, the direction of effect is not noted; most effects of course reflect impairment. Fifth, most human studies were epidemiological in design, and precisely which solvent was associated with a given effect cannot be determined. Finally, an empty cell in the table indicates only the absence of evidence for a given effect, and *not evidence for absence of effect*. For example, the lack of entries under CFA in human studies does not mean that no solvent causes a flavor aversion in humans, but rather that no studies of this possibility are known to us. This distinction is critical to appreciating this information because it indicates the limited range of our knowledge of the toxicity of these solvents, instead of confidence that the functions lacking entries are insensitive to solvent toxicity.

3.2   EFFECTS OF SOLVENTS ON MOTOR FUNCTION

The peripheral neuropathy induced by 2,5-HD leads to motor dysfunction as well as to the sensory deficits described above. Symptoms of this syndrome

TABLE 14.3

Solvents for which there exists evidence regarding effects on sensory systems

	Olfaction, taste and irritation			Vision			Hearing	Somatosensation
	Detect/ Ident	Irritation	CFA [a]	Acuity	Visual field	Color		
Animal studies		See Table 14.1	Toluene p-Xylene	CS_2 [b]	Methanol		See Table 14.2	Benzene Cumene Ethylbenzene n-Hexane Methyl butyl ketone Propyl benzene Toluene Trichloroethylene m-Xylene
Human studies	CS_2 Toluene Xylene	See Table 14.1		CS_2 Methanol	CS_2 Methanol	CS_2 Ethanol Styrene	See Table 14.2	n-Hexane Methyl butyl ketone 1,1,1-Trichloro- ethane Trichloroethylene

Detect/Ident = Detection or identification threshold
[a] CFA = Conditioned Flavor Aversion (see text)
[b] CS_2 = Carbon disulfide

TABLE 14.4

Solvents for which there exists evidence regarding effects on motor functions

	Motor activity/Arousal	Strength	Balance/Coordination/ Speed	Performance[a]
Animal studies	Benzene Cumene Ethylbenzene Propylbenzene Styrene Toluene 1,1,1-Trichloroethane Trichloroethylene m-Xylene	Benzene Cumene Ethylbenzene n-Hexane Methyl butyl ketone Propylbenzene Styrene Toluene m-Xylene	Benzene Carbon disulfide Cumene Ethylbenzene n-Hexane Methyl butyl ketone Methyl chloride Methylene chloride Propylbenzene Toluene 1,1,1-Trichloroethane Trichloroethylene m-Xylene	Carbon disulfide Ethanol Heptane Hexane Methanol Methylene chloride Methyl ethyl ketone Octane Pentane Styrene Toluene 1,1,1-Trichloroethane Trichloroethylene
Human studies	Carbon disulfide Ethanol n-Hexane Methanol Methyl butyl ketone Methylene chloride Methyl ethyl ketone Styrene Toluene 1,1,1-Trichloroethane Trichloroethylene Xylenes	n-Hexane Methyl butyl ketone Styrene 1,1,1-Trichloroethane Trichloroethylene	Carbon disulfide Ethanol n-Hexane Methyl butyl ketone Methylene chloride Methyl ethyl ketone Styrene Toluene 1,1,1-Trichloroethane Trichloroethylene Xylenes	Carbon disulfide Ethanol Toluene 1,1,1-Trichloroethane

[a] Performance of a previously-learned conditioned response.

413

TABLE 14.5
Solvents for which there exists evidence regarding effects on cognitive function

	Attention	Learning	Memory
Animal studies	Toluene Trichloroethylene	Carbon disulfide Ethanol Methylene chloride Styrene Toluene 1,1,1-Trichloro-ethane Trichloroethylene Octane	Ethanol Trichloroethylene
Human studies	Acetone Carbon disulfide Ethanol Methyl chloride Methylene chloride Styrene Toluene 1,1,1-Trichloro-ethane	Ethanol Styrene	Carbon disulfide Ethanol Styrene Toluene Trichloroethylene

include muscle weakness in the extremities, as may be indicated behaviorally by loss of grip strength and gait abnormalities. Neurological tests may be used to quantify reported motor impairment in humans, including direct measurements of hand grip strength and observations of gait. In addition, quantitative measurements of postural instability can be made by standing the subject on a platform equipped with force transducers to detect changes in the distribution of his/her weight. While the subject maintains vertical posture by muscular adjustments, changes in the distribution of his/her weight over time ("sway") is indexed by the variability observed across a series of measurements. Intoxication with solvents frequently increases sway. Comparison of sway with eyes open and closed provides information about the oculomotor and vestibulomotor control of posture, respectively.

Motor effects can also be induced in animals by exposure to 2,5-HD or to its metabolic precursors n-hexane and methyl n-butylketone. The measurement of grip strength, gait deficit assessment, landing foot splay, and rotarod performance have all been applied to the study of solvent-induced motor dysfunction in rodents. For example, gait disturbances may appear in rats after continuous exposure to n-hexane at 400 ppm for 45 days; reduced grip strength and locomotor activity appear to require higher levels of continuous exposure (1000 ppm).

In contrast to the peripheral neuropathy just described, motor behavior can be affected reversibly by acute inhalation of solvent vapors acting on the CNS. In experimental animals, locomotor activity increases from normal "baseline" levels, then with increasing vapor concentration, reaches a maximum, and then falls to zero as anesthetic levels are approached. In humans, symptoms of euphoria, giddiness, and excitement follow a similar concentration-effect relationship: little or no effect occurs at very low levels while at very high levels, the excitation is overcome by depression of the CNS and unconsciousness. Thus the activity level of experimental animals can provide a rough parallel metric of the stages of acute intoxication in humans inhaling solvents.

To measure unconditioned motor activity, an animal is placed in a test chamber which quantifies its movements. Common methods for quantifying

See Chapter 3 for a review of these methods.

414

movement include counting infrared photobeam breaks, physical movements of the test cage against a switch, disruptions of an applied electric or acoustic field, or by recording frequencies and durations of specified behavior patterns by a trained observer[38,60]. Tests of conditioned behavior (e.g., performance of an operant schedule for food reward), show that response rates (e.g., of lever pressing) increase up to a maximum and then fall with increasing solvent vapor concentration. The concentration producing maximal activity varies across solvents and species; for toluene and xylene in rats for example, it lies in the range of 1000–2000 ppm, with facilitation beginning at about 500 ppm and depression beginning at about 3000 ppm. Parallel experiments assessing motor function in humans are lacking due to ethical constraints on the concentrations of solvent vapors that may be presented to human subjects.

The relationship between concentrations of toluene in air, blood, and brain of rats has been studied by Kishi and colleagues[34]. Rats were trained to avoid electric shock by pressing a lever when a warning light was illuminated. The number of presses and percent of shocks avoided (avoidance rate) were measured before, during, and after each of several exposures to toluene vapor at concentrations ranging from 125 to 4000 ppm in air. Other rats, not trained to perform the task, were exposed to toluene under otherwise similar conditions; concentrations of toluene in brain were determined from analysis of the tissues of these animals. The biphasic effect of toluene on activity, as indexed by lever press rate, is clear in the upper panel of Figure 14.7, which shows that as toluene level rose in the brain, activity first increased and then decreased. In contrast, shock avoidance rate did not increase before falling (Figure 14.7, lower panel), indicating that the additional lever presses emitted during inhalation of 1000–2000 ppm toluene did not reduce the number of shocks the rats received.

This data illustrates two important points. First, the motoric effect of toluene (as measured by lever press rate) is related in a non-linear fashion to the level of toluene in the brain of the animal, such that low to moderate levels increase activity whereas high levels suppress activity. Second, the toluene-induced increase in activity reflects adventitious ("extra") responding not under control of the environmental contingencies of the task. This *enhancement of non-contingent responding* may be a general feature of the effects of solvents on behavior, as will be discussed next.

The mechanism by which these changes in behavioral activity occur are not fully understood, but probably involve staged action on a hierarchy of CNS systems. Behavioral excitation may result from release of subcortical motor systems from the inhibitory control of the cortex at low to moderate solvent levels. Higher levels of the solvent in the brain may then inhibit the subcortical motor centers as well, leading to suppression of all behavior, and eventually to unconsciousness. With prolonged exposure at high concentrations, the respiratory control centers in the brainstem become inhibited, and death may ensue from respiratory failure.

In addition to these overt effects of acute intoxication, subtle motor effects of solvent exposure may persist beyond the period of intoxication. For example, *p*-xylene inhalation has been shown to increase motor activity of rats 0.25 - 2 hr after a 4-hr exposure at 1600 ppm, and paradoxically to facilitate acquisition of a lever-press response for food reward as well[9]. The faster acquisition of the response was attributed to a *p*-xylene-induced loss of motor coordination which brought the intoxicated animals in to contact with the experimental contingencies of the task sooner than the controls. This paradoxical facilitation of learning associated with inhalation of *p*-xylene was abolished by increasing the amount of force necessary to press the lever, thus rendering the rats' "clumsy" explorations less effective as responses.

415

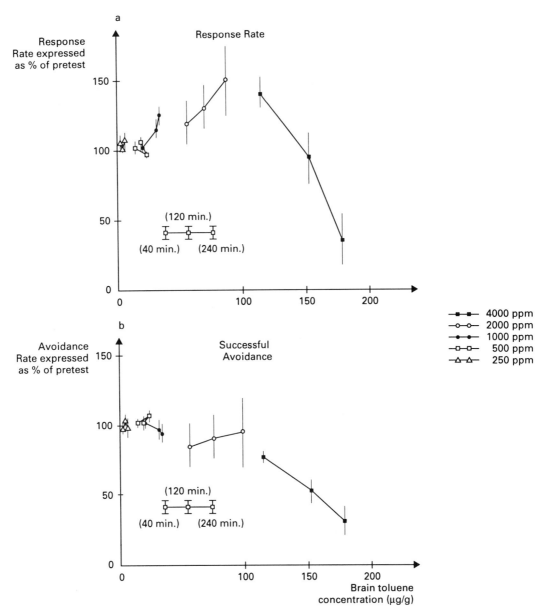

FIGURE 14.7

Relationship between levels of toluene in brain and response rate (upper panel) on a lever providing avoidance or escape from shock, and number of successful shock avoidance responses (lower panel) in rats

Each set of three connected points shows the effects of inhaling a given concentration of toluene vapor, measured after 40, 120, or 240 min. Brain levels of toluene were obtained from sedentary, untrained rats, whereas behavior was assessed in trained rats performing the avoidance task while breathing toluene. As toluene levels in the brain increased, response rate (upper panel) increased to a maximum and then fell (biphasic concentration-effect). This biphasic effect was not observed when avoidance efficiency was measured: in this case (lower panel), shock avoidance declined steadily with increasing levels of toluene. Data from 125 ppm toluene. (Adapted from: Kishi, R. et al. (1988) *Br. J. Indust. Med.* **45**: 396–408.)

These effects of *p*-xylene were observed within a few hours after inhalation of the vapor. Similar effects have been observed in rats when exposure to solvents occurred months prior to training. For example, rats with a history of exposure to trichloroethylene vapor also learned the lever-press response faster than did controls (Figure 14.8); furthermore, they showed increased rates of adventitious

416

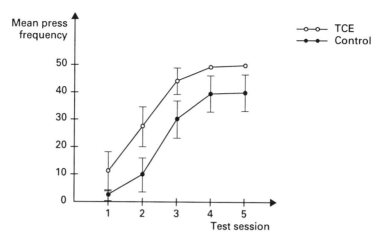

FIGURE 14.8

Facilitation of acquisition of a lever-press response by prior exposure to trichloro-ethylene (TCE)

Rats were exposed to air or TCE (4,000 ppm × 6 hr/day × 5 days/week × 6 weeks) two to four months prior to testing. During training, lever exploration was elicited by repeatedly pairing retraction of the lever with food delivery ("autoshaping"). Explorations vigorous enough to move the lever were defined as presses. TCE-exposed rats emitted an increased frequency of these vigorous responses, and thus learned the association between the lever press and reward sooner than controls. A similar effect in rats exposed to p-xylene[9] was interpreted as a reduced ability to control the motor responses involved in exploration of the lever. (Bushnell, unpublished.)

responding during subsequent training. In addition, rats dosed orally with styrene monomer showed persistent elevations of response rates in other complex conditioning tasks, which retarded learning under appropriate conditions[10].

The rats in this latter study had been dosed orally with styrene monomer at 4–6 months of age; behavioral abnormalities persisted in them for more than a year. Because these effects — as well as those following repeated inhalation of TCE — occurred so long after exposure, it is difficult to attribute them to immediate effects of the solvents on behavior; it is more likely that some persistent change occurred in the CNS which was expressed later in behavior. However, their similarity to the acute, or immediately post-exposure effects of solvents suggest a common mechanism of action. Perhaps the neural substrate(s) upon which solvents act acutely does not completely recover after a long period of intoxication or a large number of toxic episodes.

These observations in rats are generally consistent with observations of psychomotor deficits reported in humans after long-term intoxication with a variety of organic solvents. For example, reduced speed and accuracy on tests requiring reproduction of visual patterns and response slowing have been observed in workers exposed to styrene (see Figure 14.10). These results suggest the hypothesis that solvents affect the generation and control of motor responses by the nervous system. The excitation and euphoria sought by those who inhale solvents voluntarily may be an expression of this increased motor activity. Further more, it is clear (in animals at least) that this activity is not directly related to the contingencies acting in the environment of the subject; similar activity in humans is thus unlikely to lead to productive, socially acceptable behavior.

A list of motor functions and solvents which have been shown to affect them is shown in Table 14.4. All of the cautions listed for Table 14.3 apply to this table as well. In addition to these largely motor effects of solvents, changes not completely attributable to either sensory or motor dysfunction have been reported in humans; these changes instead involve *cognitive dysfunction* (i.e.,

417

disturbances of attention, learning, or memory). How could motor impairment and cognitive dysfunction be related? Given the tendency of intoxicated individuals to emit high rates of non-contingent behavior, it is conceivable that prolonged experience in the intoxicated state could lead to a deterioration of the associations between environmental stimuli and the responses that those stimuli have been conditioned to elicit. In other words, if solvent intoxication causes adventitious responses that are not controlled by the environmental cues which normally shape behavior, and the intoxicated state is repeated and/or prolonged, then the relationships between those stimuli and "normal" (i.e., previously conditioned) behavior may weaken. At some point, these normal behavior patterns may become disrupted sufficiently to generalize to functions that are not entirely motor. In this way, an initial change in motor function could develop into alterations of cognitive capacities, expressed as deficits of attention, learning, or memory.

3.3   EFFECTS OF SOLVENTS ON COGNITION

It is particularly difficult to document disruption of cognitive processes per se because they are not directly observable or measurable. Sensory and motor changes can be directly assessed, the former by psychophysical or electrophysiological determination of objectively defined thresholds, and the latter by direct measurement of muscle strength or response speed. In contrast, cognitive processes must be wholly inferred from changes in behavior, whose quantification requires behavioral assessments that rely upon both sensory input and motor output[22]. Thus if a toxicant impairs sensory, motor, and cognitive processes simultaneously, it becomes extremely difficult to parcel out its effects among these processes. If it affects cognition only, then normal performance on measures of sensory and motor function must be demonstrated along with changes in a measure of cognition, in order to discount sensory or motor dysfunction as the reason for the behavior change. This inferential process may be particularly difficult when sensory or motor changes may be involved in the development of the cognitive dysfunction. In reality, it is likely that any given syndrome of intoxication will reflect a dynamic interplay of changes in all three processes simultaneously.

Several approaches have been taken to address the problem of specifying the degree to which a behavioral change may reflect impairment of cognition. For example, the pattern of changes obtained in a battery of assessments may point toward memory loss, if little or no change in performance on other measures of sensory (e.g., visual acuity), motor (e.g., finger-tapping), or motivation (e.g., appetite or sensitivity to punishment) is observed. Alternatively, a subject may perform a memory task at low accuracy, but with normal speed, suggesting no lack of motor ability. Finally, if accuracy is high when the mnemonic load is removed (immediate recall) but low when a temporal delay is imposed before responding is permitted (delayed recall), then a conclusion of memory impairment may be justified.

As with motor dysfunction, changes in cognition may be induced acutely by solvents or after prolonged exposure. In general, acute effects tend to be transient, while effects which develop over long periods of exposure tend to persist. Quantifying acute cognitive effects of solvents is a simpler problem than assessing their long-term effects, because effects tend to be less subtle, low-level experimental exposures can be carried out in human volunteers, and acute exposure experiments cost less in time and other resources.

A common complaint from acute solvent intoxication is confusion, or loss of the ability to concentrate; these symptoms suggest that *attention* may be compromised. Several methods have been developed for assessing attention in humans. For studies of solvent intoxication, these tests have often assessed *vigilance*, the ability to maintain attention to a tedious task over time. Tests of vigilance involve monitoring, or "tracking" repetitive irrelevant events, and making a response when a specified target event occurs. For example, the Mackworth clock test requires the subject to watch a display representing a clock face on which a hand rotates in discrete 1-second steps. Occasionally and unpredictably, the target event occurs: the hand "jumps" ahead in a 2-second step, and the subject must press a key to record the event. The numbers of correct detections ("hits"), missed targets ("misses"), and incorrect detections ("false alarms") are recorded, from which sensitivity and bias may be calculated. These parameters are derived from the Theory of Signal Detection[27]. *Sensitivity* reflects the subject's accuracy in detecting target events; *bias* represents his/her tendency to respond as if a target occurred, regardless of whether or not it actually occurred.

Inhalation of toluene has been shown to impair vigilance in both human volunteers and in experimental animals. Under demanding conditions (e.g., a high event rate and with stimuli which are difficult to detect), sensitivity falls across time on the task, resulting in a *vigilance decrement*[54]. This vigilance decrement may be induced by inhalation of solvents. For example, Dick and colleagues[21] required volunteers to perform a version of the Mackworth test in an atmosphere containing either filtered air or 100 ppm toluene and found that their sensitivity declined across 2-hr blocks of time when toluene was present, but did not do so in its absence.

This result has been confirmed using rats trained to detect auditory signals[11]. At the end of each trial in this vigilance task two response levers were inserted into the test chamber: if a "signal" (a 20-ms increase in the amplitude of continuously presented white noise) had occurred within the preceding 3 s, then a response to one lever was rewarded by food; if no signal had been presented, a response to the other lever was rewarded. When rats performed this task in the presence of toluene vapor, sensitivity declined across 20-min blocks of the session (Figure 14.9). Very similar effects have also been observed with trichloroethylene.

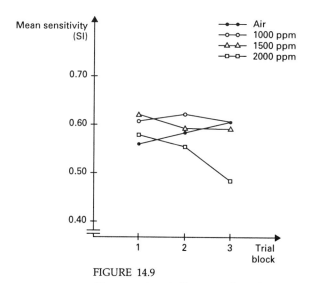

FIGURE 14.9

Changes in vigilance induced in rats by inhalation of toluene
Sensitivity (the accuracy with which rats detected suprathreshold auditory stimuli) increased slightly across blocks of trials in clean air (filled circles), but fell across blocks in atmospheres of toluene vapor (open symbols). This decrease in sensitivity across a 1-hour exposure-test session suggests that toluene interfered with maintaining attention to the demands of the task, thus inducing a vigilance decrement. (Data from Bushnell, P.J. et al. (1988) *Neurobehav. Toxicol. Teratol.* **16:** 149-160.)

Evidence that this effect reflected a change in attention was strengthened by examining other possible causes of the behavior change. Sensory deficits were shown to be unlikely because performance *after* toluene (either immediately or the next day) was normal; rats exposed to these concentrations and durations of toluene retained normal hearing in a reflex modification test; similar deficits have been observed with a visual analog of the test. The role of motor impairment in causing the deficit is more difficult to rule out because response latencies increased along with changes in the signal detection measures of sensitivity and bias. However, operations of the food-cup door did not increase in the presence of toluene, suggesting that the effects on vigilance were not associated with high levels of adventitious responding in the food cup (though other behaviors not detectable by the apparatus could have increased in response to toluene).

These studies suggest that vigilance may be compromised by concurrent inhalation of toluene. They also show that repeated, short-term exposures did not affect vigilance (or hearing, vision, or motor responding) in a permanent manner. Thus they do not provide evidence for the kind of persistent cognitive dysfunction that has been observed in humans after prolonged occupational or recreational exposure to solvents. It is probable that the intensity and duration of exposure were not adequate to cause permanent changes in behavior in this task. Yet persistent cognitive deficits do sometimes occur after prolonged exposure to solvents: what is the nature of these deficits and the means by which they are produced?

These questions are perhaps the most vexing in the field of solvent neurotoxicology today, and understanding the persistent cognitive deficits reported by persons with long term exposure to solvents presents a great challenge. Clinical studies provide evidence for motor, sensory, and emotional difficulties; in addition, cognitive impairment features prominently in the clinical literature on solvent workers[5] and abusers[62].

Documenting these changes in the clinic has usually involved application of standardized intelligence tests, including subtests of the Wechsler Adult Intelligence Scale (WAIS). Some examples include the *Block Design test*, in which the subject's ability to integrate spatial information is assessed by the speed and accuracy with which she/he arranges a set of colored blocks into a pre-determined pattern. *Recall tests* may be used to assess verbal and logical memory by asking a subject to repeat a list of words or concepts, both immediately ("short-term") and after a 30-minute retention interval ("long-term"). The *Digit-Symbol Substitution test* measures the rate at which a subject can write down a set of numbers, each of which is paired with a non-verbal symbol when given a list of the symbols. Whereas each test emphasizes a particular cognitive process, sensory and motor skills play a large role in their performance and, as with animal tests, it is difficult to ascribe observed deficits in performance solely to the cognitive process putatively under assessment.

Nevertheless, these tests provide important evidence of cognitive dysfunction, particularly when a systematic pattern of deficit appears. Thus, several studies have reported memory impairment in workers employed in the manufacture of polystyrene plastic products, including ship hulls and storage tanks. The nature of the work — laminating fiberglass cloth and styrene monomer on large surfaces — can result in high-level exposure to styrene vapor for significant periods of the work day. An early study of Finnish workers with an average occupational exposure to styrene of 4.9 years showed normal intelligence, memory, and vigilance, but a tendency toward reduced visuomotor speed as exposure duration increased[39]. A later study of Italian workers[51] reported deficits on several tests of memory (recall tests) and tests of spatial relationships among visual stimuli (block design) in workers with an average exposure to styrene of 8.6 years (Figure 14.10). Importantly, the

420

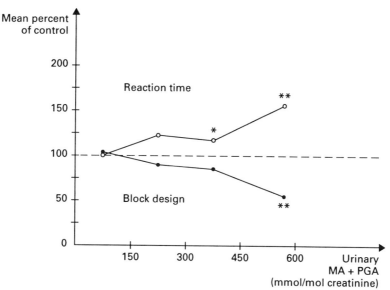

FIGURE 14.10

Effects of occupational exposure to styrene on two neurobehavioral tests in humans

Exposure to styrene was quantified by the levels of two metabolites of styrene, mandelic acid (MA) and phenylglyoxylic acid (PGA), excreted in the urine the morning after a work shift. The Block Design test (a subtest of the Wechsler Adult Intelligence Scale) was used to assess the ability of the subject to reconstruct a two-dimensional design from a drawn pattern, using blocks with partially colored top surfaces. Performance was based on both accuracy and speed of completion of the design. The Reaction Time test assessed the speed with which a subject responded to one of two buttons; the choice depended upon the color of a visual cue (choice reaction time). In general, workers performed more slowly than and less accurately as urinary styrene metabolite levels increased. Asterisks indicate significant differences from a matched control group. (Original data from Mutti, A. et al. (1984) *Am. J. Ind. Med.* **5**: 275–286.)

magnitude of several of these deficits increased both with the level of urinary metabolites of styrene as well as with duration of exposure in the workplace.

In addition to these studies of workers exposed to styrene, extensive evaluations of Danish painters, whose paints were thinned with white spirits, indicated a complex syndrome of cognitive dysfunction, called "chronic toxic encephalopathy" or "organic solvent syndrome". The core symptoms of this syndrome include poor memory and attentiveness, fatigue, irritability, and loss of initiative. In clinical studies, the prevalence of reported memory problems ranged from 46% to 100% of the painters tested[5]. Evaluation of the cognitive capacity of workers assessed with objective behavioral tests showed lower, but still disturbing prevalence rates approaching 65%.

It is clear from these reports that solvents can impair cognitive function. However, the precise nature of these impairments is not clear, for the reasons stated above. In addition, these findings have not been universally observed in other subject populations, nor have they been successfully modeled in animals. The major difficulty with characterizing the organic solvent syndrome arises from the kinds of assessments that are possible in the clinic. Patients reporting these symptoms have invariably experienced a variety of occupations and exposures to solvents which are both highly variable and very difficult to quantify in terms of either the particular chemicals or their concentrations in the air. In addition, each person brings an unique history of other physical and mental characteristics to the clinic, which combine with the solvent exposure history to produce an impenetrable array of possible sources of the symptoms. Finally, experimental studies of humans are limited ethically to conditions which are highly unlikely to cause irreversible changes in mental function, the very effects of concern. Thus scientific understanding of this debilitating condition, hence its prevention and treatment, will probably require development of an animal model of the syndrome.

Efforts to model in animals the neuropsychiatric syndrome resulting from long-term exposure to solvents have not yet succeeded in providing a measure of dysfunction that reflects reliably the symptoms reported by humans. The ototoxicity discussed above permits quantification of a sensory deficit after solvent inhalation, but it is the cognitive difficulties that predominate in the clinic[5], and for which a metric is needed in experimental animals.

Given the prevalence of memory loss in painters and workers exposed to styrene, several attempts have been made to impair memory in rodents with exposure to these agents. For example, rats were dosed orally with styrene monomer daily for up to 40 days, yielding total doses of styrene of 2.5 to 20 g/kg body weight[10], which compare reasonably with estimated *lifetime* exposures of 5 to 10 g/kg in human shipbuilders. No effects on memory were observed despite the high styrene dose rate (daily dose), training either before or after exposure to styrene, and use of a test well known to quantify working memory, a mnemonic process sensitive to many neurological insults.

Working memory can be thought of as the cognitive process by which *information which changes frequently* is acquired, used, and forgotten[29]. It is the "scratch-pad" of information processing, where information is held temporarily until used, and then replaced with new information. It contrasts with *reference memory*, which can be thought of as the processes involved in storing *information which does not change*, and which is retained for long periods of time. A telephone number that one looks up for a single use would be held in working memory until the call is complete; once memorized, one's own telephone number resides in reference memory. Working memory is more easily disrupted — by other events, by disease, by intoxication, and by aging — than is reference memory. Working memory loss characterizes many of the complaints of "forgetfulness" and "confusion" that are often reported by chronically intoxicated people.

Working memory can be quantified in humans and other animals by several means, including tests of *delayed matching*. In these tests, the subject is provided with a bit of information, the *sample* stimulus, and then after a delay, she/he is asked to compare the sample stimulus to two or more *comparison* stimuli. If she/he correctly identifies the comparison stimulus which matches the sample, a positive result (food, for an animal) occurs. The accuracy with which a subject correctly matches the sample decreases with the length of the delay; the function relating accuracy to delay is called a *retention gradient*. For rats, the sample and comparison stimuli are simply the locations of two retractable levers in an operant chamber. As long as the rat is required to do something at a location removed from the levers during the delay period, its matching accuracy falls reliably to chance with delays of less than 1 minute (see Figure 14.11). For humans, the stimuli can be colors, shapes, or patterns, and the number of comparison stimuli can be large.

Despite the proven ability of this method to detect and quantify impairment of working memory after a number of brain lesions, psychoactive drugs, and neurotoxic chemicals, exposure to styrene monomer was without effect, either before training or when given to rats previously trained to perform the task (Figure 14.11). At the same time, styrene did affect other aspects of the rats' behavior, including their ability to learn a series of spatial reversals. Thus, several months after exposure to styrene (using the same dosing protocol), the rats were trained to press either of two retractable response levers for food using an automaintenance schedule. The extension and retraction of one of the

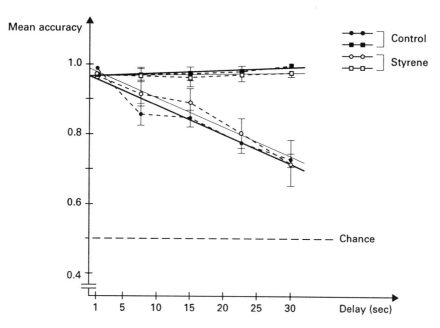

FIGURE 14.11

A test of two kinds of memory in rats previously exposed to styrene

A Delayed Matching-to-Position/Visual Discrimination (DMTP/VD) procedure was used to quantify working memory (as matching accuracy: lower curves) and reference memory (as discrimination accuracy: upper curves). Because matching required retention of information which changed from trial to trial, accuracy in this component of the test fell across retention intervals (delays); conversely, because discrimination required retention of information which remained constant across trials, accuracy in this component was independent of delay. Two groups of rats were trained to perform the task and then were given 40 oral doses (5 days/week for 8 weeks) of either styrene monomer (500 mg/kg/day) or corn oil vehicle. All animals were evaluated daily throughout the dosing period and for 1 year thereafter. These data illustrate performance averaged during the last week of dosing. The group given styrene performed as accurately as controls on both the matching and discrimination components of the test throughout the experiment. (Bushnell, unpublished.)

levers was associated with food; however, the identity of this positive lever alternated ("reversed") unpredictably two or three times per week. Under this schedule, the rats responded preferentially to the positive lever, and the rate at which their responding shifted toward it after each reversal provided a metric of *reversal learning*. Rats previously exposed to styrene exhibited high rates of adventitious responding and performed these reversals more slowly than did controls (Figure 14.12).

The long-term behavioral effects of solvents observed in rats — the TCE-induced changes in learning (Figure 14.8) and the styrene-induced impairment of reversal learning (Figure 14.12) — probably resulted from changes in motor control. However, because neither unconditioned motor activity nor simple operant discriminations appear to provide useful measures of persistent effects of either styrene[36] or white spirits[37,53], more complex, integrative aspects of motor control involved in response selection, execution, and inhibition probably underlie these deficits (see Graybiel et al[26] for a review of the role of the basal ganglia of the mammalian forebrain in adaptive motor control). In addition, a role for dopamine, a neurotransmitter prevalent in the basal ganglia, in toluene-induced changes in behavior has recently been proposed[71].

While progress is being made, no satisfactory rodent model has yet been developed of the persistent cognitive effects of solvents reported in humans. The learning deficits occurring after styrene exposure and the changes in motor function after *p*-xylene and TCE suggest possible leads for further research. In

423

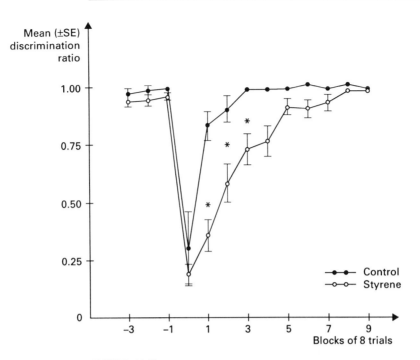

FIGURE 14.12

A reversal learning deficit in rats with a history of oral exposure to styrene monomer

The Discrimination Ratio (DR) reflects the proportion of responses made to a retractable lever whose retraction from the test chamber was associated with food delivery (positive lever), in comparison to a second (neutral) lever whose retraction occurred independently of food delivery. A DR of 0.5 indicates equal rates to both levers; a DR of 1.0 indicates responding to the positive lever only. The reversal involved an unsignalled switch in contingencies, such that the neutral lever became positive, and vice versa. Prior to the reversal (Blocks -3 to -1), both groups responded almost exclusively to the positive lever; after the reversal (beginning in Block 0), the DR dropped to about 0.25 and returned toward 1.0 as the rats shifted their responding in accordance with the change in contingencies. The rats exposed to styrene acquired the reversal more slowly than did the controls. Asterisks indicate points differing significantly from control. (Data from Bushnell, P.J. et al. (1994) *J. Am. Coll. Toxicol.* **13**: 279–300.)

addition, improvements in computerized neurobehavioral tests are presently being devised for humans[33]. Behavioral characterization of the solvent-induced cognitive impairment reported clinically by humans will provide important information regarding the type of tests that will be required to model this syndrome in animals.

It is conceivable that differences in metabolism between rodents and humans could account for the difficulty in documenting clear and persistent changes in behavior in rats. However, it is unlikely that compounds as different as styrene and TCE would yield a toxic metabolite in humans but not in rats, and toxicokinetic studies do not bear out the possibility that higher levels of solvent are achieved by humans than by rats under experimental conditions. It is more likely that the "right" test method, which will demonstrate the long-sought deficits in solvent-exposed rats, has not yet been invented or applied.

A listing of the solvents with demonstrated effects on cognitive functions is presented in Table 14.5. All of the cautions stated for Table 14.3 apply here as well. In one case, an attempt was made to determine whether or not styrene affected working memory in rats (Figure 14.11) but no effect was observed. In this case, then, there is evidence for an absence of effect; few such cases are known.

4   Summary

Solvents comprise a large class of extremely useful hydrophobic chemicals with widespread use in industrialized nations. Whereas environmental exposures are probably too low to affect the nervous system significantly, occupational exposures may be sufficiently intense to affect it. In addition, because some of their biological effects are pleasurable, they are subject to recreational use and abuse. Their anesthetic properties at high concentrations are well known. The neurotoxicity of some solvents has been recognized for decades (e.g., carbon disulfide), while that of others has just recently been documented (e.g., the ototoxicity of many solvents), or is being actively explored (e.g., the effects of trichloroethylene on attention). Neurobehavioral effects of solvents may be both transient, as from acute exposure, and persistent, as from long-term exposure, and may involve all aspects of function of the nervous system including: irritation; deficits in the olfactory, visual, auditory, and somesthetic senses; alterations in the rate, patterns and control of motion; and changes in attention, learning, and memory. Solvents may affect both the central and peripheral nervous systems. Current neurobehavioral research in solvent toxicology aims to define the mechanisms by which sensory and motor dysfunctions occur; to characterize the reinforcing properties possessed by some solvents; and to develop experimental methods (including animal models) for assessing the cognitive impairments associated with long-term occupational exposure.

References

1.   ACGIH (1990–1991) *Threshold Limit Values for Chemical Substances and Physical Agents,* American Conference of Governmental Industrial Hygienists, Inc.

2.   Alarie, Y. (1973) Sensory irritation by airborne chemicals. *Crit. Rev. Toxicol.* **2**: pp 299–363.

3.   Alarie, Y. (1981) Bioassay for evaluating the potency of airborne sensory irritants and predicting acceptable levels of exposure in man. *Food Cosmet. Toxicol.* **19**: pp 623–626.

4.   Amoore, J.E. and Hautala, E. (1983) Odor as an aid to chemical safety: odor thresholds compared with threshold limit values and volatilities for 214 industrial chemicals in air and water dilution . *Appl. Toxicol.* **3**: pp 272–290.

5.   Arlien-Søborg, P. (1992) *Solvent Neurotoxicity.* CRC Press, Boca Raton, FL.

6.   Beauchamp, Jr., R.O., Bus, J.S., Popp, J.A., Boreiko, C.J., Goldberg, L. (1983) A critical review of the literature on carbon disulfide toxicity. *Crit. Rev. Toxicol.* **11**: pp 169–278.

7.   Bleecker, M.L., Bolla, K.I., Agnew, J., Schwartz, B.S., Ford, D.P. (1991) Dose-related subclinical neurobehavioral effects of chronic exposure to low levels of organic solvents. *Am. J. Indust. Med.* **19**: pp 715–728.

8.   Broadwell, D.K., Darcey, D.J., Hudnell, H.K., Otto, D.A., Boyes, W.K. (1995) Work-site clinical and neurobehavioral assessment of solvent exposed microelectronics workers. *J. Indust. Med.* **27**: pp 677–698.

9.   Bushnell, P.J. (1988) Facilitation of autoshaping by xylene in rats. *Neurotoxicol. Teratol.* **10**: pp 569–577.

10.   Bushnell, P.J. (1994) Styrene impairs serial reversal learning in rats *J. Am. Coll. Toxicol.* **13**: pp 279–300.

11.   Bushnel,l P.J., Kelly, K.L., Crofton, K.M. (1994) Effects of inhaled toluene on detection of auditory signals in rats *Neurotoxicol. Teratol.* **16**: pp 149–160.

12.   Bushnell, P.J. and Peele, D.B. (1988) Conditioned flavor aversions from inhalation of *p*-xylene. *Neurobehav. Toxicol. Teratol.* **10**: pp 273–277.

13.   Cherry, N. (1993) Neurobehavioural effects of solvents: the role of alcohol. *Environ. Res.* **62**: pp 155–158.

14.   Crofton, K.M. (1992) Reflex modification and the assessment of sensory dysfunction In: Tilson, H.A. and Mitchell, C.L. (Eds.), *Neurotoxicology.* Raven Press, New York, pp 181–212.

15. Crofton, K.M., Janssen, R., Prazma, G., Pulvar, S., Barone, S. Jr. (1994) The ototoxicity of 33'-iminodipropionitrile: functional and morphological evidence of cochlear dysfunction. *Hearing Res.* **80**: pp 129–140.

16. Crofton, K.M., Lassiter, T.L, Rebert, C.R. (1994) Solvent-induced ototoxicity in rats: an atypical selective mid-frequency hearing deficit. *Hearing Res.* **80**: pp 25–30.

17. Crofton, K.M. and Zhao, X. (1993) Mid-frequency hearing loss in rats following inhalation exposure to trichloroethylene: evidence from reflex modification audiometry. *Neurotoxicol. Teratol.* **15**: pp 413–423.

18. De Ceaurriz, J.C., Micillino, J.C., Bonnet, .P, Guenier, J.P. (1981) Sensory irritation caused by various industrial airborne chemicals. *Toxicol. Lett.* **9**: pp 137–143.

19. de Oliveira, J.A.A. (1989) *Audiovestibular Toxicity of Drugs*, Vol. 1 and 2. CRC Press, Boca Raton, FL.

20. De Valois, R.L. and De Valois, K.K. (1988) *Spatial Vision*. Oxford University Press, New York.

21. Dick, R.B., Setzer, J.V., Wait, R., Hayden, M.B., Taylor, B.J., Tolos, B., Putz-Anderson, V. (1984) Effects of acute exposure of toluene and methyl ethyl ketone on psychomotor performance. *Intl. Arch. Occup. Environ. Health* **54**: pp 91–109.

22. Eckerman, D.A. and Bushnell, P.J. (1992) The neurotoxicology of cognition: attention learning and memory. In: Tilson, H.A. and Mitchell, C.L. (Eds.), *Neurotoxicology*. Raven Press, New York, pp 213–270.

23. Ehyai, A. and Freemon, F.R. (1985) Progressive optic neuropathy and sensorineural hearing loss due to chronic glue sniffing. *Neurol. Neurosurg. Psychiatry* **46**: pp 349–351.

24. Evans, E.B. and Balster, R.L. (1991) CNS depressant effects of volatile organic solvents. *Neurosci. Biobehav. Rev.* **15**: pp 233–241.

25. Filley, C.M., Heaton, R.K., Rosenberg, N.L. (1990) White matter dementia in chronic toluene abuse. *Neurology* **40**: p 532.

26. Graybiel, A.M., Aosaki, T., Flaherty, A.W., Kimura, M. (1994) The basal ganglia and adaptive motor control. *Science* **265**: pp 1826–1831.

27. Green, D.M. and Swets, J.A. (1974) *Signal Detection Theory and Psychophysics*, 2nd edition. R.A. Krieger, Huntington, NY.

28. Gulick, L.W., Gescheider, G.A., Frisina, R.D. (1989) *Hearing: Physiological Acoustics Neural Coding and Psychoacoustics*. Oxford Press, New York, p 409.

29. Honig, W.K. (1978) Studies of working memory in the pigeon In: Hulse, S.H., Fowler, H., Honig, W.K. (Eds.), *Cognitive Processes in Animal Behavior*. Lawrence Erlbaum Assoc., Hillsdale, NJ, pp 211–248.

30. Jaspers, R.M.A., Muijser, H., Lammers, J.H.C., Kulig, B.M. (1993) Mid-frequency hearing loss and reduction of acoustic startle responding in rats following trichloroethylene exposure. *Neurotoxicol. Teratol.* **15**: pp 407–412.

31. Johnson, A.-C. (1993) The ototoxic effect of toluene and the influence of noise acetyl salicylic acid or genotype. *Scand. Audiol.* **39**(Suppl.): pp 1–40.

32. Johnson, A.-C., Canlon, B. (1994) Progressive hair cell loss induced by toluene exposure. *Hearing Res.* **75**: pp 201–209.

33. Johnson, B.L. (Ed.) *Advances in Neurobehavioral Toxicology: Applications in Environmental and Occupational Health*. Lewis Publishers, Chelsea, MI.

34. Kishi, R., Harabuchi, I., Ikeda, T., Yokota, H., Miyake, H. (1988) Neurobehavioral effects and pharmacokinetics of toluene in rats and their relevance to man. *Br. J. Indust. Med.* **45**: pp 396–408.

35. Krastel, H., Moreland, J.D. (1991) Colour vision deficiencies in ophthalmic diseases. In: Foster, D.H. (Ed.), *Inherited and Acquired Colour Vision Deficiencies: Fundamental Aspects and Clinical Studies*. Vol. 7. *Vision and Visual Dysfunction*. CRC Press, Boca Raton, FL, pp 115–156.

36. Kulig, B.M. (1988) The neurobehavioral effects of chronic styrene exposure in the rat. *Neurotoxicol. Teratol.* **10**: pp 511–517.

37. Kulig, B.M. (1989) *Effects of White Spirits on Neurobehavioral Functioning in the Rat*. Summary Report Rijswijk, The Netherlands, Medical Biological Laboratory, TNO.

38. Kulig, B.M. (1992) Assessment of neurotoxicant-induced effects on motor function. In: Tilson, H.A. and Mitchell, C.L. (Eds.), *Neurotoxicology*. Raven Press, New York, pp 147–179.

39. Lindström, K., Härkönen, H., Hernberg, S. (1976) Disturbances in psychological functions of workers occupationally exposed to styrene. *Scand. J. Work Environ. Health* **3**: 129–139.

40. Marjot, R., McLeod, A.A. (1989) Chronic non-neurological toxicity from volatile substance abuse. *Hum. Toxicol.* **8**: p 301.

41. Maurissen, J.P.J. (1979) Effects of toxicants on the somatosensory system. Test methods for definition of effects of toxic substances on behavior and neuromotor function. *Neurotoxicol. Teratol.* (Suppl 1): pp 23–31.

42. Mergler, D. (1990) Color vision loss: a sensitive indicator of the severity of optic neuropathy In: Johnson, B.L. (ed.) *Advances in Neurobehavioral Toxicology.* Lewis Publishers, Chelsea, MI, pp 175–182.

43. Mergler, D., Beauvais, B. (1992) Olfactory threshold shift following controlled 7-hour exposure to toluene and/or xylene. *NeuroToxicology* **13**: pp 211–216.

44. Merigan, W.H. (1979) Effects of toxicants on visual systems. Test methods for definition of effects of toxic substances on behavior and neuromotor function *Neurotoxicol. Teratol.* **1** (Suppl. 1): pp 15–22

45. Merigan, W.H., Wood, R.W., Zehl, D., Eskin, TA. (1988) Carbon disulfide effects on the visual system. I. Visual thresholds and ophthalmoscopy. *Invest. Ophthalmol. Vis. Sci.* **29**: pp 512–518.

46. Metrick, S.A., Brenner, R.P. (1982) Abnormal brainstem auditory evoked potentials in chronic paint sniffers. *Ann. Neurol.* **12**: pp 553–556.

47. Miyagawa, M., Honma, T., Sato, M., Hasegawa, H. (1984) Conditioned taste aversion induced by toluene administration in rats. *Neurobehav. Toxicol. Teratol.* **6**: pp 33–37.

48. Morata, T.C. (1989) Study of the effects of simultaneous exposure to noise and carbon disulfide on workers' hearing. *Scand. Audiol.,* **18**: pp 53–58.

49. Morata, T.C., Dunn, D.E., Kretschmer, L.W., Lemasters, G.K., Keith, R.W. (1993) Effects of occupational exposure to organic solvents and noise on hearing. *Scand. J. Work Environ. Health* **19**: pp 245–254.

50. Müller, G., Spassowski, M., Henschler, D. (1975) Metabolism of trichloroethylene in man. III. Interaction of trichloroethylene and ethanol. *Arch. Toxicol.* **33**: pp 173–189.

51. Mutti, A., Mazzucchi, A., Rustichelli, P., Frigeri, G., Arfini, G., Franchini, I. (1984) Exposure-effect and exposure-response relationships between occupational exposure to styrene and neuropsychological functions. *Am. J. Ind. Med.* **5**: pp 275–286.

52. Nielsen, G.D., Alarie, Y. (1982) Sensory irritation pulmonary irritation and respiratory stimulation by airborne benzene and alkylbenzenes: prediction of safe industrial exposure levels and correlation with their thermodynamic properties. *Toxicol. Appl. Pharmacol.* **65**: pp 459–477.

53. Østergaard, G., Lam, H.R., Ladefoged, O., Arlien-Søborg, P. (1993) Effects of six months' white spirit inhalation exposure in adult and old rats. *Pharmacol. Toxicol.* **72**: pp 34–39.

54. Parasuraman, R., Warm, J.S., Dember, W.N. (1987) Vigilance: taxonomy and utility. In: Mark, L.S., Warm, J.S., Huston, R.L. (eds.) *Ergonomics and Human Factors.* Springer Verlag, New York. pp 11–39.

55. Pryor, G.T. (1994) Assessment of auditory dysfunction In: Chang, L. (ed.) *Principles of Neurotoxicology.* Marcel Dekker, New York, pp 345–371.

56. Pryor, G.T., Dickinson, J., Howd, R.A., Rebert, C.S. (1993) Transient cognitive deficits and high-frequency hearing loss in weanling rats exposed to toluene. *Neurobehav. Toxicol. Teratol.* **5**: pp 53–57.

57. Pryor, G.T., Howd, R.A. (1986) Toluene-induced ototoxicity by subcutaneous injection. *Neurobehav. Toxicol. Teratol.* **8**: pp 103–104.

58. Rebert, C.S., Pryor, G.T. (1993) *Interactive Effects of Solvents on the Rat's Nervous System.* Stanford Research Institute Project LSU-1606, prepared for the U.S. Environmental Protection Agency, CA, CR816784 (available from SRI International, 836 Ravenswood Ave, Menlo Park, CA 94025-3493).

59. Rebert, C.S., Sorenson, S.S, Howd, R.A., Pryor, G.T. (1983) Toluene-induced hearing loss in rats evidenced by the brainstem auditory-evoked response. *Neurobehav. Toxicol. Teratol.* **5**: pp 59–62.

60. Reiter, L.W., MacPhail, R.C. (1979) Motor activity: a survey of methods with potential use in toxicity testing. Test methods for definition of effects of toxic substances on behavior and neuromotor function. *Neurotoxicol. Teratol.* **1** (Suppl. 1): pp 53–66.

61. Riley, A.L. and Tuck, D. (1985) Conditioned taste aversion: a behavioral index of toxicity. *Ann. N.Y. Acad. Sci.* **443**: pp 272–292.

62. Ron, M.A. (1986) Volatile substance abuse: a review of possible long-term neurological intellectual and psychiatric sequelae. *Br. J. Psychiatry* **148**: pp 235–246.

63. Rosenberg, N.L., Spitz, M.C., Filley, C.M., Davis, K.A., Schaumburg, H.H. (1988) Central nervous system effects of chronic toluene abuse — clinical brainstem evoked response and magnetic resonance imaging studies. *Neurotoxicol. Teratol.* **10**: p 489.

64. Rybak, L.P. (1992) Hearing. The effects of chemicals. *Otolaryngol. Head Neck Surg.* **106**: pp 677–686.

65. Schwartz, B.S., Ford, D.P., Bolla, K.I. et al. (1990) Solvent-associated decrements in olfactory function in paint manufacturing workers. *Am. J. Ind. Med.* **18**: pp 697–706.

66. Segal, B., *Drugs and Behavior: Cause Effects and Treatment.* Gardner Press, New York, p 120.

67. Sharp, C.W. (1988) Clinical and medical manifestations of volatile solvents. In: Arif, E.E., Grant, M., Navaratnam, V. (Eds.) *Abuse of Volatile Solvents and Inhalants.* Papers presented at a WHO Advisory Meeting. International Monograph Series Centre for Drug Research, Universiti Sains Malaysia Minden, Pulau Pinang, Malaysia **1**: p 227.

68. Shaw, N.A. (1988) The auditory evoked potential in the rat — a review. *Prog. Neurobiol.* **31**: pp 19-45.

69. Spencer, P.S., Schaumburg, H.H., Sabri, M.I., Veronesi, B. (1980) The enlarging view of hexacarbon neurotoxicity. *Crit. Rev. Toxicol.* **7**: pp 279–356.

70. Stewart, R.D., Hake, C.L., Peterson, J.E. (1974) "Degreaser's flush": dermal response to trichloroethylene and ethanol. *Arch. Environ. Health* **29**: pp 1–4.

71. von Euler, G. (1994) Toluene and dopaminergic transmission. In: Isaacson, R.L. and Jensen, K.F. (Eds.), *The Vulnerable Brain and Environmental Risks.* Vol. 3. *Toxins in Air and Water*, Plenum Press, New York.

72. Zur, J. and Yule, W. (1990) Chronic solvent abuse. 2. Relationship with depression. *Child Care Health Dev.* **16**: p 21.

Glossary

abrupto placenta: premature detachment of the placenta, often attended by maternal systemic reactions in the form of shock, oliguria, and coagulation abnormalities.

abuse liability, see abuse potential

acaricide: substance used to kill mites and ticks.

acetylcholine (ACh): one of the major neurotransmitters of the nervous system, it is involved in the arousal of the organism.

acetylcholinesterase (AChE): enzyme that catalyzes the cleavage of acetylcholine to choline and acetate; it is found in the CNS, particularly in gray matter of nerve tissue, in red blood cells and in motor endplates of skeletal muscle.

achievement test: a measure of knowledge or proficiency. The term is usually applied to an examination on outcomes of school instruction.

achromatic threshold: the least intensity of the spectrum that produces a sensation of color; reduction of the intensity below this point produces a sensation of brightness only, without color distinction.

acquisition: usually defined in operational terms as a change in response measure and typically is used to refer to that portion of any learning process during which there is a consistent increase in responsiveness and the change has a certain permanence.

acral: affecting the extremities.

active avoidance: avoidance learning (conditioning) paradigm in which the subject must overtly make a specific response; e.g., pole jumping (see pole-climb test).

activity wheel: apparatus that can be rotated by an animal while in its cage; the number of revolutions, the time spent in the wheel, or some other measure is recorded as an index of the animal's general activity level.

Addison's disease: syndrome due to inadequate secretion of corticosteroid hormones by the adrenal glands, sometimes as a result of a tuberculous infection. Symptoms include weakness, loss of energy, low blood pressure, and dark pigmentation of the skin. Formerly fatal, the disease is now curable by replacement hormone therapy.

adiadochokinesia: a dyskinesia consisting of inability to perform the rapid alternating movements of diadochokinesia (see diadochokinesia).

agranulocytosis: sudden decrease in white blood cells.

Ah-locus: gene(s) controlling the trait of aromatic hydrocarbon (Ah) responsiveness.

Ah-receptor: a protein coded by a putative regulatory gene of the AH-locus. Binding of aromatic hydrocarbons to this cytosolic enzyme is necessary for the induction of many xenobiotic-metabolizing enzymes.

Alcohol Accepting rats: strain of rats, specially bred for their quality to consume alcohol when administered to their drinking water.

alcohol dehydrogenase: an enzyme that catalyzes the reversible oxidation of primary or secondary alcohols to aldehydes.

Alcohol non-accepting rats: strain of rats, specially bred for their quality not to consume drinking water when alcohol has been administered to it.

aldehyde dehydrogenase: an enzyme that catalyzes the oxidation of various aldehydes, using NAD^+ as an electron acceptor.

aldicarb: systemic carbamate type insecticide and nematocide for soil use, widely used on sugar beet and potatoes. Neurotoxicant by virtue of its ability to inhibit acetylcholinesterase.

alkaloids: organic substances existing in combination with organic acids in great variety in many plants, and to which many drugs owe their medicinal properties.

amino acids: organic compounds containing an amino group ($-NH_2$) and a carboxyl group (-COOH). Amino acids are the fundamental constituents of all proteins.

amnesic syndrome: an organic mental disorder characterized by the impairment of memory, both retrograde and anterograde amnesia, occurring in a normal state of consciousness. This syndrome may be caused by thiamine deficiency, but also result from any pathological process causing bilateral damage to certain structures in the medial temporal lobe and diencephalon (e.g.,the hippocampal formation). Causes include head trauma, brain tumors, infarction, cerebral hypoxia, and carbon monoxide poisoning.

amniocentesis: withdrawal of a sample of the amniotic fluid surrounding the embryo in the uterus by piercing the amniotic sac through the abdominal wall.

amphetamines: a class of drugs, including Benzedrine, Dexedrine and Methedrine that act as central-nervous-system stimulants. Amphetamines suppress appetite, increase heart rate and blood pressure, and, in larger doses, produce a feeling of euphoria and power. Therapeutically, they are used to alleviate depression, control appetite, relieve narcolepsy and paradoxically, to control childhood hyperkinesis. Amphetamine abuse is common and chronic use leads to a model paranoid psychosis.

amygdala (nucleus amygdalae): one of the basal ganglia: a rough almond shaped mass of gray matter deep inside each cerebral hemisphere. It has extensive connections with the olfactory system and sends nerve fibers to the hypothalamus, its functions are apparently concerned with mood, feeling, instinct, and possibly memory for recent events.

amyotrophic lateral sclerosis (ALS): also called Lou Gehrig's Disease. Amyotrophy refers to the neurogenic atrophy of muscle and lateral sclerosis to the hardness felt when the spinal cord is examined at autopsy. ALS is a disease marked by the progressive degeneration of the neurons that give rise to the corticospinal tract and of the motor cells of the brain stem and spinal cord, and resulting in a deficit of upper and lower motor neurons. Usually, the disease ends fatally within two to three years.

analgesic: a drug that relieves pain.

analysis of covariance: extension of the analysis of variance, used when there is improper control of one or more of the variables. The procedure allows for statistical control of the uncontrolled variations so that normal analysis techniques may be carried out without distorting the results.

angiopathy: any disease of the blood vessels.

anhedonia: condition marked by a general lack of interest in living, in the pleasures of life. A loss of the ability to enjoy things.

anorectic: drug that induces loss of appetite.

anorexia nervosa, psychological illness, most common in female adolescents, in which the patients starve themselves or use other techniques, such as vomiting or taking laxatives, to induce weight loss.

anorexia: loss of appetite

anoxia: condition in which the tissues of the body receive inadequate amounts of oxygen. This may result from low atmospheric pressure at high altitudes, a shortage of circulating blood, red blood cells or hemoglobin, or disordered blood flow, such as occurs in heart failure.

antacids: drugs that neutralize the hydrochloric acid secreted in the digestive juices of the stomach.

antenatal: the period just before birth.

antipyretic: drug that relieves fever by lowering the body temperature.

anxiety disorders: a group of disorders characterized by unrealistic fear, panic, or avoidance behavior. These disorders include (among others) panic attacks, phobias, obsessive-compulsive disorder, and generalized anxiety disorder.

area postrema: a small tongue-shaped area on the lateral wall of the fourth ventricle.

asphyxia: pathological changes caused by lack of oxygen in respired air, resulting in hypoxia and hypercapnia.

asphyxiation: the causing of, or the state of asphyxia.

assessment batteries of cognition: test battery to measure cognitive functions (see cognition).

assessment of visual fields: testing the perception of the visual fields, that is all the points in the physical environment that can be perceived by a stable eye at a given moment.

assessment: any procedure that measures or describes. Most often refers to integration of several pieces of information, in order to reach a judgment about an individual's health or fitness for an assignment. Sometimes a study of a group or a situation.

astrocyte: type of cell with numerous sheet-like processes extending from its cell-body, found throughout the central nervous system. It is one of the several different types of cells that make up the neuroglia. The cells have been ascribed the function of providing nutrients for neurons and possibly of taking part in information storage processing.

astroglia, see astrocyte.

ataxia: failure of muscular coordination, irregularity of muscular action.

athetoid: resembling or affected with athetosis.

athetosis: a form of dyskinesia marked by the ceaseless occurrence of slow, sinuous, writhing movements, especially severe in the hands, and performed involuntary.

atropine: an alkaloid derived from belladonna (from the deadly night-shade plant). It is a respiratory and circulatory stimulant and counteracts parasympathetic stimulation.

audiogenic seizures: seizures caused by a sound.

audiometric threshold testing: assessment of the least intensity of a sound of a certain frequency that still can be perceived by the subject.

auditory evoked potentials: evoked potential occurring in response to an auditory stimulus, e.g.,a peep.

auditory startle habituation: adaptation (gradual elimination) to a sudden, unanticipated stimulus.

auditory stimulus: stimulus relating to the sense of hearing.

auditory threshold: the slightest perceptible sound.

auto shaping: process in which organisms come to direct operant behavior (e.g.,key picking; lever pressing) toward stimuli which are regularly associated with the presentation of reinforcement.

aversion: a repugnance or dislike for something, an internal negative reaction.

axon: thread-like nerve cell extension functioning to conduct information. At the specialized endings, the synapses, information is transmitted to other cells.

axonopathy: axonal degeneration with the axon itself being the primary focus of injury.

Babinski's sign: loss or lessening of the Achilles tendon reflex in sciatica.

basal ganglia: several large masses of gray matter embedded deep within the white matter of the cerebrum. They include the caudate and lenticular nuclei, together known as the corpus striatum, and the amygdaloid nucleus. The lenticular nucleus consists of the putamen and globus pallidus. The basal ganglia have complex neural connections with both the cerebral cortex and thalamus: they are involved with the regulation of voluntary movements and at a subconscious level.

behavioral rating scale: any device used to assist a rater in making ratings of behavior. Rating behavior is the use of rating techniques in a fairly restricted manner such that only overt, objectively observable behaviors enter into the assessment.

benzodiazepines: a group of pharmacologically active compounds used as minor tranquilizers and hypnotics.

Betz's cells: large pyramidal ganglion cells found in the internal pyramidal layer of the cerebral cortex; also called giant pyramids and giant pyramidal cells.

bias: an inclination toward a position or conclusion: a prejudice. A distortion of the facts caused by errors in selecting or classifying the subjects of a study.

Biel maze: water maze with a multiple T pattern with six choice points present in the correct pathway. This maze forms part of the Cincinnati Psychoteratogenicity Screening Battery, and both acquisition and reversal learning of the task are evaluated.

biochemical marker: a biochemical measure that designates the presence or absence of distinctive features.

blood-brain barrier: a selectively permeable physiological 'barrier' between the circulating blood and the brain which functions by preventing some substances from reaching the brain.

bolus dose: a concentrated mass of a pharmaceutical preparation given (intravenously) within a minimal amount of time; often used for diagnostic purposes, e.g.,in the form of a radioactive isotope.

Bonferroni correction: in statistical analysis, multiple tests or comparisons greatly increase the Type I error rate compared to the single-test situation. The most direct way to obtain multiple-test error rates is to use a Bonferroni correction, in which the single-test alpha is multiplied by the number of independent (non redundant) comparisons (k) to yield a multiple-test alpha that protects against even one fortuitous association.

botulinum: protein neurotoxin produced by strains of *Clostridium botulinum*, which is responsible for the symptoms of botulism.

bradyphrenia: slowness of thought or fatigability of initiative, resulting from depression or CNS disease.

brainstem auditory evoked potentials (BAER): auditory evoked potentials originating from the brainstem.

brevetoxins: polycyclic toxins produced by the dinoflagellate *Ptychodiscus brevis*. They are highly lipid soluble and are potent neurotoxins, causing depolarization of nerve membranes.

bromocriptine: a drug derived from ergot, that has similar to those of dopamine. It is used in the treatment of parkinsonism and to prevent lactation by inhibiting the secretion of the hormone prolactin by the pituitary gland.

Buckthorn toxin: toxins found in the fruit of the spineless shrub *Karwinska humboldtonia*, a desert species found in northern Mexico and south-west Texas. These toxins produce segmental demyelination of peripheral nerves in animals and man.

bulimia nervosa: mental disorder occurring predominantly in females, with onset usually in adolescence or early adulthood, characterized by episodes of binge eating that continue until terminated by abdominal pain, sleep, or self-induced vomiting, by awareness that the binges are abnormal

cacosmia: altered olfactory function consisting of a bad smell, not related to exposure of any specific odor.

caffeine: 1,3,7-trimethylxanthine, a purine derivative present e.g.,in tea leaves and coffee beans.

cage-side observation: behavioral observations, whether or not using sensing devices, of one or more animals in its (home) cage. Often this type of observational tests is based on time sampling, at repeated intervals the location and/or the posture of the animal is recorded.

calcarine fissure (sulcus calcarinus): a sulcus on the medial surface of the occipital lobe, separating the cuneus from the lingual gyrus.

camphor: $C_{10}H_{16}O$, organic substance from the wood of the camphor tree, *Cinnamonum camphora*.

canulla: tube implanted into an organism, usually in a vein, through which drugs may be injected.

capsaicin: compound found in hot peppers that has sensory neurotoxic effects. It acts selectively on primary afferent neurons, depleting these processes of the peptide Substance P. Neonatal rodents are particularly sensitive to capsaicin, which causes a selective profound and permanent degradation of C and A fibers.

carbamates: salts and esters of carbamic acid (H2NCOOH).

carbamazepine (Tegretol™): analgesic and anticonvulsant substance.

cassava: tropical shrub or plant belonging to genus *Manihot*, producing large thick roots from which an edible starch is obtained. The edible starch obtained from the cassava is dried and used to make tapioca and bread.

cataract: opacity, partial or complete, of one or both eyes, on or in the lens or capsule, especially an opacity impairing vision or causing blindness.

catecholamines: a group of physiologically important substances, including adrenaline, noradrenaline and dopamine, having various different roles (mainly as neurotransmitters) in the functioning of the sympathetic and central nervous system. Chemically, all contain a benzene ring with adjacent hydroxyl groups (catechol) and an amine group on a side chain.

challenge: administration of a chemical substance to a patient for observation of whether the normal physiological response occurs.

Chinese Restaurant Syndrome (CRS): transient syndrome associated with arterial dilatation, due to ingestion of monosodium glutamate, which is used liberally in seasoning Chinese food. It is characterized by throbbing of the head, lightheadedness, tightness of the jaw, neck and shoulder and backache.

chi-square: a statistical test that allows tests of differences between independent samples using frequency data or between a sample and some set of expected scores.

chlordane: one of the cyclodiene insecticides, like aldrin and dieldrin, cyclodienes are chlorinated insecticides based on the cyclodiene ring structure: formerly widely used for the control of agricultural pests and for structural pest control.

chlordecone: insecticide developed for the control of ants and roaches. It possesses fungicide activity too. It is a neuro- and hepatotoxicant. As a neurotoxicant, it blocks Na^+/K^+-ATPase. In humans, it causes tremors, chest pain, weight loss, mental changes, arthralgia, skin rash, muscle weakness, incoordination and slurred speech.

chlordiazepoxide: a minor tranquilizer used in the treatment of anxiety.

choeric: characterized by chorea.

choice response: refers to tests that offer fixed response options (e.g.,multiple-choice, true-false, like-dislike…).

choline acetyl transferase (ChAT): mitochondrial enzyme that synthesizes the neurotransmitter acetylcholine from its precursors acetyl CoA and choline.

chorea: the ceaseless occurrence of a wide variety of rapid, highly complex, jerky, dyskinetic movements that appear to be well coordinated but are performed involuntary.

chorea: the jerky involuntary movement particularly affecting the head, face, or limbs.

choreiform: resembling chorea.

choreoathethosis: a condition marked by choreic and athetoid movements

chorioid plexus: a rich network of blood vessels, derived from those of the pia mater, in each of the brain's ventricles. It is responsible for the production of cerebrospinal fluid.

Cincinnatti maze: synonym for Biel maze.

circumventricular organs: several small structures located around the edges of the third and fourth ventricles, lacking the regular blood brain barrier and thus serving as significant sites for neural-endocrine interaction. They include the area postrema, the median eminence, the subcommissural organ, the subfornical organ and the organum vasculosum of the lamina terminalis.

classical conditioning: basic form of learning in which stimuli initially incapable of evoking certain responses acquire the ability to do so through repeated pairing with other stimuli that are able to elicit such responses. The learning in classical conditioning consists of acquiring responsiveness (conditioned response) to a stimulus that originally was ineffective (conditioned stimulus) by pairing it with a stimulus (unconditioned stimulus) that elicits an overt response (unconditioned response).

classification: broadly any decision in which alternative descriptive labels or courses of action are available, and one of them is chosen for the individual.

clearance (renal clearance): a quantitative measure of the rate at which waste products are removed from the blood by the kidney. It is expressed in terms of the volume of blood that could be completely cleared of a particular substance in one minute.

cliff avoidance: test that measures whether or not an animal will choose to step onto a nearby platform or floor ('shallow' floor), as compared to one perceived to be farther away ('deep' floor), used to assess depth perception.

clinical investigations: investigations founded on actual observation of patients.

clinical trial: a large scale plan for testing and evaluating the effectiveness of some drug or therapeutic procedure in human subjects.

clonic movements, see clonus.

clonic: pertaining to or of the nature of clonus.

clonidine (Catapres™; Catapresan™): an α_2-adrenergic agonist-antagonist, acting as an antagonist when the norepinephrine concentration is high; oral antihypertensive, also used as prophylaxis for migraine, and in the treatment of dysmennorrhea.

clonus: rhythmical contraction of a muscle in response to a suddenly applied and then sustained stretch stimulus. It is caused by an exaggeration of the stretch reflexes and is usually a sign of disease in the brain or spinal cord. The term is most commonly used to describe the rhythmical limb movements in convulsive epilepsy.

cognition: the mental process by which knowledge is acquired. These include perception, reasoning, acts of creativity, problem-solving, and possibly intuition.

cognitive (spatial) maps: a term describing the theoretical interpretation of the behavior of an animal learning a (spatial) maze. The original author, E.C. Tolman, argued that the animal was developing a set of spatial relationships – a cognitive 'map' – rather than merely learning a chain of overt responses.

cognitive tests: tests to measure cognitive functions (see cognition).

Cohen's kappa test: statistical method to measure the intraclass correlation between two methods; a kappa > 0.75 means that the similarity between both tests is excellent, 0.4 < kappa < 0.75 means that the similarity is moderate and a kappa < 0.40 means a bad similarity.

cohort study: epidemiological method used to search for the cause(s) of disease. Groups of people whose members differ on one or more characteristic(s) suspected of causing disease are followed over time to see if the people with the characteristic(s) get more disease.

collateral data: accompanying data.

compliance: yielding to others; overt behavior of one person that conforms to the wishes or behaviors of others. That is, they use it so that there is no notion that the compliant person necessarily believes in what he or she is doing.

computer axial tomography (CAT): brain imaging technique in which a series of X-rays of the brain is computerized and built up into a three-dimensional picture.

computer-based assessment: using computers as a medium for the presentation of psychological tests.

computerized tomography: diagnostic radiologic technique for the examination of the soft tissues of the body. It involves the recording of slices of the body with an X-ray scanner (CT-scanner); these records are then integrated by computer to give a cross-sectional image. This investigation is without risk to the patient (cf. PET). Within the skull it can be used to diagnose pathological conditions of the brain, such as tumors, abscesses and hematomas.

concordance rate: a measure of the extent to which one member of a twin pair will express a trait if the other member of the twin pair expresses that trait.

conditioned avoidance responding: the flight response evoked by a conditioned stimulus.

conditioned flavor aversion: a negative reaction towards a flavor, resulting from pairing the flavor with some painful or unpleasant stimulus.

conditioned stimulus: a stimulus which acquires the capacity to evoke particular responses through repeated pairing with another stimulus capable of eliciting such reactions.

conditioned suppression: a reduction or suppression in responding in the presence of a previously neutral stimulus. The suppression is produced by pairing the neutral stimulus with a noxious stimulus. For example, if a 1-minute light signal is repeatedly followed by a shock, the organism will come to suppress responding during the full minute that the light is on.

confabulation, see fabulation.

confidence interval: the margin of the error calculated for a risk estimate or an effect. 95% confidence intervals are the most common, meaning that there is a 95% probability that the risk, or the effect, is no higher or lower than the range of values included in this interval.

confounder: a cause of disease that is not under investigation, but that distorts the cause-effect relationship under study.

construct validity: the validity is the extent to which a model measures what it purports to measure. In the case of real construct validity the processes affected by a substance in the animal model are the same as those affected in the species in the real situation. In other words, there is a high closeness of fit between the animal and the theoretical model.

contrast sensitivity threshold: the lowest intensity of a stimulus in the presence of other stimuli that can be detected by a sense organ as an independent stimulus.

control group: a group in an experiment which is not exposed to the independent variable under investigation. The behavior of subjects in this condition is used as a baseline against which to evaluate the effects of experimental treatments.

coprine (1-cyclopropanol-1-N-glutamine): mushroom (*Coprinus atramentarius*) toxin from which the metabolite cyclopropanone hydrate elicits a disulfiram like response to alcohol consumption for up to three days after eating the mushroom.

corpus striatum: the striate body, one of the components of the basal nuclei.

correlation coefficient: a statistic which indicates, on a scale from −1.00 to +1.00, the degree of relationship between two or more variables. The larger the correlation, the stronger the observed relationship.

correlational method: a research method in which variables of interest are observed in a careful and systemic manner in order to determine whether changes in one are associated with changes in the other.

cotinine: the major urinary metabolite of nicotine

covariance analysis: an extension of the ANOVA used when there is improper control of one or more of the variables.

crack: the most common street name for freebase cocaine, probably called this because a crackling sound is made both when cocaine hydrochloride powder ('snow') is mixed with water and sodium bicarbonate to make the freebase cocaine, and when crack is smoked.

crest: a projection or a projecting structure or ridge.

criterion: standard against which the success of prediction is judged. For a diagnosis, the criterion is usually an independent diagnosis considered to be rather trustworthy but impractical for routine use.

critical flicker fusion test: test in which the point at which a flickering stimulus is no longer perceived as periodic but shifts to continuous, in this case the flicker fusion threshold is measured.

cross-fostering: technique in which pups from dams that are exposed to a certain substance during pregnancy receive a drug naive 'foster' mother, to be able to exclude possible environmental factors (e.g.,disturbed mother care behavior, exposure through breast feeding).

crossover design: variation on the double blind (see double blind) design, in which as an added control, the conditions are crossed in the middle of the experiment. For example a group of subjects receives one week placebo and the second week a drug, the control group receives the drug in the first week and the placebo during the second week.

cross-sectional study: epidemiological method used to determine whether a problem exists that warrants further study. A population of interest is identified and the individuals asked about current illnesses and current exposures.

cross-tolerance: a condition in which tolerance to a certain drug results in tolerance to another drug, mostly the latter are from the same class.

cutaneous: relating to the skin.

cyanamide (NH_2CN): a colorless crystalline unstable substance.

cyanogenic glycosides: monosaccharide or disaccharide conjugates of cyanohydrins. There is evidence that cyanohydrins are derived from amino acids.

cyanohydrins: organic compounds containing a C(OH)CN group.

cysteine: sulfur-containing amino-acid that is an important constituent of many enzymes. The disulfide links (S-S) between adjacent cysteine molecules in polypeptide chins contribute to the three-dimensional molecular structure of proteins.

cytisine: alkaloid neurotoxin present in *Cytisus laburnum*, causing nausea, emesis, sweating, dizziness and clonic convulsions.

cytokine: a generic term for non antibody proteins released by one cell population (e.g.,primed T-lymphocytes) on contact with specific antigen, which act as intercellular mediator, as in the generation of an immune response. Examples are lymphokines and monokines.

DDE (Dichlorodipheyldichloroethylene); one of the major metabolites of dichlorodiphenyltrichloroethane (DDT).

DDT (dichlorodiphenyltrichloroethane): powerful and persistent insecticide, used to control mosquitoes in countries were malaria is a problem. It is stored in the fat of higher animals.

Debrisoquine genotype: the 4-hydroxylation of the drug debrisoquine shows a bimodial distribution in the population. Most people are extensive metabolizers of this drug, but there is a small group of poor metabolizers. This bimodality in metabolism is due to a genetic polymorphism in cytochrome P-450 dependent drug oxidation.

delayed matching: test in which a subject (animal/human) is asked to match a sample stimulus with comparison stimuli that are provided after a delay (see retention gradient).

delayed recall: defer in the process of retrieving information from memory.

dementia pugilistica: boxer's dementia.

dementia: progressive decline in mental function, in memory, and in acquired intellectual skills.

demyelination: disease process selectively damaging the myelin sheets surrounding the nerve fibers in the central and peripheral nervous system. This in turn affects the functioning of the nerve fibers, which the myelin normally supports. Demyelination may be the primary disorder, as in multiple sclerosis, or it may occur after injury of the nervous system.

dendrite: thread-like nerve cell extension functioning to receive information.

depressive disorders: a group of disorders including various forms of depression and manic-depression.

detection threshold: the statistically determined point along a stimulus continuum at which the energy level is just sufficient for one to detect the presence of the stimulus.

developmental milestones: significant behaviors which are used to mark the progress of development. Examples are: saying phrases, turning pages, carrying out requests, pointing to body parts, holding a pencil, imitating a drawn circle, catching a ball.

developmental studies: tests for evaluating the developmental stage of infants and preschoolers.

diadochokinesia: the function of arresting one motor impulse and substituting for it one that is diametrically opposite, to permit sequential alternating movements, as pronation and supination of the arm.

diagnosis: narrowly, choosing one of a set of labels that best fits an individual's disorder or disability. Broadly, developing an understanding of the individual's difficulties, and insofar possible, their origins.

Diagnostic and Statistical Manual of Mental Disorders (DSM): the official system for classification of psychological and psychiatric disorders prepared and published by the American Psychiatric Association. The current version is DSM-IV.

2,4-diaminobutyric acid: one of the many compounds found in different *Lathyrus spp.* that may be responsible for the neurotoxic actions of these species and causing Lathyrism.

dieldrin: common name for an insecticidal product containing not less than 85% of $C_{12}H_8Cl_6O$. It is the epoxy derivative of aldrin, from which it is prepared by epoxidation. It has a high contact and stomach toxicity to most insects, and is very persistent. Stored in the bodies of birds and mammals and now little used.

differential rate reinforcement (DRR): in operant conditioning a term for a class of schedules of reinforcement in which the delivery of reinforcement depends on the immediately preceding rate of responding. *DRL (differential reinforcement of low rates)*: a class of schedules based on a specified rate of responses which must not be exceeded for reinforcement to occur. Thus, in DRL 5 (seconds), 5 seconds must pass between responses or no reinforcement is delivered. *DRH (differential reinforcement of high rates)*: in contrast to DRL, here the rate must exceed some set value for reinforcement to occur. DRH 5 (seconds) means that the interresponse time must be less than 5 seconds.

1-(2,5-dimethoxy-4-iodophenyl)-2-aminopropane (DOI): a non-indole 5-hydroxytryptamine-sub-2 (5-HT-sub-2) agonist.

dipsomania: 19th century term for morbid and insatiable craving for alcohol, occurring in paroxysms. Only a small proportion of alcoholics show this symptom.

discrimination learning, see discrimination training procedure.

discrimination training procedures: class of experimental procedures in which a subject learns to judge between two (or more) stimuli. In *operant-conditioning* experiments, responses in the presence of one stimulus (S1) are reinforced but responses in the presence of another (Sx) are not. Such training leads to the emitting of responses in the presence of Sx, but not in the presence of S1. In *classical conditioning*, in the presence of one stimulus the conditioned stimulus (CS) and the unconditioned stimulus (UCS) are paired, but in the presence of another stimulus they are not. This leads to elicitation of the conditioned response (CR) in the presence of the conditioned stimulus, but not in the presence of another stimulus.

discrimination: the ability to perceive differences between two or more stimuli.

dissociative anesthetic: a class of drug that was developed as anesthetics that do not cause significant respiratory depression. Although these drugs can cause hallucinations, the profile of their effects is different from drugs such as LSD. Two of these drugs are phencyclidine (PCP) and ketamine. Ketamine is used in clinical medicine and both, ketamine and phencyclidine are used as veterinary anesthetics.

disulfiram: an antioxidant which inhibits the oxidation of the acetaldehyde metabolized from alcohol, resulting in high concentrations of acetaldehyde in the body. Extremely uncomfortable symptoms occur when alcohol is ingested; therefore this drug is used as an aversion therapy for alcoholism.

divergent thinking: intellectual fluency; finding a variety of possible solutions to a problem.

DOB (2,5-dimethoxy-4-bromoamphetamine): a phenylalkylamine with hallucinogenic and psychotogenic activity

domoic acid: algal toxin responsible for the amnestic shellfish poisoning syndrome.

dopaminergic: characterizing or pertaining to pathways, fibers or neurons in which dopamine is a neurotransmitter.

dorsal root ganglion (ganglion spinale, sensory ganglion): spinal ganglion: the ganglion found on the posterior root of each spinal nerve, composed of the unipolar nerve cell bodies of the sensory neurons of the nerve.

double-blind procedure: experimental procedure in which neither the subject nor the person administering the experimental procedure knows what are considered to be the crucial aspects of the experiment. In the case of drug studies neither the subject nor the administering person knows whether the drug concerned or a placebo is administered.

DRL (differential reinforcement of low rates), see differential rate reinforcement.

DSP-4 (N-chloroethyl-N-ethyl-2-bromobenzylamine hydrochloride): a potent noradrenaline neurotoxin that only affects noradrenergic neurons. Neurobiologists employ DSP-4 to destroy specific groups of noradrenergic neurons, using stereotactic injections in specific anatomic sites.

dying back neuropathy: degeneration of an axon, beginning distally and progressing to more proximal areas.

dynamometer: instrument for measuring strength of muscular response, usually hand-grip.

dynein: large protein playing several key roles in movement associated with microtubules. Attached to the microtubules of cilia and flagella, its ATPase activity drives a cyclic interaction that moves along the tubulin subunits and produces the bending movement of cilia and flagella by alternately forming and releasing cross-bridges between adjacent tubulin subunits.

dysarthria: imperfect articulation of speech due to disturbances of muscular control which results from damage to the central or peripheral nervous system.

dyschromatopsia: disorder of color vision.

dysethesia: distortion of any sense, especially that of touch; unpleasant abnormal sensation produced by normal stimuli.

dystonia: dyskinetic movements due to disordered tonicity of muscle

Eaton-Lambert syndrome, see Lambert-Eaton syndrome.

echoencephalography: radiographic method demonstrating the intracranial fluid containing spaces using the echo obtained from beams of ultrasonic waves directed through the cranium

ectoderm: the outer layer of the three germ layers (ectoderm, mesoderm and entoderm) of the early embryo. It gives rise to the nervous system and the sense organs, the teeth and lining of the mouth and to the epidermis and its associated structures.

ectodermal origin: originating from the ectoderm.

electoconvulsive shock (ECS): brief electrical shock applied to the head that produces full-body seizure, convulsions and usually loss of consciousness. It is used in animals to study the neurobiology of memory and in patients as a therapeutic procedure for psychiatric disorders (electroconvulsive treatment, ECT)

electroencephalogram (EEG): record of the changes in electrical potential of the brain. The pattern of the EEG reflects the state of the patient's brain and his level of consciousness in a characteristic manner.

electroshock, see electoconvulsive shock (ECS)

embryo: organism in its early stages of development. In humans the first 2 months after conception.

embryogenesis: the development of the embryo.

encephalopathy: any degenerative disease of the brain.

β-endorphin: one of the endogenous endorphins; it is self-administered in animals and possesses opioid activity.

endorphins: any of three neuropeptides, amino acid residues of β-lipotropin. They bind to opioid receptors in the brain and have potent analgesic activity.

endosulfan: chlorinated cyclodiene insecticide.

enkephalin: peptide naturally occurring in the brain and having effects resembling those of opiates like morphine

entorhinal cortex: part of the cortex near the entorhinal fissure that links the neocortex with the limbic system. The entorhinal cortex receives its input from areas of the association cortex and sends its information to the hippocampus by way of the perforant path.

eosinophilia: an increase in the number of eosinophils in the blood. Eosinophilia occurs in response to certain drugs and in a variety of diseases, including allergies, parasitic infestations and certain forms of leukemia.

eosinophilia-myalgia syndrome: sometimes fatal combined syndrome of eosinophilia and severe generalized myalgia in patients ingesting L-tryptophan, occurring in the absence of infection, neoplasm or other known causes of eosinophilia: other characteristics may include subjective weakness, fever, arthralgia, shortness of breath, rash, peripheral edema, and pneumonia.

ependyma: the extremely thin membrane, composed of cells of the neuroglia (ependymal cells), that lines the ventricles of the brain and of the chorioid plexus. It helps to form the cerebrospinal fluid.

ependymal cell: neuroglial cells that line the ventricles (cavities) of the brain and the central canal of the spinal cord (see ependyma).

epidemiology: the study of how and why diseases and other conditions are distributed within the population.

epiphysis, see pineal gland.

equine: pertaining to, characteristic of, or derived from the horse.

ergot alkaloids (ergolines): alkaloids obtained primarily from Claviceps spp; they are also produced by other fungi (e.g.,Aspergillus, Penicillum, Rhizopus) and by certain plants. Natural ergot alkaloids stimulate smooth muscle, especially that of peripheral blood vessels (causing vasoconstriction) and of the pregnant uterus at term.

ergot alkaloids: alkaloids obtained primarily from Claviceps species, they are also produced by other fungi and by certain plants. Natural ergot alkaloids stimulate smooth muscle, especially that of peripheral vessels causing vasoconstriction and of the pregnant uterus at term. Certain hydrogenated derivatives have the reverse effect and cause vasodilatation. There are two classes of ergot alkaloids: the clavine alkaloids and the lysergic acid derivatives.

essential oils: natural oils obtained from plants, mostly benzene derivatives or terpenes. They are used for their flavor or odor.

etonitazene, a narcotic analgesic

event related potential (ERP): change in potential recorded from many neurons in response to a sensory stimulus. Typically lasts for several hundred milliseconds and consists of a number of positive and negative waves. The potentials are generally recorded on the scalp, and it is very difficult to localize the internal sources of these waves. The term is used when the author is referring to a specific evoked potential that occurs in response to a specific known stimulus, the event.

evoked potentials: regular pattern of electrical activity recorded from neural tissue evoked by a controlled stimulus. The stimulus may be auditory (brainstem auditory evoked potential, BAEP), visual (visual evoked potential, VEP) or somatosensory (somatosensory evoked potential, SSEP).The term specifically applies to potentials from the brain (see event related potential).

excitotoxic amino acid, see excitotoxin

excitotoxicity: toxicity induced by excitotoxin (see excitotoxin).

excitotoxin: neurotoxic substances analogous to glutamic acid, that mimic glutamic acids excitatory effect on neurons of the CNS as well as producing lesions on the perikarya.

extinction: the process through which conditioned responses are weakened and eventually eliminated.

fabulation (confabulation): unconscious filling in of gaps in memory with fabricated facts and experiences, commonly seen in organic amnestic syndromes. It differs from lying in that the patient has no intention to deceive and believes the fabricated memories to be real.

face validity: a high face validity of an animal model means that the effect of a chemical measured in the model has a high similarity with the effect the chemical causes in the human. This however does not mean that the underlying mechanisms in the animal model and the human situation are the same.

Feingold diet: diet that purports to treat many illnesses by the elimination of artificial food colorings, preservatives and salicylates from the diet. It has been recommended for the treatment of hyperkinetic syndrome, but is of unproved value.

Feingold Hypothesis: hypothesis that states that many behavioral disorders in children are caused by the presence of (artificial) food additives.

Fetal Alcohol Syndrome (FAS): cluster of abnormal developmental features of a fetus resulting from severe alcoholism in the mother. The features may include: microcephaly, growth deficiencies, mental retardation, hyperactivity, heart murmurs and skeletal malformations.

fetogenesis: development of the organism in the period between the embryonal stage and birth. In humans, the period between the second month of pregnancy until birth.

finger tapping: an automated performance test in which the subject must press a button as many times as possible within a 30 second interval, first with the index finger of each hand separately and then alternating between both hands.

fixed interval performance: response following the passage of a fixed period of time in a fixed interval schedule of reinforcement.

fixed interval schedule of reinforcement: a schedule in which the first response following the passage of a fixed interval of time yields reinforcement.

flumazenil: a benzodiazepine antagonist.

flunitrazepam: benzodiazepin hypnotic and induction agent in anesthesia.

forelimb grip strength: technique to examine the effects of chemical exposures on neuromuscular function of the forelimbs in rodents. The apparatus consists of a push-pull strain gauge attached to a t-bar positioned at the end of a specially built platform. To measure forelimb grip strength, the rat is placed with its forepaws on the t-bar and gently pulled backwards by the tail until it engages the bar. It is then pulled back further until its grip is broken. To measure hindlimb grip strength, the rat is placed on the platform facing away from the t-bar and a similar procedure is employed as in the forelimb grip strength procedure.

fornix (fornix cerebri): a triangular anatomical structure of white matter in the brain, situated between the hippocampus and hypothalamus.

free-choice paradigm: experimental situation in which a subject can choose out of more than one possibilities.

frontal lobe tests: tests to assess impairment of functions that are anatomically located in the frontal lobe of the brain. These tests mainly assess cognitive functioning, mood and personality.

frontal lobe: area of the neocortex, involved in executive functioning. Executive functioning can be conceptualized as having four components: goal formulation, planning, carrying out goal directed plans and effective performance.

Fumonisin B1 (1,2,3-propatericarboxylic acid, 1,1'-[1-(12-amino-4,9,11-trihydroxy-2-methyltridecyl)-2-(1-methylpentyl)-1,2-ethanediyl]ester): a mycotoxin contaminant of maize, produced by the corn mold *Fusarium moniliforme*. Shown to be responsible for reported outbreaks of equine leukoencephalomalacia and a pulmonary edema syndrome in swine. Structurally similar to sphingosine; fumonisin B1 is a potent inhibitor of sphingolipid biosynthesis by inhibiting sphingosine N-acetyltransferase. Mol. Wt. 721,838 $C_{34}H_{59}NO_{15}$.

generalized anxiety disorder: a disorder characterized by at least 6 months of persistent and excessive anxiety and worry not resulting from exposure to a drug or medication.

glial cells, see neuroglia.

glial scarring: the formation of scars by astrocytes in the vicinity of a neural injury. Only mature astrocytes are capable to form glial scars, embryonic or postnatal astrocytes are not. The formation of glial scars prevents the regeneration of the axon.

goiter: an enlargement of the thyroid gland, causing a swelling in the front part of the neck.

goitrogen: a goiter-producing compound.

Guanabenz: an α_2-adrenergic agonist that stimulates the α_2-adrenergic receptors of the central nervous system, resulting in a reduction of sympathetic outflow to the heart and peripheral vascular system.

Guillain Barré syndrome: disease of the peripheral nerves, in which there is numbness and weakness in the limbs. It usually develops 10-20 days after a respiratory infection that provokes an allergic response in the peripheral nerves.

gyromitrin: toxin from the mushroom *Gyromitra esculenta* (false morel).

437

habituation: decrease in the behavioral response to a repeatedly presented stimulus.

half-life (half-time): the time taken for the concentration of drug in the blood or plasma to decline to half its original value.

Haloperidol: a butyrophenone antipsychotic drug used to relieve anxiety and tension in the treatment of schizophrenia and other psychiatric disorders.

harmine: an alkaloid with hallucinogenic activity. It was first isolated from seeds of *Perganum harmala*, a plant from the Asian steppes which has spread from the Mediterranean to Southeast Asia. Seed preparations have been used traditionally for a variety of purposes. Harmine and related alkaloids are also the active principles of psychoactive preparations from *Banisteriopsis caapi* and related species used by Amazonian Indians in social and religious festivals. There is cross tolerance with lysergide and psylocybin, suggesting a common site of action. Harmine is a MAO-inhibitor. At one time it was used in the treatment of Parkinson's disease.

Hebb-Williams maze: actually a series of 12 maze problems conducted within the same apparatus. The maze itself is an enclosed rectangular field, with a start and goal box placed at diagonally opposite ends of the apparatus. Barriers can be placed at different points within this field to present different maze configurations. The maze is unique in that a series of problems can be given to the animal within the same general environment, a procedure that eliminates some of the inference with learning that arises from exploratory behavior elicited by a novel situation. Twelve different maze problems differing in complexity constitute the *Hebb-Williams test*.

hemi-paresis: muscular weakness or partial paralysis affecting one side of the body

hepatomegaly: an increase in liver size

hepatotoxic: damaging or destroying liver cells.

hexachlorobenzene(HCB): obsolete fungicidal dressing, being applied to seed grain as a dry powder. Between 1955 and 1959 an epidemic, involving 4000 people, occurred in Turkey were people consumed treated grain during times of crop failure.

High Alcohol Drinking rats: strain of rats, specially bred for their quality to consume alcohol when administered to their drinking water. The rats were selectively bred based upon their daily intake of an 8% alcohol solution.

higher order conditioning: a process in which previously established conditioned stimuli serve as the basis for further conditioning.

hippocampal formation: a curved band of cortex lying within each cerebral hemisphere: in evolutionary terms one of the brain's most primitive parts. It forms a portion of the limbic system and is involved in the complex physical aspects of behavior governed by emotion and instinct.

hippocampus: a swelling of the floor of the lateral ventricle of the brain. It contains complex foldings of cortical tissue and is involved, with other connections of the hippocampal formation, in the workings of the limbic system.

histocompatibility: the form of compatibility that depends upon tissue components, mainly specific glycoprotein antigens in cell membranes. A high degree of histocompatibility is necessary for a tissue graft or organ transplantation.

hole board: apparatus to test exploration and motor activity in animals (mice and rats). The apparatus consists of a box with spaced holes in the floor. Sometimes objects are placed in these holes. The frequency and duration that an animal dips its head in one of the holes is measured as a quantity for explorative behavior, the activity of an animal is quantified by the amount of time that the animal spents moving about on the hole-board.

home cage activity: measuring the diurnal cycle of activity displayed by rodents in their home cage by some kind of sensory device such as photocells, infrared sensors and radio-frequency recorders (see also cage observation methods).

HOME-scale: scale assessing quality of the home environment of subjects undergoing a psychological test. The home environment seems to be an important confounder in neuropsychological studies using developmental psychological tests.

homovanillic acid (HVA): a product of the catecholamine metabolism.

Horner's syndrome: a group of symptoms that are due to a disorder of the sympathetic nerves in the brainstem or cervical (neck) region. The syndrome consists of a constricted pupil, ptosis, and an absence of sweating over the affected side of the face.

human leukocyte antigen system (HLA system): This is a series of four gene families that code for polymorphic proteins expresses on the surface of most nucleated cells. Individuals inherit from each parent one gene, or set of genes, for each subdivision of the HLA system. If two individuals have identical HLA types, they are said to be histocompatible. Successful tissue transplantation requires a minimum of HLA differences between donor and recipient tissues.

Hunter-Russell syndrome, see Minamata Disease

hydrazines: organic derivatives of hydrazine (= diamine; $H_2N=NH_2$). Many of these compounds are potent carcinogens, hepatoxicants or are toxic to the reproductive system.

hyperkeratosis: thickening of the outer horny layer of the skin.

hyperkinetic syndrome: mental disorder, usually of children, characterized by a grossly excessive level of activity and a marked impairment of the ability to attend. Learning is impaired as a result, and behavior is disruptive and may be defiant or aggressive. The syndrome is more common in the intellectually subnormal, the epileptic and the brain damaged. Treatment usually involves drugs (stimulants) and behavior therapy. The terms attention deficit disorder and hyperkinetic syndrome are often erroneously used indifferently.

hypersialorrhea (ptyalkism): excessive production and excretion of saliva.

hypertonia: condition of excessive tone of the skeletal muscles

hypoesthesia (= hypestesia): a dysesthesia consisting of abnormally decreased sensitivity, particularly to touch.

hyposmia: decreased olfactory function.

hypothalamus: the region in the forebrain in the floor of the third ventricle, linked with the thalamus above and the pituitary gland below. It contains important centers controlling body temperature, thirst, hunger, and eating, water balance and sexual functions. It is, via the limbic system, closely connected with emotional activity and sleep and functions as the integration center for hormonal and autonomic nervous activity through its control of the pituitary secretions.

hypothesis: a proposition that seeks to place certain facts (or variables) within a construct that will explain or predict relationships between these facts. A prediction regarding the relationship between two variables is tested by conducting research: if the findings offer support for the hypothesis, confidence in accuracy may increase, while if findings fail to offer such support, confidence in its accuracy may be reduced.

Ibuprofen: an anti-inflammatory drug used in the treatment of arthritic conditions.

ICI-174,864: a selective δ-opioid receptor antagonist.

ICS 205-930: a 5HT3/5HT4 receptor antagonist.

ideopathic: denoting a disease or condition the cause of which is not known or that arises spontaneously.

imaging techniques: the technique of producing images of organs or tissues using radiological procedures, particularly by using scanning techniques. Examples of brain imaging techniques are: CAT (see computer axial tomography), MRI (see magnetic resonance imaging), PET (see positron emission tomography).

immediate recall: the process of bringing a memory of an immediately preceding event or activity into consciousness.

immunolabeling procedures: procedures in which endogenous molecules are labeled with immunoactive substances that have been labeled with a detectable site, e.g.,a stain or a fluorescent molecule. The procedure makes it possible to localize certain systems within the brain, e.g.,the dopaminergic system, or certain peptidergic (endorphins) systems.

impotentia coeundi: lack of copulative power in the male due to initiate an erection or to maintain an erection until ejaculation caused by psychogenic or organic dysfunction.

inbreeding: the production of offspring by parents who are closely related; for example who are first cousins or siblings.

incidence rate: a fraction expressing the rate of new cases of disease in a population over a period of time: the numerator of the rate is the number of new cases during a specified time period and the denominator is the population at risk during the period.

incidence: the rate at which a certain event occurs, e.g.,the number of new cases of a specific disease occurring during a certain period.

informed consent: permission to carry out a research or medical procedure where the subject is given information including the nature of the procedures, potential risks and benefits, any other alternative procedures that are available and acknowledgment that such consent is voluntary.

instrumental conditioning: a basic form of learning in which responses that yield positive (i.e., desirable) consequences or lead to escape from avoidance of negative (i.e., undesirable) outcomes are strengthened.

intellectual maturity: intellectual ripeness; state of full intellectual functioning

intelligence quotient (IQ): age related measure of intelligence level.

intermittent reinforcement: a term for all those schedules of reinforcement in which some of the responses made go unreinforced; that is all those schedules of reinforcement other than continuous reinforcement and extinction. Also called partial reinforcement.

intracerebroventricular: directly into the cerebral ventricles

intragastric: situated or occurring within the stomach. Intragastric administration means a form of (drug)administration in which the substance is directly administered into the stomach, using a cannula or catheter.

intramuscular, within the substance of a muscle.

intraperitoneal: within the peritoneal cavity.

inventory: a questionnaire, typically one that represents many questions about each aspect of personality that is under investigation. Directions may ask for a self-description of an acquaintance who is being assessed.

inverse agonist: a drug that produces a response that is opposite to the response of the agonist, e.g.,if the agonist increases blood pressure, the inverse agonist will decrease it.

inverted U-shaped curve: dose effect curve in which medium doses have large effects, small and high doses have relatively little effect; also called bell-shaped curve.

IPPO: isopropylbicyclophosphate, a picrotoxin ligand.

Ipsapirone: a non-benzodiazepine anxiolytic, a 5-HT$_1$ receptor agonist.

IQ tests: any test that purports to measure an intelligence quotient. Generally such tests consist of a graded series of tasks each of which have been standardized with a large, representative population of individuals.

ischemia: inadequate blood flow to a part of the body, caused by constriction or blockage of the blood vessels supplying it.

jumping down to home cage: test to measure the ability/willingness of rat pups to jump from a pedestal situated above their home cage. A positive score depends of the height of the pedestal and the age of the animals.

kainate receptor: one of the four types of NMDA receptors. It binds the glutamate agonist kainate and regulates a channel permeable to Na$^+$ and K$^+$.

Kava: fermented drink made by Polynesian people of the South Pacific from the stem and root of the shrub *Piper methysticum*; it produces mild stimulation followed by drowsiness.

ketanserine: serotonin antagonist.

kinesin: large soluble cytoplasmic protein that binds tightly to microtubules and transports vesicles and particles along them using energy from ATP hydrolysis.

kinetics: mathematical study of the changes of the concentration of a substance or its metabolites in the human or animal body after exposure or administration.

Korsakoff syndrome: syndrome of anterograde and retrograde amnesia; currently used synonymously with 'amnestic syndrome' or, more narrowly, to refer to the amnestic component of the Wernicke-Korsakoff syndrome.

lacrimation: excess production of tears.

Lambert Eaton Myasthenic Syndrome, see Lambert-Eaton syndrome.

Lambert-Eaton syndrome: myasthenia-like syndrome in which the weakness usually affects the limbs, and ocular and bulbar muscles are spared; there is reduced muscle action potential on stimulation of its nerve but with repetitive stimulation it becomes augmented.

landing foot splay: simple test of motor function. To measure landing foot splay, the hindpaws of a rat are inked and the rat is dropped from a horizontal position, onto a blank sheet of paper covering a cushion. The distance between the two resulting ink spots on the paper provide a measure of 'landing foot splay' which tends to be increased in rats with motor dysfunction.

Lashley III maze: relatively complex maze. It is an enclosed, rectangular chamber that consists of four parallel alleys, a start box, and a goal box. The start box and the goal box are located on opposite external walls of the apparatus. The correct path from the start box to the goal is described by a zigzag pattern through doorways in each of the walls of the four interior alleys. The ends of the four alleyways form eight culs-de-sac and contribute to the complexity of the task.

lathyrism (spastic paraparesis): disease characterized by muscular weakness and paralysis, found among people whose staple diet contains mostly of large quantities of Lathyrus sativus, a kind of chicken pea. Except in mild cases, complete recovery does not occur, despite administration of an adequate diet and physiotherapy.

L-dopa (levodopa): a naturally occurring amino acid administered orally to treat parkinsonism.

learned helplessness: in learning experiments, a subject's passive response to stress after being placed in situations in which there is no way to avoid an electric shock.

leukoencephalomalacia: softening of the brain, especially affecting the white matter of the cerebral hemispheres.

Lewy bodies: concentrically laminated, round bodies found in vacuoles in the cytoplasm of some of the neurons of the midbrain in paralysis agitans (Parkinson's disease).

limbic system: complex system of nerve pathways and networks in the brain, involving several different nuclei, that is involved in the expression of instinct and mood in activities of the endocrine and motor systems of the body. Among the brain regions involved are the amygdala, hippocampal formation, and hypothalamus.

linamarin: cyanogenic glycoside in the leaves and roots of the cassava plant, Manihot utilissima, from which hydrogen cyanide is released by enzymatic hydrolysis. Much of the cyanide is removed by peeling the fresh roots, washing them in running water, and then drying them in the sun. Remaining free HCN is readily votalized on boiling. However, if the cooking utensil is covered with a lid, condensation occurs and the cooking water become heavily contaminated. Inadequate preparation of cassava can lead to acute poisoning characterized by abdominal pains and vomiting, followed by mental fusion, generalized muscular weakness and respiratory distress. Chronic poisoning arising from prolonged exposure to small amounts of cyanide may lead to neurological symptoms associated with demyelinating of nerve fibers, including those of the optic nerve.

β-lipotropin: a precursor molecule for different endogenous peptides, such as the endorphins and ACTH.

lithium carbonate: a white granular powder used in the treatment of acute manic states and in the prophylaxis of recurrent affective disorders manifested by depression or mania only.

lithium, see lithium carbonate

liver transaminases: a sub-subclass of enzymes of the transferase class within the liver that catalyze the transfer of an amino group from a donor (usually an amino acid) to an acceptor (generally a 2-keto acid). Also called aminotransferases.

locomotor activity tests: observational tests involving an environment in which the activity of an animal is recorded as it moves from place to place.

locus coeruleus: small pigmented region in the floor of the fourth ventricle of the brain.

Lolitrem-B: a tremorgenic mycotoxin isolated from perennial ryegrass and produced by the endophyte Acremomium lolii. Associated with the neurological 'ryegrass staggers' in livestock. Mol. wt. 685,899 $C_{42}H_{55}NO_7$.

long term potentiation (LTP): increased excitatory synaptic potential in postsynaptic neurons of the hippocampus caused by a brief high-frequency train of stimuli to any one of the three afferent pathways to the hippocampus.

long-term exposure: continuous or repeated exposure to a substance over a long period of time, usually several years in man, and of the greater part of the total life-span in animals.

l-tryptophan: amino acid existing in proteins from which it is set free by triptic digestion. It is essential for optimal growth in infants and for nitrogen equilibrium in human adults. It is also the precursor for serotonin.

lupinine: hepatotoxic toxin present in the seeds of leguminous herbs of the genus Lupinus. The toxin produces lupinosis, a morbid, often fatal condition characterized by acute atrophy of the liver and affecting domestic animals, including cattle, sheep, goats, and horses.

Lytigo Bodeg: ALS-variant endemic to many Pacific areas, particularly studied on the island of Guam. It exists on Guam in two main clinical forms which also encompass dementia and extrapyramidal manifestations. The first form which resembles ALS has declined in recent years, the second form, which is characterized by dementia and extrapyramidal manifestations has not.

magnetic resonance imaging (MRI): in this technique, the brain is bombarded with radio waves; molecules in the neurons respond by producing radio waves of their own, and these are recorded, computerized, and assembled into a three-dimensional picture.

major depressive episode: a period of at least 2 weeks during which there is either depressed mood or the loss of interest or pleasure in nearly all activities.

major histocompatibility complex: a set of linked genetic loci which dominates the control of modification of the immune response. The genes and their products are designated by prefixes (e.g., HLA in man, ChLA in chimpanzee) MHC gene products occur on cell surfaces and serve as markers which help to distinguish 'self' from 'non-self' tissue.

mania: a period of abnormally and persistently elevated, expansive, or irritable mood.

manic-depressive: characterized by alternating manic and depressive episodes.

Marchiafava-Bignami syndrome (disease): A rare condition, originally described in Italian drinkers of crude red wine, occurs occasionally in other alcoholic patients. It is characterized clinically by disorders of emotional control and cognitive function followed by variable delirium, fits, tremors, rigidity, and paralysis; most patients eventually become comatose and die within a few months. Symmetrical demyelination with subsequent cavitation and axonal destruction is found in the corpus callosum and often, in varying degree, in the central white matter of the cerebral hemispheres, the optic chiasm, and in the middle cerebellar peduncles.

maze: any experimental apparatus consisting of intersecting paths used in intelligence tests and in demonstrating learning in experimental animals. There is a variety of models, from the most simple T-maze to the complex Hampton Court Maze. The most common is the multiple T-maze. There are also water mazes (Morris maze, Biel maze) that must be swum through, elevated mazes that consist of narrow ramps, etc.

MDI, mental development index (see Bayley Scales of Infant Development).

mean: the arithmetic average of a set of data. The mean is computed by adding all of the data together and then dividing by the number of scores in the set.

medial forebrain bundle: a pathway in the limbic system leading from the precommissural fornix and the olfactory bulb through the preoptic mesencephalon. The pathway has been found to produce highly reinforcing in self-stimulation experiments.

median: the midpoint of a set of scores. Fifty percent of the scores fall above the median, 50 percent below.

Meibomian glands: small sebaceous glands that lie under the conjunction of the eyelids.

mental age: a measure of intellectual ability obtained on early experimenters' tests of intelligence. An individual's mental age was assumed to reflect his or her level of intellectual maturity.

mental arithmetic: assessment test for measuring skills necessary for academic achievement. In the WAIS-R mediocre value as measures of general ability, assesses knowledge and ability to apply arithmetic operations only.

mesoderm: the middle germ layer of the early embryo. It gives rise to cartilage, muscle, bone, blood, kidneys, gonads and their ducts, and connective tissue. It separates into two layers, an outer somatic and an inner splanchnic mesoderm, separated by a cavity. The dorsal somatic mesoderm becomes segmented into a number of somites.

mesodermal origin: originating from the mesoderm.

meta-analysis: systematic method that uses statistical analysis to integrate the data from a number of independent studies.

β-N-methylaminoalanine (BMAA): neurotoxin found in the seeds of cycad (*Cycas circinalis*) and chemically and pharmacologically related to [Beta]-N-oxylaminoalanine. BMAA causes chromatolysis of giant Betz's cells and smaller pyramidal cells in the cynomolgus monkey.

N-methyl-D-aspartate (NMDA): a neurotransmitter similar to glutamate, found in the central nervous system; a synthetic preparation is used experimentally to study the excitatory mechanisms of glutamate transmitters.

methyl mercury: main organic mercury compound to which humans are exposed. Exposure occurs mainly via the oral pathway. In nature, methyl mercury can be formed by biomethylation in aquatic environments; it is also used as a fungicide. Some familiar epidemics of methyl mercury poisoning have occurred in Japan (fish from polluted waters) and Iraq (seed for sowing). Because of the lipophilic character of the compound, the central nervous system forms the main target organ. Symptoms are visual disturbance, sensory distortion, coordination and speech disturbance, and deafness.

methylenedioxyamphetamine (MDA): a psychoactive substance, chemically related to mescaline and amphetamine, first synthesized in the 1930s. MDA is obtained from safrol, a psychoactive oil found in nutmeg that is chemically related to amphetamine.

Methylphenidate: a central stimulant used in the treatment of hyperkinetic children, various types of depression and narcolepsy.

metronidazole: an antiprotozoal and antibacterial drug effective against obligate anaerobes.

Meynert's nucleus (nucleus basalis of Meynert): a group of neurons situated in the basal part of the forebrain that has wide projections to the neocortex and is rich in acetylcholine and choline acetyltransferase. It undergoes degeneration in Parkinson's disease and in Alzheimer's disease.

MHC: major histocompatibility complex.

microcephalic: small headed, this term is reserved for those cases in which the abnormality is so great that retardation results.

441

microcyte: one of the cells of the neuroglia, responsible for phagocytizing the waste products of nerve tissue.

microdialysis: technique used to locally measure the rate of synthesis and/or secretion of neurotransmitters in the brain.

microglia, see microcyte.

micrognathia: unusual or undue smallness of the jaws

microsomal ethanol-oxidizing system: one of the 3 enzymatic systems that metabolize ethanol in the body. The other two are catalase and alcohol dehydrogenase.

mid-air righting: simple tests of motor function in which the rat is held in a reclining position and dropped from a height of approximately 40 cm on to a cushion. A rat with normal motor function will right itself in mid-air to land on all four paws, while animals with motor dysfunction may be unable to do so.

Miller Fisher syndrome: a variant of the Lambert-Eaton syndrome (see Lambert-Eaton syndrome). characterized by areflexia, ataxia and ophthalmoplegia.

Minamata disease: severe neurological disorder, formerly known as Hunter-Russell syndrome, usually characterized by peripheral and circumoral paresthesia, ataxia, dysarthria and loss of peripheral vision and leading to permanent disabilities or death. The disease is caused by alkyl mercury poisoning and was prevalent between 1953 and 1958 among those who ate seafood from a bay in Japan that was polluted with alkyl mercury compounds.

minimal brain dysfunction (MBD): cover term for a variety of behavioral, cognitive and affective abnormalities observed in young children. The term is typically reserved for such cases in which the patterns of thought and action are such that one would expect to find some organic abnormality but none is apparent. The term is often used as though there were an identifiable MBD syndrome, a collection of fairly specific disorders that could be taken as hallmarks of some underlying neurological causal mechanism. However, the evidence to support a single MBD syndrome is largely unconvincing.

miosis: constriction of the pupil of the eye to contract.

M-maze: M-shaped maze, often used as a water maze in developmental toxicology (see maze).

mnemonic: relating to memory.

monosodium glutamate (MSG): flavoring agent used as a food additive. Prepared from natural or synthetic L-glutamic acid.

mood disorders: depressive and manic disorders.

mood questionnaire: questionnaire to assess the state of mood.

mood: a relatively short-lived, low intensity emotional state.

morphine: principal alkaloid of opium. Morphine and its salts are very valuable analgesic drugs, but are highly addictive. In addition to suppression of pain, morphine causes constipation, decreases pupillary size and depresses respiration. Only the (+)-stereoisomer is biologically active.

Morris maze: maze used to assess memory for spatial location. In this task, the rat must swim to a submerged platform in a large tank filled with opaque water; escape from the water is the reward, so no food deprivation is required. Starting position in the tank varies from trial to trial, and latency to reach the platform, as well as directness of the route, can be measured.

motivation: a hypothetical internal process that provides the energy for behavioral and directs it towards a specific goal.

motive: an acquired motivational system.

motor latency: the time between the onset of a stimulus and the overt response to that stimulus.

MPTP: 1-methyl-4-phenyl-1,2,3,4-tetrahydropyridine, a byproduct in the manufacture of MPPP due to inadequate technique. It produces severe and irreversible Parkinson's disease.

multiple correlation: the relationship between one dependent variable and two or more independent variables. A multiple correlation coefficient yields an estimate of the combined influence of the independent variables on the dependent.

multiple FI-FR schedules: a schedule of reinforcement (see schedule of reinforcement), in which a fixed-interval (see fixed interval (FI) schedule of reinforcement) and a fixed ratio schedule (see fixed ratio (FR) schedule of reinforcement) are combined into a compound form.

multiple regression method: a statistical technique that makes it possible to make predictions about performance on one variable (= the criterion variable) based on performance on two or more other variables (= predictor variables).

multiple regression: statistical technique that is an extension of simple regression; allows predictions about performance on one variable (called the criterion variable) based on performance on two or more variables (called the predictor variables). If the regression equation is in standard score form then the relative weights or contributions of each of the predictor variables may be assessed (see multiple correlation).

multiple sclerosis: chronic disease of the nervous system affecting young and middle aged adults. The myelin sheaths surrounding nerves in the brain and pinal cord are damaged, which affects the functioning of the nerves involved. The underlying cause of the myelin sheath damage remains unknown. The illness affects different parts of the CNS and spinal cord.

multivariate analysis: any statistical test that is designed to analyze data from more than one variable.

myalgia: pain in the muscles.

myasthenia gravis: chronic disease marked by abnormal fatigability and weakness of skeletal muscles, relieved by rest or anticholinesterase drugs. The degree of fatigue is so extreme, that these muscles are temporarily paralyzed. Other symptoms include drooping of the upper eye lid (ptosis), double vision and dysarthria. The cause is uncertain, but appears to be associated with impaired ability of the neurotransmitter acetylcholine to induce muscular contraction. It chiefly affects adolescents and young adults (usually women) and adults over 40.

mycotoxin: fungal toxin.

myelin: the substance of the cell membrane of Schwann's cells that coils to form the myelin sheath; it has a high proportion of lipid to protein and serves as an electrical insulator.

myoclonus: shock-like contractions of a portion of a muscle, an entire muscle, or a group of muscles, restricted to one area of the body or appearing synchronously or asynchronously in several areas.

myristicin: naturally occurring methylenedioxyphenyl compound found in nutmeg. It has been suggested that it may be (partly) responsible for the toxicity of nutmeg. Nutmeg poisoning causes symptoms similar to atropine poisoning: skin flushing, tachycardia, absence of salivation and excitation of the CNS. Euphoria and hallucinations have given rise to abuse of this material.

negative geotaxis: a rat turns to face upward when it is placed on an incline, head pointing downward; this postural reaction is called negative geotaxis.

negative reinforcement: the process by which a person learns to avoid behavior that causes unpleasant sensations.

negative reinforcer: stimulus that increases the frequency of behavior that prevents its presentation.

nerve growth factor (NGF): one of the neurotrophic factors that control neuronal growth during nervous system development and during regeneration (see neurotrophic factor).

neuralgia: severe burning or stabbing pain often following the course of a nerve

neurofibrillary tangles: bundles of abnormal filaments within neurons. The filaments of these cytoskeletal abnormalities, as seen in Alzheimer's disease, consist of two thin filaments arranged in a helix (paired helical filament).These structures do not resemble normal cytoskeletal proteins of neurons.

neuroglia: the supporting structure of the nervous system. It consists of a fine web of tissue made up of modified ectodermal elements, in which are enclosed peculiar branched cells known as neuroglial cells or glial cells. The neuroglial cells are of three types: astrocytes and oligodendrocytes (astroglia and oligodendroglia), which appear to play a role in myelin formation, transport of material to neurons, and maintenance of the ionic environment of neurons, and microcytes (microglia), which phagocytize waste products of nerve tissue.

neuroglial cells, see neuroglia.

neuropathy: a functional disturbance or pathological change in the peripheral nervous system.

neurotensin: a tridecapeptide found in the small intestine and in the brain. It induces vasodilatation and hypotension and in the brain it is a neurotransmitter.

neurotrophic factor: proteins that ensure the survival and maintenance of defined populations of neurons. In addition, neurotrophic factors have the competence to promote the outgrowth of neurites. To date, three neurotrophic proteins have been purified: nerve growth factor (NGF), brain-derived neurotrophic factor (BDNF) and ciliary neurotrophic factor (CNTF).

3-nitropropionate (3-nitropropionic acid): a plant neurotoxic amino acid derived from *Indigofera endecaphylla*. An irreversible inhibitor of succinic dehydrogenase and a competitive inhibitor of fumerase; an excitotoxin shown to cause lateral caudate putamen brain lesions similar to those of Huntington's disease. Mol. Wt. 119,08 $C_3H_5NO_4$.

NMDA receptor: N-methyl-D-aspartate receptor.

NMDA, see N-methyl-D-aspartate

nociception: perception of painful/damaging stimuli.

non contingent responding: responding not under the control of the environmental contingencies of the task.

no-observed-effect level (NOEL): experimentally determined dose at which no statistically or biologically significant indication of the toxic effect of concern is observed.

norepinephrine: synaptic neurotransmitter, also called noradrenaline. It is one of the monoamines; lowered levels are associated with depression. Norepinephrine is also released from the adrenal medulla as part of the peripheral arousal response.

nucleus accumbens (NA): collection of pleomorphic cells in the caudal part of the anterior horn of the lateral ventricle of the olfactory tubercle, lying between the head of the caudate nucleus and the anterior perforated substance.

numbness: anesthesia.

nystagmus: an involuntary, rapid, rhythmic movement of the eyeball, which may be horizontal, vertical, rotatory, or mixed.

object reversal learning: discrimination reversal learning task in which a subject is required to make initial discrimination between two stimulus objects, in which the correct object is of a particular color and shape. After the initial discrimination, the original stimulus object is changed for another.

objective test: a personality test involving structured items and a limited set of responses (such as TRUE - FALSE)

obsessive compulsive behavior: characteristic behavior as performed in an obsessive-compulsive disorder.

obsessive compulsive disorder: a subclass of anxiety disorders with two essential characteristics: recurrent and persistent thoughts, ideas and feelings and repetitive, ritualized behaviors. Attempts to resist a compulsion produce mounting tension and anxiety, which are relieved immediately by giving into it. The term is not properly used for behaviors like excessive drinking, gambling, eating etc. on the grounds that the compulsive gambler actually derives considerable pleasure from gambling, it is the losing that hurts.

odor identification: the ability to discriminate between different olfactory stimuli (smells).

olfaction: the sense of smell.

olfactory perception threshold: the minimal concentration of a substance that evokes smell.

oligodendrocyte: one of the cells of the neuroglia, responsible for producing myelin sheaths of the neurons of the central nervous system and therefore equivalent to the Schwann cells of the peripheral nerves.

oligodendroglia, see oligodendrocyte.

Olton spatial maze, see radial arm maze.

Ondansetron: an antiemetic used in conjunction with cancer chemotherapy.

one-way active avoidance test: conditioned avoidance task in which an animal must learn to avoid an unpleasant stimulus (e.g., electric shock) by moving away from it (e.g., by jumping in a pole [see pole-climb test] or on a platform). Just before that stimulus (electric shock; the unconditioned stimulus) a tone or light (the conditioned stimulus) is presented.

443

open field activity, see open field.

open field behavior: the activities performed by an animal in an open field test.

open field rearing behaviors: typical postures of a rat in an open field, rearing middle and rearing field. The rat is standing on its hindlimbs and is facing the wall (= rearing wall) or standing on its hindlimbs in the middle of the field an sniffing in the air.

open field: observational test method to measure the behavior of animals, especially rats. An animal is placed in the center of a field and number of squares crossed in a limited time is measured. Other posture/activities that can be measured are: rearing (in the middle and against the wall), defecations, urination etc. The test has been automated and devices are available that record locomotion, rearing, and even grooming, as separate acts. Many modifications of this test are introduced, such as introducing (unknown) objects in the field.

operant conditioning: instrumental conditioning. Learning by association of an organism's own behavior with a subsequent reinforcing or punishing environmental event.

operant paradigm: paradigm in which the actions of the subject are voluntary and are not elicited by a discrete, identifiable stimulus.

operant response: response, for example a behavior, which is voluntary and is not elicited by a discrete, identifiable stimulus.

ophthalmoplegia: paralysis of the muscles of the eye.

orbital edema: edema in the bony cavity that contains the eye ball.

organophosphates: any organic salt of a phosphorous oxyacid; generally applied to organic salts of H_3PO_4.

organophosphorus compounds: substances containing C-P bounds but often extended to esters and thio-esters. The most important application of organophosphates is in the field of pesticides. Among the best known are malathion, parathion, schradan and dimefox.

organum subfornicale, see subfornical organ.

outbreeding: the mating of totally unrelated individuals, which frequently results in the production of offspring that show more vigor, as measured in terms of growth, survival, and fertility, than the parents.

β-N-oxylaminoalanine: neurotoxic substance extracted from Lathyrus sativus. It produces signs of neurotoxicity when injected systemically into young animals at an age when the blood brain barrier is deficient, but it is without effect in adult animals. When chronically injected into the CSF of adult monkeys by lumbar puncture, it produces the typical signs of human lathyrism, associated with destruction of nerve cells in the gray matter of the cord and a proliferation of microglia.

palpebral closure: closure of the eyelids.

pandemic: epidemic, so widely spread that vast numbers of people in different countries are affected.

panic disorder: a disorder characterized by sudden, unexpected, and persistent episodes of intense fear, accompanied by a sense of imminent danger and an urge to escape.

Paralytic Shellfish Poisoning (mussel poisoning): a severe, often fatal condition resulting from the consumption of mussels or clams, which have themselves consumed neurotoxin-forming dinoflagellates and accumulated the toxin (e.g.,saxitoxin) in their tissues. Symptoms include diarrhea, fatigue, and a tingling sensation beginning around the face and spreading to the arms, fingers and toes; later numbness may develop, followed by weakness, paralysis and death from respiratory paralysis.

paraparesis: partial paralysis of the lower extremities.

Paregoric: preparation of powdered opium, anise oil, benzoic acid, camphor, diluted alcohol and glycerin, each 100 ml. of which yields 35-45 mg. of anhydrous morphine; used as antiperistaltic in the treatment of diarrhea.

paresis: muscular weakness caused by disease of the nervous system. It implies a lesser degree of weakness than paralysis, although these two words are often used interchangeably.

paresthesia: abnormal touch sensation, such as burning, prickling, or formication, often in the absence of an external stimulus.

parkinsonism: clinical picture characterized by tremor, rigidity and a poverty of spontaneous movements. The commonest symptom is tremor, which often affects one hand, spreading first to the leg on the same side and then to the other limbs. It is most pronounced in resting limbs, interfering with such actions as holding a cup. The patient has an expressionless face, an unmodulated voice, an increasing tendency to bend, and a shuffling walk. Parkinsonism is a disease process affecting the basal ganglia of the brain and associated with a deficiency of the neurotransmitter dopamine. Sometimes, a distinction is made between *Parkinson's disease*, a degenerative disorder associated with aging, and parkinsonism due to other causes. For example, it may be induced by the long-term use of antipsychotic drugs and uncommonly, it can be attributed to the late effects encephalitis or coal-gas poisoning. Relief of the symptoms may be obtained with anticholinergic drugs, dopamine-receptor agonists and levodopa.

paroxysm: (1) sudden violent attack, especially a spasm or convulsion; (2) the abrupt worsening of symptoms or recurrence of disease.

passive avoidance: avoidance learning (conditioning) paradigm in which the subject must refrain from making a response which will produce the aversive stimulus.

peer rating: a classmate, fellow soldier, fellow student, or other acquaintance marks a rating scale or inventory to describe the target person. Usually, the average of several reports is taken as the target's peer rating.

penitrem-A: a mycotoxin contaminant isolated from *Penicillium cycolpium*. Produces a neurotoxicity syndrome characterized by sustained tremors and accompanied by other neurological signs such as limb weakness, ataxia and convulsions. Mol. Wt. 634,210 $C_{37}H_{44}NO_6Cl$.

perceptual learning: the learning of an organism to detect the physical differences between two stimuli.

performance test: among ability tests the term is usually applied to those where the respondent is to execute an appropriate physical action – tracing a maze path for example; although it can also be verbal as in the Wechsler Performance section. Among personality measures, the term is usually applied to observations of response in a standardized situation, usually a situation that arouses strong motives.

pergolide mesylate: drug that stimulates dopamine receptors in the brain and is used in the treatment of parkinsonism

periamygdalian cortex: neocortex situated near to the nucleus amygdalae.

perikaryon: the cell body of a neuron, as distinguished from the nucleus and the processes.

perinatal: period before birth, birth and just after birth. In man pertaining to the period extending from the 28th week of gestation to the 28th day after birth.

personality indices: the items in questionnaires and inventories that are used to assess a subject's personality.

personality questionnaire: questionnaire to assess a subject's personality.

personality: the characteristic way in which a person thinks, feels, and behaves; the relatively stable and predictable part of a person's thought and behavior; it includes conscious attitudes, values, and styles as well as unconscious conflicts and defense mechanisms.

pharmacokinetic phase: process in which a drug is absorbed, distributed, biotransformed in, and excreted out of the body.

pharmacotreatment: any form of medical therapy in which medicines are used.

phenacetin: derivative of *p*-aminophenol; a large portion of phenacetin is metabolized to acetaminophen (= paracetamol), which has lower toxicity. Phenacetin and acetaminophen have the same therapeutic indications as the salicylates (analgesic, antipyretic), except that they do not have anti-inflammatory activity. Besides its effects on the blood, phenacetin has been indicated in the production of kidney damage.

phenazone (= antipyrine): pyrazolon derivative with weak antipyretic or analgesic activity, but high anti-inflammatory activity. It may cause serious blood disorders (agranulocytosis) and skin eruptions.

phenelzine sulfate: a monoamine oxidase inhibitor used as an antidepressant, administered orally.

phenobarbital: sedative barbiturate with low reinforcing properties

Phentermine: an adrenergic isomeric with amphetamine, used as an anorexic.

phenylketonuria (PKU): the most severe manifestation of hyperphenylalaninemia due to a phenylalanine 4-monooxygenase deficiency, with accumulation and excretion of phenylalanine, phenylpuryvic acid, and related compounds. It is inherited as an autosomal recessive trait. Characterized by severe mental retardation, tumors, seizures, hypopigmentation of skin and hair, eczema and mousy odor, all preventable by early restriction of dietary phenylalanine.

Phi-coefficient: an index of the relationship between any two sets of scores, provided that both sets of scores can be represented on ordered, binary dimensions, e.g.,male-female; married-single.

pica: the indiscriminate eating of non-nutritious or harmful substances, such as dirt, clay, gravel, grass, stones, or clothing. It is common in early childhood, but may also be found in mentally handicapped and psychotic persons.

pilocarpine: an alkaloid, extracted from jaborandi, *Pilocarpus pinnatifolius,* and used in medicine to stimulate sweating.

piloerection: erection of the hair.

Pimozide: antipsychotic drug used in the treatment of Gilles de la Tourette syndrome and schizophrenia.

pineal gland or pineal body (epiphysis): a pea-sized mass of nerve tissue attached by a stalk to the posterior wall of the third ventricle of the brain, deep between the cerebral hemispheres at the back of the skull. It functions as a gland, secreting the hormone melatonin. The gland becomes calcified as age progresses.

pirenperone: serotonin antagonist

pivoting: circular motion of young rats between 5 and 13 days of age, resulting from the fact that their hindlimbs do not support the movements of the forelimbs.

placebo effect: changes in behavior stemming from conditions or procedures which accompany, but are not directly related to, independent variables in an experiment. For example, changes in behavior following injections of a specific drug may result from the act of being injected, rather than from the drug itself.

pole-climb test: an active avoidance task, in which a rat is learned to jump in a pole in the presence of a conditioned stimulus to avoid an electrical shock; thus, the animal can escape the shock. Failure to acquire avoidance while adequately developing escape performance indicates learning deficits.

pole-jumping, see pole-climb test.

polybrominated biphenyls (PBB's): mixture of brominated biphenyls with an average bromide content of six atoms per molecule. PBBs have been used as flame retardants, but in view of environmental and health problems, such use is being curtailed.

polychlorinated biphenyl (PCB): a derivative of biphenyl $(C_6H_5)_2$ in which some of the hydrogen atoms on the two benzene rings have been substituted by chlorine atoms. These highly poisonous and carcinogenic compounds are used in synthetic resins, particularly as electrical insulators. Their increased usage has caused concern as they tend to accumulate in the food chain.

polydipsia: drinking of large quantities of fluids.

polyneuropathy: neuropathy of several peripheral nerves simultaneously; also called multiple or peripheral neuropathy.

poor metabolizer: large variations occur in drug metabolism in humans. This often is due to a genetic predisposition. Persons in which the rate of metabolism is slow are called poor metabolizers (see debrisoquine genotype).

445

positive reinforcement: the process by which a person learns to repeat rewarding behavior.

positive reinforcer: stimulus that increases the frequency of behavior that leads to its presentation.

positron emission tomography (PET): imaging technique involving the injection of radioactive glucose in the bloodstream. Glucose is used as energy source by brain cells, the most active cells taking up more glucose. The radioactive particles emitted by the glucose are picked up by an array of detectors around the head, and after computer-analysis give an overall picture of brain cell activity.

post-translational modification: modification of a protein after it has once been manufactured at the ribosomes.

predictive value: the validity is the extent to which a model measures what it purports to measure. In the case of high predictive value a model demonstrates concordance between predictions based on the effects of a chemical agent on the animal model and effects of the agent on humans, in essence an ex post facto validation for the model.

prefrontal cortex: the superstructure above all other parts of the cortex, situated in the anterior part of the frontal lobe of the brain.

prenatal: period before birth. In man pertaining to the period between the last menstrual cycle and birth of the child.

Preyers's reflex: involuntary movement of the ears produced by auditory stimulation.

progressive supranuclear palsy: also known as Steele-Richardson-Olszewski syndrome, progressive neurological disorder having onset at about 60 years of age, characterized by supranuclear ophtalmoplegia, especially paralysis of the downward gaze, pseudobulbar palsy, dysarthria, dystonic rigidity of the neck and trunk and dementia.

projective test: a personality test involving ambiguous stimuli. A subject's responses are supported to reveal aspects of the unconscious.

proportional hazards model: statistical procedure to determine the proportion of a sample that develops adverse effects.

proprioceptive sensation, see proprioceptor.

proprioceptor: specialized sensory nerve ending that monitors internal changes in the body brought about by movement and muscular activity. Proprioceptors located in muscles and tendons transmit information that is used to coordinate muscular activity.

prospective study, see cohort study.

pseudobulbar palsy, see supranuclear paralysis.

psychiatric rating scale: device used to assist a psychiatrist in making a diagnosis of a patient.

psychodysleptics: substance inducing a dreamlike or delusional state of mind (see psychedelic).

psychomotor: refers to abilities that require coordinated adaptation of muscular actions. (Example: aiming at a target that moves irregularly.)

psychotaraxics: mind disrupting hallucinogenic drug.

ptosis: drooping of the upper eyelid.

punisher: stimulus that decreases the frequency of behavior that leads to its presentation.

putamen: part of the lenticular nucleus, one of the basal ganglia.

quantitative trait loci (QTL): the collection of genes that are positively correlated with a certain deviation or illness.

Quinpirole (= SKF-38393): selective dopamine D-2 receptor agonist.

racemic: made up of two enantiomorphic isomers and therefore optically inactive.

Raclopride: an atypical neuroleptic; selective dopamine D-2 receptor antagonist.

radial arm maze (RAM): maze, developed for the analysis of memory function in the rodent. The RAM utilizes spatial cues as part of the paradigm. Generally, the maze is open to the room. The typical RAM consists of an open central platform, from which 8 arms radiate like spokes of a wheel. Food deprived rats are allowed to choose freely from among the 8 arms, but only 1 piece of food is present at the end of each.

radicle: any of the smallest branches of a vessel or nerve.

radioimmunoassay: any assay procedure that employs an immune reaction, in which either the antigen or the antibody is labeled with a radionuclide to permit accurate quantification.

radioreceptor assay: a radioimmunoassay to localize specific receptors.

range: the difference between the highest and lowest scores in a distribution, or set of data.

raphe nuclei: a system of nuclei in the brain that controls sleep.

rate-independent threshold procedure: reward threshold levels are determined in intracranial self-stimulation designs (ICSS), using a discrete trial task in which the presentation of stimulus 1 (S1) signals the availability of an identical electrical pulse of the same intensity. An operant response within 7.5 sec. after S1 results in the immediate delivery of the second stimulus (S2), which is the reinforcement. The rate independent procedure includes that responses made during an interval institute a 30 sec. delay or time-out period before the onset of the next trial to punish unsolicited responding. Therefore, it is to the subjects advantage to make responses (discrete wheel turns) only in response to stimuli that are rewarding.

reaction time (RT) paradigms: tests to measure the minimum time between the presentation of a stimulus and the subject's response to it. One of the oldest dependent variables in experimental psychology, specialized types have been studied. The introduction of computerized assessment has facilitated the recording and calculating of reaction times. (1) choice reaction time is the reaction time between the onset of one of a set of possible stimuli and the completion of the set of responses associated with the stimulus. (2) simple reaction time is the time between the onset of a stimulus and the completion of a response to it. Only one stimulus and one response are possible.

reactive gliosis: the transformation of resting astrocytes to their reactive form as a response to brain injury, the hallmark of this process is the increase in glial fibrillary acidic protein (GFAP), the major intermediate filament protein of this cell type. Reactive astrocytes synthesize many neuropeptides, transmitters, cytokines and growth factors. The expression of these factors is generally restricted by brain region and to subsets of the reactive astrocytes in the affected region. The role of the glia-derived factors in brain injury are not well understood, but potentially, these factors could contribute to observed pathological changes, or alternatively might be important in recovery from brain injury. It has been argued that inflammatory cytokines and other micro-glia derived factors account for neurotoxicity in gliosis, while reactive-astrocyte products tend to be neuroprotective.

recall tests: tests to measure memory, in which previously learned items must be reproduced (as by listing them or repeating them).

receptor antagonist: a substance that binds to a specific receptor site and blocks the effect of the agonist, e.g.,a neurotransmitter.

receptor binding studies: studies to measure the binding of certain substances to a specific receptor.

reference memory: processes involved in storing information which does not change and which is retained for long periods of time.

reflex modification: whole body startle response elicited by a sudden, intense sensory stimulus. The reflex modification refers to a perceptible and preceding change in the sensory environment.

reinforcer: *(primary)* a stimulus whose reinforcing effect appears to be inherent (unconditioned, i.e. does not have to be acquired through conditioning or learning); *(secondary)* a stimulus whose reinforcing effect has been acquired through its previous association with a primary reinforcer (also called conditioned reinforcer).

repeated measure design: statistical design of a study in which a certain variable is measured more than one time in the same subject, e.g.,at different time intervals.

residue: part of substance that remains in the food after manufacturing.

respiration depression (RD50): measure to quantify the irritant potency of solvents.

retention gradient: in a delayed matching test, the correct matching of a sample stimulus with the comparison stimuli decreases with the length of the delay, the function relating accuracy to delay is the retention gradient (see delayed matching).

retention: the process of holding onto or retaining a thing. Most commonly used with respect to issues surrounding the retention of information, where the basic presumption is that some 'mental content' persists from the time of initial exposure to the material or initial learning of a response until some later request for recall or re-performance.

reticular activating system (RAS): the system of nerve pathways in the brain concerned with the level of consciousness, from the state of sleep, drowsiness and relaxation to full alertness and attention. The system integrates information from all of the senses and from the cerebrum and cerebellum and determines the overall activity of the brain and the autonomic nervous system and patterns of behavior during waking and sleeping.

reversal learning: this term covers a variety of learning situations during which at some point in discrimination training (see discrimination training procedure) the original cues for correct responding are modified, so that they no longer serve as indicators for the original correct response. For example, a triangle may serve as a cue for the right turn in a T-maze and a circle for a left. After learning has reached some criterion (e.g.,80% correct responses) the cues are reversed so that now right turns to circles are reinforced and left turns to triangles.

reverse dialysis: brain dialysis involves the implantation of a dialysis tube into the brain tissue. Substances in the extracellular fluid will diffuse into the perfusate, enabling monitoring of levels of neurotransmitters and their metabolites in the extracellular fluid *in vivo*. In addition, substances included in the perfusate can diffuse into the tissue ('reverse').

rhinophyma: hypertrophy of the nose with follicular dilatation

rhinorrhoea: persistent watery mucus discharge from the nose, as in the common cold.

riboflavin (vitamin B2): a vitamin of the B-complex that is a constituent of the coenzymes FAD (flavine adenine dinucleotide) and FMN (flavine mononucleotide).

righting reflex, see mid-air righting.

Risperidone: a putative atypical neuroleptic; a 5Ht-2/D-2 receptor antagonist.

Ritanserin, a trademark for preparations of methylphenidate hydrochloride

RO 15-4513: GABA antagonist and benzodiapine partial inverse agonist; also ethanol antagonist; anxiogenic.

rodenticide: chemical for destroying rodents.

rotorod test: test to assess motor coordination in animals. A rat or mouse is placed on a motor-driven rotating rod and the time it takes the animal to lose its footing and fall of is measured.

saxitoxin: potent heterocyclic (nitrogen containing) neurotoxin produced by e.g.,the dinoflagellates *Gonyaulax catenella* and *Gonyaulax tamarensis* (red tide organisms) and apparently also by the cyanobacterium *Aphanizomenon flos-aquae*. In warm blooded animals, saxitoxin blocks the transmission of nerve impulses, causing paralysis and eventually death by asphyxiation.

SCH-23390: a selective D-1 receptor antagonist.

schedule induced polydipsia: an organism can be induced to drink enormous quantities of water simply by delivering small quantities of food on a regular basis. The food may be a reinforcement for responding on a FI schedule, or it may not be contingent on any behavior.

schedule of reinforcement: program determining relationship between responses and the occurrence of reinforcing stimuli.

schedule-controlled behavior: behavior that has been placed on a schedule of reinforcement. Following an animal's responses with reinforcing stimuli according to an explicit schedule produces orderly patterns of behavior.

sciatica: a syndrome characterized by pain radiating from the back into the buttock and into the lower extremity along its posterior or lateral aspect, and most commonly caused by protrusion of a low lumbar intervertebral disk.

second messenger: organic molecule acting within a cell to initiate the response to a signal carried by a chemical messenger (e.g.,a neurotransmitter or a hormone) that does not itself enter the cell. Examples of second messengers are inositol triphosphate and cyclic AMP.

second-order schedule: a schedule of a schedule. The defining feature of a second order schedule is that the component schedule (the first order schedule) is treated as if it were a single response by the second order schedule. For example, a second order FR4 (FI 1 min.) schedule would require the completion of four 1-min. Fixed Intervals prior to delivery of the reinforcer. The FI schedule is the first-order schedule and the FR schedule is the second-order schedule.

segment studies: in reproductive toxicology, studies of the effect of substances on various stages of the reproductive cycle in animal models. A distinction can be made between tests of fertility and reproduction *(segment I studies)*, tests of embryo-toxicity and teratogenicity *(segment II studies)*, perinatal and postnatal tests *(segment III studies)* and *(multi-)generation* tests.

self-administration model: model in which subjects can self-administer drugs and in which they have control over the infusion apparatus. Frequently animals are trained to lever press in a food-reward set-up before being tested for drug self-administration. Different protocols may be engaged, for instance, intravenous, intracerebral or oral and animals may be drug naive or drug dependent.

self-administration, see drug self-administration.

senile plaques: pathological hallmark of Alzheimer's disease formed by an extracellularly accumulation of amyloid and irregular, loosely arranged aggregate of neuronal and glial processes. They are found most characteristically in the gray matter of the neocortex and hippocampus, but also occur in the basal ganglia, thalamus and cerebellum.

sensitivity: the degree of responsiveness to weak stimuli; having a low threshold.

sensitization: increased response after a challenge induced by previous exposure to the substance.

sensorimotor tests: tests to assess basic central and peripheral nervous system integrity of sensory and motor nerves.

sensory ganglion, see dorsal root ganglion.

sensory latency: the time between the onset of a stimulus and the actual perception of the stimulus.

septum pellucidum (= septum lucidum): triangular double membrane separating the anterior horns of the lateral ventricles of the brain. It is part of the limbic system and controls aggressive behavior.

septum, see septum pellucidum.

Serax™, see oxazepam

sexual dimorphic area (SDN): nucleus in the hypothalamus for which a difference in cell number has been found between men and women

Shellfish Poisoning: form of food poisoning which results from the consumption of shellfish (mollusks, crustacea) contaminated with toxins or pathogens.

shock-elicited fighting: agonistic behavior by a rat that is stressed by an electric shock given through his paws.

short term memory: a memory system subject to decay and in which information must be maintained by rehearsal.

shuttle-box: apparatus employed in the investigation of escape or avoidance conditioning. Usually, organisms are placed in one side of the apparatus and must jump to the other in order to escape or avoid painful electric shocks.

side-effects: actions of a drug other than those desired for a beneficial pharmacological effect.

Sidman avoidance (temporal avoidance conditioning): an operant-conditioning procedure in which an aversive stimulus is presented at regular intervals. When the organism makes the proper response the shock is delayed some fixed amount of time. The procedure thus has two independent variables, the shock-shock and the shock-interval. Typically, good temporal avoidance conditioning develops and the animal learns to avoid the noxious events without the presence of any external stimulus.

single-blind procedure: an experimental technique in which the subjects are ignorant of the experimental conditions, but the researcher running the experiment knows them.

SKF-38393 [Quinpirole]: a dopamine D-1 receptor agonist.

SLC-90: 90-item symptom checklist, provides a measure of mood and psychopathological condition.

smack: one of the street names for heroin.

solanine: glycoalkaloid present in the common potato. It is a potent irritant of the intestinal mucosa and a cholinesterase inhibitor. Poisoning results in gastrointestinal and neurological symptoms. Gastrointestinal symptoms include vomiting and diarrhea and neurological symptoms irritability, confusion, delirium and respiratory failure which may ultimately result in death. Poisoning is often accompanied by high fever. In general, the glycoalkaloid content in potatoes do not pose harmful effects in humans.

span of apprehension: in memory testing, the maximum number of items that can be perceived and remembered in a single glance.

spatial discrimination learning: discrimination reversal learning task in which a subject is required to make initial discrimination between two (or more) stimulus objects, in which the correct object occupies a particular spatial position.

spatial orientation: the ability to relate to the position, direction, or movements of objects or points in space; deficits are associated with damages to different areas of the brain and involve different functions.

spatial reversal learning: spatial discrimination learning paradigm, in which after the initial discrimination has been learned the original stimulus object is shifted to another location.

spatial: refers to a problem where success depends mostly on imaging how objects or diagrams will appear after rotation, combination, or other transformation.

specificity: preciseness, uniqueness. The term is used to the connotation that the thing so characterized is relevant to but a single phenomenon or event.

spider naevi: telangiectatic arterioles in the skin with radiating capillary branches simulating the legs of a spider

spinal ganglion, see dorsal root ganglion.

Spiperone: a tranquilizer used in the treatment of schizophrenia.

spontaneous alternation: the tendency of animals, especially rats, to alternate left and right turns in a T-maze. A form of learning that depends on the cholinergic system.

St. Vitus dance, see Sydenham's chorea.

startle response: behavioral response as a consequence of an unexpected frightening stimulus, e.g.,an acoustic sound.

stimulus discrimination: the ability to tell the difference between two or more stimuli.

stimulus property of a drug: the ability of a drug to act as a stimulant to induce or change a behavior.

stimulus: any action or agent that causes or changes an activity in an organism, organ, or part.

strychnine: principal alkaloid of *Nux vomica* and some other plants. It stimulates all parts of the nervous system and in large doses produces convulsion It is used for killing vermin.

Student t-test: any of a number of statistical tests of significance based upon the t statistic.

subarachnoid: situated or occurring between the arachnoid and the pia mater.

subchronic study: animal experiment serving to study the effects produced by the test material when administered in repeated doses (or continually in food, drinking water, air) over a period of up to about 90 days.

subependymal: situated beneath the ependyma.

subfornical organ (organum subfornicale): a group of specialized ependymal cells, similar to those of the subcommissural organ, projecting toward the cavity of the third ventricle from its anterior wall between the columns of the fornix.

substantia innominata: nerve tissue immediately caudate to the anterior perforated substance and ventral to the globus pallidus and ansa lenticularis.

substantia nigra: a midbrain nucleus, the origin of the nigrostriatal dopamine pathway whose degeneration causes Parkinson's disease.

succinylcholine: synthetic muscle relaxant with a curare-like activity, it blocks acetylcholine receptors.

Sudden Infant Death syndrome (SIDS): the sudden and unexpected death of an apparently healthy infant, typically occurring between the ages of 3 weeks and 5 months and not explained by careful post mortem studies; also called crib or cot death.

suicide inhibitor (metabolite inhibitory complexes): compound that in its metabolized form functions as an enzyme inhibitor.

supersensitivity: abnormally increased sensitivity; increased activity of a neural pathway following chronic exposure to an antagonist caused by changes in postsynaptic receptors.

supranuclear paralysis (pseudobulbar p.): spastic weakness of the muscles innervated by the cranial nerves, i.e., the muscles of the face, pharynx and tongue, due to bilateral lesions of the corticospinal tract. Symptoms include dysphagia, dysarthria, and spastic facial jerks.

surface righting: the ability of a rat pup to right himself from a supine position.

surveys: a method of research in which the opinions of a large number of individuals are measured. Provided the sampling is genuinely representative, the results of surveys can often be employed to predict product sales, public reactions to various events, or the results of political elections.

survival analysis: statistical technique for the analysis of longitudinal data on the occurrence and timing of events.

sustained attention: the ability to stay alert during a longer lasting test or situation, e.g.,in the Mackworth clock test (see Mackworth clock test).

Sydenham's chorea: an acute, usually self-limited disorder of early life, between the ages of five and fifteen, or during pregnancy, and closely linked with rheumatic fever. It is characterized by involuntary movements that gradually become severe, affecting all motor activities including gait, arm movements and speech.

tachyphylaxis: rapidly decreasing response to a drug or physiologically active agent after administration of a few doses.

tachypnea: excessive rapidity of respiration.

tail-flick test: rodent test to measure changes in nociception (pain sensitivity) by placing the distal part of an animal's (mouse or rat) tail in warm water and noting the emergence of the tail from the water.

tapping speed, see finger tapping.

tardive dyskinesia: iatrogen disorder due to long-term treatment with dopamine antagonists (phenothiazines, butyrophenones), manifested by abnormal involuntary movements especially of the face and tongue, usually temporary, but it can be permanent. The cause is an alteration in dopaminergic receptors causing hypersensitivity to dopamine and dopamine agonists.

Tegretol™, see carbamazepine

terpenes ($(C_5H_8)_n$):class of hydrocarbons occurring in many fragrant essential oils of plants. They are colorless liquids, generally with a pleasant smell.

test battery: a collection of tests the results of which can be combined to produce a single score. Such batteries are used on the assumption that the errors inherent in each separate test will cancel each other out and the single score obtained will be maximally valid. In neurobehavioral toxicology it is often a collection of behavioral tests which makes it possible to test for more than one functional behavioral disturbance.

tetrachlorodibenzo-*p*-dioxin(TCDD): a compound from a group of about 75 structurally related compounds (dioxins). They have a skeleton of 3 linked rings with 1 to 18 chlorine atoms. In molecules with the same number of chlorine atoms, these can occur in different places (isomers). The central hexagon, with two oxygen atoms opposite one another, is called a dioxane ring. The full name of dioxins is polychlorodibenzo-paradioxins, PCDD for short. The Seveso dioxin contained 4 chlorine atoms in the positions 2,3,7,8; its official name is 2,3,7,8-tetrachlorodibenzo-paradioxine, abbreviated 2,3,7,8-TCDD. The toxicity results from the presence of 4 chlorine atoms in a rectangle of 4 by 10 angstroms. The related PCDF (polychlorodibenzofuran) compound 2,3,7,8-TCDF is only slightly less toxic. TCDD is mainly formed as undesirable contaminant in the production of trichlorophenols, and the pesticide 2,4,5-T or the bactericide hexachlorophene, which are manufactured from them.

tetrodotoxin: toxin contained in the ovary and liver of the puffer fish, in the eggs of the Californian newt and in some other animals. It is used as a chemical tool for the study of ion channels. By blocking the sodium channels, it is a toxicant of nerves and muscles, causing nerve and muscle paralysis. Symptoms include weakening of voluntary muscles, respiratory failure from diaphragm paralysis and hypotension.

thallium: naturally occurring heavy metal used in organic syntheses, to form alloys with other metals, as a rodenticide and in superconductor research. Acute toxicity includes nausea, vomiting, diarrhea, polyneuritis, coma, convulsions and death. Chronic toxicity includes neural, hepatic and renal damage, as well as deafness and loss of vision. The mechanism is thought to involve complexing of thallium with sulfhydryl groups in mitochondria and consequent interference with oxidative phosphorylation.

THBC: a serotonin agonist used as analgetic.

threshold limit value (TLV): occupational exposure limit recommended by the American Conference of Governmental Industrial Hygienists (ACGIH).

thujone: colorless oil the smell of which resembles that of menthol. This dicyclic ketone occurs in thuja, tansy, wormwood and many other oils.

time weighted average: the average, of e.g.,the concentration, of a substance to which a person is exposed in the ambient air over a period. Usually this period is 8 hours.

TIQ-challenge: a challenge treatment with tetrahydroisoquinoline

T-maze, see maze.

tolerance: a decreasing response to repeated constant doses of a drug or the need for increasing doses to maintain a constant response.

tonic movement: a movement marked by continuous tension (contraction) of muscles; spasm.

toxic equivalency factor (TEF): factor used in risk assessment to estimate the toxicity of a complex mixture, most commonly a mixture of chlorinated dibenzo-p-dioxins, furans and biphenyls: in this case, TEF is based on relative toxicity to 2,3,7,8-tetrachlorodibenzo-p-dioxin.

tracking task: task in which a target must be followed.

translation: the manufacture of proteins in a cell, which takes place at the ribosomes.

tranylcypromine: monoamine oxidase inhibiter structurally related to amphetamine.

tremor: an involuntary trembling or quivering.

tremorgenic mycotoxin: fungal toxin that induces tremor.

tremulousness: behavior characterized by tremor.

tri-ortho-cresylphosphate (TOCP): industrial compound, used as a plasticizer in lacquers and varnishes, as an additive in lubricants and gas oil, and as a fire retardant. It has a low acute toxicity but causes serious neuropathies after bio-activation. It is a neurotoxic esterase and choline esterase inhibitor.

trophic: of or pertaining to nutrition.

tropic: turning toward.

tropical ataxic neuropathy: toxic, distal axonopathy probably due to chronic consumption of improperly prepared cassava. Clinical features comprise central and peripheral manifestations of a central-peripheral distal axonopathy, characterized by painful paresthesiae of the feet with numbness of the hand, with sight and hearing loss when the disease progresses.

tropical spastic paraparesis: chronic progressive myelopathy.

t-tests: any of a number of statistical tests of significance based upon the t statistic.

twin studies: studies carried out on monozygotic and dizygotic twins raised together or apart from each other. The focus of these studies is to sort out the relative contributions of heredity and environment to human behavior.

two-way active avoidance task: conditioned avoidance test in which the animal must return after it first has escaped from a shock or other unpleasant stimulus. For this paradigm, often a shuttle box is used in which the rat can jump from one side of the cage to another and vice versa.

tyramine: amine, naturally occurring in cheese. It has a similar effect in the body to that of epinephrine (adrenaline). This can be dangerous in patients taking antidepressants of the MAO-inhibitor type.

unavoidable shock procedure: a procedure in which a subject (mostly a rat) receives an electrical shock that it cannot avoid. The procedure is often used to study stress or stress-related phenomena.

U-shaped curve: dose effect curve in which small and high doses have relatively large effects and medium doses have relatively low effects or no effect at all, the curve has the form of an 'u'.

vagus nerve: the 10th cranial nerve which supplies motor nerve fibers to the muscles of swallowing and parasympathetic fibers to the heart and organs of the crest cavity and abdomen. Sensory branches of the vagus carry impulses from the viscera and the sensation of taste from the mouth.

vas deferens (ductus deferens): the excretory duct of the testis, which unites with the excretory duct of the seminal vesicle to form the ejaculatory duct.

ventral pallidum: ventral part of the pallidum, one of the large subcortical nuclei that make up the basal ganglia.

ventral striatum: the ventral part of the corpus striatum.

ventral tegmental area (VTA): area in the central nervous system from which two main dopaminergic pathways start, the mesocortical dopamine pathway and the mesolimbic pathway.

ventriculomegaly: a dilation of the heart ventricle

verruculogen: a tremorgenic mycotoxin produced e.g., by *Penicillium piscarium* and *P. estinogeum.*

video tracking system: automated system to measure activity in an open field (see open field).

vigilance: alertness, watchfulness.

vision acuity: sharpness of vision.

visual discrimination learning: discrimination learning using visual stimuli (see discrimination learning).

Wallerian-type degeneration: fatty degeneration of a nerve fiber which has been severed from its nutritive centers.

water T-maze, see maze.

working memory task: equivalent to short-term memory.

Zimeldine: a trademark for zimelidine hydrochloride, a serotonergic antidepressant.

Z-scores: a statistical score which has been standardized by being expressed in terms of relative position in the full distribution of scores. A z-score is always expressed relative to the mean of the distribution and in standard deviation units.

Table of (neuro)psychological tests

Alberts Famous Faces Test: a test of remote memory using faces of famous personalities as the test stimuli.

Bayley's Scales of Infant Development: a developmental scale for assessing the status of infants from 2 months to 2 years of age. There are separate scales for mental development (perception, memory, learning, simple problem solving, etceteras) (*MDI = mental development index*), motor development (sitting, standing, walking, stair climbing, manipulatory skills) and an infant behavior record (personality development, social behavior, attention span, persistence and goal directness) (*PDI = personality development index*).

behavioral rating scale: any device used to assist a rater in making ratings of behavior. Rating behavior is the use of rating techniques in a fairly restricted manner such that only overt, objectively observable behaviors enter into the assessment.

Bender gestalt-test (= Bender Visual Motor Gestalt Test): test originally conceptualized as a maturational test for use with children and as a device for exploring regression, retardation and possible brain damage. Nowadays it is predominantly used to detect organic cerebral damage. The typical procedure is to present the testee with 9 relatively simple standard designs and ask him/her to copy them. Results are interpreted on the basis of the quality of the reproduction, the manner of organization of each copy and the pattern of spatial errors.

Benton Visual Recognition test: test to assess visual memory and perception and visuoconstructive abilities in children from 8 years old to adult. It is an individual test and the patient has to reproduce geometric figures from memory.

block design test: any performance (i.e. nonverbal) test using colored blocks or tiles in which the subject is required to match or copy given designs. In the WAIS-R test this part of the test on visuoconstructive reasoning and ability requires the replication of visual designs made with red and white blocks.

block design: experimental design in which subjects are distributed into distinct groups (blocks), so that each is representative of the population. Each block of subjects is then subjected to particular experimental manipulations. The name originates from agricultural research in which a large area was divided into sub-plots or 'blocks'.

Boston Naming Test: a test of dysphasic speech. The subject is shown 85 separate drawings of common objects. The subject's task is to name each of the objects. Word finding difficulties on this type of confrontation naming test is manifested as literal and verbal paraphasias, circumlocutions, and neologisms.

Bourdon-Wiersma test: test for the assessment of attention, using dots, either 3, 4 or 5, as stimuli. Subjects have to check each row as quickly as possible, crossing out groups of 4 dots.

Brief Psychiatric Rating Scale: scale providing a measure for a psychiatric status of a patient.

Brazelton's Neonatal Behavioral Assessment Scale (BNBAS): method for assessing infant (ages 3 days to 4 weeks) behavior by its response to environmental stimuli.

British Ability Scales (BASC or BAS): test to assess mental ability of children from 2.5-17.5 years of age; process areas included are: speed of problem solving, reasoning, spatial imagery, perceptual matching, short term memory and retrieval and application of knowledge.

California verbal learning test: a neuropsychological test of learning and memory ability in adults; it tests verbal and nonverbal learning, short- and long-term memory.

Child behavior checklist: a rating scale for internalizing and externalizing behavior disorders in children between 7 and 16 years of age.

choice response: refers to tests that offer fixed response options (e.g., multiple-choice, true-false, like-dislike...).

clinical trial: a large scale plan for testing and evaluating the effectiveness of some drug or therapeutic procedure in human subjects.

cognitive tests: tests to measure cognitive functions (→ cognition).

computer-based assessment: using computers as a medium for the presentation of psychological tests.

Continuous Performance Test: an automated test of sustained attention in which the subject is required to press a button whenever a given letter, e.g., X, appears on a video monitor screen. Deficits in attention and perceptual motor speed are detected by this test.

Controlled Word Association Test: a test of verbal fluency in which the patient must generate words beginning with specific letters.

contrast sensitivity threshold: the lowest intensity of a stimulus in the presence of other stimuli that can be detected by a sense organ as an independent stimulus.

critical flicker fusion test: test in which the point at which a flickering stimulus is no longer perceived as periodic but shifts to continuous, in this case the flicker fusion threshold, is measured.

delayed matching: test in which a subject (animal/human) is asked to match a sample stimulus with comparison stimuli that are provided after a delay.

Denver Developmental Screening Test: assessment test for developmental delays in children aged 0-6 years. The test concerns four areas: motor, social and/or language delays.

developmental mile stones: significant behaviors which are used to mark the progress of development. Examples are: saying phrases, turning pages, carrying out requests, pointing to body parts, holding a pencil, imitating a drawn circle, catching a ball.

diagnosis: narrowly choosing one of a set of labels that best fits an individual's disorder or disability. Broadly, developing an understanding of the individual's difficulties, and insofar possible, their origins.

Diagnostic and Statistical Manual (Diagnostic and Statistical Manual of Mental Disorders; DSM): the official system for classification of psychological and psychiatric disorders prepared and published by the American Psychiatric Association. The current version is DSM-IV.

Digit Span: test assessing immediate and short-term memory disturbances. In the test, the subject is asked to repeat sequences of random digits forward and backward. (Forms part of the Wechsler scales).

Digit Symbol: coding test requiring the subject to match as many symbols as possible to their corresponding digits in 90 seconds. The tests assesses motor speed, sustained attention and visuomotor skills.

Digit-Symbol Substitution Test: coding test that requires the subject to match as many symbols as possible to their corresponding digits within 90 sec. Motor speed, sustained attention and visuomotor skills play an important role in this test.

Fagan Visual Recognition Memory Test: test of visual recognition memory in young children (< 1 year of age) based on differential fixation to novel and previously seen targets. Fagan claims that positive results from this test correlate with a high IQ in later life.

Farnsworth-Munsell 100-Hue and dichotomous Test for Color Vision: a test to assess color perception and recognition. In this test colors are represented on small counters which have to be arranged in sequence, involving quite subtle discrimination. The test provides a sensitive, non-verbal assessment for a purely sensory disorder, as well as the material for assessing color agnosias through verbal response or by other tasks.

Finger Tapping Test: an automated performance test in which the subject must press a button as many times as possible within a 30 second interval, first with the index finger of each hand separately and then alternating between both hands.

Finnegan Scoring System: clinical assessment test for neonatal abstinence symptoms.

frontal lobe tests: tests to assess impairment of functions that are anatomically located in the frontal lobe of the brain. These tests mainly assess cognitive functioning, mood and personality.

Functional Observational test Battery (FOB): battery consisting of standardized observational ratings designed to assess different aspects of nervous system function as well as several noninvasive tests to evaluate sensory, motor and autonomic dysfunction.

Groningen Behavioral Observation Scale: assessment scale for detecting attention deficit disorder.

Hamilton search task: specific test to measure spatial memory.

Hänninen neurobehavioral test battery: one of the earliest neuropsychological screening batteries in toxicology. This Finnish test battery, includes tests for motor performance, visuoconstrictive ability, and visual and auditory memory, and was developed to assess neurotoxicological effects due to occupational exposure.

453

Halstead-Reitan Neuropsychological Battery: a brain functioning test that tests cerebral functioning and organic brain damage in patients from 6 years of age and older. The test consists of various subtests that measure aspects of cerebral functioning.

International Classification of Diseases (ICD): system of classification of diseases developed under the supervision of the World Health Organization. Over the years, many revisions have been made, the most recent system is the 10th (ICD-10). The ICD has an extensive section on psychiatric and psychological disorders, which is in wide use in many countries.

inventory: a questionnaire, typically one that represents many questions about each aspect of personality that is under investigation. Directions may ask for a self-description of an acquaintance who is being assessed.

IQ tests: any test that purports to measure an intelligence quotient. Generally such tests consist of a graded series of tasks each of which have been standardized with a large, representative population of individuals.

Kaufmann Assessment Battery for Children (K-ABC): test battery for assessment of cognition (sequential and simultaneous processing) and intelligence in children 2.5-12.5 years of age. In this test intelligence and achievement are measured separately.

Lanthony D-15 test: variant of the Farnsworth-Munsell 100 Hue test for color vision. In this test color stimuli are displayed on small moveable caps, which are presented to the subject in a random arrangement. The subject is asked to arrange the caps in sequence between two extremes. Subjects with normal color vision order the colors correctly around the color circle, while those with dyschromatopsia arrange the caps with specific errors, characteristic of vision loss along the blue-yellow and red-green axis.

Mackworth clock test: a classical test constructed to assess sustained attention. A subject is placed in a small room for two hours, watching a clock hand jerking round in regular jumps, one jump a second, one hundred jumps per revolution. The signal to press a key, consists of a jump of twice the usual distance. The interval between successive signals varies from 0.75 min. to 10 min. Each half an hour, 12 signals will be presented. Recorded are: number of correct detections ('hits'), missed targets ('misses') and incorrect detections ('false alarms'). From these, sensitivity and bias are calculated.

McCarthy Scales of Children's Abilities: a set of developmental scales for assessing the abilities of preschoolers between ages of 2 and 8. There are 18 separate tests combined into six distinct scales: verbal, perceptual, quantitative, memory, motor and general cognitive. The general cognitive scale is based on all but three of the tests and is generally treated as a measure of intellectual development.

MDI, mental development index (See: Bayley Scales of Infant Development).

mental age: a measure of intellectual ability obtained on early experimenters' tests of intelligence. An individual's mental age was assumed to reflect his or her level of intellectual maturity.

mental arithmetic: assessment test for measuring skills necessary for academic achievement. In the WAIS-R mediocre value as measures of general ability, assesses knowledge and ability to apply arithmetic operations only.

Milner Facial Recognition Test: a short-term recognition memory task using unfamiliar faces as the test stimuli.

MMPI (Minnesota Multiphasic Personality Inventory): an objective test in questionnaire form which compares a particular subject's responses to those of individuals in various diagnostic categories. The 400-item true/false questionnaire provides a profile of personal and social adjustment.

mood questionnaire: questionnaire to assess the state of mood.

Neurobehavioral Core Test Battery (NCTB): a core battery of tests put forward by a committee of the World Health Organization in 1985 covering specific nervous system domains which are sensitive to neurotoxic damage. The intention of this Core Test Battery was to standardize testing in neurobehavioral toxicology. This battery includes the following tests: aiming, simple reaction time, Santa Ana Dexterity Test, Digit Symbol Test, Benton Visual Retention Test, Digit Span and Profile of Mood States.

Neurobehavioral Evaluation System (NES): neurobehavioral test battery developed to bring about greater unity in the neurobehavioral test methods applied in occupational toxicology. The NES contains more than 20 computerized neuropsychological tests, some intended for use in

454

epidemiological studies in industry, others for experimental studies in laboratory settings. Most of the tests are analogs of traditional tests used in neurobehavioral research.

object reversal learning: discrimination reversal learning task in which a subject is required to make initial discrimination between two stimulus objects, in which the correct object is of a particular color and shape. After the initial discrimination, the original stimulus object is changed for another.

objective test: a personality test involving structured items and a limited set of responses (such as TRUE - FALSE)

odor identification: the ability to discriminate between different olfactory stimuli (smells).

olfactory perception threshold: the minimal concentration of a substance that evokes smell.

Peabody Individual Achievement Test (PIAT): test to assess academic achievement for children ages 5-18. It consists of 6 subtests on: general information, reading recognition, reading comprehension, spelling, mathematics, written expression. The test is thought to be especially appropriate for use with culturally disadvantaged, mentally handicapped, distractible children, and those with learning disabilities.

peer rating: a classmate, fellow soldier, fellow student, or other acquaintance marks a rating scale or inventory to describe the target person. Usually, the average of several reports is taken as the target's peer rating.

performance test: among ability tests the term is usually applied to those where the respondent is to execute an appropriate physical action-tracing a maze path for example; although it can also be verbal as in the Wechsler Performance section. Among personality measures, the term is usually applied to observations of response in a standardized situation, usually a situation that arouses strong motives.

Personality Development Index (PDI): See: Bayley Scales of Infant Development.

personality indices: the items in questionnaires and inventories that are used to assess a subject's personality.

personality questionnaire: questionnaire to assess a subject's personality.

PIAT: Peabody Individual Achievement Test.

Preyers's reflex: involuntary movement of the ears produced by auditory stimulation.

Profile Of Mood States (POMS): a questionnaire containing adjectives describing different mood states, e.g. tired, happy, etceteras. The subject indicates on a four point scale the degree which the appropriate description of his or her own feelings.

projective test: a personality test involving ambiguous stimuli. A subject's responses are supported to reveal aspects of the unconscious.

psychiatric rating scale: device used to assist a psychiatrist in making a diagnosis of a patient.

Raven Progressive Matrices: a nonverbal test to assess mental abilities in children from 5 years and older and adults.

reaction time (RT) paradigms: tests to measure the minimum time between the presentation of a stimulus and the subject's response to it. One of the oldest dependent variables in experimental psychology, specialized types have been studied. The introduction of computerized assessment has facilitated the recording and calculating of reaction times. (1) *choice reaction time* is the reaction time between the onset of one of a set of possible stimuli and the completion of the set of responses associated with the stimulus. (2) *simple reaction time* is the time between the onset of a stimulus and the completion of a response to it. Only one stimulus and one response are possible.

recall tests: tests to measure memory, in which previously learned items must be reproduced (as by listing them or repeating them).

Rennel Developmental Language Scale: assessment scale to detect language development in children.

reversal learning: this term covers a variety of learning situations during which at some point in discrimination training the original cues for correct responding are modified, so that they no longer serve as indicators for the original correct response. For example, a triangle may serve as a cue for the right turn in a T-maze and a circle for a left. After learning has reached some criterion (e.g. 80% correct responses) the cues are reversed so that now right turns to circles are reinforced and left turns to triangles.

455

Rutter Child Behavior Test: assessment test for behavioral problems in preschool children, using a parental questionnaire.

Santa-Ana Sensory Motor Coordination test: a text of manual speed and dexterity which requires the subject to turn white- and black-colored pegs 180 in a form board first with each hand separately and subsequently with both hands. The task has been used extensively in neurotoxicity studies.

sensorimotor tests: tests to assess basic central and peripheral nervous system integrity of sensory and motor nerves.

Serial Digit learning: See: Digit Symbol Test.

serial reaction time task: test of memory in which items are considered correct only if they are recalled in their correct order.

SLC-90: a 90-item symptom checklist; provides a measure of mood and psychopathological condition.

Snellen eye chart: commonest chart to test sharpness of distant vision, often used by the oculist.

Stanford-Binet IQ test (or scale): test of intelligence, originally prepared in 1916, as a revision of the Binet-Simon Scale of 1911. Several revisions have appeared throughout the years, which incorporate a large number of modifications, including extensive norms and the scales for measuring adult IQ.

stimulus discrimination: the ability to tell the difference between two or more stimuli.

Stroop test: procedure to study verbal processing. The test consist of a series of color name words (blue, red, green) printed in non-matching colors. That is the word blue is printed in red. Most people find it extremely difficult to attend to the ink color alone when asked to name the color in which each word is printed, because of an automatic tendency to read the words, which produces interfering information.

surveys: a method of research in which the opinions of a large number of individuals are measured. Provided the sampling is genuinely representative, the results of surveys can often be employed to predict product sales, public reactions to various events, or the results of political elections.

Swedish Performance Evaluation System (SPES): computerized neurotoxicological test battery, mainly used in occupational field testing.

test battery: a collection of tests the results of which can be combined to produce a single score. Such batteries are used on the assumption that the errors inherent in each separate test will cancel each other out and the single score obtained will be maximally valid. In neurobehavioral toxicology it is often a collection of behavioral tests which makes it possible to test for more than one functional behavioral disturbance.

Test of Language Development (TOLD): assessment test for language development in children between 4 and 13 years of age. The test measures pronunciation and skills on sound, word and sentence level.

tracking task: task in which a target must be followed.

Trail Making Test: a simple two-part test to measure visual conceptual and visuomotor tracking. In part A, the patient must draw lines to connect consecutively numbered circles. In part B, the patient must connect the same number of consecutively numbered and lettered circles by alternating between the two sequences.

visual reproduction test: assessment test for memory, in which a subject must draw from memory three figures which have been presented individually for 10 sec. A delay condition of 30 minutes is often included. The subject may also be asked to copy the figures in order to differentiate memory difficulties from visual perceptual deficits. This test forms part of the Wechsler Memory Scale.

WAIS-R (Wechsler Adult Intelligence Scale-Revised): a group of tests for assessment of intellectual functioning in adults. The output of this test is the performance on 10 subtests (verbal and performance tests) measuring various dimensions of intellectual functioning.

Wechsler Intelligence Scale for Children – Revised (WISC-R): a group of tests for assessment of intellectual functioning, thought processes and ego functioning in children ages 5 to 15. The output of this test is the performance on 10 subtests measuring various dimensions of intellectual functioning.

Wechsler Memory Scale: a psychiatric examination of memory ability to assess verbal and nonverbal memory in adults. It is a battery of tests designed to asses different aspects of verbal and visual short-term memory functions in brain-damaged populations consisting of 7 subtests: information/orientation, mental control, logical memory, memory span, visual reproduction and paired associates.

Wechsler Preschool and Primary Scale of Intelligence (WPPSI): assessment scale for a variety of cognitive abilities in children between 4 and 6 years of age.

Werry-Weiss-Peters Activity Scale: activity rating scale to assess (hyper)activity in children; often used to test hyperactivity and aggression in child psychiatry.

Wide Range Achievement Test (WRAT): assessment test for reading, spelling and arithmetic skills in children from 5 years on and in adults.

Wisconsin Card Sorting Test: a neuropsychological test of card sorting to a defined concept. It tests the abstracting ability. The test is especially sensitive to anterior frontal lobe functioning. The output gives concept learning ability measures of cognitive flexibility.

working memory task: equivalent to short-term memory.

writing sample: the patient is asked to write name and simple phrases to dictation.

List of commonly used abbreviations

AA, alcohol accepting strain

ABP, auditory brainstem potential

3-AP, 3-acetylpyridine

ACGIH, the American Conference of Governmental Industrial Hygienists

ACh, acetylcholine.

AChE, acetylcholinesterase.

ACTH, adrenocorticotrophic hormone

AD, Alzheimer's disease

ADD, attention deficit disorder

ADH, alcohol dehydrogenase

ADHD, Attention-Deficit Hyperactivity Disorder

ADI, acceptable daily intake

AETT, acetylethyl tetramethyl tetralin

Ah receptor, aromatic hydrocarbon receptor

ALA, aminolevulinic acid

ALAT, alanine transaminase

ALDH, aldehyde dehydrogenase

ALS, Amyotrophic lateral sclerosis (ALS) 5.4.1.4

AMPA, alpha-amino-3-hydroxy-t-methyl-isoxazole-4-pripionic acid

AN, arcuate nucleus

ANA, alcohol non-accepting strain

ANOVA, analysis of variance

AP, amphetamine

APO, apomorphine

ARBD, Alcohol Related Birth Defects,

ASAT, aspartate aminotransferase

ASPD, antisocial personality disorder

AST, aspartate amine transferase, a liver enzyme.

BAER, brainstem auditory evoked reponse

BDNF, brain-derived neurotrophic factor

BHA, butylated hydroxyanisole

B-Hg, Hg-concentrations in blood

BHT, butylated hydroxytoluene

BMAA, β-N-methyl-amino-L-alanine

BNAS, Brazelton Neonatal Assessment Scale,

BOAA, beta-N-oxalyl-amino-L-alanine

BOAA, β-N-oxalylamino-L-alanine

BPRS, Brief Psychiatric Rating Scale

CCK, cholecystokinin

CFF, critical flicker fusion

CFSAN, Center for Food Safety and Applied Nutrition of the FDA (US)

cGMP, cyclisch guanosine monophosphate

ChAT, choline acetyl transferase

CIDI, Composite International Diagnostic Interview

CNS, central nervous system

CNTF, ciliary neurotrophic factor

CPP, conditioned place preference

CRF, corticotropin-releasing factor

CRS, Chinese Restaurant Syndrome

CSF, cerebrospinal fluid

CT, computerized tomography

CTA, conditioned taste aversion

DA, dopamine

DDE, dichlorodiphenyldichloroethylene

DDT, dichlorodiphenyltrichloroethane

2-DG, 2-[^{14}C]deoxyglucose

DMT, dimethyltryptamine

DOB, 2,5-dimethoxy-4-bromoamphetamine

DOI, 1-(2,5-dimethoxy-4-iodophenyl)-2-aminopropane

DPOE, distortion product otoacustic emission

DSA, delayed spatial alternation

DSM, Diagnostic and Statistical Manual

DSP-4, N-chloroethyl-N-ethyl-2-bromobenzylamine hydrochloride

DZ, dizygotic twins

EAA, excitatory amino acid

ECS, electoconvulsive shock

ECT, electroconvulsive treatment

EPA, Environmantal Protection Agency (US)

ER, endoplasmic reticulum

ERP, event-related potentials

ESM, Experience Sampling Methodology

ETZ, etonitizine

FAE, Fetal Alcohol Effects,

FAS, fetal alcohol syndrome

FDA, Food and Drug Administration (US)

FI, fixed interval

FIFRA, Federal Insecticide, Fungicide and Rodenticide Act (US)

FR, fixed ratio schedule

FSS, Finnegan Scaling System

G × E, genotype by environmental interaction

GABA, γ-amino butyric acid

GAD, glutamate decarboxylase

gamma-GT, gamma glutamyl transferase

GD, gestational day

GFAP, glial fibrillary acidic protein.

GGT, γ-glutamyltransferase

GRAS, generally recognized as safe

GTP, guanosine triphosphate

HAD, high alcohol drinking strain

HCB, hexachlorobenzene

2,5-HD, 2,5-hexanedione

5-HIAA, 5-Hydroxyindoleacetic acid

HPA-axis, hypothalamic-pituitary-adrenal-axis

5-HT, 5-hydroxytryptamine (serotonine)

HVA, homovanillic acid

i.c.v., intracerebroventricular

i.g., intragastric

i.m., intramuscular

i.p., intraperitoneal

ICH, International Commission for Harmonization

IDPN, β,β'-Iminodiproprionitrile
IP$_3$, phosphoinositol triphosphate
IPPO, isopropylbicyclophosphate
IUGR, Intra Uterine Growth Retardation,
LAD, low alcohol drinking strain
LB, Lewis bodies
LC, locus coeruleus
LHRH, luteinizing hormone-releasing hormone
LS, Long Sleep strain
LTP, long term potentiation
MAG, myelin associated glycoprotein
MAM, methyl azoxymethanol
MAO, monoamine oxidase
MAOI, monoamino oxidase inhibitor
MBD, minimal brain dysfunction
MBP, myelin basic protein
MCV, mean corpuscular volume
MDA, methylenedioxyamphetamine
MDMA, 3,4-Methylene-dioxy-methyl-amphetamine (ecstasy, XTC)
METH, Methamphetamine
MHC, major histocompatibility complex.
MMC, methylmercury
MMH, monomethylhydrazine
MMT, methylcyclopentadienyl manganese tricarboxyl
Mn, manganese
MND, motor neuron disease
MOA, monoamine oxidase
MPA, mercaptopropionic acid
MPP$^+$, 1-methyl-4-phenylpyridinium
MPPP, 1-methyl-4-phenylpropionoxypiperidine
MPTP, 1,2,3,6-tetrahydro-1-methyl-4-phenyl-pyridine1.4.3
MSG, monosodium glutamate
MT, microtubules
MZ, monozygotic twins
NA, nucleus accumbens
NBAS, Neonatal Behavioral Assessment Scale
NCAM, neural cell adhesion molecule
NCTB, Neurobehavioral Core Test Battery
NDMA, N-methyl-D-aspartate
NES, Neurobehavioral Evaluation System
NF, neurofilaments
NFT, neurofibrillary tangles
NGF, nerve growth factor
NMDA, N-methyl-D-aspartate
NOAEL, no observed adverse effect level
NP, alcohol non preferring strain
3-NPA, 3-Nitroproprionic acid
β-ODAP, β,β'-N-oxalyl-L-α,β-diaminoproprionic acid
P, alcohol preferring strain
PBB, polybrominated biphenyls
PCAG, pentobarbital-chlorpromazine alcohol group of ARCI
PCB, polychlorinated biphenyl
PCDD, polychlorodibenzo-paradioxins
PCDF, polychlorodibenzo furans

PCP, phencyclidine
PD, Parkinson's disease
PD, postnatal day
PET, positon emission tomography
PKU, phenylketonuria
PLP, proteolipid protein
PNS, peripheral nervous system
POMC, pro-opiomelanocortine
POMS, Profile of Mood States
PR, progressive ratio (schedule)
PSP, paralytic shellfish poisoning
PTZ, pentylenetetrazol
QTL, quantitative trait loci
RAS, reticular activating system
REM, rapid eye movement
RI, recombinant inbred strain
s.c., subcutaneous
SDH, succinate dehydrogenase
SER, smooth endoplasmic reticulum
SGA, Small for Gestational Age,
-SH group, thiol group
SIDS, Sudden Infant Death Sybdrome,
SIHL, see solvent induced hearing loss
SN, substantia nigra
SOD, superoxide dismutase
SP, senile plaque
SPES, Swedish Performance Evaluation System
SRIF, somatotropin-release inhibiting factor
SS, Short Sleep strain
TCA, tricarboxylic acid
TEF, toxic equivalency factor
TEQ, toxic equivalents
TET, tetraethlyl lead
THC, Δ^9-tetrahydrocannabinol
THC, tetrahydrocannabinol
TILE, total integrated lifetime exposure
TLV, threshold limit value
TMT, trimethyl tin
TOCP, tri-ortho cresyl phosphate
TRH, tyrotropin-releasing hormone
TSCA, Toxic Substance Control Act (US)
U-Hg, Hg-concentrations in urine
VAS, Visual Analog Scales
VTA, ventral tegmental area
WHO, World Health Organization
WSP, withdrawal seizure prone strain
WSR, withdrawal seizure resistant strain
ZPP, zinc protoporphyrin

Index